Lecture Notes in Mathematics   1952

Editors:
J.-M. Morel, Cachan
F. Takens, Groningen
B. Teissier, Paris

Roman Mikhailov · Inder Bir Singh Passi

Lower Central and Dimension Series of Groups

 Springer

Roman Mikhailov
Steklov Mathematical Institute
Department of Algebra
Gubkina 8
Moscow 119991
Russia
romanvm@mi.ras.ru

Inder Bir S. Passi
Centre for Advanced Study in Mathematics
Panjab University
Chandigarh 160014
India
ibspassi@yahoo.co.in

QA
3
.L28
no.1952

ISBN: 978-3-540-85817-1   e-ISBN: 978-3-540-85818-8
DOI: 10.1007/978-3-540-85818-8

Lecture Notes in Mathematics ISSN print edition: 0075-8434
                        ISSN electronic edition: 1617-9692

Library of Congress Control Number: 2008934461

Mathematics Subject Classification (2000): 18G10, 18G30, 20C05, 20C07, 20E26, 20F05, 20F14, 20F18, 20J05, 55Q40,

© 2009 Springer-Verlag Berlin Heidelberg
This work is subject to copyright. All rights are reserved, whether the whole or part of the material is concerned, specifically the rights of translation, reprinting, reuse of illustrations, recitation, broadcasting, reproduction on microfilm or in any other way, and storage in data banks. Duplication of this publication or parts thereof is permitted only under the provisions of the German Copyright Law of September 9, 1965, in its current version, and permission for use must always be obtained from Springer. Violations are liable to prosecution under the German Copyright Law.

The use of general descriptive names, registered names, trademarks, etc. in this publication does not imply, even in the absence of a specific statement, that such names are exempt from the relevant protective laws and regulations and therefore free for general use.

Cover design: SPi Publishing Services

Printed on acid-free paper

9 8 7 6 5 4 3 2 1

springer.com

To Olga and Surinder

Preface

A fundamental object of study in the theory of groups is the lower central series of groups whose terms are defined for a group G inductively by setting

$$\gamma_1(G) = G, \quad \gamma_{n+1}(G) = [G, \gamma_n(G)] \quad (n \geq 1),$$

where, for subsets H, K of G, $[H, K]$ denotes the subgroup of G generated by the commutators $[h, k] := h^{-1}k^{-1}hk$ for $h \in H$ and $k \in K$. The lower central series of free groups was first investigated by Magnus [Mag35]. To recall Magnus's work, let F be a free group with basis $\{x_i\}_{i \in I}$ and $\mathcal{A} = \mathbb{Z}[[X_i \mid i \in I]]$ the ring of formal power series in the non-commuting variables $\{X_i\}_{i \in I}$ over the ring $\mathbb{Z}$ of integers. Let $\mathcal{U}(\mathcal{A})$ be the group of units of $\mathcal{A}$. The map $x_i \mapsto 1 + X_i$, $i \in I$, extends to a homomorphism

$$\theta : F \to \mathcal{U}(\mathcal{A}), \tag{1}$$

since $1 + X_i$ is invertible in $\mathcal{A}$ with $1 - X_i + X_i^2 - \cdots$ as its inverse. The homomorphism θ is, in fact, a monomorphism (Theorem 5.6 in [Mag66]). For $a \in \mathcal{A}$, let a_n denote its homogeneous component of degree n, so that

$$a = a_0 + a_1 + \cdots + a_n + \cdots .$$

Define
$$\mathcal{D}_n(F) := \{f \in F \mid \theta(f)_i = 0, \ 1 \leq i < n\}, \quad n \geq 1.$$

It is easy to see that $\mathcal{D}_n(F)$ is a normal subgroup of F and the series $\{\mathcal{D}_n(F)\}_{n \geq 1}$ is a central series in F, i.e., $[F, \mathcal{D}_n(F)] \subseteq \mathcal{D}_{n+1}(F)$ for all $n \geq 1$. Clearly, the intersection of the series $\{\mathcal{D}_n(F)\}_{n \geq 1}$ is trivial. Since $\{\mathcal{D}_n(F)\}_{n \geq 1}$

is a central series, we have $\gamma_n(F) \subseteq \mathcal{D}_n(F)$ for all $n \geq 1$. Thus, it follows that the intersection $\bigcap_{n\geq 1}\gamma_n(F)$ is trivial, i.e., F is residually nilpotent.

Let G be an arbitrary group and R a commutative ring with identity. The *group ring* of G over R, denoted by $R[G]$, is the R-algebra whose elements are the formal sums $\sum \alpha(g)g$, $g \in G$, $\alpha(g) \in R$, with only finitely many coefficients $\alpha(g)$ being non zero. The addition and multiplication in $R[G]$ are defined as follows:

$$\sum_{g\in G}\alpha(g)g + \sum_{g\in G}\beta(g)g = \sum_{g\in G}\big(\alpha(g) + \beta(g)\big)g,$$

$$\sum_{g\in G}\alpha(g)g \sum_{h\in G}\beta(h)h = \sum_{x\in G}\bigg(\sum_{gh=x}\alpha(g)\beta(h)\bigg)x.$$

The group G can be identified with a subgroup of the group of units of $R[G]$, by identifying $g \in G$ with $1_R g$, where 1_R is the identity element of R, and it then constitutes an R-basis for $R[G]$. The map

$$\epsilon : R[G] \to R, \quad \sum\alpha(g)g \mapsto \sum\alpha(g),$$

is an algebra homomorphism and is called the *augmentation map*; its kernel is called the *augmentation ideal* of $R[G]$; we denote it by $\Delta_R(G)$. In the case when R is the ring $\mathbb{Z}$ of integers, we refer to $\mathbb{Z}[G]$ as the integral group ring of G and denote the augmentation ideal also by $\mathfrak{g}$, the corresponding Euler fraktur lowercase letter.

The augmentation ideal $\Delta_R(G)$ leads to the following filtration of $R[G]$:

$$R[G] \supseteq \Delta_R(G) \supseteq \Delta_R^2(G) \supseteq \ldots \supseteq \Delta_R^n(G) \supseteq \ldots . \tag{2}$$

Note that the subset $G \cap \big(1 + \Delta_R^n(G)\big)$, $n \geq 1$, is a normal subgroup of G; this subgroup is called the nth *dimension subgroup* of G over R and is denoted by $D_{n,R}(G)$. It is easy to see that $\{D_{n,R}(G)\}_{n\geq 1}$ is a central series in G, and therefore $\gamma_n(G) \subseteq D_{n,R}(G)$ for all $n \geq 1$. In the case when R is the ring $\mathbb{Z}$ of integers, we drop the suffix $\mathbb{Z}$ and write $D_n(G)$ for $D_{n,\mathbb{Z}}(G)$. The quotients $\Delta_R^n(G)/\Delta_R^{n+1}(G)$, $n \geq 1$, are $R[G]$-modules with trivial G-action. There then naturally arise the following problems about dimension subgroups and augmentation powers.

Problem 0.1 *Identify the subgroups* $D_{n,R}(G) = G \cap \big(1 + \Delta_R^n(G)\big)$, $n \geq 1$.

Problem 0.2 *Describe the structure of the quotients* $\Delta_R^n(G)/\Delta_R^{n+1}(G)$, $n \geq 1$.

Problem 0.3 *Describe the intersection* $\bigcap_{n\geq 1}\Delta_R^n(G)$; *in particular, characterize the case when this intersection is trivial.*

In the case when F is a free group, then, for all $n \geq 1$, $\gamma_n(F) \subseteq D_n(F) \subseteq \mathcal{D}_n(F)$. The homomorphism $\theta : F \to \mathcal{A}$, defined in (1), extends by linearity to the integral group ring $\mathbb{Z}[F]$ of the free group F; we continue to denote the extended map by θ:

$$\theta : \mathbb{Z}[F] \to \mathcal{A}.$$

Let $\mathfrak{f}$ be the augmentation ideal of $\mathbb{Z}[F]$; then, for $\alpha \in \mathfrak{f}^n$, $\theta(\alpha)_i = 0$, $i \leq n-1$. With the help of free differential calculus, it can be seen that the intersection of the ideals $\mathfrak{f}^n$, $n \geq 1$, is zero and the homomorphism $\theta : \mathbb{Z}[F] \to \mathcal{A}$ is a monomorphism (see Chap. 4 in [Gru70]). A fundamental result about free groups ([Mag37], [Gru36], [Wit37]; see also [Röh85]) is that the inclusions $\gamma_n(F) \subseteq D_n(F) \subseteq \mathcal{D}_n(F)$ are equalities:

$$\gamma_n(F) = D_n(F) = \mathcal{D}_n(F), \text{ for all } n \geq 1. \tag{3}$$

This result exhibits a close relationship among the lower central series, the dimension series, and the powers of the augmentation ideal of the integral group ring of a free group. Thus, for free groups, Problems 1 and 3 have a definitive answer in the integral case. Problem 2 also has a simple answer in this case: for every $n \geq 1$, the quotient $\mathfrak{f}^n/\mathfrak{f}^{n+1}$ is a free abelian group with the set of elements $(x_{i_1} - 1)\dots(x_{i_n} - 1) + \mathfrak{f}^{n+1}$ as basis, where x_{i_j} range over a basis of F (see p. 116 in [Pas79]).

The foregoing results about free groups naturally raise the problem of investigation of the relationship among the lower central series $\{\gamma_n(G)\}_{n\geq1}$, dimension series $\{D_{n,R}(G)\}_{n\geq1}$, and augmentation quotients $\Delta_R^n(G)/\Delta_R^{n+1}(G)$, $n \geq 1$, of an arbitrary group G over the commutative ring R. While these series have been extensively studied by various authors over the last several decades (see [Pas79], [Gup87c]), we are still far from a definitive theory. The most challenging case here is that when R is the ring $\mathbb{Z}$ of integers, where a striking feature is that, unlike the case of free groups, the lower central series can differ from the dimension series, as first shown by Rips [Rip72].

Besides being purely of algebraic interest, lower central series and augmentation powers occur naturally in several other contexts, notably in algebraic K-theory, number theory, and topology. For example, the lower central series is the main ingredient of the theory of Milnor's $\bar{\mu}$-invariants of classical links [Mil57]; the lower central series and augmentation powers come naturally in [Cur71], [Gru80], [Qui69], and in the works of many other authors.

The main object of this monograph is to present an exposition of different methods related to the theory of the lower central series of groups, the dimension subgroups, and the augmentation powers. We will also be concerned with another important related series, namely, the derived series whose terms are defined, for a given group G, inductively by setting

$$\delta_0(G) = G, \quad \delta_{n+1}(G) = [\delta_n(G), \delta_n(G)] \text{ for } n \geq 0.$$

Our focus will be primarily on homological, homotopical, and combinatorial methods for the study of group rings. Simplicial methods, in fact, provide new possibilities for the theory of groups, Lie algebras, and group rings. For example, the derived functors of endofunctors on the category of groups come into play. Thus, working with simplicial objects and homotopy theory suggests new approaches for studying invariants of group presentations, a point of view which may be termed as "homotopical group theory." By homological group theory one normally means the study of properties of groups based on the properties of projective resolutions over their group rings. In contrast to this theory, by homotopical group theory we may understand the study of groups with the help of simplicial resolutions. From this point of view, homological group theory then appears as an abelianization of the homotopical one.

We now briefly describe the contents of this monograph.

Chapter 1: Lower central series   We discuss examples and methods for investigating the lower central series of groups with a view to examining residual nilpotence, i.e., the property that this series intersects in the identity subgroup. We begin with Magnus's theorem [Mag35] on residual nilpotence of free groups and Gruenberg's [Gru57] result on free polynilpotent groups. Mal'cev's observation [Mal68] on the adjoint group of an algebra provides a method for constructing residually nilpotent groups. We consider next free products and describe Lichtman's characterization [Lic78] for the residual nilpotence of a free product of groups. If a group G is such that the augmentation ideal $\mathfrak{g}$ of its integral group ring $\mathbb{Z}[G]$ is residually nilpotent, i.e., $\bigcap_{n \geq 0} \mathfrak{g}^n = 0$, then it is easily seen that G is residually nilpotent. This property, namely, the residual nilpotence of $\mathfrak{g}$, has been characterized by Lichtman [Lic77]. Our next object is to discuss residual nilpotence of wreath products. We give a detailed account of Hartley's work, along with Shmelkin's theorem [Shm73] on verbal wreath products. For HNN-extensions we discuss a method introduced by Raptis and Varsos [Rap89]. Turning to linear groups, we give an exposition of recent work of Mikhailov and Bardakov [Bar07]. An interesting class of groups arising from geometric considerations is that of braid groups; we discuss the result of Falk and Randell [Fal88] on pure braid groups.

If $1 \to R \to F \to G \to 1$ is a free presentation of a group G, then the quotient group $R/[R, R]$ of R can be viewed as a G-module, called a "relation module." The relationship between the properties of $F/[R, R]$ and those of F/R has been investigated by many authors [Gru70], [Gru], [Gup87c]. We discuss a generalization of this notion. Let R and S be normal subgroups of the free group F. Then the quotient group $(R \cap S)/[R, S]$ is abelian and it can be viewed, in a natural way, as a module over F/RS. Clearly the relation modules, and more generally, the higher relation modules $\gamma_n(R)/\gamma_{n+1}(R)$, $n \geq 2$, are all special cases of this construction. Such modules are related to the second homotopy module $\pi_2(X)$ of the standard complex X associated to the free presentation $G \simeq F/R$. We discuss here the work in [Mik06b]. The main point of investigation here is the faithfulness of the F/RS-module $(R \cap S)/[R, S]$.

We next turn to k-central extensions, namely, the extensions

$$1 \to N \to \tilde{G} \to G \to 1,$$

where N is contained in the kth central subgroup $\zeta_k(\tilde{G})$ of $\tilde{G}$. We examine the connection between residual nilpotence of G and that of $\tilde{G}$. In general, neither implies the other.

The construction of the lower central series $\{\gamma_n(G)\}_{n \geq 1}$ of a group G can be extended in an obvious way to define the transfinite terms $\gamma_\tau(G)$ of the lower central series for infinite ordinals τ. Let ω denote the first non-finite ordinal. The groups G whose lower central series has the property that $\gamma_\omega(G) \neq \gamma_{\omega+1}(G)$, called groups with long lower central series, are of topological interest. We discuss methods for constructing such groups.

It has long been known that the Schur multiplicator is related to the study of the lower central series. We explore a similar relationship with generalized multiplicators, better known as Baer invariants of free presentations of groups. In particular, we discuss generalized Dwyer filtration of Baer invariants and its relation with the residual nilpotence of groups. Using a generalization of the Magnus embedding, we shall see that every 2-central extension of a one-relator residually nilpotent group is itself residually nilpotent [Mik07a].

The residually nilpotent groups with the same lower central quotients as some free group are known as para-free groups. Non-free para-free groups were first discovered by Baumslag [Bau67]. We make some remarks related to the para-free conjecture, namely, the statement that a finitely generated para-free group has trivial Schur multiplicator.

Next we study the nilpotent completion $Z_\infty(G)$, which is the inverse limit of the system of epimorphisms $G \to G/\gamma_n(G)$, $n \geq 1$, and certain subgroups of this completion. We study the Bousfield–Kan completion $R_\infty X$ of a simplicial set X over a commutative ring R, and homological localization (called HZ-localization) functor $L : G \mapsto L(G)$ on the category of groups, due to Bousfield, in particular, the uncountability of the Schur multiplicator of the free nilpotent completion of a non abelian free group. After discussing the homology of the nilpotent completion, we go on to study transfinite para-free groups. Given an ordinal number τ, a group G is defined to be τ-para-free if there exists a homomorphism $F \to G$, with F free, such that

$$L(F)/\gamma_\tau\big(L(F)\big) \simeq L(G)/\gamma_\tau\big(L(G)\big).$$

Our final topic of discussion in this chapter is the study of the lower central series and the homology of crossed modules. These modules were first defined by Whitehead [Whi41]. With added structure to take into account, the computations naturally become rather more complicated. It has recently been shown that the cokernel of a non aspherical projective crossed module with a free base group acts faithfully on its kernel [Mik06a]; we give an exposition of this and some other related results.

Chapter 2: Dimension subgroups   In this chapter we study various problems concerning the dimension subgroups. The relationship between the lower central and the dimension series of groups is highly intriguing.

For every group G and integer $n \geq 1$, we have

$$\gamma_n(G) \subseteq D_n(G) := G \cap (1 + \mathfrak{g}^n).$$

As first shown by Rips [Rip72], equality does not hold in general. We pursue the counter example of Rips and the subsequent counterexamples constructed by N. Gupta, and construct several further examples of groups without the dimension property, i.e., groups where the lower central series does not coincide with the integral dimension series. We construct a 4-generator and 3-relator example of a group G with $D_4(G) \neq \gamma_4(G)$ and show further that, in a sense, this is a minimal counter example by proving that every 2-relator group G has the property that $D_4(G) = \gamma_4(G)$. At the moment it seems to be an intractable problem to compute the length of the dimension series of a finite 2-group of class 3. However, we show that for the group without the dimension property considered by Gupta and Passi (see p. 76 in [Gup87c]), the fifth dimension subgroup is trivial. Examples of groups with $D_n(G) \neq \gamma_n(G)$ with $n \geq 5$ were first constructed by Gupta [Gup90]. In this direction, for each $n \geq 5$, we construct a 5-generator 5-relator group $\mathfrak{G}_n$ such that $D_n(\mathfrak{G}_n) \neq \gamma_n(\mathfrak{G}_n)$. We also construct a nilpotent group of class 4 with non trivial sixth dimension subgroup. We hope that these examples will lead to a closer understanding of groups without the dimension property.

For each $n \geq 4$, in view of the existence of groups with $\gamma_n(G) \neq D_n(G)$, the class $\mathcal{D}_n$ of groups with trivial nth dimension subgroup is not a variety. This class is, however, a quasi-variety [Plo71]. We present an account of our work [Mik06c] showing that the quasi-variety $\mathcal{D}_4$ is not finitely based, thus answering a problem of Plotkin.

We next review the progress on the identification of integral dimension subgroups and on Plotkin's problems about the length of the dimension series of nilpotent groups.

Related to the dimension subgroups are the Lie dimension subgroups $D_{[n]}(G)$ and $D_{(n)}(G)$, $n \geq 1$, with $D_{[n]}(G) \subseteq D_{(n)}(G) \subseteq D_n(G)$ (see Sect. 2.10 for definitions). We show that, for every natural number s, there exists a group of class n such that $D_{[n+s]}(G) \neq \gamma_{n+s}(G) \neq 1$. In contrast with the integral case, many more definitive results are known about the dimension subgroups and the Lie dimension subgroups over fields. We review these results and their applications, in particular, to the study of Lie nilpotency indices of the augmentation ideals.

Chapter 3: Derived series   We study the derived series of free nilpotent groups. Combinatorial methods developed for the study of the dimension subgroups can be employed for this purpose. We give an exposition of the work of Gupta and Passi [Gup07].

Homological methods were first effectively applied by Strebel [Str74] for the investigation of the derived series of groups. We give an exposition of Strebel's work and then apply it to study properties analogous to those studied in Chap. 1 for the lower central series of groups; in particular, we study the behaviour of the transfinite terms of the derived series. We show that this study has an impact on Whitehead's asphericity question which asks whether every sub complex of an aspherical two-dimensional complex is itself aspherical.

Chapter 4: Augmentation powers The structure of the augmentation powers $\mathfrak{g}^n$ and the quotients $\mathfrak{g}^n/\mathfrak{g}^{n+1}$ for the group G is of algebraic and number-theoretic interest. While the augmentation powers have been investigated by several authors, a solution to this problem, when G is torsion free or torsion abelian group, has recently been given by Bak and Tang [Bak04]. We give an account of the main features of this work. We then discuss transfinite augmentation powers $\mathfrak{g}^\tau$, where τ is any ordinal number. We next give an exposition of some of Hartley's work on the augmentation powers.

A filtration $\{P_n H^2(G, \mathbb{T})\}_{n \geq 0}$ of the Schur multiplicator $H^2(G, \mathbb{T})$ arising from the notion of the polynomial 2-cocycles is a useful tool for the investigation of the dimension subgroups. This approach to dimension subgroups naturally leads to relative dimension subgroups $D_n(E, N) := E \cap (1 + \mathfrak{e}^n + \mathfrak{n}\mathfrak{e})$, where N is a normal subgroup of the group E. The relative dimension subgroups provide a generalization of dimension subgroups, and have been extensively studied by Hartl [Har08]. Using the above-mentioned filtration of the Schur multiplicator, Passi and Stammbach [Pas74] have given a characterization of para-free groups. These ideas were further developed in [Mik04, Mik05b]. We give here an account of this (co)homological approach for the study of subgroups determined by two-sided ideals in group rings.

In analogy with the notion of HZ-localization of groups, we study the Bousfield HZ localization of modules. Given a group G, a G-module homomorphism $f : M \to N$ is said to be an HZ-map if the maps $f_0 : H_0(G, M) \to H_0(G, N)$ and $f_1 : H_1(G, M) \to H_1(G, N)$ induced on the homology groups are such that f_0 is an epimorphism and f_1 is an isomorphism. In the category of G-modules, we examine the localization of G-modules with respect to the class $\mathcal{HZ}$ of HZ maps. We discuss the work of Brown and Dror [Bro75] and of Dwyer [Dwy75] where the relation between the HZ localization of a module M and its $\mathfrak{g}$-adic completion $\varprojlim M/\mathfrak{g}^n M$ has been investigated.

Chapter 5: Homotopical aspects After recalling the construction of certain functors, we give an exposition of the work of Curtis [Cur63] on the lower central series of simplicial groups. Of particular interest to us are his two spectral sequences of homotopy exact couples arising from the application of the lower central and the augmentation power functors, γ_n and Δ^n, respectively, to a simplicial group. Our study is motivated by the work Stallings [Sta75] where he suggested a program and pointed out the main problem in its pursuit:

Finally, Rips has shown that there is a difference between the "dimension subgroups" and the terms of the lower central series .... The problem would be, how to compute with the Curtis spectral sequence, at least to the point of going through the Rips example in detail?

To some extent, this program was realized by Sjögren [Sjo79]. However, in general, it is an open question whether there exists any homotopical role of the groups without the dimension property. The simplicial approach for the investigation of the relationship between the lower central series and the augmentation powers of a group has been studied by Gruenenfelder [Gru80]. We compute certain initial terms of the Curtis spectral sequences and, following [Har08], discuss applications to the identification of dimension subgroups. We show that it is possible to place some of the known results on the dimension subgroups in a categorical setting, and thus hope that this point of view might lead to a deeper insight. One of the interesting features of the simplicial approach is the connection that it provides to the derived functors of non-additive functors in the sense of Dold and Puppe [Dol61]. Finally, we present a number of homotopical applications; in particular, we give an algebraic proof of the computation of the low-dimensional homotopy groups of the 2-sphere. Another application that we give is an algebraic proof of the well-known theorem, due to Serre [Ser51], about the p-torsion of the homotopy groups of the 2-sphere.

Chapter 6: Miscellanea   In this concluding chapter we present some applications of the group ring construction in different, rather unexpected, contexts. As examples, we may mention here (1) the solution, due to Lam and Leung [Lam00], of a problem in number theory which asks, for a given natural number m, the computation of the set $W(m)$ of all possible integers n for which there exist mth roots of unity $\alpha_1, \ldots, \alpha_n$ in the field $\mathbb{C}$ of complex numbers such that $\alpha_1 + \cdots + \alpha_n = 0$, and (2) application of dimension subgroups by Massuyeau [Mas07] in low-dimensional topology.

Appendix   At several places in the text we need results about simplicial objects. Rather than interrupt the discussion at each such point, we have preferred to collect the results needed in an appendix to which the reader may refer as and when necessary. The material in the appendix is mainly that which is needed for the group-theoretic problems in hand. For a more detailed exposition of simplicial methods, the reader is referred to the books of Goerss and Jardine [Goe99] and May [May67].

To conclude, we may mention that a crucial point that emerges from the study of the various series carried out in this monograph is that the least transfinite step in all cases is the one which at the moment deserves most to be understood from the point of view of applications.

Acknowledgement

It is a pleasure to acknowledge the helpful suggestions and encouraging comments that we received from a number of experts who looked at an earlier draft of this work; in particular, we would like to thank Valerij Bardakov, Viktor Bovdi, Karl Gruenberg, Manfred Hartl, Eliyahu Rips, Ralph Stöhr, John Wilson, and the anonymous referee to whom the book was referred by the LNM series Editor at Springer. Our special thanks are due to Karl Gruenberg whose pioneering work on residual properties of groups and (co)homological approach to group theory has been a continuing motivation for our research. The reader will undoubtedly notice an imprint of his influence on the present text. Karl's untimely demise has deprived us of extensive discussions with him on some of the contents of this monograph. We are thankful to L. Breen, J. Mostovoy, B. I. Plotkin, E. Rips, and A. L. Shmel'kin for their interest in this work and for useful discussions with the first author.

The support provided by Steklov Mathematical Institute (Moscow), Harish-Chandra Research Institute (Allahabad), Indian Institute of Science Education and Research (Mohali), Indian National Science Academy (New Delhi) and Panjab University (Chandigarh) is gratefully acknowledged.

Contents

Notation

$\mathbb{Z}$	the ring of rational integers.
$\mathbb{Z}_n$	the ring of integers mod n.
$\mathbb{Q}$	the field of rational numbers.
$\mathbb{R}$	the field of real numbers.
$\mathbb{C}$	the field of complex numbers.
$\mathbb{T}$	the additive group of rationals mod 1, i.e., $\mathbb{Q}/\mathbb{Z}$.
$\mathbb{T}_p$	the p-torsion subgroup of $\mathbb{T}$.
Σ_n	the symmetric group of degree n.
$\langle S \rangle$	the subgroup (of G) generated by a subset S (of G).
$\langle S \rangle^G$	the normal subgroup of G generated by S.
G'	derived subgroup of G.
x^y	$y^{-1}xy$, for elements x, y of a group.
$[x, y]$	$x^{-1}y^{-1}xy$, the commutator of x and y.
$[x, {}_n y]$	$[[\dots [x, \underbrace{y], \dots y]]}_{n \text{ terms}}$
$[H, K]$	$\langle [h, k] \mid h \in H,\ k \in K \rangle$.
$[H, {}_n K]$	$[\underbrace{\dots [[H, K], K], \dots, K]}_{n \text{ terms}}$.
$H \triangleleft G$	H is a normal subgroup of the group G.
$\{\gamma_n(G)\}_{n \geq 1}$	the lower central series of the group G.
$\gamma_\omega(G)$	$\bigcap_{n \geq 1} \gamma_n(G)$.
$\{\delta_n(G)\}_{n \geq 0}$	the derived series of the group G.
$\delta_\omega(G)$	$\bigcap_{n \geq 1} \delta_n(G)$.
$\gamma_{\omega+1}(G)$	$[\gamma_\omega(G), G]$
$\delta_{(\omega)}(G)$	$\bigcap_{n \geq 1} [\delta_\omega(G), \delta_n(G)]$.
$\delta_{\omega+1}(G)$	$[\delta_\omega(G), \delta_\omega(G)]$.
$\mathcal{P}(G)$	the perfect radical of the group G.
K_P	the standard 2-complex associated with the group presentation $\mathsf{P} = (X, \varphi, \mathcal{R})$.
$gd(G)$	the geometric dimension of the group G.
$cd(G)$	the cohomological dimension of the group G.

$hd(G)$ the homological dimension of the group G.

$|X|$ the cardinality of the set X.

$R^{\oplus n}$ the direct sum of n copies of the ring R.

$\pi_n(X)$ the nth homotopy group of the topological space X.

$R[G]$ the group ring of the group G over the ring R.

$\mathcal{U}(R)$ the group of units of the ring R.

$\Delta_R(G)$ augmentation ideal of the group ring $R[G]$.

$\mathcal{U}_1(R[G])$ the group of units of augmentation 1 in $R[G]$.

$\delta^+(G)$ the subgroup generated by the torsion elements $g \in G$ which have only finitely many conjugates in G.

$\mathfrak{g}$ the augmentation ideal of the integral group ring $\mathbb{Z}[G]$.

$\mathfrak{N}$ the class of nilpotent groups

$\mathfrak{N}_\tau$ the class of groups G satisfying $\gamma_\tau(G) = 1$.

$\mathfrak{N}_0$ the class of torsion-free nilpotent groups.

$\overline{\mathfrak{N}}_p$ the class of nilpotent p-groups of finite exponent.

$\mathrm{r}\mathfrak{C}$ the class of residually $\mathfrak{C}$-groups.

Top   the category of topological spaces.

Mod_R category of (right)R-modules.

Gr   category of groups.

$\mathcal{H}o_n$ category of homotopy n-types.

Alg_R category of R-algebras.

Lie_R category of R-Lie algebras.

Ab   category of abelian groups.

$\mathfrak{S}$ the class of all solvable groups.

$\mathfrak{F}$ the class of all finite groups.

$\mathcal{L}_i T$ ith derived functor of the functor $T : \mathsf{Gr} \to \mathsf{Gr}$.

$\mathfrak{L}_i T$ ith derived functor $L_i T(-, 0)$ at level zero of the functor $T : \mathsf{Ab} \to \mathsf{Ab}$ (in the sense of Dold-Puppe).

G_{ab} $G/\gamma_2(G)$, the abelianization of the group G.

$\mathcal{L}_R(M)$ the free R-Lie algebra over the R-module M.

$\mathcal{L}_R^n(M)$ nth component of $\mathcal{L}_R(M)$.

$\mathcal{U}_R(L)$ the universal enveloping algebra of the R-Lie ring L.

$\mathfrak{S}(A)$ the symmetric algebra over the abelian group A.

$\mathfrak{T}_R(M)$ the tensor algebra over the R-module M.

$L : G \mapsto L(G)$ homological localization functor.

$\widetilde{\mathcal{J}}_k$ the class of residually nilpotent groups groups G for which every k-central extension $1 \to N \to \widetilde{G} \to G \to 1$ has the property that $\widetilde{G}$ is residually nilpotent.

$\widetilde{\mathcal{J}}$ the class $\widetilde{\mathcal{J}}_1$

$\mathrm{tor}(A)$ the torsion subgroup of the abelian group A.

Z_∞ nilpotent completion of G.

X^+ the plus-construction of the space X.

$\mathcal{CM}$ the category of crossed modules.

$GL(R)$ the general linear group over R.

$R_\infty X$ the R-Bousfield-Kan localization of the space X.

$D_n(G)$	nth integral dimension subgroup of the group G.
$\mathcal{H}o_n$	the category of homotopy n-types.
$\mathcal{L}(X)$	the fundamental cat^1-group of the space X.
$T^n(A)$	the nth tensor power of the abelain group A.
$\mathrm{SP}^n(A)$	the nth symmetric power of the abelian group A.
$\wedge^n(A)$	the nth exterior power of the abelian group A.
$J^n(A)$	the metablian Lie functor J^n applied to the abelian group A.
$\mathcal{U}_R(L)$	universal enveloping algebra of the R-Lie algebra L.
$\mathcal{A}_R(G)$	the associated graded ring of $R[G]$.
$Z_n{:}\mathsf{Gr}\to\mathsf{Gr}$	the functor $G \to G/\gamma_n(G)$.
$\mathcal{C}h(R)$	the category of chain complexes of R-modules.
$\mathcal{SC}$	the category of simplicial objects in the category $\mathcal{C}$.
$N_*(X)$	the Moore complex of the simplicial object X.
$\overline{W}$	classifying space functor on the category of simplicial groups.
$\Gamma(A)$	the divided power ring over the abelian group A.
$\mathcal{D}_n$	the quasi-variety of groups with trivial nth dimension subgroup.
$\Box$	signifies end (also omission) of proof.

Chapter 1
Lower Central Series

The lower central series $\{\gamma_\alpha(G)\}$ of a group G, where α varies over all ordinal numbers, is defined by setting $\gamma_1(G) = G, \gamma_{\alpha+1}(G) = [G, \gamma_\alpha(G)]$ and for a limit ordinal τ, $\gamma_\tau(G) = \bigcap_{\alpha<\tau} \gamma_\alpha(G)$, where for subsets H, K of G, $[H, K]$ denotes the subgroup of G generated by all commutators $[h, k] := h^{-1}h^k = h^{-1}k^{-1}hk$, $h \in H$, $k \in K$. The group G is said to be *transfinitely nilpotent* if $\gamma_\alpha(G) = 1$ for some ordinal α, or simply *nilpotent* if α is a finite ordinal. In particular, if $\gamma_\omega(G) = 1$, where ω is the least infinite ordinal, then G is said to be *residually nilpotent*. In this Chapter we present various constructions and methods for studying the residual nilpotence of groups. Our aim is primarily to investigate the transfinite terms of the lower central series. The results discussed here can be viewed as an attempt to develop a 'limit theory' for lower central series.

1.1 Commutator Calculus

Since we will have frequent occasion to work with commutators, we begin with recording here some of the basic identities concerning them.

Let G be a group and let $x_1, x_2, \ldots$, be elements of G. A *simple commutator* $[x_1, \ldots, x_n]$ of weight $n \geq 1$ is defined inductively by setting

$$[x_1] = x_1 \text{ and } [x_1, \ldots, x_n] = [[x_1, \ldots, x_{n-1}], x_n] \text{ for } n \geq 2.$$

The simple commutator $[x_1, \underbrace{x_2, \ldots x_2}_{n \text{ terms}}]$ will be denoted by $[x_1, {}_n x_2]$.

Commutator Identities

The following identities, which are straight-forward to verify, hold for arbitrary elements x, y, z in every group G.

R. Mikhailov, I.B.S. Passi, *Lower Central and Dimension Series of Groups*,
Lecture Notes in Mathematics 1952,
© Springer-Verlag Berlin Heidelberg 2009

(i) $[x, y] = [y, x]^{-1}$.

(ii) $[xy, z] = [x, z]^y [y, z]$ and $[x, yz] = [x, z][x, y]^z$.

(iii) $[x, y^{-1}] = \left([x, y]^{y^{-1}}\right)^{-1}$ and $[x^{-1}, y] = \left([x, y]^{x^{-1}}\right)^{-1}$.

(iv) $[x, y^{-1}, z]^y [y, z^{-1}, x]^z [z, x^{-1}, y]^x = 1$. (Hall-Witt identity)

(v) If $\gamma_2(G)$ is abelian, then

(a) $[x, y, z][y, z, x][z, x, y] = 1$.

(b) $[x_1, x_2, x_3, \ldots, x_n] = [x_1, x_2, x_{\sigma(3)}, \ldots, x_{\sigma(n)}]$, where σ is any permutation of the set $\{3, \ldots, n\}$.

(c) For $n \geq i \geq 3$, $[x_1, \ldots, x_{i-1}, x_i, x_i^{-1}, x_{i+1}, \ldots, x_n] = [x_1, \ldots, x_{i-1}, x_i, x_{i+1}, \ldots x_n]^{-1}[x_1, \ldots, x_{i-1}, x_i^{-1}, x_{i+1}, \ldots x_n]^{-1}$.

1.2 Residually Nilpotent Groups

Free Groups

The first result about residual nilpotence of groups is due to W. Magnus [Mag35].

Theorem 1.1 *Every free group is residually nilpotent.*

Proof. Let F be a free group and $X = \{x_i \mid i \in I\}$ a basis of F. Let $\mathcal{A} = \mathbb{Z}[[X]]$ be the ring of formal power series, in non-commuting indeterminates X_i, $i \in I$, with coefficients in the ring $\mathbb{Z}$ of integers. Since $1 + X_i$, $i \in I$, is an invertible element in $\mathcal{A}$ with inverse the power series $1 - X_i + X_i^2 - \cdots$, the map $X \to \mathcal{U}(\mathcal{A})$, $x_i \mapsto 1 + X_i$, where $\mathcal{U}(\mathcal{A})$ is the group of units of $\mathcal{A}$, extends to a homomorphism $\theta : F \to \mathcal{U}(\mathcal{A})$. The homomorphism θ is a monomorphism. For, let $w = x_{i_1}^{\epsilon_{i_1}} \ldots x_{i_n}^{\epsilon_{i_n}}$, with $x_{i_j} \in X$, $\epsilon_{i_j} \in \mathbb{Z}$, $1 \leq j \leq n$, be a reduced word over X. Then, in the power series $\theta(w)$, the coefficient of the monomial $X_{i_1} \ldots X_{i_n}$ is $\epsilon_{i_1} \ldots \epsilon_{i_n}$. Thus $\theta(w) = 1$ if and only if $w = 1$. A simple induction on $n \geq 1$ shows that for $f \in \gamma_n(F)$, $n \geq 1$, the homogeneous terms of degree i, $1 \leq i \leq n - 1$, in $\theta(f)$ are all zero. It thus follows that if $w \in \bigcap_{n \geq 1} \gamma_n(F)$, then $\theta(w) = 1$, and hence $w = 1$. Consequently F is residually nilpotent. $\square$

Free Poly-nilpotent Groups

If $\mathcal{P}$ is a group property, then a group G is said to be a *residually $\mathcal{P}$-group* if, for every non-identity element $x \in G$, there exists a normal subgroup N, depending on x, such that (i) $x \notin N$ and (ii) the quotient group G/N has the property $\mathcal{P}$.

Given a class $\mathfrak{C}$ of groups, we refer to groups in the class $\mathfrak{C}$ as $\mathfrak{C}$-groups, and denote by $\mathfrak{r}\mathfrak{C}$ the class of residually $\mathfrak{C}$-groups. Let us also adopt the following notation for certain classes of groups:

$\mathfrak{F}_p$ the class of finite p-groups;
$\mathfrak{N}$ the class of nilpotent groups;
$\mathfrak{N}_0$ the class of torsion-free nilpotent groups.
$\overline{\mathfrak{N}}_p$ the class of nilpotent p-groups of finite exponent.

It is easy to see that a group $G \in \mathfrak{r}\mathfrak{N}$ if and only if $\gamma_\omega(G) = \bigcap_{n \geq 1} \gamma_n(G) = 1$. Clearly, if a property $\mathcal{P}$ implies the property $\mathcal{Q}$, then a residually $\mathcal{P}$-group is also a residually $\mathcal{Q}$-group. In particular, since prime-power groups are nilpotent, $\mathfrak{r}\mathfrak{F}_p \subseteq \mathfrak{r}\mathfrak{N}$.

Residual properties of groups were extensively studied by K. W. Gruenberg ([Gru57], [Gru62]).

Theorem 1.2 (Gruenberg [Gru57]). *Every free nilpotent group* $F/\gamma_n(F)$, $n \geq 1$, *is in the class* $\mathfrak{r}\mathfrak{F}_p$, *where* p *is any given prime.*

A free nilpotent group $G = F/\gamma_n(F)$, $n \geq 1$, in fact, has the property that the augmentation ideal of its group algebra $\mathbb{F}_p[G]$ over the field $\mathbb{F}_p$ of p elements is residually nilpotent; equivalently, $G \in \mathfrak{r}\overline{\mathfrak{N}}_p$, which is in turn equivalent to the intersection $\bigcap_{m,r \geq 1} \gamma_m(G)G^{p^r}$ being trivial (see Passi [Pas79]).

Theorem 1.3 (Gruenberg [Gru57]). *Let* $c_1, \ldots, c_m$ *be natural numbers, and* F *a free group. Then the free poly-nilpotent group* $F/\gamma_{c_m}(\cdots(\gamma_{c_1}(F))) \in \mathfrak{r}\mathfrak{F}_p$ *for any prime* $p > \max\{c_1, \ldots, c_{m-1}\}$.

Thus, in particular, free poly-nilpotent groups are in the class $\mathfrak{r}\mathfrak{N}$. We will see later (Theorem 1.20) that these groups are, in fact, in the class $\mathfrak{r}\mathfrak{N}_0$.

Adjoint Groups of Residually Nilpotent Algebras

The theory of associative algebras was applied by A. I. Mal'cev [Mal68] to study residually nilpotent groups. Given an algebra A, consider the *adjoint product* (also called *circle operation*) which is defined by setting

$$g \circ h = g + h - gh, \quad g,\ h \in A. \tag{1.1}$$

Let $J(A)$ be the set of elements $g \in A$ such that there exists an element $\bar{g}$ satisfying $g \circ \bar{g} = 0 = \bar{g} \circ g$. Then $J(A)$ can be viewed as a group with the binary operation defined by (1.1) having the zero element of the algebra as the identity element; this group is called the *adjoint group* of A. Recall that an associative algebra A is called *residually nilpotent* if $\bigcap_i A^i = \{0\}$. A simple relationship between residual nilpotence of algebras and that of groups, observed in [Mal68], is that, *for any residually nilpotent algebra* A, *the group $J(A)$ is residually nilpotent.* Thus a group G which embeds in the adjoint group of a residually nilpotent algebra is necessarily a residually nilpotent group.

It may, however, be noted that the residual properties of algebras and those of groups are, in fact, quite different. For instance, Mal'cev [Mal68] proved that *a free product of residually nilpotent algebras is residually nilpotent*; on the other hand, it is easy to see that the free product $\mathbb{Z}_2 * \mathbb{Z}_3$ of the cyclic groups of orders 2 and 3 is *not* residually nilpotent.

As a consequence of his method of proving a group to be residually nilpotent, Mal'cev obtained a sufficient condition for the free product of two residually nilpotent groups to be residually nilpotent:

> *If two groups G, H can be embedded into adjoint groups of residually nilpotent algebras over fields of the same characteristic, then their free product $G * H$ is residually nilpotent.*

As simple examples of residually nilpotent groups of this type, we may mention the groups $\mathbb{Z} * \mathbb{Z}_p$, $\mathbb{Z}_{p^n} * \mathbb{Z}_{p^m}$ for primes p.

Free Products

Necessary and sufficient conditions for the free product of a family of groups to be residually nilpotent have been obtained by A. I. Lichtman [Lic78]. Recall that an element $g \in G$ is called *an element of infinite p-height*, p a prime, if, given any two natural numbers i and j, there exist elements $x \in G$, $y \in \gamma_i(G)$, such that $x^{p^j} = gy$. Lichtman's characterization is as follows:

Theorem 1.4 (Lichtman [Lic78]). *The free product of a family $\{G_j : j \in J\}$ of nontrivial groups with $|J| \geq 2$ is residually nilpotent if and only if one of the following conditions holds:*

(i) All the groups G_j, $j \in J$, are in $\mathrm{r}\mathfrak{N}_0$.

(ii) For any finite set of non-identity elements $g_1, \ldots, g_k \in \bigcup_{j \in J} G_j$, there exists a prime p, depending on $g_1, \cdots, g_k$, such that none of these elements is of infinite p-height in the corresponding G_j.

Unit Groups of Group Rings

Let G be a group, $\mathbb{Z}[G]$ its integral group ring, $\epsilon : \mathbb{Z}[G] \to \mathbb{Z}$ the augmentation map and $\mathfrak{g} = \ker \epsilon$, the augmentation ideal of $\mathbb{Z}[G]$. Let $\mathcal{U}_1 := \mathcal{U}_1(\mathbb{Z}[G])$ be the group of units $\alpha \in \mathbb{Z}[G]$ satisfying $\epsilon(\alpha) = 1$. Residual nilpotence of $\mathcal{U}_1$ has been studied by Musson-Weiss [Mus82]. It is easy to see that

$$\gamma_n(\mathcal{U}_1) \subseteq 1 + \mathfrak{g}^n \qquad (n \geq 1).$$

Thus, if the augmentation ideal $\mathfrak{g}$ is residually nilpotent, then the unit group $\mathcal{U}_1$ is a residually nilpotent group. It is, however, possible for the unit group to be residually nilpotent without the augmentation ideal $\mathfrak{g}$ being residually nilpotent; for example, if G is the cyclic group of order 6.

The residual nilpotence of augmentation ideals has been characterized by A. I. Lichtman [Lic77]. To state the result, recall that if $\mathcal{C}$ is a class of groups, then a group G is said to be *discriminated by the class* $\mathcal{C}$ if for every finite subset $g_1, \ldots, g_n$ of distinct elements of G, there exists a group H in the class $\mathcal{C}$ and a homomorphism $\varphi : G \to H$ such that $\varphi(g_i) \neq \varphi(g_j)$ for $i \neq j$.

Theorem 1.5 (Lichtman [Lic77]). *Let G be a group. Then $\mathfrak{g}$ is residually nilpotent if and only if* either $G \in r\mathfrak{N}_0$, *or G is discriminated by the class* $\bigcup_{i \in I} r\overline{\mathfrak{N}}_{p_i}$, *where $\{p_i \,|\, i \in I\}$ is some set of primes.*

Lower Central Series of a Lie Ring

Let L be a Lie ring. The lower central series $\{L_n\}_{n \geq 1}$ of L is defined inductively by setting $L_1 = L$, $L_{n+1} = [L, L_n]$, the Lie ideal generated by the elements $[a, b]$ with $a \in L$ and $b \in L_n$, for $n \geq 1$. Let A be an associative ring with identity. We view A as a Lie ring under the operation $[a, b] = ab - ba$ $(a, b \in A)$, and define $A^{[n]}$, $n \geq 1$, to be the two-sided ideal of A generated by A_n. Let $\mathcal{U}(A)$ denote the group of units of A. With these notations, we have

Theorem 1.6 (Gupta-Levin [Gup83]). $\gamma_n(\mathcal{U}(A)) \subseteq 1 + A^{[n]}$ *for all $n \geq 1$.*

An immediate consequence of the preceding result is that if the associative algebra A is such that $\bigcap_{n \geq 1} A^{[n]} = 0$, then $\mathcal{U}(A)$ is a residually nilpotent group. To apply this result to the case of integral group rings, let $\mathfrak{K}_p$ be the class of groups $G \in \mathfrak{N}$ with $\gamma_2(G) \in r\overline{\mathfrak{N}}_p$, $\mathfrak{K}_0$ the class of groups $G \in \mathfrak{N}$ with $\gamma_2(G) \in \mathfrak{N}_0$, and $\mathfrak{K} = \bigcup_p \mathfrak{K}_p$, where p ranges over all primes.

Theorem 1.7 *If G is a group which satisfies* either

    (*i*) $G \in r\mathfrak{K}_0$, or

    (*ii*) G is discriminated by the class $\mathfrak{K}$,

then the unit group $\mathcal{U}(\mathbb{Z}[G]) \in r\mathfrak{N}$.

Proof. Under the hypothesis of the theorem, $\bigcap_{n \geq 1} \mathbb{Z}[G]^{[n]} = 0$ [Bha92b], and hence $\mathcal{U}(\mathbb{Z}[G])$ is residually nilpotent. $\square$

Wreath Product

An important construction in the theory of groups is the wreath product of groups (see [Rob95]). Let H and K be permutation groups acting on sets X

and Y respectively. The wreath product of H and K, denoted by $H \wr K$, is a group of permutations of the set $Z := X \times Y$, which is defined as follows:

If $\gamma \in H$, $y \in Y$ and $\kappa \in K$, define permutations $\gamma(y)$ and κ^* of Z by the rules

$$\gamma(y) : \begin{cases} (x, \; y) \mapsto (x\gamma, \; y), \\ (x, \; y_1) \mapsto (x, \; y_1) \text{ if } y_1 \neq y, \; x \in X, \end{cases}$$

and

$$\kappa^* : (x, \; y) \mapsto (x, \; y\kappa), \; x \in X.$$

The functions $\gamma \mapsto \gamma(y)$, and $\kappa \mapsto \kappa^*$ are monomorphisms from H and K respectively into $\mathcal{S}_Z$, the symmetric group of Z; let their images be $H(y)$ and K^* respectively. Then the *wreath product* of $H \wr K$ is the subgroup of $\mathcal{S}_Z$ generated by the subgroups K^* and $H(y)$, $y \in Y$. The subgroup of $H \wr K$ generated by the subset $\{H(y) \,|\, y \in Y\}$ is called the *base group* of the wreath product; note that the base group is the direct product of the subgroups $H(y)$, $y \in Y$. If H, K are arbitrary groups, we can view them as permutation groups via right regular representation and form the wreath product $H \wr K$; this group is called the *standard wreath product* of H and K.

It is not hard to see that the investigation of the residual nilpotence of wreath products is directly related to the residual nilpotence of augmentation ideals, a topic briefly discused above; the reader may refer to (Passi [Pas79]) for more details. In particular, note that, for any group G and integer $n \geq 0$, the wreath product $C_n \wr G$, where $C_n = \mathbb{Z}/n\mathbb{Z}$, is isomorphic to the semi-direct product $\mathbb{Z}_n[G] \rtimes G$, with G acting on $\mathbb{Z}_n[G]$ by right multiplication. As a consequence it may be noted that the wreath product $C_n \wr G$, $n \geq 0$, is residually nilpotent if and only if the augmentation ideal $\Delta_{\mathbb{Z}_n}(G)$ is residually nilpotent. While a characterization of the residual nilpotence of $\mathfrak{g}$ is provided by Theorem 1.5, recall that the residual nilpotence of $\Delta_{\mathbb{Z}_p}(G)$ is equivalent to G being in the class $r\overline{\mathfrak{N}}_p$ (see Passi [Pas79], p. 90).

Let A and G be two nontrivial groups and let $W = A \wr G$ be their standard wreath product. Necessary and sufficient conditions for W to be nilpotent are given by the following result.

Theorem 1.8 (Baumslag [Bau59]). *The wreath product $W = A \wr G$ is nilpotent if and only if, for some prime p, $A \in \overline{\mathfrak{N}}_p$ and $G \in \mathfrak{F}_p$.*

Let Ab denote the class of abelian groups. It is interesting to note ([Har70], Lemma 1) that if $W = A \wr G$ is residually nilpotent and G is infinite, then $A \in$ Ab. For, *if G is infinite and x, y are two non-commuting elements of A, then one can easily show that $[x, y] \in \gamma_\omega(W)$.*

Necessary and sufficient conditions for the residual nilpotence of the wreath product $W = A \wr G$ of two nontrivial groups have been given by B. Hartley [Har70] *in the case when G is nilpotent*. Hartley's work [Har70], combined with Lichtman's characterization for the residual nilpotence of the augmentation ideal $\mathfrak{g}$, provides the following characterization for the residual nilpotence of a wreath product $W = A \wr G$.

Theorem 1.9 (Hartley [Har70], Lichtman [Lic77]). *The wreath product $W = A \wr G$ of two nontrivial groups A and G, of which G is nilpotent, is residually nilpotent if only if one of the following three statements holds:*

(i) For some prime p, $G \in \mathfrak{F}_p$ and $A \in \overline{r\mathfrak{N}}_p$.

(ii) G is infinite but not torsion-free, $A \in \mathsf{Ab}$, and for some prime p, both A and G belong to $\overline{r\mathfrak{N}}_p$.

(iii) G is torsion-free, $A \in \mathsf{Ab}$, $G \in \overline{r\mathfrak{N}}_p$ whenever A contains an element of order p. Furthermore, if A is not a torsion group, then either $G \in r\mathfrak{N}_0$, or G is discriminated by the class $\bigcup_{i \in I} \overline{\mathfrak{N}}_{p_i}$, where $\{p_i \mid i \in I\}$ is some set of primes.

Proof. Suppose $W = A \wr G \in r\mathfrak{N}$ and both G and A are nontrivial.
Case 1: G finite. Let $1 \neq x \in G$ be an element of order p. Then the subgroup $A \wr \langle x \rangle$ of W is residually nilpotent. Note that

$$[a^{p^n}, x] \in \gamma_{n(p-1)+1}(W) \tag{1.2}$$

for all $a \in A$ and $n \geq 1$. Let $A_n = A^{p^n}(\gamma_{n(p-1)+1}(W) \cap A)$. Since W is residually nilpotent, the equation (1.2) implies that $\cap A_n = 1$ and hence $A \in r\overline{\mathfrak{N}}_p$.
Case 2: G infinite. In this case, as mentioned above, A must be abelian.

Suppose A has an element y of prime order p. From the residual nilpotence of the subgroup $\langle y \rangle \wr G$ of W, it follows that the augmentation ideal $\Delta_{\mathbb{F}_p}(G)$ of the group algebra $\mathbb{F}_p[G]$, where $\mathbb{F}_p$ is the field of p elements, is residually nilpotent, and consequently $G \in r\overline{\mathfrak{N}}_p$ (see [Pas79]).

If there exists an element $1 \neq x \in G$ of prime order q, say, then $A \wr \langle x \rangle$ is residually nilpotent. By case 1, $A \in r\overline{\mathfrak{N}}_q$. Thus, if A has an element of order p, then q must be p and, in this case both A and G belong to $r\overline{\mathfrak{N}}_p$. If A is torsion-free, then it follows that the wreath product $Z \wr G \in r\mathfrak{N}$, where Z is an infinite cyclic group. Therefore $\mathfrak{g}$ is residually nilpotent, and, since G has an element of order p, it must be in the class $r\overline{\mathfrak{N}}_p$ (see Theorem 1.5).

Finally, if G is torsion-free, then, as already noted, $G \in r\overline{\mathfrak{N}}_p$ for every prime p for which A has a p-torsion element. If A has an element a, say, of infinite order, then $\langle a \rangle \wr G \in r\mathfrak{N}$ and so $\mathfrak{g}$ is residually nilpotent; therefore, by Lichtman's Theorem 1.5, we have the remaining part of statement (iii).

In the direction of proving that each of the three conditions (i)-(iii) is sufficient for the residual nilpotence of the wreath product W, the following lemma for wreath products helps to simplify the argument.

Lemma 1.10 (Hartley [Har70], Lemma 9). *(i) If $\mathfrak{C}$ is a class of groups, $A \in r\mathfrak{C}$, and G is an arbitrary group, then $A \wr G \in r(\mathfrak{C} \wr G)$.
(ii) Let $\mathfrak{C}$ be a class of groups and $\mathfrak{D}$ a class of groups closed under taking subgroups and finite direct products. If $A \in r\mathfrak{C}$ and $G \in r\mathfrak{D}$, then $A \wr G \in r(\mathfrak{C} \wr \mathfrak{D})$.* □

Let $G \in \mathfrak{F}_p$ and $A \in r\overline{\mathfrak{N}}_p$. First consider the case when $A \in \mathfrak{N}_p$. In this case, W is a group of p-power exponent and, by Theorem 1.8, it is nilpotent.

Hence $W \in r\overline{\mathfrak{N}}_p$. Now observe, using Lemma 1.10, that if $A \in r\overline{\mathfrak{N}}_p$, then $W \in r(\overline{\mathfrak{N}}_p \wr \mathfrak{F}_p)$. Hence $W \in r\overline{\mathfrak{N}}_p$.

Suppose G is infinite but not torsion-free, A is abelian, and for some prime p, G and A both belong to the class $r\overline{\mathfrak{N}}_p$. Observe that the hypothesis on A implies that A is residually a cyclic p-group. Therefore, by Lemma 1.10, in order to show that $W \in r\overline{\mathfrak{N}}_p$, we can assume that A is a cyclic p-group and $G \in r\overline{\mathfrak{N}}_p$. That $W \in r\overline{\mathfrak{N}}_p$ then follows from the residual nilpotence of the augmentation ideal $\Delta_{\mathbb{Z}_p}(G)$ (see Passi [Pas79], p. 84).

Finally, suppose $G \neq 1$ is a torsion-free group, $A \in \mathsf{Ab}$, and for all primes $p \in \pi(A)$, the set of primes p for which A has p-torsion, $G \in r\overline{\mathfrak{N}}_p$. Further, if A has an element of infinite order, then $\mathfrak{g}$ is residually nilpotent. These conditions suffice to show that $W \in r\mathfrak{N}$ in case A is finitely generated. For an arbitrary abelian group A, a reduction argument is needed to reduce to the finitely generated case; see (Hartley [Har70]) for details. $\square$

Remark 1.11 *In the preceding theorem, the hypothesis on G being nilpotent is required only for showing the sufficiency of condition (iii). In fact, in general, condition (iii) is not sufficient for the residual nilpotence of the wreath product.*

Example 1.12
Let A be the quasi-cyclic 2-group and

$$G = \langle a, b \mid aba^{-1}b = 1 \rangle$$

be the fundamental group of the Klein bottle. Observe that G is torsion-free and $G \in r\widetilde{\mathfrak{F}}_2$. Thus the pair (A, G) of groups satisfies the hypothesis of Theorem 1.9 (iii). However, $W = A \wr G \notin r\mathfrak{N}$, since $1 \neq [A, b] \subseteq \gamma_\omega(W)$.

Complete answer to following problem remains open.

Problem 1.13 *Find a set of neessary and sufficient conditions for the residual nilpotence of the wreath product $A \wr G$ of two nontrivial groups.*

In case A is a torsion-free abelian group and $G \in r\mathfrak{N}_0$, a stronger assertion than residual nilpotence holds for $A \wr G$, namely:

Theorem 1.14 (Hartley [Har70]). *If A is a torsion-free abelian group and $G \in r\mathfrak{N}_0$, then $W = A \wr G \in r\mathfrak{N}_0$ or, equivalently, the augmentation ideal $\Delta_{\mathbb{Q}}(W)$ of the rational group algebra $\mathbb{Q}[W]$ is residually nilpotent.*

The main steps in the proof of the above result are:

(i) By Lemma 1.10 it suffices to consider only the case when $G \in \mathfrak{N}_0$.

(ii) A reduction argument shows that one may restrict to the case when A is finitely generated.

(iii) Invoke the result that $\Delta_{\mathbb{Q}}(G)$ is residually nilpotent when $G \in r\mathfrak{N}_0$.

Example 1.15
Consider the wreath product $G = Z \wr Z$ of two copies of the infinite cyclic group. This group has the following presentation:

$$G = \langle a, b \mid [a, a^{b^i}] = 1, \ i \in \mathbb{Z} \rangle. \tag{1.3}$$

Theorem 1.14 implies that $G \in r\mathfrak{N}_0$.

Verbal Wreath Product

Let us recall Shmel'kin's definition of verbal wreath product of groups (resp. Lie algebras).

Let $\mathcal{V}$ be a variety of groups (resp. Lie algebras over a commutative ring k with identity), A and B two groups (resp. Lie k-algebras) with $A \in \mathcal{V}$. Then the *verbal $\mathcal{V}$-wreath product*, denoted by $A \operatorname{wr}_\mathcal{V} B$, is defined as follows

$$A \operatorname{wr}_\mathcal{V} B := F/V(A^F),$$

where F is the free product $F = A * B$, $V(A^F)$ is the verbal subgroup (resp. ideal), corresponding to the variety $\mathcal{V}$, of the normal subgroup (resp. ideal) A^F generated by A in F.

Let $\mathcal{V}$ be a variety of Lie algebras over a field of characteristic 0. Let A be a $\mathcal{V}$-free Lie algebra with basis $X = \{x_i\}_{i \in I}$, B a nilpotent Lie algebra, and $W = A \operatorname{wr}_\mathcal{V} B$ their verbal $\mathcal{V}$-wreath product.

Lemma 1.16 (Shmel'kin [Shm73]). *The wreath product $W = A \operatorname{wr}_\mathcal{V} B$ is a residually nilpotent Lie algebra.*

In view of Lemma 1.16, the Lie algebra W embeds in its nilpotent completion $\widetilde{W}$, which again is residually nilpotent. Let W° be the group with $\widetilde{W}$ as the underlying set and the binary operation defined by means of the Campbell-Hausdorff formula

$$x \circ y = x + y + \frac{1}{2}[x, y] + \dots . \tag{1.4}$$

Let B° be the subgroup of W° consisting of the elements of B.

Since $\widetilde{W}$ is residually nilpotent, the group $W^0 \in r\mathfrak{N}_0$. Thus any group G that embeds in such a group W° must necessarily belong to the class $r\mathfrak{N}_0$.

A variety $\mathcal{V}$ of groups is said to be of *Lie type* if its free groups lie in the class $r\mathfrak{N}_0$. For example, the variety of all groups is a variety of Lie type.

Proposition 1.17 (see Andreev [And68], [And69]). *If A is a $\mathcal{V}$-free Lie algebra with basis $X = \{x_i\}_{i \in I}$, and A° is the group $(\widetilde{A}, \circ)$, arising from its nilpotent completion under the operation (1.4), then the subgroup of A° generated by X is a $\mathcal{V}^\circ$-free group relative to a certain variety $\mathcal{V}^\circ$ of Lie type.*

Theorem 1.18 (Shmel'kin [Shm73]). *The subgroup of W° generated by the elements X and B° is a group V°-wreath product of the V°-free group generated by X and B°.*

Theorem 1.19 (Shmel'kin [Shm73]). *Let V be a group variety of Lie type, A a free group in V, and B a group in the class $\mathfrak{rN}_0$. Then their V-wreath product A wr $_V B \in \mathfrak{rN}_0$.*

Let F be a free group, $R \lhd F$. Suppose F/R and $R/V(R)$ are both in $\mathfrak{rN}_0$. Then the group $F/V(R)$ can be embedded in a group of the type W° considered in the preceding theorem (see Shmel'kin [Shm65, Shm67]). Thus we have:

Theorem 1.20 *If F is a free group, and $R \lhd F$ with both F/R and $R/V(R)$ in $\mathfrak{rN}_0$, then $F/V(R) \in \mathfrak{rN}_0$.*

Let F be the free group with basis $\{a,\ b\}$ and let $R = \langle a \rangle^F$. It is easy to see that the group $G = Z \wr Z$ considered in Example 1.15 is a free abelian extension of the infinite cyclic group, and, in fact, has the presentation:

$$G = F/[R,\ R].$$

Thus the assertion that $G \in \mathfrak{rN}_0$ follows also from Theorem 1.20.

HNN-extensions

The residual properties of amalgamated products and HNN-extensions have been studied extensively by G. Higman, G. Baumslag, E. Raptis and D. Varsos, D. I Moldavanskii and others. Here we mention the main results of Raptis-Varsos [Rap89] about the residual nilpotence of an HNN-extension of a finitely generated abelian group (see also [Rap91]).

Let K be a finitely generated abelian group, A and B isomorphic subgroups of K with isomorphism $\varphi : A \to B$. Define

$$M_0 = A \cap B,\ M_1 = \varphi^{-1}(M_0) \cap M_0 \cap \varphi(M_0)$$

and inductively

$$M_{i+1} = \varphi^{-1}(M_i) \cap M_i \cap \varphi(M_i),\ i \geq 1.$$

Then define $H = H_{K,A,B,\varphi} = \bigcap_{i \geq 1} M_i$. Clearly, $\varphi(H) = H$ and furthermore, H is the largest subgroup such that $\varphi(H) = H$, i.e. for any subgroup L of K, such that $\varphi(L) = L$, one has $L \subseteq H$. Let λ be the least integer such that M_λ and $M_{\lambda+1}$ have equal torsion-free rank, and set $D = i_K(M_\lambda)$, the isolated closure of the subgroup M_λ.

Proposition 1.21 (Raptis-Varsos [Rap89]). *If the HNN-extension*

$$G = \langle t, \ K \mid t^{-1}at = \varphi(a), \ a \in A \rangle$$

is residually nilpotent, then either $K = A$ or $K = B$ or $|D/H|$ is a prime-power.

While the above result gives only a necessary condition for residual nilpotence of an HNN-extension of a finitely generated abelian group, in certain cases necessary and sufficient conditions are available.

Define the endomorphism $f : H_{K,A,B,\varphi} \to H_{K,A,B,\varphi}$ (or in the case $K = A$ the endomorphism $f : K \to K$) by setting

$$f : a \mapsto a^{-1}\varphi(a), \ a \in H_{K,A,B,\varphi}.$$

Theorem 1.22 (Raptis-Varsos [Rap89]). *Let K be a finitely generated abelian group, A and B subgroups of K, $\varphi : A \to B$ an isomorphism.*

(i) If $K = A$ or $K = B$, then the HNN-extension $G = \langle t, K \mid t^{-1}At = B, \varphi \rangle$ is residually nilpotent if and only if the only f-invariant subgroup of K is the trivial subgroup.

(ii) The HNN-extension $G = \langle t, K \mid t^{-1}At = A, \varphi \rangle$ is residually nilpotent if and only if the torsion subgroup of K/A is a prime-power group and the only f-invariant subgroup of A is the trivial subgroup.

Example 1.23
Let p be a prime, and K the group defined by the presentation

$$K = \langle a, b \mid [a, b] = 1, a^p = 1 \rangle.$$

Set $A = B = \langle b \rangle$ and define the isomorphism $\varphi : A \to B$ by setting $\varphi : b \mapsto b^{-1}$. Then $f : H \to H$ is defined by

$$f : b^k \mapsto b^{-2k}, \ k \in \mathbb{Z}.$$

Clearly, the only f-invariant element of H is the trivial element. Since $K/A = K/B = \mathbb{Z}/p\mathbb{Z}$, Theorem 1.22 implies that the HNN-extension

$$G = \langle a, b, t \mid [a, b] = 1, \ a^p = 1, \ t^{-1}btb = 1 \rangle$$

is residually nilpotent. On the other hand, if p, q are two different primes, then we conclude by Theorem 1.22 that the group defined by the presentation

$$\langle a, b, t \mid [a, b] = 1, \ a^{pq} = 1, \ t^{-1}btb = 1 \rangle$$

is *not* residually nilpotent.

Linear Groups

Recall that the *derived series* of a group G is the series defined by (transfinite) induction as follows:

$$\delta_1(G) = G, \ \delta_{\alpha+1}(G) = [\delta_\alpha(G), \delta_\alpha(G)],$$

and for a limit ordinal τ, $\delta_\tau(G) = \bigcap_{\alpha<\tau} \delta_\alpha(G)$. A group G is called *transfinitely solvable* if $\delta_\alpha(G) = 1$ for some ordinal α and simply *solvable* if $\delta_\alpha(G) = 1$ for some finite α. Let $\mathfrak{S}$ denote the class of all solvable groups. Note that $G \in r\mathfrak{S}$ if and only if $\delta_\omega(G) = \bigcap_{n\geq 1} \delta_n(G) = 1$. Let $\mathfrak{F}$ denote the class of finite groups.

Theorem 1.24 (Mal'cev [Mal65]; see also [Weh73, Theorem 4.7], and [Mer87, Theorem 51.2.1]). *If G is a finitely generated subgroup of the general linear group $GL_n(F)$, where F is a field of characteristic zero, then, for almost all primes p, there exists a normal subgroup N of finite index which is a residually finite p-group.*

The above theorem indeed provides a method of showing that a given finitely generated linear group belongs to the class $r\mathfrak{F}_p$. For, if for some prime p, it is possible to choose a normal subgroup $N \in r\mathfrak{F}_p$ such that $G/N \in \mathfrak{F}_p$, then, by the following result, G itself would be in $r\mathfrak{F}_p$.

Proposition 1.25 (Gruenberg [Gru57]). *Let $\mathcal{F}$ be the class $\mathfrak{S}$, $\mathfrak{F}$ or $\mathfrak{F}_p$ for some prime p. If $P \in \mathcal{F}$ and $H \in r\mathcal{F}$, then, for any group extension*

$$1 \to H \to G \to P \to 1,$$

the group $G \in \mathcal{F}$.

However, when we have a specific group G, the choice of the prime p can be quite a problem. Hence it would be desirable to find more constructive conditions for a linear group G to belong to the class $r\mathfrak{F}_p$ which does not use Mal'cev's theorem.

Let G be a subgroup of $GL_n(F)$ generated by the finite set A, where F is a field of characteristic zero. Consider the subring

$$K = \mathbb{Z}[\{a_{ij}, 1 \leq i, j \leq n, \mid a = (a_{ij}) \in A \cup A^{-1}\}]$$

of the field F, generated over $\mathbb{Z}$ by entries of matrices from A and A^{-1}, where A^{-1} is the set of inverses of the matrices in A. The ring K then is an integral domain and we can consider the group G as a subgroup of the group $GL_n(K)$.

By ([Mer87], Lemma 51.1.3), the ring K is residually finite. Furthermore, there exists an infinite set $\pi(K) = \{I_1, I_2, \ldots\}$ of ideals such that the quotient K/I_j is a field of characteristic p_j and the family of homomorphisms

$$\varphi_{j,n} : K \longrightarrow K/(I_j)^n, \quad j, \, n = 1, \, 2, \, \ldots,$$

is residually finalizing for any ideal I_j, i.e.,

$$|\mathrm{im}\,\varphi_{j,n}| < \infty, \quad \bigcap_{n=1}^{\infty} \ker \varphi_{j,n} = 0.$$

Let $\varphi_j = \varphi_{j,1}$. The homomorphism $\varphi_{j,n}$ can be extended to the homomorphism

$$\widetilde{\varphi}_{j,n} : GL_n(K) \longrightarrow GL_n(K/(I_j)^n).$$

We then have the following

Theorem 1.26 (Bardakov-Mikhailov [Bar07]). *Let G be a finitely generated subgroup of the group $GL_n(F)$, where F is a field of characteristic zero, K a finitely generated sub-domain of F, such that $G \subseteq GL_n(K)$. If for some ideal $I_j \in \pi(K)$ the group $\widetilde{\varphi}_j(G) \in \mathfrak{F}_{p_j}$ (resp. $\mathrm{r}\mathfrak{S}$), then $G \in \mathrm{r}\mathfrak{F}_{p_j}$ (resp. $\mathrm{r}\mathfrak{S}$).*

Proof. Let I_j be an ideal from the set $\pi(K)$. There exists the following exact sequence

$$1 \longrightarrow GL_n(K, \, I_j) \longrightarrow GL_n(K) \overset{\widetilde{\varphi}_j}{\longrightarrow} GL_n(K/I_j) \longrightarrow 1,$$

where $GL_n(K, \, I_j) = \ker(\widetilde{\varphi}_j)$ is a congruence subgroup. It follows from the proof of Mal'cev's Theorem ([Mer87], 51.2.1) that $GL_n(K, \, I_j)$ is a residually finite p_j-group. Now we have the following short exact sequence:

$$1 \longrightarrow G \cap GL_n(K, \, I_j) \longrightarrow G \longrightarrow \widetilde{\varphi}_j(G) \longrightarrow 1, \qquad (1.5)$$

where the group $G \cap GL_n(K, \, I_j)$ is a residually finite p_j-group. Since $\widetilde{\varphi}_j(G)$ is a finite p_j-group (resp. solvable group), G is a residually finite p_j-group (resp. residually solvable group) by Lemma 1.25 applied to the extension (1.5). $\square$

Let $\mathfrak{o}_F$ be the ring of integers of the field F. In the same way as Theorem 1.26, one can prove the following result.

Proposition 1.27 *Let G be a finitely generated subgroup of the group $\mathrm{PSL}_n(\mathfrak{o}_F)$. If for some prime p the image of G in $\mathrm{PSL}_n(\mathfrak{o}_F/p\mathfrak{o}_F)$ is a finite p-group, then G is a residually finite p-group.*

Consider the simplest case of Proposition 1.27, namely the case $F = \mathbb{Q}[i] = \mathbb{Q}[\sqrt{-1}]$. In this case the ring of integers is the ring $\mathbb{Z}[i] = \mathbb{Z} + i\mathbb{Z}$ of Gaussian integers .

Corollary 1.28 *Let G be a finitely generated subgroup of the Picard group* $\mathrm{PSL}_2(\mathbb{Z}[i])$. *If the image of G in* $\mathrm{PSL}_2(\mathbb{Z}_p[i])$ *is a finite p-group, then G is a residually finite p-group.*

Let us now have some examples of linear groups for which residual nilpotence can be proved by the above-mentioned method.

Consider subgroups of the Picard group of indices 12 and 24 respectively. The presentations of these groups can be found in [Wie78, Kru86] and [Bru84].

Example 1.29
Let
$$G_1 = \langle x,\, y \mid [x,\, yxy^{-2}xy] = 1\rangle.$$
This group has a faithful representation $f_1 : G_1 \to \mathrm{PSL}_2(\mathbb{Z}[i])$ given by
$$x \longmapsto \begin{pmatrix} 1 & 0 \\ 1 & 1 \end{pmatrix}, \quad y \longmapsto \begin{pmatrix} i & 1+i \\ -1 & -1 \end{pmatrix},$$
and the group $f_1(G_1)$ has index 12 in $\mathrm{PSL}_2(\mathbb{Z}[i])$ [Bru84]. Let $a = \varphi_2(f_1(x))$, $b = \varphi_2(f_1(y))$. Then $a^2 = b^2 = 1$. It is easy to see that
$$[a,\, b] = \begin{pmatrix} i & 1+i \\ 0 & i \end{pmatrix}$$

(where we use the same notation for elements in $\mathbb{Z}[i]$ and their images in $\mathbb{Z}_2[i]$). Thus we have $[a,\, b]^2 = 1$, and therefore the order of the group $\varphi_2(G_1)$ is 8. Thus, by Corollary 1.28, it follows that G_1 is a residually finite 2-group, and is therefore residually nilpotent.

Example 1.30
The unique normal subgroup of index 24 in $\mathrm{PSL}_2(\mathbb{Z}[i])$ is the group of Borromean rings:
$$G_2 = \langle x_1,\, x_2,\, x_3 \mid [x_3^{-1}, x_2, x_1] = 1,\ [x_1, x_3, x_2] = 1\rangle.$$

A faithful representation $f_2 : G_2 \to \mathrm{PSL}_2(\mathbb{Z}[i])$ of this group in the Picard group is given by:
$$x_1 \longmapsto \begin{pmatrix} 1 & -2i \\ 0 & 1 \end{pmatrix}, \quad x_2 \longmapsto \begin{pmatrix} 1 & 0 \\ -1 & 1 \end{pmatrix}, \quad x_3 \longmapsto \begin{pmatrix} 2+i & 2i \\ -1 & -i \end{pmatrix}$$

[Bru84]. We have $\varphi_2(x_1) = 1$. Let $z_i = \varphi_2(x_i)$, $i = 2, 3$. Then $z_i^2 = 1$ and $[z_2,\, z_3] = 1$; thus the group $\varphi_2(G_2) \simeq \mathbb{Z}_2 \oplus \mathbb{Z}_2$ is of order 4 and it follows from Corollary 1.28 that G_2 is a residually finite 2-group and is therefore residually nilpotent.

Pure Braid Groups

Let $n \geq 2$, $\mathbb{C}^n$ n-dimensional complex space and

$$\Delta = \{(z_1, \ldots, z_n) \mid z_i = z_j \text{ for some } i < j\} \subset \mathbb{C}^n.$$

There is an obvious permutation action of the symmetric group Σ_n on $\mathbb{C}^n \backslash \Delta$. The fundamental group

$$B_n = \pi_1((\mathbb{C}^n \backslash \Delta)/\Sigma_n)$$

of the quotient space is called the nth *braid group*. Clearly, every element of B_n can be viewed as an n-strand object which connects two collections of marked points in two planes in the 3-dimensional space. Braid groups B_n, $n \geq 2$, were introduced in [Art25] by E. Artin, where further details about these groups can be found.

For a fixed $n \geq 2$, the group B_n is known to have the following presentation:

$$\langle \sigma_1, \ldots, \sigma_{n-1} \mid \sigma_i \sigma_j = \sigma_j \sigma_i, \ |i - j| > 1,$$
$$\sigma_i \sigma_j \sigma_i = \sigma_j \sigma_i \sigma_j, \ |i - j| = 1 \rangle.$$

It is easy to see that, for every $n \geq 2$, there exists an epimorphism

$$\nu_n : B_n \to \Sigma_n,$$

where Σ_n is the symmetric group of degree n, defined by setting $\sigma_i \mapsto (i, i+1)$. The kernel $\ker(\nu_n)$ of this epimorphism is called the nth *pure braid group*, and denoted by P_n. Clearly, elements of pure braid groups can be viewed as braids which preserve the fixed order of the marked points on the plane. Also $P_n = \pi_1(\mathbb{C}^n \backslash \Delta)$. Furthermore, $\mathbb{C}^n \backslash \Delta$ has trivial homotopy groups in dimensions greater than one; hence, $\mathbb{C}^n \backslash \Delta$ is the classifying space of the pure braid group P_n.

The group P_n is generated by elements $a_{i,j}$, $1 \leq i < j \leq n$, where

$$a_{i-1,i} = \sigma_{i-1}^2, \ 2 \leq i \leq n,$$
$$a_{i,j} = \sigma_{j-1}\sigma_{j-2}\ldots\sigma_{i+1}\sigma_i^2\sigma_{i+1}^{-1}\ldots\sigma_{j-2}^{-1}\sigma_{j-1}^{-1}, \ i+1 < j \leq n$$

and has the following presentation:

$$\langle a_{i,j}, \ 1 \leq i < j \leq n \mid a_{i,k}^{-\varepsilon}a_{k,j}a_{i,k}^{\varepsilon} = (a_{i,j}a_{k,j})^\varepsilon a_{k,j}(a_{i,j}a_{k,j})^{-\varepsilon},$$
$$a_{k,m}^{-\varepsilon}a_{k,j}a_{k,m}^\varepsilon = (a_{k,j}a_{m,j})^\varepsilon a_{k,j}(a_{k,j}a_{m,j})^{-\varepsilon}, \ m < j,$$
$$a_{i,m}^{-\varepsilon}a_{k,j}a_{i,m}^\varepsilon = [a_{i,j}^{-\varepsilon}, a_{m,j}^{-\varepsilon}]^\varepsilon a_{k,j}[a_{ij}^{-\varepsilon}, a_{m,j}^{-\varepsilon}]^{-\varepsilon}, \ i < k < m,$$
$$a_{i,m}^{-\varepsilon}a_{k,j}a_{i,m}^\varepsilon = a_{k,j}, \ k < m, \ m < j, \text{ or } m < k, \ \varepsilon = \pm 1 \rangle.$$

It is a well known result of Artin that the pure braid group P_n is an iterated semidirect product of free groups. To be precise, the group P_n is the semi-direct product of the normal subgroup U_n, generated by elements $a_{1,n}, a_{2,n}, \ldots, a_{n-1,n}$ and the group naturally isomorphic to P_{n-1}. The subgroup U_n is free. By iterating this process, we get the decomposition of P_n as the following semi-direct product:

$$P_n = U_n \rtimes U_{n-1} \rtimes \cdots \rtimes U_3 \rtimes U_2, \qquad (1.6)$$

where U_i is the free subgroup of P_n, generated by elements $a_{1,i}, \ldots, a_{i-1,i}$.

Theorem 1.31 (Falk and Randell [Fal88]). *For all $n \geq 2$, the pure braid group P_n is residually torsion-free nilpotent.*

Proof. It follows from the form of defining relations of the group P_n that

$$[a_{i,n}, a_{k,m}] \in \gamma_2(U_n), \ 1 \leq i \leq n-1, \ 1 \leq k < m \leq n.$$

Thus, by induction on $k \geq 1$, it follows that the kth term of the lower central series of P_n can be decomposed as

$$\gamma_k(P_n) = \gamma_k(U_n) \rtimes \gamma_k(P_{n-1}), \ k \geq 2,$$

and induction on $n \geq 2$ yields the assertion. $\square$

It may be observed that, in general, it is easy to construct an example of a semidirect product of free groups which is not residually nilpotent.

Example 1.32

Let F be a 2-generated free group with generators a, b. Consider an automorphism φ of F given by

$$\varphi : a \mapsto a^2 b,$$
$$b \mapsto ab.$$

Consider the semidirect product $G = F \rtimes \mathbb{Z}$, where the generator of $\mathbb{Z}$ acts on F as the above automorphism. Then G has the following presentation:

$$G = \langle a, b, t \mid a^t = a^2 b, \ b^t = ab \rangle \simeq \langle b, t \mid [t, b^{-1}]^t = [t, b^{-1}]^2 b \rangle,$$

and we see that the element b lies in the intersection of lower central series of G; hence G is not residually nilpotent, since F is a subgroup in G and b is non-trivial.

1.3 Generalized Relation Modules

Free Presentations with Abelian Kernel

Recall that, given a ring R, a right R-module M is said to be *faithful* if $\operatorname{Ann}_R(M) := \{r \in R \mid M.r = 0\} = 0$. Given a group G, and an $R[G]$-module M, we say that G *acts faithfully* on M, if $G \cap (1 + \operatorname{Ann}_{R[G]}(M)) = 1$.

If $1 \to N \to \Pi \xrightarrow{\theta} G \to 1$ is an exact sequence of groups in which N is abelian, then N can be viewed as a $\mathbb{Z}[G]$-module via conjugation in Π, i.e., under the action induced by setting

$$n.g = x^{-1}nx, \ n \in N, \ x \in \Pi, \ \theta(x) = g.$$

It is easy to see that

$$N.\mathfrak{g}^n \subseteq \gamma_{n+1}(\Pi), \ n \geq 1,$$

and the following result is therefore an immediate consequence.

Theorem 1.33 *Let $1 \to N \to \Pi \xrightarrow{\theta} G \to 1$ be an exact sequence of groups in which N is abelian and Π is residually nilpotent.*

(i) *If N is a faithful $\mathbb{Z}[\Pi]$-module, then $\mathfrak{g}^\omega = 0$.*

(ii) *If the action of G on N is faithful, then*

$$G \cap (1 + \mathfrak{g}^\omega) = 1.$$

In particular, in either case, G is residually nilpotent.

The technique provided by the above Theorem has been successfully employed to characterize the residual nilpotence of certain groups.

Theorem 1.34 (Passi ([Pas75a], see also Lichtman [Lic77])). *Let F be a non-cyclic free group and R' the derived subgroup of a normal subgroup R of F. Then F/R' is residually nilpotent if and only if the augmentation ideal of the integral group ring $\mathbb{Z}[F/R]$ is residually nilpotent.*

Let us consider a case more general than the one to which the above result applies. Let F be a group and R, S its normal subgroups. Then the quotient group $\frac{R \cap S}{[R, S]}$ is an abelian group which can be viewed as a right $\mathbb{Z}[F/RS]$-module via conjugation in F, i.e., with the F/RS-action given by

$$w[R, S].(fRS) = f^{-1}wf[R, S] \quad (w \in R \cap S, \ f \in F).$$

We call this module a *generalized relation module* over the group F/RS. The study of such modules is of interest; for, observe that if F is a free group and $R = S$, then $\frac{R \cap S}{[R, S]}$ is a *relation module* for the group F/R and for $S = \gamma_n(R)$, $n > 1$, these modules are the *higher relation modules*. From the preceding discussion it is clear that the faithfulness of $\frac{R \cap S}{[R, S]}$ is related to the residual nilpotence of F/RS. By Theorem 1.33, we have

Theorem 1.35 *Let R and S be normal subgroups of a free group F such that $F/[R, S]$ is residually nilpotent. Let $G = F/RS$.*

(i) *If $\frac{R \cap S}{[R, S]}$ is a faithful $\mathbb{Z}[G]$-module, then $\mathfrak{g}^\omega = 0$.*

(ii) *If the action of G on $\frac{R \cap S}{[R, S]}$ is faithful, then*

$$G \cap (1 + \mathfrak{g}^\omega) = 1.$$

In particular, in either case, G is residually nilpotent.

Let V be a normal subgroup of F which is contained in $[R, S]$ and let

$$C = \{ f \in RS \mid [r, f] \in V \text{ for all } r \in R \cap S \}.$$

With the above notations we have:

Theorem 1.36 (Mikhailov-Passi [Mik06b]). *Suppose F/RS acts faithfully on $\frac{R \cap S}{[R, S]}$ and $\gamma_\omega(F/V) = 1$, then $\gamma_\omega(F/C) = 1$.*

Proof. Consider the following semi-direct product

$$(R \cap S)/V \rtimes F/C = \{ (rV, hC) \mid r \in R \cap S, h \in F \},$$

with binary composition given by

$$(h_1 C, r_1 V)(h_2 C, r_2 V) = (h_1 h_2 C, r_1^{h_2} r_2 V).$$

Note that

$$[\tfrac{R \cap S}{V}, \gamma_n(F/C)] \subseteq \tfrac{R \cap S}{V} \cap \gamma_{n+1}(F/V), \text{ for all integers } n \geq 1.$$

Let $f \in F$ be such that $fC \in \gamma_\omega(F/C)$. Then the hypothesis implies that $[\frac{R \cap S}{V}, fC] = 1$. Furthermore, since the action of F/RS on $\frac{R \cap S}{[R, S]}$ is assumed to be faithful, it follows that $f \in RS$ and consequently $f \in C$. $\square$

Observe that, in the case $R = S$, C/V is exactly the centre of R/V. Since the F/R-action on $R/\gamma_2(R)$ is known to be faithful (see Theorem 1.55), we have:

Corollary 1.37 (Mital [Mit71], Theorem 4.19). *Let $V \subset R$ and let C/V be the centre of R/V. Then $\gamma_\omega(F/V) = 1$ implies that $\gamma_\omega(F/C) = 1$.*

The investigation of the modules $\frac{R \cap S}{[R, S]}$ is also motivated from topological considerations, particularly in view of the relationship of such modules with the second homotopy modules which we discuss next.

The Second Homotopy Module

Let

$$\mathsf{P} = (X, \varphi, \mathcal{R}) \tag{1.7}$$

be a free presentation of the group G, $F = F(X)$ the free group with basis X, R the normal subgroup of F generated, as a normal subgroup, by $\mathcal{R}$; then we have an exact sequence $1 \to R \to F \xrightarrow{\varphi} G \to 1$. For a given presentation (1.7), we can construct the cellular model of (1.7) called the *standard 2-complex of* P, i.e., the 2-complex $K = K_{\mathsf{P}}$ which is constructed in the following way:

The complex K has a single 0-cell, the 1-skeleton of K is a bouquet of circles, one circle corresponding to each generator $x \in X$, which is oriented and labelled x. The words over $X \cup X^{-1}$ are then in bijective correspondence with the edge loops in the 1-skeleton. The 2-cells of K correspond bijectively to the relators $r \in \mathcal{R}$. The 2-cell corresponding to the relator $r = x_1 x_2 \ldots x_n$, $x_i \in X^{\pm 1}$ is attached to the 1-skeleton along the loop defined by $x_1 x_2 \ldots x_n$. The fundamental group $\pi_1(K)$ is known to be naturally isomorphic to the group G; this follows from the Seifert-van Kampen theorem (see Rotman [Rot88], p. 178). The group G then acts on the second homotopy group $\pi_2(K)$, via the standard action of the fundamental group of K (see Rotman [Rot88], p. 342).

Example 1.38
1. Let $\mathsf{P} = \langle x \mid xx^{-1}x \rangle$ be a presentation of the trivial group. Then K_{P} is the so called *Dunce hat*; it is a contractible space.

2. Let $\mathsf{P} = \langle x \mid x^2 \rangle$ be a presentation of the cyclic group of order 2. Then K_{P} is homotopically equivalent to the 2-dimensional real projective plane. Its second homotopy module is the infinite cyclic group $\mathbb{Z}$ with the generator of $\mathbb{Z}_2$ acting by change of sign: $1 \mapsto -1$.

3. For the one-relator presentation $\mathsf{P} = \langle x_1, \ldots, x_{2n} \mid \prod_{i=1}^{n} [x_{2i-1}, x_{2i}] \rangle$, $n \geq 1$, the standard 2-complex K_{P} is homotopically equivalent to the oriented surface of genus n. Its second homotopy module $\pi_2(K)$ is zero (see Example 1.43).

The first two of the above examples follow easily from Theorem 1.39 for which we now prepare to state.

Let $\mathsf{P} = (X, \varphi, \mathcal{R})$ be a free presentation of the group G and $K = K_{\mathsf{P}}$ the corresponding standard 2-complex. The homomorphism $\varphi \colon F \to G$ induces, by linear extension, a ring homomorphism $\varphi^* \colon \mathbb{Z}[F] \to \mathbb{Z}[G]$. While going from $\mathbb{Z}[F]$ to $\mathbb{Z}[G]$ under the homomorphism $\varphi^* \colon \mathbb{Z}[F] \to \mathbb{Z}[G]$, we will often drop writing φ^*.

Consider the module $\mathbb{Z}[G]^{\oplus |X|}$ (resp. $\mathbb{Z}[G]^{\oplus |\mathcal{R}|}$), namely, the direct sum of $|X|$ (resp. $|\mathcal{R}|$) copies of the integral group ring $\mathbb{Z}[G]$. Recall that $\mathbb{Z}[G]^{\oplus |X|} \simeq \mathfrak{f}/\mathfrak{f}\mathfrak{r}$, where $\mathfrak{f}$ (resp. $\mathfrak{r}$) is the augmentation ideal of $\mathbb{Z}[F]$ (resp. $\mathbb{Z}[R]$). Define

$$\kappa : \mathbb{Z}[G]^{\oplus |X|} \to \mathbb{Z}[G]$$

by setting

$$\kappa : (\alpha_x)_{x \in X} \mapsto \sum_{x \in X} (x - 1)\alpha_x, \ \alpha_x \in \mathbb{Z}[G].$$

Note that there are only a finite number of non-zero terms in the sequence $(\alpha_x) \in \mathbb{Z}[G]^{\oplus |X|}$, and as such the sum $\sum_{x \in X}(x - 1)\alpha_x$ is well-defined. Define

$$\tau : \mathbb{Z}[G]^{\oplus |\mathcal{R}|} \to \mathbb{Z}[G]^{\oplus |X|} \tag{1.8}$$

by setting

$$\tau : (\beta_r)_{r \in \mathcal{R}} \mapsto \sum_{r \in \mathcal{R}} (J_{rx}\beta_r)_{x \in X}, \quad \beta_r \in \mathbb{Z}[G],$$

where J_{rx} is the image in $\mathbb{Z}[G]$ of the right partial derivative $\frac{\partial r}{\partial x}$, $r \in \mathcal{R}$, $x \in X$ (see later the section on free differential calculus, p. 31).

The second homotopy module $\pi_2(K)$ can be viewed as the kernel of the map τ. This was first observed by K. Reidemeister [Rei50]. More precisely, we have the following result.

Theorem 1.39 *There is an exact sequence of $\mathbb{Z}[G]$-modules:*

$$0 \to \pi_2(K) \to \mathbb{Z}[G]^{\oplus |\mathcal{R}|} \xrightarrow{\tau} \mathbb{Z}[G]^{\oplus |X|} \xrightarrow{\kappa} \mathbb{Z}[G] \xrightarrow{\varepsilon} \mathbb{Z} \to 0, \tag{1.9}$$

where ε is the augmentation map. $\square$

Let $\widetilde{K}$ be the universal cover of the two-dimensional CW-complex K. Then the singular chain complex of $\widetilde{K}$ gives the following exact sequence of $\mathbb{Z}[\pi_1(K)]$-modules:

$$0 \to H_2(\widetilde{K}) \to C_2(\widetilde{K}) \to C_1(\widetilde{K}) \to C_0(\widetilde{K}) \to \mathbb{Z} \to 0,$$

which, in fact, is exactly the sequence (1.9). For more details see [Sie93] .

Note that $\mathrm{im}(\tau)$ is precisely the relation module R/R' arising from the presentation (1.7), the embedding

$$i : R/R' \to \mathbb{Z}[G]^{\oplus |X|} \tag{1.10}$$

is the well-known Magnus embedding

$$i : rR' \mapsto \left(\frac{\partial r}{\partial x}\right)_{x \in X}, \quad r \in R,$$

and we have the following exact sequence

$$0 \to R/R' \xrightarrow{i} \mathbb{Z}[G]^{\oplus |X|} \xrightarrow{\kappa} \mathfrak{g} \to 0. \tag{1.11}$$

Identity Sequences

There is an illustrative interpretation of elements of the second homotopy modules of standard complexes in terms of the so-called identity sequences (see, for example, [Pri91]). Let c_i, $i = 1, \ldots, m$, be words in the free group F, which are conjugates of elements from $\mathcal{R}$, i.e. $c_i = t_i^{\pm w_i}$, $t_i \in \mathcal{R}$, $w_i \in F$. Then the sequence

$$c = (c_1, \ldots, c_m) \tag{1.12}$$

is called an *identity sequence* if the product $c_1 \ldots c_m$ is the identity element in F. For an identity sequence (1.12), define its inverse c^{-1} by setting

$$c^{-1} = (c_m^{-1}, \ldots, c_1^{-1}).$$

For $w \in F$, the conjugate c^w is the sequence:

$$c^w = (c_1^w, \ldots, c_m^w),$$

which clearly is again an identity sequence. Define the following operations, called *Peiffer operations*, on the class of identity sequences:

(i) replace each w_i by any word equal to it in F;

(ii) delete two consecutive terms in the sequence if one is equal identically to the inverse of the other;

(iii) insert two consecutive terms in the sequence one of which is equal identically to the inverse of the other;

(iv) replace two consecutive terms c_i, c_{i+1} by terms $c_{i+1}, c_{i+1}^{-1} c_i c_{i+1}$;

(v) replace two consecutive terms c_i, c_{i+1} by terms $c_i c_{i+1} c_i^{-1}, c_i$.

Two identity sequences are called *equivalent* if one can be obtained from the other by a finite number of Peiffer operations. This defines an equivalence relation in the class of identity sequences. Denote by E_P the set of equivalence classes of identity sequences for a given group presentation (1.7) . Then E_P can be viewed as a group, with the binary operation defined to be the class obtained by the juxtaposition of two sequences:

For identity sequences c_1, c_2 and their equivalence classes $\langle c_1 \rangle, \langle c_2 \rangle \in E_P$, $\langle c_1 \rangle + \langle c_2 \rangle = \langle c_1 c_2 \rangle$. The inverse element of the class $\langle c \rangle$ is $\langle c^{-1} \rangle$ and the identity in E_P is the empty sequence.

It is easy to see that E_P is an abelian group. For two identity sequences $c = (c_1, \ldots, c_m)$ and $d = (d_1, \ldots, d_k)$, we have

$$\langle cd \rangle = \langle (c_1, \ldots, c_m, d_1, \ldots, d_k) \rangle = \langle (d_1, \ldots, d_k, c_1^{d_1 \ldots d_k}, \ldots, c_m^{d_1 \ldots d_m}) \rangle$$

by the relation (iv). Since $d_1 \ldots d_k = 1$ in F, we have

$$\langle cd \rangle = \langle (d_1, \ldots, d_k, c_1, \ldots, c_m) \rangle = \langle dc \rangle.$$

Furthermore, E_P is a $\mathbb{Z}[F]$-module, where the action of F is given by

$$\langle c \rangle \circ f = \langle c^f \rangle, \quad f \in F.$$

It is easy to show that

$$\langle c \rangle \circ r = \langle c \rangle, \quad r \in R,$$

i.e. the subgroup R acts trivially on E_P. To see this, let $r = r_1^{\pm v_1} \ldots r_k^{\pm v_k}$, $r_i \in \mathcal{R}$, $v_i \in F$. Then for any identity sequence $c = (c_1, \ldots, c_m)$, by (ii)-(iv), we have

$$\langle (c_1, \ldots, c_m) \rangle = \langle (c_1, \ldots, c_m, r_1^{\pm v_1}, \ldots, r_k^{\pm v_k}, r_k^{\mp v_k}, \ldots, r_1^{\mp v_1}) \rangle =$$
$$\langle (r_1^{\pm v_1}, \ldots, r_k^{\pm v_k}, c_1^r, \ldots, c_m^r, r_k^{\mp v_k}, \ldots, r_1^{\mp v_1}) \rangle = \langle (c_1^r, \ldots, c_m^r) \rangle.$$

Thus E_P can be viewed as a $\mathbb{Z}[G]$-module.

Now it is not hard to show that for a given presentation $\mathcal{P}$, the three $\mathbb{Z}[G]$-modules:

(1) the second homotopy module $\pi_2(K_P)$,

(2) the identity sequence module E_P, and

(3) the module $\pi_1(\mathrm{sk}^1 S(X, \mathcal{R}))$ (see Appendix, equation A.9)

are naturally isomorphic. The isomorphisms are defied as follows.

Let

$$\langle c \rangle = \langle (c_1^{\pm w_1}, \ldots, c_m^{\pm w_m}) \rangle \in E_P, \quad c_i \in \mathcal{R}, \; w_i \in F.$$

For a given $r_j \in \mathcal{R}$, $j \in J$, let $w(c)_j = \pm w_{k_1} \ldots \pm w_{k_l}$, where $(r_j^{\pm w_{k_1}}, \ldots, r_j^{\pm w_{k_l}})$ is a subsequence of c with $c_{k_i} = r_j$. Define

$$\psi_P : E_P \to \mathbb{Z}[G]^{\oplus |\mathcal{R}|}$$

by setting

$$\langle c \rangle \mapsto (\varphi^*(w(c)_1), \ldots, \varphi^*(w(c)_i), \ldots).$$

It is easy to show that $\mathrm{im}(\psi) \subseteq \ker(\tau)$, where τ is the map (1.8). Therefore, ψ can be viewed as a map

$$\psi : E_P \to \pi_2(K_P);$$

this map is an isomorphism of the $\mathbb{Z}[G]$-modules E_P and $\pi_2(K_P)$.

Recall (see Appendix, equation A.9) that

$$\pi_1(\mathrm{sk}^1 S(X, \mathcal{R})) = \frac{\langle y_j, \; j \in J \rangle^{F_1} \cap \langle y_j r_j^{-1}, \; j \in J \rangle^{F_1}}{[\langle y_j, \; j \in J \rangle^{F_1}, \langle y_j r_j^{-1}, \; j \in J \rangle^{F_1}]}, \qquad (1.13)$$

where $F_1 = F * F(y_j \mid j \in J)$. Let $f \in \langle y_j, \; j \in J \rangle^{F_1} \cap \langle y_j r_j^{-1}, \; j \in J \rangle^{F_1}$ with natural projection map $\partial_0 : F_1 \to F$. Then f can be written as

$$f = (y_{k_1} r_{k_1}^{-1})^{\pm u_1} \ldots (y_{k_m} r_{k_m}^{-1})^{\pm u_m}, \ u_i \in F_1.$$

Define the map

$$\xi : \mathrm{sk}^1(S(X, \mathcal{R})) \to E_\mathsf{P}$$

by setting

$$f[\langle y_j, \ j \in J \rangle^{F_1}, \ \langle y_j r_j^{-1}, \ j \in J \rangle^{F_1}] \mapsto (r_{k_1}^{\pm \partial_0(u_1)}, \ \ldots, \ r_{k_m}^{\pm \partial_0(u_m)}). \qquad (1.14)$$

Since $f \in \langle y_j \mid j \in J \rangle^{F_1}$, clearly the sequence in (1.14) is an identity sequence. It can be checked that ξ is an isomorphism of $\mathbb{Z}[G]$-modules.

Remark. Analogously, there are different ways to define the second homotopy module for a given free presentation of a Lie algebra. For a free Lie algebra F over a field k, with generating set $X = \{x_i\}_{i \in I}$ and a subset $\mathcal{R} = \{r_j \in F\}_{j \in J}$, form the two-sided ideal R as the closure of $\mathcal{R}$ in F. Then one can define the second homotopy module of the Lie algebra presentation $L := F/R = \langle x_i, \ i \in I \mid r_j, \ j \in J \rangle$ as follows (see Appendix, equation A.10):

$$\pi_1(\mathrm{sk}^1 S(X, \mathcal{R})) = \frac{(y_j, \ j \in J)F \cap (y_j - r_j, \ j \in J)F}{[(y_j, \ j \in J)F, \ (y_j - r_j, \ j \in J)F]},$$

which clearly becomes an $\mathcal{U}(L)$-module, where $\mathcal{U}(L)$ is the universal enveloping algebra of L.

There is also a *Fox derivative* version of the definition of the second homotopy module for Lie algebras. Let $A = A(x_i \mid i \in I)$ be a free associative algebra with the set of generators $\{x_i\}_{i \in I}$. Then A is the universal enveloping algebra $\mathcal{U}(F)$ of F. There is the augmentation map $\varepsilon : A \to k$, defined as $\varepsilon(x_i) = 0, \ i \in I$. Then every element $u \in \ker(\varepsilon)$ can be uniquely expressed as

$$u = \sum_{i \in I} \frac{\partial u}{\partial x_i} x_i, \ \frac{\partial u}{\partial x_i} \in A.$$

Analogous to the case of groups, the elements $\frac{\partial u}{\partial x_i}$ can be called *Fox derivatives*. One can extend to the map $\frac{\partial}{\partial x_i} : A \to A$, by setting $\frac{\partial 1}{\partial x_i} = 0, \ i \in I$. In analogy with the map τ defined in (1.8), define the map

$$\tau^{lie} : \mathcal{U}(L)^{\oplus |\mathcal{R}|} \to \mathcal{U}(L)^{\oplus |X|},$$

by setting

$$\tau^{lie} : (\beta_j)_{j \in J} \mapsto \sum_{j \in J} (J_{ji}^{lie} \beta_j)_{i \in I}, \ \beta_j \in \mathcal{U}(L),$$

where J_{ji}^{lie} is the image in $\mathcal{U}(L)$ of the element $\frac{\partial r_j}{\partial x_i} \in A$. Then it is not hard to see that there exists an isomorphism of $\mathcal{U}(L)$-modules:

$$\ker(\tau^{lie}) = \pi_1(\mathrm{sk}^1 S(X, \mathcal{R})).$$

Comparison of Different Presentations

Let $1 \to R \to F \to G \to 1$ and $1 \to S \to E \to G \to 1$ be two free presentations of the group G and R/R', S/S' the corresponding relation modules. Then, by Schanuel's lemma (see e.g. [Lut02]) there are free $\mathbb{Z}[G]$-modules P, Q such that there exists an isomorphism of $\mathbb{Z}[G]$-modules:

$$R/R' \oplus P \simeq S/S' \oplus Q. \qquad (1.15)$$

If, furthermore, $E = F$, then we can take $P = Q$.

In view of the sequence (1.9), the following statements are equivalent:

(i) $\pi_2(K) = 0$;

(ii) the relation module is a free $\mathbb{Z}[G]$-module, with basis

$$\{rR' \, ; \, r \in \mathcal{R}\}.$$

A presentation satisfying the statement (i) (or, equivalently, (ii)) is called an *aspherical presentation*.

Example 1.40

Let

$$P = \langle x_1, \ldots, x_n \mid r_1, \ldots, r_n \rangle \qquad (1.16)$$

be a balanced presentation of the trivial group and K its standard 2-complex (a presentation is said to be *balanced* if the number of generators equals the number of defining relators). Then, $\pi_1(K) = 1$ and the sequence (1.9) has the following form:

$$0 \to \pi_2(K) \to \mathbb{Z}^{\oplus n} \xrightarrow{\tau} \mathbb{Z}^{\oplus n} \xrightarrow{\kappa} \mathbb{Z} \xrightarrow{\varepsilon} \mathbb{Z} \to 0,$$

where the augmentation map ε is the identity map. Therefore, $\pi_2(K) = 0$ and the presentation $\mathcal{P}$ is aspherical. It may be further observed that the complex K is contractible.

Problem 1.41 *Does there exist a group-theoretical description of the class of groups which have balanced presentations?*

Consider the following transformations on a given presentation

$$P = \langle x_1, \ldots, x_m \mid r_1, \ldots, r_n \rangle :$$

(i) Replace any relator r_i by $r_i^{\pm f}$, where $f \in F_m = \langle x_1, \ldots, x_m \mid \emptyset \rangle$.

(ii) Replace any relator r_i by $r_i r_j$, where r_j is another relator of the given presentation and $i \neq j$ (and the converse operation).

(iii) Add a new generator t and a new relator t (and the converse operation).

The transformations (i) - (iii) are called *Andrews-Curtis moves*. The following problem was introduced by J. J. Andrews and M. L. Curtis in 1965 [And65] and is of great interest in topology as well as group theory.

Problem 1.42 Andrews-Curtis Conjecture: *Any balanced presentation of the trivial group can be transformed by Andrews-Curtis moves into the presentation $\langle x \mid x \rangle$.*

Example 1.43

One-relator presentations [Lyn50]. Let $\langle X \mid r \rangle$ be a one-relator presentation of a group G. Then the second homotopy module of its standard complex K is

$$\pi_2(K) \simeq (\bar{r} - 1)\mathbb{Z}[G], \tag{1.17}$$

where $\bar{r}$ is a root of r in the free group F with X as basis. Further, if r is not a proper power in the free group F, then $\pi_2(K) = 0$.

Remark. Let a group G admit an aspherical free presentation. Then, it follows from (1.15) that for any other free presentation $1 \to R \to F \to G \to 1$, the relation module R/R' is a projective $\mathbb{Z}[G]$-module.

Example 1.44

(M. J. Dunwoody) [Dun72]. Consider the trefoil group G given by the presentation $\langle a, b \mid a^2 = b^3 \rangle$. By (1.17), this presentation is aspherical. The group G is known to admit a 2-generator, 2-relator presentation $\langle x, y \mid v, w \rangle$, such that the corresponding relation module cannot be generated by a single element. It follows from (1.15) that the corresponding relation module S/S' is a projective $\mathbb{Z}[G]$-module, which, however, is not a free $\mathbb{Z}[G]$-module [Ber79]. Thus it is possible for a group to have two presentations such that one of them is aspherical and the other is not.

The following result is due to Gutierrez and Ratcliffe [Gut81].

Theorem 1.45 *If K is a connected 2-dimensional CW-complex, and K_1 and K_2 are sub-complexes such that $K = K_1 \cup K_2$ and $K_1 \cap K_2$ is the 1-skeleton $K^{(1)}$ of K, then there is an exact sequence*

$$0 \to i_1\pi_2(K_1) \oplus i_2\pi_2(K_2) \xrightarrow{\alpha} \pi_2(K) \to \frac{R \cap S}{[R, S]} \to 0,$$

of $\mathbb{Z}[\pi_1(K)]$-modules, where α is induced by inclusion, R is the kernel of $\pi_1(K^{(1)}) \to \pi_1(K_1)$, S is the kernel of $\pi_1(K^{(1)}) \to \pi_1(K_2)$ and the action of $\pi_1(K) \simeq \pi_1(K^{(1)})/RS$ on $\frac{R \cap S}{[R, S]}$ is induced by conjugation. □

Let $\mathsf{P} = \langle X \mid \mathbf{t} \rangle$ be a free presentation of a group G, and let K be the standard 2-complex associated to this presentation. Suppose $\mathbf{t} = \mathbf{r} \cup \mathbf{s}$, where $\mathbf{r}$, $\mathbf{s}$ are some sets of words over X. Let $\mathsf{P}_1 = \langle X \mid \mathbf{r} \rangle$, $\mathsf{P}_2 = \langle X \mid \mathbf{s} \rangle$. It follows from Theorem 1.45 that there exists a short exact sequence of $\mathbb{Z}[G]$-modules:

$$0 \to j_1(\pi_2(K_1)) \oplus j_2(\pi_2(K_2)) \to \pi_2(K) \to \frac{R \cap S}{[R,\,S]} \to 0, \qquad (1.18)$$

where R, S are the normal closures of $\mathbf{r}$ and $\mathbf{s}$ respectively in the free group F with basis X, $G = F/RS$ and j_i is the homomorphism induced by the inclusion of the standard 2-complex K_i into K, $i = 1, 2$.

The sequence (1.18) can be easily viewed by using the interpretation of $\pi_2(K)$ in terms of identity sequences [Pri91]. For an identity sequence $c :=$ $(c_1, \ldots, c_m)$ with $c_i \in \mathbf{r} \cup \mathbf{s}$, choose a subsequence $(c_{k_1}, \ldots, c_{k_l})$ in c with $c_{k_i} \in \mathbf{r}$. Then the map

$$\kappa : \pi_2(K) \to \frac{R \cap S}{[R,\,S]} \qquad (1.19)$$

is defined by setting

$$\kappa : c \mapsto c_{k_1} \ldots c_{k_l}[R,\,S].$$

Clearly, $c_{k_1} \ldots c_{k_l} \in R$, by construction, but also $c_{k_1} \ldots c_{k_l} \in S$, since the sequence $(c_1, \ldots, c_m)$ is an identity sequence, i.e. $c_1 \ldots c_m = 1$ in F. In particular, if P is aspherical, then $R \cap S = [R,\,S]$. It is thus clear from the exact sequence (1.18) that the study of $\frac{R \cap S}{[R,\,S]}$ as a $\mathbb{Z}[F/RS]$-module is of interest for the investigation of the second homotopy modules of the standard 2-complexes.

The structure of the second homotopy module for one-relator presentations (1.17) gives another kind of exact sequence for second homotopy modules:

Theorem 1.46 [Gut81]. *Let* $G = \langle X \mid \mathcal{N} \rangle$ *be a group presentation with relation module* $N/\gamma_2(N)$. *Let* K_r *be the standard 2-complex for a presentation* $\langle X \mid r \rangle$, $r \in \mathcal{N}$, *and let* s_r *be the root of* r *in the free group* $F(X)$. *Then there is an exact sequence of* $\mathbb{Z}[G]$-*modules:*

$$0 \to \oplus_{r \in \mathcal{N}} i_* \pi_2(K_r) \xrightarrow{\alpha} \pi_2(K) \to \oplus_{r \in \mathcal{N}} \mathbb{Z}[G]/(s_r - 1)\mathbb{Z}[G] \xrightarrow{\gamma} N/\gamma_2(N) \to 0,$$

where α *is induced by inclusion and* γ *maps* $1 + (s_r - 1)\mathbb{Z}[G]$ *onto* $r\gamma_2(N)$ *for each* $r \in \mathcal{N}$. $\qquad \square$

Applying Theorem 1.46 to the case of two-relator presentations, and using the sequence (1.18), we immediately have the following result.

Corollary 1.47 (see also [Har91a]). *If* F *is a free group,* r *and* s *words in* F, $R = \langle r \rangle^F$, $S = \langle s \rangle^F$, *then the group* $\frac{R \cap S}{[R,\,S]}$ *is a free abelian group.* $\square$

Consider the following homotopy pushout of topological spaces:

$$
\begin{array}{ccc}
K(F,1) & \longrightarrow & K(F/R,1) \\
\downarrow & & \downarrow \\
K(F/S,1) & \longrightarrow & X,
\end{array}
\qquad (1.20)
$$

where $K(G,1)$ denotes the classifying space of G and the maps between the classifying spaces are induced by the natural projections of the corresponding groups. By Seifert-van Kampen theorem (see [Mas67], Theorem 2.1) the fundamental group $\pi_1(X)$ is isomorphic to F/RS. Furthermore, we have:

Theorem 1.48 [Bro84] *The second homotopy module* $\pi_2(X)$ *is isomorphic to* $\frac{R \cap S}{[R,S]}$ *as* $\mathbb{Z}[F/RS]$-*module.* $\square$

Remark. The map κ defined on page 26 can be obtained as a map between π_2 terms in the following diagram of homotopy pushouts:

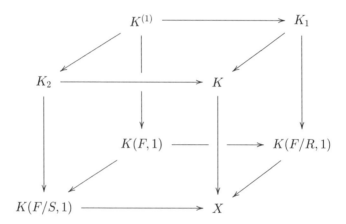

Example 1.49
Let G be a group, and Y a $K(G,1)$-space and $Y^{(1)}$ its 1-skeleton. Let $F = \pi_1(Y^{(1)})$ and R its normal subgroup such that $G = F/R$. Then the second homotopy module of the space X obtained from Y by contracting its 1-skeleton into a single point is $R/[F, R]$, viewed as a trivial G-module.

If $F = RS$, then the Mayer-Vietoris homology sequence applied to (1.20) immediately gives the following:

Theorem 1.50 (Brown [Bro84]). *There exists the following exact sequence:*

$$
H_2(F) \to H_2(F/R) \oplus H_2(F/S) \to \frac{R \cap S}{[R,S]} \xrightarrow{q}
$$
$$
H_1(F) \to H_1(F/R) \oplus H_1(F/S) \to 0. \quad (1.21)
$$

$\square$

If $F = RS$ is a normal subgroup of a group E, then it is easy to see that the long exact sequence (1.21) is a sequence of $\mathbb{Z}[E/F]$-modules, via conjugation. In the case when E is a non-cyclic free group, Hartley and Kuz'min constructed an exact sequence like (1.21), which gives an analog of the Magnus embedding (see also 1.10 for details), namely the embedding

$$R/R' \hookrightarrow \mathfrak{e}/\mathfrak{e}\mathfrak{r} \simeq \mathbb{Z}[E/R]^{\oplus rank(E)}, \; rR' \mapsto r - 1 + \mathfrak{e}\mathfrak{r}, \; r \in R \triangleleft E. \quad (1.22)$$

Theorem 1.51 (Hartley-Kuzmin [Har91a]). *There exists the following exact sequence of $\mathbb{Z}[E/F]$-modules:*

$$0 \to H_2(F/R) \oplus H_2(F/S) \to \frac{R \cap S}{[R, S]} \to W \xrightarrow{\nu}$$
$$\Delta(E/R \cap S) \otimes_{\mathbb{Z}[E/R \cap S]} \mathbb{Z}[E/F] \to 0, \quad (1.23)$$

where W is a free $\mathbb{Z}[E/F]$-module on a basis in one to one correspondence with a basis of E.

[Recall that $\Delta(G)$, as also $\mathfrak{g}$, denotes the augmentation ideal of the integral group ring $\mathbb{Z}[G]$.]

Proof. Let $V = \Delta(E/R \cap S) \otimes_{\mathbb{Z}[E/R \cap S]} \mathbb{Z}[E/F]$. Applying to the sequence

$$0 \to \Delta(E/R \cap S) \to \mathbb{Z}[E/R \cap S] \to \mathbb{Z} \to 0$$

the functor $\otimes_{\mathbb{Z}[E/R \cap S]} \mathbb{Z}[E/RS]$, we get the following exact sequence:

$$0 \to \mathrm{Tor}_1^{\mathbb{Z}[E/R \cap S]}(\mathbb{Z}, \mathbb{Z}[E/RS]) \to V \to \Delta(E/F) \to 0.$$

Applying the homological functor $H_*(E/R \cap S, -)$ to the augmentation sequence

$$0 \to \Delta(F/R \cap S)\mathbb{Z}[E/R \cap S] \to \mathbb{Z}[E/R \cap S] \to \mathbb{Z}[E/F] \to 0,$$

we conclude that

$$\mathrm{Tor}_1^{\mathbb{Z}[E/R \cap S]}(\mathbb{Z}, \mathbb{Z}[E/F]) = H_1(E/R \cap S, \mathbb{Z}[E/F]) \simeq \frac{\Delta(F/R \cap S)\mathbb{Z}[E/R \cap S]}{\Delta(F/R \cap S)\Delta(E/R \cap S)}.$$

Observe that the last term is naturally isomorphic to $H_1(F/R \cap S)$.

To construct the map $\nu : W \to V$, let us apply the functor $\otimes_{\mathbb{Z}[E/R \cap S]} \mathbb{Z}[E/F]$ to the Magnus embedding

$$0 \to R \cap S/\gamma_2(R \cap S) \to \mathbb{Z}[E/R \cap S]^{\oplus rank(E)} \xrightarrow{\kappa} \Delta(E/R \cap S) \to 0,$$

and define

$$\nu := \kappa \otimes_{\mathbb{Z}[E/R \cap S]} \mathrm{id}_{\mathbb{Z}[E/F]}.$$

We thus have the following commutative diagram:

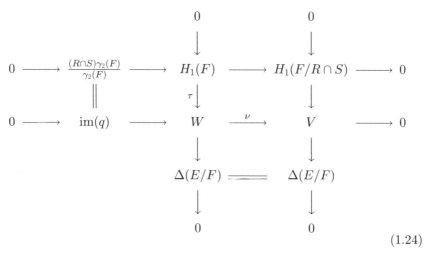

$$(1.24)$$

where τ is the Magnus embedding and q is the same as in (1.21). The assertion then follows from the commutative diagram (1.24) and exact sequence (1.21). □

Remark.
Theorems 1.45 and 1.48 are generalized for the case of three normal subgroups in [Bau] and [Ell08]; in particular, the following result is given in [Bau]:

Let K be a two-dimensional CW-complex with subcomplexes K_1, K_2, K_3 such that $K = K_1 \cup K_2 \cup K_3$ and $K_1 \cap K_2 \cap K_3$ is the 1-skeleton K^1 of K. There is a natural homomorphism of $\pi_1(K)$-modules

$$\pi_3(K) \to \frac{R_1 \cap R_2 \cap R_3}{[R_1, R_2 \cap R_3][R_2, R_3 \cap R_1][R_3, R_1 \cap R_2]},$$

where $R_i = \ker\{\pi_1(K^1) \to \pi_1(K_i)\}$, $i = 1, 2, 3$ and the action of $\pi_1(K) = F/R_1R_2R_3$ on the right hand abelian group is defined via conjugation in F.

Faithfulness of $\frac{R \cap S}{[R, S]}$

We need the following well-known result.

Lemma 1.52 *If $z \in \mathbb{Z}[G]$ is a nonzero element and $(g - 1)z = 0$ for all $g \in G$, then G must be of finite order.* □

For a group G, denote by $\delta^+(G)$ the subgroup generated by the torsion elements $g \in G$ which have only finitely many conjugates in G. Recall (see [Pas77b]) that

(i) $\delta^+(G)$ is a locally finite normal subgroup of G;

(ii) $\delta^+(G) = 1$ if and only if G does not contain a nontrivial finite normal subgroup.

An ideal I in the group algebra $k[G]$, where k is a field, is said to be *controlled* by a normal subgroup H of G if

$$I = (k[H] \cap I)k[G].$$

An *annihilator ideal* I in $k[G]$ is, by definition, a two-sided ideal such that there exists a subset $X \subseteq k[G]$ with

$$I = \mathrm{Ann}_{k[G]}(X) := \{\alpha \in k[G] \mid X.\alpha = 0\}.$$

Observe that if a $k[G]$-module M embeds in a free $k[G]$-module, then $\mathrm{Ann}_{k[G]}(M)$ is an annihilator ideal in $k[G]$. We need a result, due to M. Smith [Smi70], about semi-prime group algebras, i.e. group algebras which do not have any nonzero two-sided ideal with square zero. If k is a field of characteristic 0, then the group algebra $k[G]$ is always semi-prime ([Pas77b]).

Theorem 1.53 (Smith [Smi70]). *In a semi-prime group algebra $k[G]$, in particular if $char(k) = 0$, the annihilator ideals are controlled by the normal subgroup $\delta^+(G)$.*

As an immediate consequence we have:

Corollary 1.54 *If a $\mathbb{Z}[G]$-module M admits a nontrivial homomorphism into a free $\mathbb{Z}[G]$-module and $\delta^+(G) = 1$, then M is a faithful $\mathbb{Z}[G]$-module.*

Let F be a non-cyclic free group and R a proper normal subgroup of F. Then the quotients $\gamma_n(R)/\gamma_{n+1}(R)$ $(n \geq 1)$ of the lower central series of R can be regarded as right modules over the group $G := F/R$ via conjugation in F. It has been proved by Passi [Pas75a], using Theorem 1.53, that the relation module $R/\gamma_2(R)$ is always faithful. It had been earlier shown by Mital-Passi [Mit73] that the faithfulness of relation modules implies that of all the higher relation modules $\gamma_n(R)/\gamma_{n+1}(R)$ $(n > 1)$ as well. We thus have the following

Theorem 1.55 (Mital-Passi [Mit73], [Pas75a]). *If R is a proper normal subgroup of a non-cyclic free group F, then the F/R-modules $\gamma_n(R)/\gamma_{n+1}(R)$ $(n \geq 1)$ are all faithful.*

To present the proof of the above result, following the work of Mital and Passi, we need the basic notions of Fox's free differential calculus [Fox53].

Free Differential Calculus

Let G be a group, k a commutative ring with identity and M a right $k[G]$-module. A (right) *derivation* $d : k[G] \to M$ is a k-homomorphism which satisfies

$$d(xy) = d(x)y + d(y)$$

for all x, $y \in G$. Let $\epsilon : k[G] \to k$ be the augmentation map. It is easy to see that every derivation $d : k[G] \to M$ satisfies the following equations:

$$d(uv) = d(u)v + \epsilon(u)v \text{ for all } u, \ v \in k[G]; \qquad (1.25)$$

$$d(\lambda) = 0 \text{ for all } \lambda \in k; \qquad (1.26)$$

$$d(g^{-1}) = -d(g)g^{-1}, \text{ for all } g \in G. \qquad (1.27)$$

Let F be a free group with basis $X = \{x_i\}_{i \in I}$. Then, for every $i \in I$, we have a unique derivation $d_i : F \to \mathbb{Z}[F]$, called the *partial derivation* with respect to the generator x_i (also denoted by $\frac{\partial}{\partial x_i}$), which satisfies

$$d_i(x_j) = \begin{cases} 1, \text{ for } i = j, \\ 0 \text{ for } i \neq j \end{cases}.$$

If $d : \mathbb{Z}[F] \to \mathbb{Z}[F]$ is an arbitrary derivation and $d(x_i) = h_i$, then

$$d(u) = \sum_{i \in I} h_i d_i(u), \text{ for all } u \in \mathbb{Z}[F]. \qquad (1.28)$$

Let $1 \to R \to F \xrightarrow{\alpha} G \to 1$ be a noncyclic free presentation of a group G, with $R \neq 1$. Extend α to a ring homomorphism $\mathbb{Z}[F] \to \mathbb{Z}[G]$ by linearity. We need two formulas which are both easy to verify.

Let n, i be integers with $n > i \geq 0$, and $u \in \mathfrak{f}^i$, $v \in \mathfrak{f}^{n-i-1}\mathfrak{r}$. Then, for arbitrary derivations $D_1, \ldots, D_n$,

$$\alpha D_n \cdots D_1(uv) = \epsilon D_i \cdots D_1(u)\alpha D_n \cdots D_{i+1}(v), \qquad (1.29)$$

where $\epsilon : \mathbb{Z}[F] \to \mathbb{Z}$ is the augmentation map. Since F is residually nilpotent, there exists an integer $c \geq 1$ such that

$$R \subseteq \gamma_c(F) \text{ but } R \not\subseteq \gamma_{c+1}(F).$$

Let $r \in \gamma_i(R)$, $s \in \gamma_j(R)$, $i + j = n$. Then, for arbitrary derivations, $D_1, \ldots, D_{c(n-1)+1} : \mathbb{Z}[F] \to \mathbb{Z}[F]$, we have

$$\alpha D_{c(n-1)+1} \cdots D_1([r, s]) = \epsilon D_{ci} \cdots D_1(r)\alpha D_{c(n-1)+1} \cdots D_{ci+1}(s)$$
$$- \epsilon D_{cj} \cdots D_1(s)\alpha D_{c(n-1)+1} \cdots D_{cj+1}(r). \qquad (1.30)$$

For all natural numbers n, we have a monomorphism

$$\theta_n : \gamma_n(R)/\gamma_{n+1}(R) \to \mathfrak{f}\mathfrak{r}^{n-1}/\mathfrak{f}\mathfrak{r}^n, \ r\gamma_{n+1}(R) \mapsto r - 1 + \mathfrak{f}\mathfrak{r}^n, \ r \in \gamma_n(R)$$

of right G-modules. Let K_i be the set consisting of all elements

$$\alpha D_{c(i-1)+1} \cdots D_1(r) \in \mathbb{Z}[G]$$

with $r \in \gamma_i(R)$ and $D_1, \cdots, D_{c(i-1)+1} : \mathbb{Z}[F] \to \mathbb{Z}[F]$ arbitrary derivations.

Lemma 1.56 $\mathrm{Ann}_{\mathbb{Z}[G]}(\gamma_n(R)/\gamma_{n+1}(R)) \subseteq \mathrm{Ann}_{\mathbb{Z}[G]}(K_n)$.

Proof. Let $z \in \mathrm{Ann}_{\mathbb{Z}G}(\gamma_n(R)/\gamma_{n+1}(R))$ and let $u \in \mathbb{Z}[F]$ be a preimage of u under α. Then, applying θ_n, we have

$$(r - 1)u \in \mathfrak{f}\mathfrak{r}^n \subseteq \mathfrak{f}^{c(n-1)+1}\mathfrak{r}, \text{ for all } r \in \gamma_n(R).$$

Successive differentiation shows that

$$\alpha D_{c(n-1)+1} \cdots D_1(r).z = 0$$

for all $r \in \gamma_n(R)$ and derivations $D_1, \cdots, D_{c(n-1)+1} : \mathbb{Z}[F] \to \mathbb{Z}[F]$. $\square$

Since $\mathfrak{f}/\mathfrak{f}\mathfrak{r}$ is a free $\mathbb{Z}[G]$-module with basis $x_i - 1 + \mathfrak{f}\mathfrak{r}$, it follows that

$$\mathrm{Ann}_{\mathbb{Z}[G]}(R/\gamma_2(R)) = \mathrm{Ann}_{\mathbb{Z}[G]}(K_1),$$

and thus $\mathrm{Ann}_{\mathbb{Z}[G]}(R/\gamma_2(R))$ is an annihilator ideal. Consequently, by Theorem 1.53, the ideal $\mathbb{Q}\,\mathrm{Ann}_{\mathbb{Z}[G]}(R/\gamma_2(R))$ of the rational group algebra $\mathbb{Q}[G]$ is controlled by $\delta^+(G)$. It then follows that, to prove the vanishing of $\mathrm{Ann}_{\mathbb{Z}[G]}(R/\gamma_2(R))$, it suffices to consider the case when G is finite. It is, however, known [Gas54] that, in this case, the $\mathbb{Q}[G]$-module $\mathbb{Q} \otimes R/\gamma_2(R)$ contains $\mathbb{Q}[G]$ as a direct summand and so has trivial annihilator. Therefore $\mathrm{Ann}_{\mathbb{Z}[G]}(R/\gamma_2(R))$ is trivial when G is finite, and hence it is so in general.

We next consider higher relation modules. Let i be the least natural number such that

$$\mathrm{Ann}_{\mathbb{Z}[G]}(\gamma_n(R)/\gamma_{n+1}(R)) \subseteq \mathrm{Ann}_{\mathbb{Z}[G]}(K_i).$$

If $i = 1$, then $\mathrm{Ann}_{\mathbb{Z}[G]}(\gamma_n(R)/\gamma_{n+1}(R)) = 0$, since

$$\mathrm{Ann}_{\mathbb{Z}[G]}(K_1) = \mathrm{Ann}_{\mathbb{Z}[G]}(R/\gamma_2(R)) = 0.$$

Suppose $i \geq 2$. In this case we assert that

$$\gamma_{i-1}(R) \not\subseteq \gamma_{c(i-1)+1}(F). \tag{1.31}$$

For, suppose $\gamma_{i-1}(R) \subseteq \gamma_{c(i-1)+1}(F)$. Choose $r \in R \backslash \gamma_{c+1}(F)$. Then there exist partial derivations $d_{i_1}, \cdots, d_{i_c} : F \to \mathbb{Z}[F]$ such that

$$\epsilon d_{i_c} \cdots d_{i_1}(r) \neq 0. \tag{1.32}$$

For arbitrary $t \in \gamma_{i-1}(R)$ and derivations $D_{c+1}, \cdots, D_{c(i-1)+1}$, we then have $[r, t] \in \gamma_i(R)$ and therefore

$$\alpha D_{c(i-1)+1} \cdots D_{c+1} d_{i_c} \cdots d_{i_1}([r, t]).z = 0$$

for all $z \in \mathrm{Ann}_{\mathbb{Z}[G]}(\gamma_n(R)/\gamma_{n+1}(R))$. On successive differentiation, and observing that $t - 1 \in \mathfrak{f}^{c(i-1)+1}$, it follows that

$$\alpha D_{c(i-1)+1} \cdots D_{c+1}(t).z = 0,$$

i.e., $z \in \mathrm{Ann}_{\mathbb{Z}[G]}(K_{i-1})$. This contradicts the choice of i, and so our claim (1.31) is established.

Lemma 1.57

$\mathrm{Ann}_{\mathbb{Z}[G]}(\gamma_n(R)/\gamma_{n+1}(R)) \subseteq \mathrm{Ann}_{\mathbb{Z}[G]}\{\alpha D(r) \,|\, r \in R \cap \gamma_{c+1}(F), \, D \text{ is a derivation}\}.$

Proof. Let $r \in R \cap \gamma_{c+1}(F)$, $z \in \mathrm{Ann}_{\mathbb{Z}[G]}(\gamma_n(R)/\gamma_{n+1}(R))$, and $D : \mathbb{Z}[F] \to \mathbb{Z}[F]$ a derivation. Choose $s \in \gamma_{i-1}(R) \backslash \gamma_{c(i-1)+1}(F)$. Then there exist partial derivations $d_{i_1}, \cdots, d_{i_{c(i-1)}}$ such that $\epsilon d_{i_{c(i-1)}} \cdots d_{i_1}(s) \neq 0$. Now $[r, s] \in \gamma_i(R)$ and therefore

$$\alpha D d_{i_{c(i-1)}} \cdots d_{i_1}([r, s]).z = 0$$

for arbitrary derivation D. It then easily follows on simplification that $\alpha D(r).z = 0$. $\square$

Let $z \in \mathrm{Ann}_{\mathbb{Z}[G]}(\gamma_n(R)/\gamma_{n+1}(R))$, $r \in R$ and $f_1, \cdots, f_c \in F$. Then

$$[\cdots[[r, f_1], \cdots], f_c] \in R \cap \gamma_{c+1}(F).$$

Therefore, by Lemma 1.57,

$$\alpha D([\cdots[[r, f_1], \cdots], f_c]).z = 0$$

for every derivation $D : \mathbb{Z}[F] \to \mathbb{Z}[F]$. Now observe that

$$\alpha D([\cdots[[r, f_1], \cdots], f_c])) = \alpha D(r)(\alpha(f_1) - 1) \cdots (\alpha(f_c) - 1).$$

Hence

$$\alpha D(r)(\alpha(f_1) - 1) \cdots (\alpha(f_c) - 1).z = 0.$$

It thus follows that

$$\mathfrak{g}^c. \mathrm{Ann}_{\mathbb{Z}[G]}(\gamma_n(R)/\gamma_{n+1}(R)) \subseteq \mathrm{Ann}_{\mathbb{Z}[G]}(R/\gamma_2(R)).$$

Hence
$$\mathfrak{g}^c. \mathrm{Ann}_{\mathbb{Z}[G]}(\gamma_n(R)/\gamma_{n+1}(R)) = 0.$$

If G is infinite, then, by Lemma 1.52, it immediately follows from the above equation that $\mathrm{Ann}_{\mathbb{Z}[G]}(\gamma_n(R)/\gamma_{n+1}(R)) = 0$.

Finally, suppose G is finite. Let x, y be two distinct elements of the free basis X of F and d_x, d_y the corresponding partial derivations on F. Since G is finite, there exists a natural number m such that $x^m \in R$. By Lemma 1.56 we have

$$\alpha d_y d_x^{n-1}([\cdots [[r, \underbrace{x^m], x^m], \cdots, x^m}_{n-1 \text{ terms}}].z = 0 \qquad (1.33)$$

for all $r \in R$ and $z \in \mathrm{Ann}_{\mathbb{Z}[G]}(\gamma_n(R)/\gamma_{n+1}(R))$. Iterative simplification of the equation (1.33) yields $\alpha d_y(r).z = 0$. Hence it follows that

$$\mathrm{Ann}_{\mathbb{Z}[G]}(\gamma_n(R)/\gamma_{n+1}(R)) \subseteq \mathrm{Ann}_{\mathbb{Z}[G]}(R/\gamma_2(R)) = 0,$$

completing the proof of Theorem 1.55. $\square$

We next note that, in general, the group F/RS does not act faithfully on $\frac{R \cap S}{[R, S]}$.

Example 1.58

Let $F = \langle a, b \mid \emptyset \rangle$ be the free group of rank 2, $R = \langle a, b^{-1}ab, b^2 \rangle$, $S = \langle b^2 \rangle^F$. Then $S \subset R$ and $S/[R, S] = \langle b^2[R, S] \rangle$ is an infinite cyclic group on which $G := F/R$ acts trivially. Thus $\mathrm{Ann}_{\mathbb{Z}[G]}(S/[R, S]) = \mathfrak{g}$, the augmentation ideal of the integral group ring $\mathbb{Z}[G]$.

Theorem 1.59 (Mikhailov-Passi [Mik06c]). *Let F be a free group of finite rank $j > 1$. If $\frac{RS}{(R \cap S)\gamma_2(RS)}$ is finite, then $\frac{R \cap S}{[R, S]}$ is a faithful $\mathbb{Z}[F/RS]$-module.*

Furthermore, if $F/R \cap S$ is finite, then there is an isomorphism of $\mathbb{Q}[F/RS]$-modules:

$$\frac{R \cap S}{[R, S]} \otimes \mathbb{Q} \simeq \mathbb{Q} \oplus \underbrace{\mathbb{Q}[F/RS] \oplus \cdots \oplus \mathbb{Q}[F/RS]}_{} \quad (j - 1 \text{ terms}), \qquad (1.34)$$

Proof. Consider the following exact sequence with the obvious homomorphisms:

$$1 \to \frac{R \cap S \cap \gamma_2(RS)}{[R, S]} \otimes \mathbb{Q} \to \frac{R \cap S}{[R, S]} \otimes \mathbb{Q} \to RS/\gamma_2(RS) \otimes \mathbb{Q} \to$$

$$\frac{RS}{(R \cap S)\gamma_2(RS)} \otimes \mathbb{Q} \to 1, \qquad (1.35)$$

where the tensor products are over $\mathbb{Z}$. Since $\frac{RS}{(R \cap S)\gamma_2(RS)}$ is a finite abelian group, the last term in (1.35) is trivial. It follows that $\frac{R \cap S}{[R, S]} \otimes \mathbb{Q}$ maps onto $RS/\gamma_2(RS) \otimes \mathbb{Q}$ which is a faithful $\mathbb{Q}[F/RS]$-module [Pas75a]. Therefore, $\frac{R \cap S}{[R, S]}$ is a faithful $\mathbb{Z}[F/RS]$-module.

Next observe that $[R \cap S, RS] = [R \cap S, R][R \cap S, S] \subseteq [R, S]$; therefore, there exists a natural epimorphism

$$H_2(RS/(R \cap S)) \simeq \frac{R \cap S \cap \gamma_2(RS)}{[R \cap S, RS]} \rightarrow \frac{R \cap S \cap \gamma_2(RS)}{[R, S]},$$

where $H_2(-)$ denotes the second integral homology group. In case $F/R \cap S$ is finite, then $H_2(RS/(R \cap S))$ is finite, and consequently $\frac{R \cap S \cap \gamma_2(RS)}{[R, S]}$ is finite. Thus, from (1.35), we have

$$\frac{R \cap S}{[R, S]} \otimes \mathbb{Q} \simeq RS/\gamma_2(RS) \otimes \mathbb{Q}$$

and the structure as asserted in (1.34) follows from [Gas54]. $\square$

Theorem 1.60 (Mikhailov [Mik05a]). *If F is a non-cyclic free group, $\{1\} \neq S \subseteq R$ are its normal subgroups, and $\delta^+(F/R) = 1$, then $S/[R, S]$ is a faithful $\mathbb{Z}[F/R]$-module.*

Proof. Since F is residually nilpotent, there exists $n \geq 1$ such that

$$S \subseteq \gamma_n(R), \ S \nsubseteq \gamma_{n+1}(R), \tag{1.36}$$

and so the natural projection of $\mathbb{Z}[F/R]$-modules:

$$\varphi_1 : S/[S, R] \rightarrow \frac{S\gamma_{n+1}(R)}{\gamma_{n+1}(R)}, \tag{1.37}$$

is nontrivial. Let $\mathfrak{r}$ be the kernel of the natural projection $\mathbb{Z}[F] \rightarrow \mathbb{Z}[F/R]$. Since $\frac{S\gamma_{n+1}(R)}{\gamma_{n+1}(R)}$ is a submodule of the higher relation module $\gamma_n(R)/\gamma_{n+1}(R)$, which embeds into the free $\mathbb{Z}[F/R]$-module $\mathfrak{r}^n/\mathfrak{r}^{n+1}$, we have a nontrivial homomorphism

$$S/[S, R] \rightarrow \mathfrak{r}^n/\mathfrak{r}^{n+1}$$

of $\mathbb{Z}[F/R]$-modules. Our conclusion therefore follows from Corollary 1.54. $\square$

In general, the residual nilpotence of $F/[R, S]$ does not imply the residual nilpotence of F/RS. We give an example that requires small cancellation theory for which the reader is referred to [Ols91].

Example 1.61
Let F be a free group of rank $2n+2$, $n\geq 3$, with basis $X=\{a,\,b,\,x_1,\,\ldots,\,x_{2n}\}$. Consider the words

$$r = a[b,\,x_1][b,\,x_2]\ldots[b,\,x_n],\quad s = b[a,\,x_{n+1}][a,\,x_{n+2}]\ldots[a,\,x_{2n}],$$

and let $R = \langle r\rangle^F$, $S = \langle s\rangle^F$. Then the presentation $\langle X \mid r,\,s\rangle$ satisfies the small cancellation condition $C(6)$. Since the relators r, s are not proper powers in F, the presentation $\langle X \mid r,\,s\rangle$ is aspherical (see [Ols91], Theorem 13.3 and Corollary 31.1). Consequently $R\cap S = [R,\,S]$ (see (1.18)), and so we have the following embedding

$$F/[R,\,S] \subseteq F/R \times F/S.$$

Observe that each of F/R and F/S is a free group of rank $2n+1$. Therefore $F/R\times F/S$ is residually nilpotent, and consequently so is $F/[R,\,S]$. From the definitions of the relators r and s, it is clear that the elements aRS, bRS of F/RS lie in $\gamma_\omega(F/RS)$. We assert that aRS and bRS are both nontrivial.

Consider the natural epimorphic image of F/RS defined by the presentation

$$\langle a,b,x_1,x_{n+1} \mid a[b,x_1],\ b[a,x_{n+1}]\rangle \equiv \langle a,x_1,x_{n+1} \mid a[[x_{n+1},a],x_1]\rangle.$$

By the Freiheitssatz for one-relator groups ([Mag66] Theorem 4.10), the subgroup generated by $\{a,x_1\}$ in the group defined by the right hand presentation is a free group of rank 2. Hence aRS is a nontrivial element of F/RS, and therefore F/RS is not residually nilpotent.

The foregoing analysis helps to provide some instances in which the residual nilpotence of $F/[R,\,S]$ yields information about the quotient F/RS.

Theorem 1.62 (Mikhailov-Passi, [Mik06b]). *If $F/R\cap S$ is finite and $F/[R,\,S]$ is residually nilpotent, then F/RS is a finite p-group.*

Proof. By Theorems 1.59 and 1.35 it follows that $\Delta^\omega(F/RS) = 0$. It follows from the hypothesis that F/RS is finite; therefore it is a finite p-group (see [Pas79], p. 100). $\square$

Theorem 1.63 (Mikhailov-Passi, [Mik06b]). *Let F be a free group of finite rank, and let R, S be normal subgroups of F such that*

(i) *F/RS is finite,*
(ii) *$[F,\,RS] \subseteq (R\cap S)\gamma_2(RS)$ and*
(iii) *$F/[R,\,S]$ is residually nilpotent.*

Then F/RS is nilpotent.

Proof. The hypothesis implies that F/RS acts faithfully on $\frac{R\cap S}{[R,\,S]}$ ([Mik06b], Theorem 6). The assertion therefore follows from Theorem 1.35. $\square$

1.4 k-central Extensions

Given a group G, let $\zeta_n(G)$ denote its nth centre:

$$\zeta_n(G) := \{g \in G \mid [\ldots[g, x_1], x_2], \ldots, x_n] = 1, \ \forall\, x_i \in G, \ i = 1, 2, \ldots, n\},$$

for $n \geq 1$ and $\zeta_0(G) = 1$. The series $\{\zeta_n(G)\}_{n\geq 0}$ is an ascending central series, called the *upper central series* of the group G. Let

$$1 \to N \xrightarrow{i} \widetilde{G} \xrightarrow{\pi} G \to 1 \tag{1.38}$$

be an exact sequence of groups. We say that $\widetilde{G}$ is a *k-central extension* of G, $k \geq 1$, if the image $i(N)$ is contained in $\zeta_k(\widetilde{G})$. In particular, a 1-central extension is just a *central extension*.

A natural question to study is the relationship between residual properties of G and its k-central extensions $\widetilde{G}$. For residual nilpotence this question is quite nontrivial even for the case of central extensions. In fact, it is not hard to construct examples of central extensions $1 \to N \xrightarrow{i} \widetilde{G} \xrightarrow{\pi} G \to 1$ in which $\widetilde{G}$ (resp. G) is a residually nilpotent group, but G (resp. $\widetilde{G}$) is not residually nilpotent.

Example 1.64
Let A be a free abelian group of infinite rank with basis $\{a_i\}_{i\geq 1}$, and G_i the 2-generated free nilpotent group of nilpotency class $i + 1$ with generators x_i, y_i. Let $H = A \oplus \prod_{i\geq 1} G_i$. Set $\beta_i := [x_i, {}_iy_i]$, $i \geq 1$, and let R be the central subgroup of H generated by elements $a_i a_{i+1}^{-1}\beta_i\beta_{i+1}^{-1}$, $i \geq 1$. Let $G = H/R$.

First let us show that the group G is residually nilpotent. Let R_1 be the subgroup in G generated by the central elements β_i, $i \geq 1$. Then G/R_1 is a direct product of nilpotent groups and hence G/R_1 is residually nilpotent. Therefore, $\gamma_\omega(G) \subseteq R_1$. Let $1 \neq x \in \gamma_\omega(G)$. Then x can be written as

$$x = \beta_1^{k_1} \ldots \beta_n^{k_n}, \tag{1.39}$$

for some $n \geq 1$ and some integers $k_i, i = 1, \ldots, n$ at at least one k_i not equal to zero. Consider the group

$$G(n) = (\langle a_i, \ i = 1, \ldots, n\rangle \oplus \prod_{i=1}^{n} G_i)/\{a_i a_{i+1}^{-1}\beta_i\beta_{i+1}^{-1}, \ i < n\}.$$

There exists the natural epimorphism $G \to G(n)$, whose kernel is the subgroup in G generated by elements x_i, y_i, $i > n$. Observe that the image of x is nontrivial in the group $G(n)$. Since $G(n)$ is nilpotent of class $n + 1$ it follows that $x \notin \gamma_{n+1}(G)$, a contradiction. Hence G is residually nilpotent.

It is easy to see that the group G/A is not residually nilpotent, whereas the group G/AR_1 is residually nilpotent.

The following central extensions thus provide examples of the type asserted above:

$$1 \to A \to G \to G/A \to 1,$$

$$1 \to R_1A/A \to G/A \to G/R_1A \to 1;$$

for, we have

$$\gamma_\omega(G) = 1, \ \gamma_\omega(G/A) \neq 1, \ \gamma_\omega(G/R_1A) = 1.$$

Notation. Let G be a residually nilpotent group. We say that G lies in the class $\widetilde{\mathcal{J}}_k$, $k \geq 1$, if for any k-central extension (1.38), the group $\widetilde{G}$ is also residually nilpotent; we will denote the class $\widetilde{\mathcal{J}}_1$ by $\widetilde{\mathcal{J}}$.

Groups with Long Lower Central Series

We say that a group H has *long lower central series* if $\gamma_\omega(H) \neq \gamma_{\omega+1}(H)$. The following example shows the existence of such groups, and in fact more generally, the existence of transfinitely nilpotent groups of arbitrarily long transfinite lower central series.

Example 1.65 (Hartley [Har70]).
Let α be an ordinal number and A an abelian group with the property that $A_{\lambda+1} = 0$ and $A_\lambda \neq 0$, where the series $\{A_\mu\}$ is defined by setting $A_0 = A$, $A_{\mu+1} = pA_\mu$ and, for a limit ordinal β, $A_\beta = \bigcap_{\mu<\beta} A_\mu$. Let C_p be the cyclic group of order p. Then the wreath product $W = A \wr C_p$ is a transfinitely nilpotent group with $\gamma_\lambda(W) \neq 1$.

It is clear that for any group G with $\gamma_\omega(G) \neq \gamma_{\omega+1}(G)$, $G/\gamma_\omega(G) \notin \widetilde{\mathcal{J}}$. On the other hand, if G is a residually nilpotent group not belonging to $\widetilde{\mathcal{J}}$ and $\widetilde{G}$ is a central extension (1.38) of G which is not residually nilpotent, then $\gamma_\omega(\widetilde{G}) \neq \gamma_{\omega+1}(\widetilde{G}) = 1$, and so $\widetilde{G}$ is a group with long lower central series. Thus examples of groups with long lower central series provide examples of residually nilpotent groups which do not belong to the class $\widetilde{\mathcal{J}}$, and conversely. Note that the Stallings-Stammbach five-term sequence ([Hil71], p. 202) gives, for every group G, the following exact sequence:

$$H_2(G) \to H_2(G/\gamma_\omega(G)) \to \gamma_\omega(G)/\gamma_{\omega+1}(G) \to 1. \qquad (1.40)$$

Therefore, the stabilization of the lower central series at the ω-term, i.e., $\gamma_\omega(G) = \gamma_{\omega+1}(G)$, occurs if and only if the map $H_2(G) \to H_2(G/\gamma_\omega(G))$, induced by the natural projection $G \to G/\gamma_\omega(G)$, is an epimorphism.

The following example of a finitely-presented group G with $\gamma_\omega(G) \neq \gamma_{\omega+1}(G)$ is due to J. Levine [Lev91]; it is the first such example:

Example 1.66
Let G be the group given by the following presentation:

$$\langle t, x, y, z \mid [x, y] = 1, \, txt^{-1}x = 1, \, tyt^{-1} = y^{-1}, \, z^{-1}x[z, t] = 1 \rangle.$$

Then $\gamma_\omega(G)/\gamma_{\omega+1}(G) = \mathbb{Z}_3$.

The following example is due to Cochran and Orr [Coc98]. It gives a nice illustration of homological methods used in the construction of groups with long lower central series.

Example 1.67 [Coc98]. *Let $G = \langle a, b, c \mid aba^{-1}b = c^3 = 1 \rangle$. Then $\gamma_\omega(G) \neq \gamma_{\omega+1}(G)$.*

Proof. It is easy to see that the element b is a generalized 2-element: $b^{2^{n+1}} \in \gamma_n(G)$, $n \geq 1$. Therefore, c being an element of order 3, the nontrivial element $[b, c]$ lies in the intersection $\gamma_\omega(G)$ of the lower central series. On the other hand, the group

$$H := \langle a, b, c \mid aba^{-1}b = [b, c] = c^3 = 1 \rangle$$

is residually nilpotent by Example 1.23 (case $p = 3$). Therefore, $\gamma_\omega(G) = \langle [b, c] \rangle^G$ and $G/\gamma_\omega(G) = H$.

Let us compute the second integral homology group of H. Note that H is an HNN-extension of the group $H_1 = \langle b, c \mid [b, c] = c^3 = 1 \rangle \simeq \mathbb{Z} \oplus \mathbb{Z}_3$ over the cyclic subgroup generated by element b. Therefore, we have the following Mayer-Vietoris exact sequence

$$0 \to H_2(\mathbb{Z} \oplus \mathbb{Z}_3) \to H_2(H) \to \mathbb{Z} \to \cdots,$$

showing that $H_2(H)$ contains the cyclic subgroup $\mathbb{Z}_3 \simeq H_2(\mathbb{Z} \oplus \mathbb{Z}_3)$. Now observe that the group G is the free product of the one-relator group

$$\langle a, b \mid aba^{-1}b = 1 \rangle$$

and the cyclic group $\langle c \mid c^3 = 1 \rangle$. Therefore its second integral homology group $H_2(G)$ is trivial. The assertion thus follows from the sequence (1.40). $\square$

Example 1.68 [Coc98].
Let p, q be different primes, X_p the 3-manifold, obtained from S^3 by $(p, p, 0)$ surgery on the Borromean rings. Consider the connected sum M of X_p with q-lens space (or any other 3-manifold with fundamental group isomorphic to $\mathbb{Z}_q$). In [Coc98] it is proved that the fundamental group of M has the following presentation

$$\pi_1(M) = \langle x, y, z, s \mid [x, y] = 1, x^p = [y, z^{-1}], y^p = [x, z], s^q = 1 \rangle,$$

and $\gamma_\omega(\pi_1(M)) \neq \gamma_{\omega+1}(\pi_1(M))$.

Let F be a free group with basis x_1, x_2, x_3, x_4. A striking result of C. K. Gupta [Gup73] is that the free centre-by-metabelian group

$$F/[[F, F], [F, F], F]$$

has 2-torsion. More precisely, it has been proved that the element

$$w = [[x_1^{-1}, x_2^{-1}], [x_3, x_4]][[x_1^{-1}, x_4^{-1}], [x_2, x_3]][[x_1^{-1}, x_3^{-1}], [x_4, x_2]] \cdot$$
$$[[x_4^{-1}, x_2^{-1}], [x_1, x_3]][[x_2^{-1}, x_3^{-1}], [x_1, x_4]][[x_3^{-1}, x_4^{-1}], [x_1, x_2]].$$

of F is such that $w \notin [[F, F], [F, F], F]$, whereas $w^2 \in [[F, F], [F, F], F]$.

We need the description of torsion in the second integral homologies of free abelian extensions given by Kuz'min.

Theorem 1.69 (Kuz'min [Kuz87], [Kuz06, Cor. 8.6, p. 235]). *If F is a free group and N its normal subgroup, such that F/N is 2-torsion free, then*

$$\mathsf{tor}(H_2(F/[N, N])) \simeq H_4(F/N, \mathbb{Z}_2),$$

where for an abelian group A, $\mathsf{tor}(A)$ denotes the torsion subgroup of A.

[For a more general result, see [Stö87].]

From Theorem 1.69, it follows that for a free group F of rank ≥ 4, the group $F/[[F, F], [F, F], F]$ contains the abelian 2-group

$$H_4(F/[F, F], \mathbb{Z}_2) \simeq \mathbb{Z}_2^{\binom{rank(F)}{4}};$$

thus it follows that there is a 2-torsion in the free centre-by-metabelian group $F/[[F, F], [F, F], F]$.

Example 1.70 (Mikhailov [Mik02]). *The group*

$$G = \langle a, b \mid [a, b^3] = [a, b, a]^2 = 1 \rangle$$

is such that $\gamma_\omega(G) \neq \gamma_{\omega+1}(G)$.

Proof. First note that $\gamma_2(G)^3 \subseteq \gamma_3(G)$; therefore, $[a, b, a] \subseteq \gamma_\omega(G)$. Let F be the free group with basis a, b, and N the normal closure of the elements a, b^3 in F. Then the group

$$H := G/\langle[a, b, a]\rangle^G = \langle a, b \mid [a, b^3] = [a, b, a] = 1\rangle$$

is a free abelian extension of the cyclic group $\mathbb{Z}_3$:

$$1 \to N/[N, N] \to F/[N, N](\simeq H) \to F/N(\simeq \mathbb{Z}_3) \to 1.$$

Note that H is residually nilpotent. Therefore $\gamma_\omega(G) = \langle[a, b, a]\rangle^G$. Consider the homomorphism $\pi : H_2(G) \to H_2(H)$ induced by the natural projection $G \to H$. By Theorem 1.69 we have

$$\mathsf{tor}(H_2(H)) = H_4(F/N, \mathbb{Z}_2) = 0,$$

i.e., $H_2(H)$ is a free abelian group. Therefore the element

$$e := [a, b, a][[N, N], F]$$

is not in the image of π in case it is not trivial in $H_2(H)$. Suppose the element e is trivial, i.e., $[a, b, a] \in \langle[a, b^3], [a, b, a, f], f \in F\rangle^F$. Then $[a, b, a] \in \langle b^3, [a, b, a, f], f \in F\rangle^F$. By symmetry we conclude that

$$[a, b, a], [b, a, b] \in \langle a^3, b^3, \gamma_4(F)\rangle^F;$$

therefore, $\gamma_3(S) = \gamma_4(S)$ for the group $S = \langle a, b \mid a^3 = b^3 = 1\rangle \simeq \mathbb{Z}_3 * \mathbb{Z}_3$. However, the group S is residually nilpotent, and not nilpotent. We thus have a contradiction. Consequently π is not an epimorphism and hence $\gamma_\omega(G) \neq \gamma_{\omega+1}(G)$. $\square$

Problem 1.71 *Does there exist a finitely-presented group G with $\gamma_\omega(G)/\gamma_{\omega+1}(G)$ infinite cyclic?*

Problem 1.72 *Let G be a group of Example 1.70. Consider the group $H = G/\gamma_{\omega+1}(G) * \mathbb{Z}$. Is it true that $\gamma_\alpha(H) \neq \gamma_{\alpha+1}(H)$ for $\alpha < \omega^2$?*

Baer Invariants

As seen in the preceding section, group homology plays a role in the investigation of residual nilpotence of central extensions. In the case of k-central extensions a similar role is played by Baer invariants [Bae45] which are defined as follows:

Let G be a group and

$$1 \to N \to F \to G \to 1 \tag{1.41}$$

a free presentation of G. Then the abelian group

$$M^{(n)}(G) := \frac{N \cap \gamma_{n+1}(F)}{[N, {}_nF]},\tag{1.42}$$

where

$$[N, {}_0F] = N, \; [N, {}_{n+1}F] = [[N, {}_nF], F], \; \text{for } n \geq 1,$$

does not depend on the choice of the presentation (1.41); the group $M^{(n)}(G)$ is called the nth *Baer invariant* of the group G. That the group $M^{(n)}(G)$ does not depend on the choice of a presentation (1.41) follows from the fact that it is the first derived functor of the functor $G \mapsto G/\gamma_n(G)$ on the category Gr of groups (see Appendix, Example A.16, for details). Note that the invariant $M^{(1)}(G)$ is the Schur multiplicator of G, and is isomorphic to $H_2(G)$, the second integral homology group of G; this is the reason that the Baer invariants are sometimes called *generalized* (or *n-nilpotent*) *multiplicators*. Baer invariants have been studied by J. Burns and G. Ellis ([Bur97], [Bur98], [Ell02]), and several other authors.

A direct analog of the Stallings-Stammbach five-term sequence is the following exact sequence, which can be proved by an immediate application of the presentation (1.42).

Theorem 1.73 *Let $k \geq 1$, G a group and N a normal subgroup of G. Then there is the following exact sequence of groups:*

$$M^{(k)}(G) \to M^{(k)}(G/N) \to N/[N, {}_kG] \to G/\gamma_{k+1}(G) \to G/N\gamma_{k+1}(G) \to 1.$$

In particular, for $k \geq 1$, $n \geq k+1$, there exists the following exact sequence of abelian groups:

$$M^{(k)}(G) \to M^{(k)}(G/\gamma_n(G)) \to \gamma_n(G)/\gamma_{n+k}(G) \to 1.\tag{1.43}$$

We need the following result.

Theorem 1.74 (Ellis [Ell02]) *If G is a group such that $H_2(G)$ is a torsion group, then $M^{(n)}(G)$ is also a torsion group for all $n \geq 1$.*

For a proof of the above theorem, see Proposition 1.139 or Theorem 5.14.

Generalized Dwyer Filtration

For any $k \geq 1$, $m \geq k+1$ and group $G \simeq F/N$, define the *generalized Dwyer filtration*

$$M^{(k)}(G) = \varphi_{k+1}^{(k)}(G) \supseteq \varphi_{k+2}^{(k)}(G) \supseteq \dots$$

by setting

$$\varphi_m^{(k)}(G) := \ker\{M^{(k)}(G) \to M^{(k)}(G/\gamma_{m-k}(G))\}.$$

For $k = 1$ this is the filtration of the second integral homology group $H_2(G)$ defined by Dwyer [Dwy75]. In view of (1.42), it is easy to show that the terms of the generalized Dwyer filtration can be expressed as

$$\varphi_m^{(k)}(G) = \frac{(N \cap \gamma_m(F))[N, {}_kF]}{[N, {}_kF]} = \frac{N \cap \gamma_m(F)}{[N, {}_kF] \cap \gamma_m(F)}, \quad k \geq 1, \ m \geq k+1. \tag{1.44}$$

Remark. Note that the analog of the classical Dwyer filtration, i.e., $\{\varphi_m^{(1)}(G)\}$ can be naturally defined for homology of topological spaces. Let X be a topological space. Define $\varphi_m^{(1)}(X)$ to be the kernel of the composite map

$$H_2(X) \rightarrow H_2(\pi_1(X)) \rightarrow H_2(\pi_1(X)/\gamma_{m-1}(\pi_1(X))). \tag{$*$}$$

Recall that, for every topological space X, we have a natural map $X \rightarrow K(\pi_1(X), 1)$; the first map in $(*)$ is the one obtained by applying the functor H_2 to this map. The second map in $(*)$ is induced the natural projection $\pi_1(X)/\gamma_{m-1}(\pi_1(X))$.

We may mention that, in the case when X is a 4-dimensional topological manifold, the filtration $\{\varphi_m^{(1)}(X)\}$ plays an important role in the surgery theory ([Fre95], [Kru03]).

From (1.44) we immediately have the following:

Proposition 1.75 If $m \geq k+1 \geq l+2$, then, for any normal subgroup N of a free group F, there exists the following short exact sequence of abelian groups:

$$0 \rightarrow \varphi_m^{(l)}(F/[N, {}_{k-l}F]) \rightarrow \varphi_m^{(k)}(F/N) \rightarrow \varphi_m^{(k-l)}(F/N) \rightarrow 0.$$

We now give some results from [Mik07a] which extend results from [Dwy75].

Theorem 1.76 Let $f : G \rightarrow H$ be a group homomorphism which induces an isomorphism $G/\gamma_{k+1}(G) \rightarrow H/\gamma_{k+1}(H)$ for some $k \geq 1$. Then for any $m \geq k+1$, the following statements are equivalent:

(i) f induces an isomorphism $M^{(k)}(G)/\varphi_m^{(k)}(G) \rightarrow M^{(k)}(H)/\varphi_m^{(k)}(H)$.

(ii) f induces an isomorphism $f_m : G/\gamma_m(G) \rightarrow H/\gamma_m(H)$.

Proof. It follows from Theorem 1.73 that, for any $m \geq k+1$, there exists the following commutative diagram:

$$
\begin{array}{ccccccc}
M^{(k)}(G)/\varphi_m^{(k)}(G) & \rightarrowtail & M^{(k)}(G/\gamma_{m-k}(G)) & \longrightarrow & \gamma_{m-k}(G)/\gamma_m(G) & \longrightarrow & 1 \\
\downarrow & & \downarrow & & \downarrow & & \\
M^{(k)}(H)/\varphi_m^{(k)}(H) & \rightarrowtail & M^{(k)}(H/\gamma_{m-k}(H)) & \longrightarrow & \gamma_{m-k}(H)/\gamma_m(H) & \longrightarrow & 1
\end{array}
$$

induced by f. The assertion then follows immediately by induction on $m \geq k+1$ with an iterated application of the above diagram. $\square$

Corollary 1.77 *Let $G/\gamma_{c+1}(G)$ be a free nilpotent group for some $c \geq 1$ and suppose that $M^{(c)}(G) = 0$. Then $M^{(k)}(G) = 0$, $k \leq c$.*

Proof. Let $\{g_i\}_{i \in I}$ be elements in G such that $\{g_i \gamma_{c+1}(G)\}_{i \in I}$ is a basis of the free nilpotent group $G/\gamma_{c+1}(G)$. Consider the homomorphism $f : F \to G$, where F is a free group with basis $\{f_i\}_{i \in I}$, given by setting $f_i \mapsto g_i, i \in I$. Then f induces an isomorphism $F/\gamma_{c+1}(F) \simeq G/\gamma_{c+1}(G)$ and an epimorphism

$$M^{(c)}(F)/\varphi_k^{(c)}(F) \to M^{(c)}(G)/\varphi_k^{(c)}(G), \ k \geq c+1.$$

Proposition 1.76 implies that f induces an isomorphism $F/\gamma_n(F) \simeq G/\gamma_n(G)$ for all $n \geq 1$. Since $M^{(c-1)}(F) = 0$, we again apply Proposition 1.76 to get $M^{(c-1)}(G) = \varphi_k^{(c-1)}(G)$, $k \geq c$. Now the needed statement follows from the epimorphism $M^{(c)}(G) \to \varphi_c^{(c-1)}(G)$ (see Proposition 1.75). $\square$

Theorem 1.78 *Let*

$$1 \to N \to \widetilde{G} \overset{p}{\to} G \to 1 \tag{1.45}$$

be a k-central group extension for a given $k \geq 1$. Then for any $m \geq k+1$ there exists the following exact sequence of abelian groups

$$\varphi_m^{(k)}(\widetilde{G}) \overset{p*}{\to} \varphi_m^{(k)}(G) \to N \cap \gamma_m(\widetilde{G}) \to 1, \tag{1.46}$$

where the homomorphism $p*$ is induced by the epimorphism p.

Proof. Consider free presentations of groups from (1.45):

$$
\begin{array}{ccccccccc}
1 & \longrightarrow & N & \longrightarrow & \widetilde{G} & \overset{p}{\longrightarrow} & G & \longrightarrow & 1 \\
  & & \| & & \| & & \| & & \\
1 & \longrightarrow & R/S & \longrightarrow & F/S & \overset{p}{\longrightarrow} & F/R & \longrightarrow & 1,
\end{array}
$$

where F is a free group and

$$[R, \,_k F] \subseteq S. \tag{1.47}$$

It follows from the presentation (1.44) that the homomorphism $p*$ can be viewed as a map

$$p* : \frac{(S \cap \gamma_m(F))[S, \,_k F]}{[S, \,_k F]} \to \frac{(R \cap \gamma_m(F))[R, \,_k F]}{[R, \,_k F]},$$

induced by the inclusion $S \to R$. The cokernel of $p*$ is naturally presentable as

$$\mathrm{coker}(p*) \simeq \frac{(R \cap \gamma_m(F))[R, \, _kF]}{(S \cap \gamma_m(F))[R, \, _kF]}.$$

Define a homomorphism

$$f : \frac{(R \cap \gamma_m(F))[R, \, _kF]}{(S \cap \gamma_m(F))[R, \, _kF]} \to R/S \cap \gamma_m(F/S) = \frac{R \cap \gamma_m(F)S}{S} \simeq N \cap \gamma_m(\widetilde{G})$$

by setting

$$f : r(S \cap \gamma_i(F))[R, \, _kF] \mapsto rS, \ r \in R \cap \gamma_m(F).$$

The k-centrality condition (1.47) then implies that f is an isomorphism. Thus the exact sequence (1.46) is established. $\square$

Absolutely Residually Nilpotent Groups

Definition. A residually nilpotent group G is called *absolutely residually nilpotent* if, for every integer $k \geq 1$ and k-central extension

$$1 \to N \to \widetilde{G} \to G \to 1, \qquad\qquad (1.48)$$

the group $\widetilde{G}$ is also residually nilpotent.

Theorem 1.78 makes it possible to give simple conditions under which residual nilpotence is preserved under k-central extensions.

Proposition 1.79 *Let G be a group such that $\varphi_m^{(k)}(G) = 0$ for some $k \geq 1$ and $m \geq k + 1$. Then, for any k-central extension (1.48), the group $\widetilde{G}$ is residually nilpotent if and only if the group G is residually nilpotent.*

Proof. By Theorem 1.78, applied to the k-central extension (1.48), we have $N \cap \gamma_m(\widetilde{G}) = 0$; hence, for any $l \geq m$, there exists the natural isomorphism $\gamma_l(\widetilde{G}) \simeq \gamma_l(G)$, which immediately implies the assertion. $\square$

The above Proposition immediately yields the following

Corollary 1.80 *If G is a residually nilpotent group with $M^{(n)}(G) = 0$ for all $n \geq 1$, then G is an absolutely residually nilpotent.*

We will prove later (see Proposition 1.139 or Theorem 5.14) that *if, for a given group G, $H_1(G)$ is torsion-free and $H_2(G) = 0$, then $M^{(n)}(G) = 0$ for all $n \geq 1$.*

Example 1.81 *Let G be a group given by the following presentation:*

$$G = \langle a, b, c \mid a = [c^{-1}, a][c, b] \rangle.$$

Then G is absolutely residually nilpotent group.

Proof. It has been shown by Baumslag that the group G is residually nilpotent [Bau67]. Observe that $H_1(G)$ is torsion-free and $H_2(G) = 0$. Hence $M^{(n)}(G) = 0$ for all $n \geq 1$ and therefore, by Corollary 1.80, G is absolutely residually nilpotent. $\square$

Theorems 1.74 and 1.78 also provide simple conditions for the residual nilpotence of a given k-central extension of a group G to imply the residual nilpotence of the group G itself.

Proposition 1.82 *Let G be a finitely presented group with $H_2(G)$ finite. Then for any k-central extension*

$$1 \to N \to \widetilde{G} \to G \to 1,$$

$k \geq 1$, the residual nilpotence of $\widetilde{G}$ implies the residual nilpotence of G.

Proof. Consider the inverse limit of the short exact sequences

$$1 \to N \cap \gamma_n(\widetilde{G}) \to \gamma_n(\widetilde{G}) \to \gamma_n(G) \to 1, \ n \geq 2.$$

By (Appendix, Proposition A.34) we get the inclusion

$$\gamma_\omega(G) \to \varprojlim_n^1 (N \cap \gamma_n(\widetilde{G})). \tag{1.49}$$

Since G is finitely presented and $H_2(G)$ is finite, $M^{(n)}(G)$ is finite for all $n \geq 1$, by Theorem 1.74. By Theorem 1.78, $N \cap \gamma_n(\widetilde{G}), n \geq k+1$ are also finite, hence the $\varprojlim^1$ term in (1.49) vanishes, and consequently G is residually nilpotent. $\square$

Free k-central Extensions

Let G be a group given by a free presentation

$$1 \to N \to F \to G \to 1. \tag{1.50}$$

For a given $k \geq 1$, consider the following induced free k-central extension of G:

$$1 \to N/[N, \ _kF] \to F/[N, \ _kF] \to G \to 1.$$

Clearly we have the following short exact sequence:

$$1 \to M^{(k)}(G) \to \gamma_{k+1}(F)/[R, \ _kF] \to \gamma_{k+1}(G) \to 1,$$

and hence the group $\gamma_{k+1}(F)/[R, \ _kF]$ is an invariant of the group G, i.e. it does not depend on the choice of the presentation (1.50) for G. (In fact, the group $\gamma_{k+1}(F)/[R, {}_kF]$ is the 0-th derived functor of the endofunctor $G \mapsto \gamma_{k+1}(G)$ on the category Gr of groups, see Appendix, Section A.13). Thus it follows that the residual nilpotence of the group $F/[N, {}_kF]$ depends only on the quotient group $F/N \simeq G$, and not on the choice of the presentation (1.50). More precisely, we have the following result.

Proposition 1.83 *Let G be a residually nilpotent group with the presentation (1.50). Then*

$$\gamma_\omega(F/[N, {}_kF]) \simeq \bigcap_{m \geq k+1} \varphi_m^{(k)}(G)$$

for all $k \geq 1$.

Proof. Since the group F/N is residually nilpotent, we have

$$\gamma_\omega(F/[N, {}_kF]) \subseteq N/[N, {}_kF].$$

For every $n \geq 1$, we have the inclusion

$$\gamma_\omega(F/[N, {}_kF]) \subseteq \gamma_n(F/[N, {}_kF]) = \gamma_n(F)[N, {}_kF]/[N, {}_kF].$$

Using the presentation (1.44) of the generalized Dwyer filtration terms, we conclude that

$$\gamma_\omega(F/[N, {}_kF]) \subseteq \bigcap_{m \geq k+1} \varphi_m^{(k)}(G).$$

On the other hand, the presentation (1.44) gives the inclusion

$$\varphi_m^{(k)}(G) \subseteq \gamma_m(F/[N, {}_kF]), \ \ k \geq 1, m \geq k+1,$$

which implies that for every $k \geq 1$

$$\bigcap_{m \geq k+1} \varphi_m^{(k)}(G) \subseteq \gamma_\omega(F/[N, {}_kF]),$$

and the proof is complete. $\square$

Denote by $\mathcal{J}_k$ the class of residually nilpotent groups G, such that $F/[N, {}_kF]$ is residually nilpotent for any presentation (1.50), and let $\mathcal{J}_1 = \mathcal{J}$. It is clear that the class $\widetilde{\mathcal{J}}_k$ defined on page 38 is contained in the class $\mathcal{J}_k$. The converse, however, is not true (see Example 1.85).

For any $c \geq 2$ and arbitrary normal subgroup N of a free group F, there exists a canonical homomorphism

$$\eta_c : F/[\gamma_c(N), F] \to \mathcal{M}_{c+1, \mathbb{Z}}(F/N),$$

called the *Gupta representation*, where $\mathcal{M}_{c+1,\mathbb{Z}}(F/N)$ denotes the group of $(c+1) \times (c+1)$ matrices over a certain ring (for a complete description of this ring see [Gup78]).

Proposition 1.84 *Let p be a prime. Suppose that the group F/N is p-torsion free, $\Delta^\omega(F/N) = 0$ and $H_4(F/N, \mathbb{Z}_p) = 0$. Then $\Delta^\omega(F/[\gamma_p(N), F]) = 0$. In particular, $F/\gamma_p(N) \in \mathcal{J}$.*

Proof. On invoking [Gup87a, Theorem 3.1] it follows that

$$\Delta^\omega(F/[\gamma_p(N), F]) \subseteq \Delta(\ker \eta_p)\mathbb{Z}[F/[\gamma_p(N), F]]. \tag{1.51}$$

Stöhr [Stö87] has shown that, for any normal subgroup N of F, the kernel $\ker \eta_p$ of the Gupta representation is contained in the torsion subgroup of $F/[\gamma_p(N), F]$. The torsion subgroup of $F/[\gamma_p(N), F]$, however, lies in $\gamma_p(N)/[\gamma_p(N), F]$. In the case when F/N is p-torsion free, Stöhr [Stö87] has given a complete description of this torsion [Har91b]:

$$\mathsf{tor}(\gamma_p(N)/[\gamma_p(N), F]) \simeq H_4(F/N, \mathbb{Z}_p),$$

where $\mathbb{Z}_p$ is viewed as a trivial F/N-module. The hypothesis $H_4(F/N, \mathbb{Z}_p) = 0$ therefore implies that $F/[\gamma_p(N), F]$ is torsion free and $\ker \eta_p = 0$. The asserted statement thus follows from (1.51). $\square$

The following example shows that the inclusion $\widetilde{\mathcal{J}} \subset \mathcal{J}$ is proper.

Example 1.85
Consider the group

$$G = \langle a, b \mid [b^3, a] = [a, b, a] = 1 \rangle.$$

It has already been observed that $G \notin \widetilde{\mathcal{J}}$ (see Example 1.70).

Note that G is a free abelian extension of the cyclic group of order three:

$$1 \to N/[N, N] \to F/[N, N] \; (\simeq G) \to \mathbb{Z}_3 \to 1,$$

where $F = \langle a, b \rangle, N = \langle b^3, a, a^b, a^{b^2} \rangle$. We have

$$
\begin{aligned}
(i) \quad & \Delta^\omega(F/N) = \Delta^\omega(\mathbb{Z}_3) = 0 \\
(ii) \quad & H_4(F/N, \mathbb{Z}_2) = H_4(\mathbb{Z}_3, \mathbb{Z}_2) = 0.
\end{aligned}
$$

Therefore, by Proposition 1.84, $F/[[N, N], F]$ is a residually nilpotent group and hence G belongs to the class $\mathcal{J}$. $\square$

Proposition 1.86 *Let G be a residually nilpotent group and*

$$1 \to N \to \widetilde{G} \to G \to 1 \tag{1.52}$$

a k-central extension with $\gamma_\omega(\widetilde{G}) \neq 1$ and

$$\varprojlim{}^1 \varphi_m^{(k)}(\widetilde{G}) = 0. \tag{1.53}$$

Then $G \notin \mathcal{J}_k$.

Proof. By Theorem 1.78, for every $m \geq k+1$, there exists the following exact sequence:

$$\varphi_m^{(k)}(\widetilde{G}) \xrightarrow{t_m} \varphi_m^{(k)}(G) \to N \cap \gamma_m(\widetilde{G}) \to 1. \tag{1.54}$$

Applying the inverse limit functor, we have the following exact sequence

$$0 \to \bigcap_{m \geq k+1} \mathrm{im}(t_m) \to \bigcap_{m \geq k+1} \varphi_m^{(k)}(G) \to N \cap \gamma_\omega(\widetilde{G}) \to \varprojlim{}^1 \mathrm{im}(t_m). \tag{1.55}$$

The epimorphisms $\varphi_m^{(k)}(\widetilde{G}) \to \mathrm{im}(t_m)$ induce the following epimorphism of $\varprojlim{}^1$-functors:

$$\varprojlim{}^1 \varphi_m^{(k)}(\widetilde{G}) \to \varprojlim{}^1 \mathrm{im}(t_m);$$

thus, in view of our hypothesis, we get $\varprojlim{}^1 \mathrm{im}(t_m) = 0$. Since $N \cap \gamma_\omega(\widetilde{G}) \neq 0$, it follows from (1.55) that $\bigcap_{m \geq k+1} \varphi_m^{(k)}(G) \neq 0$. Hence $G \notin \mathcal{J}_k$ by Proposition 1.83. $\square$

Proposition 1.87 *Let G be a finitely-presented group with $\gamma_\omega(G) \neq \gamma_{\omega+1}(G)$ and $H_2(G)$ finite. Then, for any $k \geq 1$ and any free presentation*

$$1 \to R \to F \to G/\gamma_\omega(G) \to 1,$$

the group $\Pi_k := F/[R, {}_kF]$ is not residually nilpotent. Furthermore, if $H_2(G/\gamma_\omega(G))$ is finite, then for any free presentation

$$1 \to S \to E \to \Pi_k/\gamma_\omega(\Pi_k) \to 1,$$

the group $E/[S, {}_lE]$ is not residually nilpotent for $1 \leq l \leq k$.

Proof. Let $H = G/\gamma_{\omega+k}(G)$. The projection $f : G \to H$ induces isomorphisms $f_n : G/\gamma_n(G) \to H/\gamma_n(H)$ for all $n \geq 1$. Therefore, by Theorem 1.76, we have the following isomorphisms

$$M^{(k)}(G)/\varphi_m^{(k)}(G) \simeq M^{(k)}(H)/\varphi_m^{(k)}(H), \ m \geq k+1. \tag{1.56}$$

Since $H_2(G)$ is finite, Theorem 1.74 implies that $M^{(k)}(G)$ is also a finite group. Thus for any $k \geq 1$ the filtration $\varphi_m^{(k)}(G)$ stabilizes from some stage m_k, say, onward:

$$\varphi_{m_k}^{(k)}(G) = \varphi_m^{(k)}(G) \ \text{for all } m \geq m_k.$$

Isomorphism (1.56) implies that the generalized Dwyer's filtration of the group H must also stabilize:

$$\varphi_{m_k}^{(k)}(H) = \varphi_{m_k+1}^{(k)}(H). \tag{1.57}$$

Consider now the k-central extension

$$1 \to \gamma_\omega(G)/\gamma_{\omega+k}(G) \to H \to G/\gamma_\omega(G) \to 1, \tag{1.58}$$

in which, because of our hypothesis, $\gamma_\omega(G)/\gamma_{\omega+k}(G) \neq 1$. The stabilization (1.57) gives

$$\varprojlim{}^1 \varphi_m^{(k)}(H) = \varprojlim{}^1 \varphi_{m_k}^{(k)}(H) = 0;$$

thus we can apply Proposition 1.86 to the k-central extension (1.58). Consequently we have

$$\bigcap_{m \geq k+1} \varphi_m^{(k)}(G/\gamma_\omega(G)) \neq 0;$$

therefore, Π_k is not residually nilpotent by Proposition 1.83.

Suppose now that the group $H_2(F/R)$ is finite. Consider the following epimorphisms

$$\pi_l : \Pi_k \to R_l := \Pi_k/\gamma_{\omega+l}(\Pi_k), \ l = 1, \ldots, k.$$

Since π_l induces isomorphisms on lower central quotients, Theorem 1.76 implies that there are isomorphisms

$$M^{(l)}(\Pi_k)/\varphi_m^{(l)}(\Pi_k) \simeq M^{(l)}(R_l)/\varphi_m^{(l)}(R_l), \ m \geq l+1. \tag{1.59}$$

In view of Proposition 1.75, we have an embedding

$$\varphi_m^{(l)}(\Pi_k) \subseteq \varphi_m^{(l+k)}(F/R), \ l \geq 1.$$

Since by assumption $H_2(F/R)$ is finite, Theorem 1.74 implies that $M^{(l+k)}(F/R)$ is finite, and hence $\varphi_m^{(l)}(\Pi_k)$ is finite for any $l \geq 1$. Thus, for any $1 \leq l \leq k$, $\varphi_m^{(l)}(\Pi_k)$ stabilizes from some finite step and it follows from (1.59) that $\varphi_m^{(l)}(R_k)$ also stabilize from some finite step; hence

$$\varprojlim{}^1 \varphi_m^{(l)}(R_l) = 0, \ 1 \leq l \leq k. \tag{1.60}$$

Consider now the l-central extension:

$$1 \to \gamma_\omega(\Pi_k)/\gamma_{\omega+l}(\Pi_k) \to R_l \to \Pi_k/\gamma_\omega(\Pi_k) \to 1. \tag{1.61}$$

Since F/R is residually nilpotent, $\gamma_{\omega+k}(\Pi_k) = 1$, and hence the fact that the group Π_k is not residually nilpotent implies that $\gamma_\omega(R_l) \neq \gamma_{\omega+1}(R_l)$. In view of the Proposition 1.86, applied to the extension (1.61), the condition

(1.60) implies that $\bigcap_{m \geq l+1} \varphi_m^{(l)}(\Pi_k/\gamma_\omega(\Pi_k)) \neq 0$. The asserted statement thus follows from Proposition 1.83. $\square$

We next give an example of a finitely-presented residually nilpotent group H such that, for every integer $k \geq 1$ and every free presentation

$$1 \to R \to F \to H \to 1,$$

the group $F/[R, {}_kF]$ is not residually nilpotent.

Example 1.88

Consider the free product of the fundamental group of the Klein bottle and the cyclic group of order 3:

$$G = \langle a, b, c \mid aba^{-1}b = c^3 = 1 \rangle.$$

As shown in Example 1.67, $\gamma_\omega(G) = \langle [b, c] \rangle^G \neq \gamma_{\omega+1}(G)$. Since the homologies of a free product are equal to the sum of homologies, we conclude that $H_2(G) = 0$. Thus, by Proposition 1.87, we see that the finitely-presented group

$$H := G/\gamma_\omega(G) = \langle a, b, c \mid aba^{-1}b = c^3 = [b, c] = 1 \rangle,$$

which is an HNN-extension of the abelian group $\mathbb{Z} \oplus \mathbb{Z}_3$, is an example of a group with the property that for every integer $k \geq 1$ and every free presentation

$$1 \to R \to F \to H \to 1, \tag{1.62}$$

the group $F/[R, {}_kF]$ is not residually nilpotent.

Remark.

For all integers $k \geq 1$, the groups $\Pi_k := F/[R, {}_kF]$ arising from (1.62) are finitely-presented transfinitely nilpotent, but not residually nilpotent groups. It follows from Mayer-Vietoris exact sequence of homologies of HNN-extensions that $H_2(H) \simeq \mathbb{Z}_3$, and hence, by Proposition 1.87, for $1 \leq l \leq k$, and any free presentation

$$1 \to S \to E \to \Pi_k/\gamma_\omega(\Pi_k) \to 1,$$

the group $E/[S, {}_lE]$ is not residually nilpotent.

One-relator Groups

For the analysis of the generalized Dwyer filtration for one-relator groups, we will use the generalized Magnus embedding which we first recall.

Generalized Magnus Embedding

Let F be a free group with basis X. For a given integer $m \geq 2$, denote by Ω_m the ring of polynomials $\mathbb{Z}[\lambda_{i,\,i+1}(X)]$ over independent commuting indeterminates $\lambda_{i,\,i+1}(x)$, $1 \leq i \leq m-1$, $x \in X$.

Define a ring homomorphism

$$
\mu_m : \mathbb{Z}[F] \to
\begin{pmatrix}
1 & \Omega_m & \dots & \Omega_m \\
0 & 1 & \dots & \Omega_m \\
& & \dots & \\
0 & 0 & \dots & 1
\end{pmatrix}
$$

from the integral group ring $\mathbb{Z}[F]$ to the ring $\mathcal{M}_m(\Omega_m)$ of $m \times m$ matrices over Ω_m by setting

$$
\mu_m : x \mapsto
\begin{pmatrix}
1 & \lambda_{12}(x) & 0 & \dots & & 0 \\
0 & 1 & \lambda_{23}(x) & \dots & & 0 \\
& & & \dots & & \\
0 & 0 & & \dots & 1 & \lambda_{m-1,\,m}(x) \\
0 & 0 & & \dots & 0 & 1
\end{pmatrix}, \quad x \in X.
$$

The homomorphism μ_m is a particular case of the generalized Magnus embedding studied in [Gup87c]; its kernel is equal to $\mathfrak{f}^m$, the mth power of the augmentation ideal of the group ring $\mathbb{Z}[F]$:

$$
\ker(\mu_m) = \mathfrak{f}^m, \ m \geq 1. \tag{1.63}
$$

For $m = 2$, we get the simplest case of the Magnus embedding of the free abelian group in the ring of uni-triangular 2×2 matrices.

Consider the multiplicative subgroup of $GL_m(\Omega_m)$ generated by elements $\mu_m(x)$, $x \in X$. It follows from (1.63) that this subgroup is isomorphic to $F/\gamma_m(F)$:

$$
F/\gamma_m(F) \simeq \langle \mu_m(x), \ x \in X \rangle.
$$

Thus we have a matrix representation of the free nilpotent group $F/\gamma_m(F)$.

Remark. It may be noted that, for $m \geq 3$, the sub-representation to $(m-1) \times (m-1)$ matrices, obtained from μ_m by deleting the first row and the first column (or the mth row and the mth column), coincides with the representation μ_{m-1}, where the ring Ω_{m-1} is the ring $\mathbb{Z}[\lambda_{i,\,i+1}(X)]$, $i = 2, \dots, m-1$ (resp. $\mathbb{Z}[\lambda_{i,\,i+1}(X)]$, $i = 1, \dots, m-2$).

Lemma 1.89 *Let $m \geq 3$ and*

$$\mu_m(f) = \begin{pmatrix} 1 & b_{12} & b_{13} & \cdots & b_{1,m} \\ 0 & 1 & b_{23} & \cdots & b_{2,m} \\ & & & \cdots & \\ 0 & 0 & 0 & \cdots & 1 \end{pmatrix}$$

for some $f \in F$. Then

$$\mu_m((f-1)^{m-1}) = \begin{pmatrix} 0\,0 & \cdots & 0 & b_{12} \ldots b_{m-1,\,m} \\ 0\,0 & \cdots & 0 & 0 \\ & & \cdots & \\ 0\,0 & \cdots & 0 & 0 \end{pmatrix}.$$

Proof. Induct on $m \geq 3$ and use the preceding Remark. $\square$

Proposition 1.90 *Let*

$$u \in \gamma_m(F) \setminus \gamma_{m+1}(F), \quad m \geq 1, \tag{1.64}$$

and $f \in F$ be such that

$$[u,\ f] \in \gamma_{m+2}(F). \tag{1.65}$$

Then

(i) $f \in \gamma_2(F)$ *in case* $m \geq 2$;
(ii) $u = f^k d$ *with* $k \in \mathbb{Z}$ *and* $d \in \gamma_2(F)$, *in case* $m = 1$.

Proof. This is a particular case of ([Mag66], Cor. 5.12(iii)). We give below a proof using the generalized Magnus embedding.

Consider the images of elements u and f under the generalized Magnus map μ_{m+2}:

$$\mu_{m+2} : u \mapsto \begin{pmatrix} 1 & 0 & 0 & \cdots & 0 & a_{1,m+1} & a_{1,m+2} \\ 0 & 1 & 0 & \cdots & 0 & 0 & a_{2,m+2} \\ 0 & 0 & 1 & \cdots & 0 & 0 & 0 \\ & & & \cdots & & & \\ 0 & 0 & 0 & \cdots & 0 & 0 & 1 \end{pmatrix}$$

$$\mu_{m+2} : f \mapsto \begin{pmatrix} 1 & b_{12} & b_{13} & \cdots & b_{1,m+2} \\ 0 & 1 & b_{23} & \cdots & b_{2,m+2} \\ & & & \cdots & \\ 0 & 0 & 0 & \cdots & 1 \end{pmatrix}$$

Since $\ker(\mu_{m+2}) = \gamma_{m+2}(F)$, the condition (1.65) is equivalent to the condition

$$\mu_{m+2}(u)\mu_{m+2}(f) = \mu_{m+2}(f)\mu_{m+2}(u).$$

After multiplication of matrices, and comparing the $(1, m+2)$-entries, we see that the condition (1.65) is equivalent to the following:

$$b_{12}\, a_{2,\,m+2} = a_{1,\,m+2}\, b_{m+1,\,m+2}. \tag{1.66}$$

Since $u \notin \gamma_{m+1}(F)$ and $f \notin \gamma_2(F)$, we have

$$b_{12} \neq 0, \quad b_{m+1,2} \neq 0, \quad a_{2,m+2} \neq 0, \quad a_{1,m+1} \neq 0.$$

For any $i = 1, \ldots, m+1$, we have $b_{i,i+1} \in \mathbb{Z}[\lambda_{i,i+1}(X)]$. Since the indeterminates $\lambda_{12}(X)$ do not enter into the construction of the element $a_{2,m+2}$ and indeterminates $\lambda_{m+1,m+2}(X)$ do not enter into the construction of $a_{1,m+1}$, we can conclude that there exists $z \in \mathbb{Z}[\lambda_{i,i+1}(X)]$, $i = 2, \ldots, m$, such that

$$a_{2,m+2} = b_{m+1,m+2}z, \quad a_{1,m+1} = b_{12}z. \tag{1.67}$$

If $a_{1,m+1}$ can be written as a polynomial

$$a_{1,m+1} = F(\lambda_{1,2}(X), \ldots, \lambda_{m,m+1}(X)), i = 1, \ldots, m-2,$$

then the element $a_{2,m+2}$ is a polynomial of the same form, but over another set of indetrminates, i.e.

$$a_{2,m+2} = F(\lambda_{2,3}(X), \ldots, \lambda_{m+1,m+2}(X)).$$

Then (1.67) implies that

$$z = b_{23}z_1, \quad z_1 \in \mathbb{Z}[\lambda_{i,i+1}], \quad i = 3, \ldots, m.$$

Repeating the same argument m times, we get

$$a_{1,m+1} = c \cdot b_{12}b_{23} \ldots b_{m,m+1}, \tag{1.68}$$

$$a_{2,m+1} = c \cdot b_{23} \ldots b_{m,m+1}b_{m+1,m+2}, \quad c \in \mathbb{Z}. \tag{1.69}$$

By Lemma 1.89, we have that the element $(\mu_{m+2}(f) - 1)^m$ has the following form:

$$(\mu_{m+2}(f) - 1)^m = \begin{pmatrix} 0\,0 \ldots & 0 & b_{12} \ldots b_{m,m+1} & * \\ 0\,0 \ldots & 0 & 0 & b_{23} \ldots b_{m+1,m+2} \\ 0\,0 \ldots & 0 & 0 & 0 \\ & \ldots & & \\ 0\,0 \ldots & 0 & 0 & 0 \end{pmatrix}.$$

Thus the matrix

$$B := \mu_{m+2}(u) - 1 - c \cdot (\mu_{m+2}(f) - 1)^m \tag{1.70}$$

consists of zero elements, with the possible exception of the $(1, m+2)$-entry:

$$B = \mu_{m+2}(u - 1 - c(f - 1)^m) = \begin{pmatrix} 0\,0 \ldots 0\,\alpha \\ 0\,0 \ldots 0\,0 \\ \ldots \\ 0\,0 \ldots 0\,0 \end{pmatrix},$$

$\alpha \in \mathbb{Z}[\lambda_{i,\,i+1}(X)]$. It follows from (1.63) that

$$u - 1 - c(f-1)^m \in \mathfrak{f}^{m+1}. \qquad (1.71)$$

Case (i): $u \in \gamma_m(F) \setminus \gamma_{m+1}(F), m \geq 2$. Going modulo $\gamma_2(F)$ in (1.71), we have:

$$c(f\gamma_2(F) - 1)^m \in \Delta^{m+1}(F/\gamma_2(F)).$$

Therefore $f \in \gamma_2(F)$.

Case (ii): $m = 1$, $u \notin \gamma_2(F)$. Then (1.71) has the form:

$$uf^{-c} - 1 \in \mathfrak{f}^2,$$

but then we get $u = f^k d$, $d \in \gamma_2(F)$, $k = c$. $\square$

Theorem 1.91 (Mikhailov[Mik05a],[Mik07a]). *Let r be a non-identity element of the free group F, and let R be the normal closure of r in F. Then for $k = 1$, 2, there exists a natural number m (which depends on k and r), such that $\varphi_{k+m}^{(k)}(F/R) = 0$.*

Proof. Let us first consider the case $k = 1$. In this case the required natural number m can be taken to be the one for which

$$r \in \gamma_m(F) \setminus \gamma_{m+1}(F).$$

Observe that the group $R/[R,\,F]$ is infinite cyclic. If $m = 1$, then the asserted statement is a well-known fact. In fact, in this case we have the following short exact sequence

$$0 \to H_2(F/R) \to R/[R,\,F](\simeq \mathbb{Z}) \to R\gamma_2(F)/\gamma_2(F) \to 1.$$

Since $F/\gamma_2(F)$ is torsion-free and $r \notin \gamma_2(F)$, it follows that $R\gamma_2(F)/\gamma_2(F)$ is an infinite cyclic group; therefore, $H_2(F/R) = 0$.

Suppose $m \geq 2$. Then $H_2(F/R) = R/[F,R] \simeq \mathbb{Z}$ and we have the following exact sequence of abelian groups:

$$0 \to \varphi_{m+1}^{(1)}(F/R) \to H_2(F/R)(\simeq \mathbb{Z}) \xrightarrow{h} \gamma_m(F)/\gamma_{m+1}(F),$$

where the homomorphism h is induced by the inclusion $r \in \gamma_m(F)$. Since the abelian group $\gamma_m(F)/\gamma_{m+1}(F)$ is torsion-free, it follows that

$$\varphi_{m+1}^{(1)}(F/R) = 0. \qquad (1.72)$$

Consider next the case $k = 2$. Suppose that $r \in \gamma_l(F) \setminus \gamma_{l+1}(F)$, $l \geq 1$. Then, by the case $k = 1$, which we have already proved, from (1.72) we conclude that $\varphi_{l+1}^{(1)}(F/R) = 0$, i.e.

$$R \cap \gamma_{l+1}(F) = [R, F] \cap \gamma_{l+1}(F). \tag{1.73}$$

Every element of $[R, F]/[R, F, F]$ can be expressed in the form $[r, f][R, F, F]$ for some $f \in F$. Thus, it follows from (1.73) that for any $m \geq l$, $\varphi_{m+2}^{(2)}(F/R) \neq 0$ implies the existence of $f \in F$, such that

$$[r, f] \in \gamma_{m+2}(F), \; [r, f] \notin [R, F, F]. \tag{1.74}$$

Since for any $f \in \gamma_2(F)$ we have $[r, f] \in [R, F, F]$, (1.74) implies that

$$f \notin \gamma_2(F). \tag{1.75}$$

If $l \geq 2$, the case (i) of Lemma 1.90 implies that for $m > l$ the condition (1.74) implies $f \in \gamma_2(F)$, thus, in view of (1.75), we get $\varphi_{l+2}^{(2)}(F/R) = 0$.

Now let $r \notin \gamma_2(F)$. If there exists $f \in F \setminus \gamma_2(F)$ such that $[r, f] \in \gamma_3(F)$, then, by case (ii) of Proposition 1.90, the following equation has a solution:

$$r = f^k d, \; k \in \mathbb{Z}, \; d \in \gamma_2(F). \tag{1.76}$$

If $k = \pm 1$, then $[r, f] \in [R, F, F]$, which is not the case. Thus we have $k \neq -1, 0, 1$, and therefore, $F/R\gamma_2(F)$ has torsion. The torsion in $F/R\gamma_2(F)$ is a cyclic group of order $N \; (> 1)$, say ; denote its generator by $wR\gamma_2(F)$, $w \in F$. Thus

$$r = w^N y, \; y \in \gamma_2(F).$$

From (1.76), we then have $f = w^{N_0} d_0$ for some $N_0 > 1$, $d_0 \in \gamma_2(F)$. Therefore

$$[r, f] = [r, w^{N_0} d_0] \equiv [r, w^{N_0}] \equiv [w^N y, w^{N_0}] \mod [R, F, F].$$

Let $q \geq 2$ be such that $y \in \gamma_q(F) \setminus \gamma_{q+1}(F)$. By Proposition 1.90 (case (i)), we see that if $[y, w^{N_0}] \in \gamma_{q+2}(F)$, then $w^{N_0} \in \gamma_2(F)$, which is not the case by construction. Thus (1.74) implies that $m < q$ and therefore we get

$$\varphi_{q+2}^{(2)}(F/R) = 0. \; \square$$

Remark. It is easy to see from the proof of Theorem 1.78 that if the element $r \in F$ is such that $F/R\gamma_2(F)$ is torsion free, then for the number m satisfying $r \in \gamma_m(F) \setminus \gamma_{m+1}(F)$, and $k = 1, \; 2$, we have $\varphi_{m+k}^{(k)}(F/R) = 0$.

The following example shows that the condition on torsion is necessary.

Example 1.92 Let $G = \langle a, b \mid aba^{-1}b = 1 \rangle$, then $M^{(2)}(G) \neq 0$.

Proof. Rewrite the relator as $r = b^2[b, a]$ and denote by F the free group with basis $\{a, b\}$ and R the normal closure of r in F. Then $[r, b] \in R \cap \gamma_3(F)$. Suppose that $[r, b] \in [R, F, F]$. Then

$$[b, a, b] \in [R, F, F] \subseteq \langle b^2, \gamma_4(F) \rangle^F.$$

Making the same construction with a change of symbols a and b, we conclude that $[a, b, a] \in \langle a^2, \gamma_4(F) \rangle^F$. This means that in the group

$$H = \langle a, b \mid a^2 = b^2 = 1 \rangle,$$

the elements $[a, b, a]$ and $[b, a, b]$ lie in the fourth term of the lower central series. However, these elements generale $\gamma_3(H)$ as a normal subgroup; consequently $\gamma_3(H) = \gamma_4(H)$. This is a contradiction, since H is residually nilpotent, but not nilpotent. We thus conclude that $[r, b] \notin [R, F, F]$. Hence $M^{(2)}(G) \neq 0$. $\square$

Lemma 1.93 *Let* $1 \to N \to \widetilde{G} \overset{\pi}{\to} G \to 1$ *be a* k-*central extension. Then* $[\pi^{-1}(\gamma_\omega(G)), {}_k\widetilde{G}] \subseteq \gamma_\omega(\widetilde{G})$.

Proof. Let $x \in \gamma_\omega(G)$ and $g \in \widetilde{G}$ be such that $\pi(g) = x$. Then for all $m \geq 1$, we can write

$$g = f_m r_m, \text{ for some } f_m \in \gamma_m(\widetilde{G}), \; r_m \in N.$$

The k-centrality condition $[N, {}_k\widetilde{G}] = 1$, implies that, for any $g_1, \ldots, g_k$ in $\widetilde{G}$, and $m \geq 1$, we have

$$[f_m r_m, g_1, \ldots, g_k] \in \gamma_m(\widetilde{G}).$$

Therefore $[g, g_1, \ldots, g_k] \in \gamma_\omega(\widetilde{G})$. $\square$

Theorem 1.94 (Mikhailov [Mik05a]). *Let* $k \geq 1$, G *a one-relator group and*

$$1 \to N \to \widetilde{G} \overset{\pi}{\to} G \to 1 \tag{1.77}$$

a k-*central extension with* $\widetilde{G}$ *residually nilpotent. Then* G *is also residually nilpotent.*

Proof. Suppose that G is not residually nilpotent. Let $x \in \gamma_\omega(G)$ and $g \in \widetilde{G}$ be such that $\pi(g) = x$. By Lemma 1.93, $[g, {}_k\widetilde{G}] \subseteq \gamma_\omega(\widetilde{G}) = 1$. Consequently $[x, {}_kG] = 1$. Due to the fact that x was an arbitrary element of $\gamma_\omega(G)$, it follows that

$$\gamma_{\omega+k}(G) = 1. \tag{1.78}$$

It follows from (1.78) that G has a nontrivial centre. To see this, note that $\gamma_{\omega+k}(G)$ lies in the centre of G. Suppose that $\gamma_{\omega+k-1}(G) = 1$, then the previous term in the transfinite lower central series lies in the centre. In the case all terms $\gamma_{\omega+l}(G)$, $l \geq 1$ are trivial, then $1 \neq \gamma_\omega(G)$ lies in the centre of G.

The remaining part of the proof follows by the scheme of the proof of the main theorem in [McC96]. We have that G is a one-relator group with nontrivial centre. Therefore it follows from [Pie74] that G can be defined by one of the following presentations:

$$G = \langle a_1, \ldots, a_m \mid a_1^{p_1} = a_2^{q_1}, \ldots, a_{m-1}^{p_{m-1}} = a_m^{q_{m-1}} \rangle, \qquad (1.79)$$

where p_i, $q_i \geq 2$, and $(p_i, q_j) = 1$, $i > j$; or

$$G = \langle a, a_1, \ldots, a_m \mid a a_1 a^{-1} = a_m, a_1^{p_1} = a_2^{q_1}, \ldots, a_{m-1}^{p_{m-1}} = a_m^{q_{m-1}} \rangle, \quad (1.80)$$

where p_i, $q_i \geq 2$, $p_1 \ldots p_{m-1} = q_1 \ldots q_{m-1}$ $(p_i, q_j) = 1, i > j$.

Consider first the case when the group G is given by the presentation (1.79). Then G is an amalgamated free product of cyclic groups and contains subgroups

$$G_i = \langle a_i, a_{i+1} \mid a_i^{p_i} = a_{i+1}^{q_i} \rangle, \ i = 1, \ldots, m-1.$$

Since the group G is transfinitely nilpotent by (1.78), subgroups G_i are also transfinitely nilpotent. If for some $1 \leq i \leq m-1$, the pair p_i, q_i is not a pair of powers of some prime, then $[a_i, a_{i+1}] \in \gamma_3(G_i)$, and we have a stabilization of the lower central series of G_i, and transfinite nilpotence for G_i is equivalent to commutativity, which is not possible for $p_i, q_i \geq 2$. Thus, for any $1 \leq i \leq m-1$, the numbers p_i and q_i are powers of a prime: $p_i = P_i^{s_i}$, $q_i = P_i^{t_i}$. If $m = 2$, we have the residually nilpotent group $G = \langle a_1, a_2 \mid a_1^{P_1^{s_1}} = a_2^{P_2^{t_2}} \rangle$. Hence we can assume that $m \geq 3$. The condition $(p_i, q_j) = 1$, $i > j$ implies that P_i are different primes for different i. Consider the subgroup H in G generated by elements $a_1^{P_2}$ and $a_3^{P_1}$. The subgroup H is not abelian by the construction of a free product. We have the following congruences in H:

$$[a_1^{P_2}, a_3^{P_1}]^{P_1} \equiv [a_1^{P_2 P_1}, a_3^{P_1}] \equiv [a_2^{P_2}, a_3^{P_1}] \equiv 1 \mod \gamma_3(H),$$

$$[a_1^{P_2}, a_3^{P_1}]^{P_2} \equiv [a_1^{P_2}, a_3^{P_1 P_2}] \equiv [a_1^{P_2}, a_2^{P_1}] \equiv 1 \mod \gamma_3(H),$$

which imply that $\gamma_2(H) = \gamma_3(H)$. Hence, for $m \geq 3$, transfinite nilpotence of G is not possible. We thus have a contradiction in this case.

Finally, let us consider the case when G has the presentation (1.80). In this case the group G is an HNN-extension of the group given by the presentation (1.79) with some additional conditions on coefficients p_i, q_i. As we have seen, transfinite nilpotence of the base group of the HNN-extension is possible only in the case $m = 2$. Thus G has the following presentation

$$G = \langle a, a_1, a_2 \mid a a_1 a^{-1} = a_2, a_1^{P^{s_1}} = a_2^{P^{t_1}} \rangle$$

for some prime P. However, such a group is residually nilpotent (see, for example, [McC96]), which is again a contradiction. $\square$

In the reverse direction we have the following

Theorem 1.95 (Mikhailov [Mik07a]) *Let G be a one-relator residually nilpotent group and*

$$1 \to N \to \widetilde{G} \to G \to 1 \tag{1.81}$$

a k-central extension, with $k \in \{1, 2\}$. Then $\widetilde{G}$ is residually nilpotent.

Proof. By hypothesis, the group G has a free presentation

$$1 \to R \to F \to G \to 1,$$

where R is the normal closure in F of a single element $r \in F$. By Theorem 1.91, there exists a natural number m such that

$$\varphi_{m+k}^{(k)}(G) = 0.$$

Theorem 1.78, applied to the k-central extension (1.81) then implies that

$$N \cap \gamma_{m+k}(\widetilde{G}) = 1;$$

hence $\widetilde{G}$ is residually nilpotent. $\square$

We proceed next to generalize Theorems 1.91 and 1.95 to free products of one-relator groups. For this we need more detailed analysis of the generalized Magnus embedding.

Theorem 1.96 *Let G_i, $i = 1, \ldots, n$ be a family of one-relator groups and $G = G_1 * \cdots * G_n$ their free product. Then, for $k \in \{1, 2\}$, there exists m such that $\varphi_m^{(k)}(G) = 0$.*

Proof. Let G_i, $i = 1, \ldots, n$ be defined by the presentation

$$1 \to R_i \to F_i \to G_i \to 1,$$

where the subgroups R_i is the normal closure in F_i of the element $r_i \in F_i$. Then the group G has the free presentation

$$1 \to R \to F \to G \to 1,$$

where $F = F_1 * \cdots * F_n$, $R = \langle r_1, \ldots, r_n \rangle^F$.

Let us first consider the case $k = 1$. Let l_i $(i = 1, \ldots, n)$ be natural numbers such that

$$r_i \in \gamma_{l_i}(F_i) \setminus \gamma_{l_i+1}(F_i), \tag{1.82}$$

and $l = \max_i(l_i)$. We assert that $\varphi_{l+1}^{(1)}(G) = 0$. Suppose that $\varphi_{l+1}^{(1)}(G) \neq 0$. Then there exist $k_1, \ldots, k_n$ such that

$$r_1^{k_1} \ldots r_n^{k_n} \in \gamma_{l+1}(F), \quad r_1^{k_1} \ldots r_n^{k_n} \notin [R, F].$$

We can then choose r_j $(1 \le j \le n)$ such that $r_j^{k_j} \notin [R_j, F_j]$. Consider the image of the element $r_j^{k_j}$ under the natural projection $d_i : F \to F_i$. We have $d(r_j^{k_j}) \in \gamma_{l+1}(F_j) \setminus [R_j, F_j]$, and consequently $\varphi_{l+1}^{(1)}(F_j/R_j) \neq 0$. On the other hand, by Theorem 1.91, we have $\varphi_{l+1}^{(1)}(F_j/R_j) = 0$. This is a contradiction, showing that we must have $\varphi_{l+1}^{(1)}(G) = 0$.

Next let us consider the case $k = 2$. By elementary commutator calculus, it is easy to show that if for some $m \ge 3$, $\varphi_m^{(2)}(G) \neq 0$, then there exist $f_j \in F_j$, such that:

$$[r_1, f_1] \ldots [r_n, f_n] \in \gamma_m(F) \setminus ([R, F, F] \cap \gamma_m(F)). \qquad (1.83)$$

In view of Theorem 1.91, we can assume that $n \ge 2$. Consider first the case when $n = 2$. Let

$$w := [r_1, f_1][r_2, f_2] \in \gamma_m(F), \quad [r_1, f_1][r_2, f_2] \notin [R, F, F]. \qquad (1.84)$$

We can then clearly assume that $f_1, f_2 \notin \gamma_2(F)$, because otherwise the consideration can be reduced to the case $n = 1$, which follows from Theorem 1.91. Let $f_1 = g_1 g_2 g_3$, $g_1 \in F_1$, $g_2 \in F_2$, $g_3 \in \gamma_2(F)$, $f_2 = g_4 g_5 g_6$, $g_4 \in F_1$, $g_5 \in F_1$, $g_6 \in \gamma_2(F)$. If $[r_1, g_1] \notin [R_1, F_1, F_1]$, then projecting the element $[r_1, g_1]$ on F_1 by the natural projection $d_1 : F \to F_1$, we get $d_1(w) \in \gamma_m(F_1)$. But by Theorem 1.91, there exists m_0, depending on r_1, such that $\varphi_{m_0}^{(2)}(F_1/R_1) = 0$; then necessarily, $m < m_0$ in (1.84). Thus, we can assume that

$$f_1 \in F_2 \setminus (F_2 \cap \gamma_2(F)), \quad f_2 \in F_1 \setminus (F_1 \cap \gamma_2(F)). \qquad (1.85)$$

Suppose $l_1 \neq l_2$, where l_i are defined as in (1.82). We can assume without loss of generality that $l_1 > l_2$. Then for $m > l_1$, (1.84) implies that $[r_2, f_2] \in \gamma_{l+1}(F)$. If $l_2 > 1$, then, by Lemma 1.90 (case (i)), we have $f_2 \in \gamma_2(F)$, which contradicts (1.85). If $l_1 = 1$, by Lemma 1.90 (case (ii)), we have $r_2 d = f_2^C$, $d \in \gamma_2(F)$; but, in view of (1.85, this is also impossible since $r_2 \in F_2$, $f_2 \in F_1$. Thus it is enough to consider the case $l_1 = l_2$.

Consider the Magnus embedding μ_{l_1+2}. By definition, we then have

$$\mu_{l_1+2} : r_1 \mapsto \begin{pmatrix} 1\,0\,0 \ldots 0 & s_{1,l_1+1} & s_{1,l_1+2} \\ 0\,1\,0 \ldots 0 & 0 & s_{2,l_1+2} \\ 0\,0\,1 \ldots 0 & 0 & 0 \\ & \cdots & \\ 0\,0\,0 \ldots 0 & 0 & 1, \end{pmatrix}$$

$$r_2 \mapsto \begin{pmatrix} 1\,0\,0 \ldots 0 & t_{1,l_1+1} & t_{1,l_1+2} \\ 0\,1\,0 \ldots 0 & 0 & t_{2,l_1+2} \\ 0\,0\,1 \ldots 0 & 0 & 0 \\ & \cdots & \\ 0\,0\,0 \ldots 0 & 0 & 1 \end{pmatrix},$$

$$f_1 \mapsto \begin{pmatrix} 1 & q_{12} & q_{13} & \cdots & q_{1,l_1+2} \\ 0 & 1 & q_{23} & \cdots & q_{2,l_1+2} \\ & & & \cdots & \\ 0 & 0 & 0 & \cdots & 1 \end{pmatrix}, f_2 \mapsto \begin{pmatrix} 1 & p_{12} & p_{13} & \cdots & p_{1,l_1+2} \\ 0 & 1 & p_{23} & \cdots & p_{2,l_1+2} \\ & & & \cdots & \\ 0 & 0 & 0 & \cdots & 1 \end{pmatrix},$$

for some

$$s_{1,l_1+1} \in \mathbb{Z}[\lambda_{i,i+1}(X_1)], \ i = 1, \ldots, l_1,$$
$$s_{2,l_1+2} \in \mathbb{Z}[\lambda_{i,i+1}(X_1)], \ i = 2, \ldots, l_1+1,$$
$$t_{1,l_1+1} \in \mathbb{Z}[\lambda_{i,i+1}(X_2)], \ i = 1, \ldots, l_1,$$
$$t_{2,l_1+2} \in \mathbb{Z}[\lambda_{i,i+1}(X_2)], \ i = 2, \ldots, l_1+1,$$
$$q_{ij} \in \mathbb{Z}[\lambda_{i,i+1}(X_2)], \ i = 1, \ldots, l_1+1,$$
$$p_{ij} \in \mathbb{Z}[\lambda_{i,i+1}(X_1)], \ i = 1, \ldots, l_1+1,$$

where X_1 and X_2 are bases of free groups F_1 and F_2 respectively. Then

$$\mu_{l_1+2} : w \mapsto \begin{pmatrix} 1 & 0 & 0 & \cdots & 0 & \alpha \\ 0 & 1 & 0 & \cdots & 0 & 0 \\ & & & \cdots & & \\ 0 & 0 & 0 & \cdots & 0 & 1 \end{pmatrix},$$

where

$$\alpha = q_{12}s_{2,l_1+2} - s_{1,l_1+1}q_{l_1+1\,l_1+2} + p_{12}t_{2\,l_1+2} - t_{1\,l_1+1}p_{l_1+1\,l_1+2}.$$

Therefore, in view of (1.63), $w \in \gamma_{l_1+2}(F)$ if and only if $\alpha = 0$. Since $q_{12}s_{2\,l_1+2} - t_{1\,l_1+1}p_{l_1+1\,l_1+2}$ does not depend on $\lambda_{12}(X_1)\,\lambda_{l_1+1\,l_1+2}(X_2)$, condition (1.85) implies then that $0 \neq p_{12} \in \mathbb{Z}[\lambda_{12}(X_1)]$, $0 \neq q_{l_1+1\,l_1+2} \in \mathbb{Z}[\lambda_{l_1+1\,l_1+2}]$, and it follows that $\alpha = 0$ if and only if

$$q_{12}s_{2\,l_1+2} - t_{1\,l_1+1}p_{l_1+1\,l_1+2} = 0, \tag{1.86}$$

and

$$p_{12}t_{2\,l_1+2} - s_{1\,l_1+1}q_{l_1+1\,l_1+2} = 0. \tag{1.87}$$

Since $q_{12}\,t_{1\,l_1+1} \in \mathbb{Z}[\lambda_{i\,i+1}(X_2)]$, $s_{2\,l_1+2}\,p_{l_1+1\,l_1+2} \in \mathbb{Z}[\lambda_{i\,i+1}(X_1)]$, it follows from (1.86) that $C_1q_{12} = C_2t_{1\,l_1+1}$, $C_1\,C_2 \in \mathbb{Z}$. By the Remark on page 52, we have $C_1q_{23} = C_2t_{2\,l_1+2}$. Then (1.87) implies that there exist C_3, $C_4 \in \mathbb{Z}$, such that $C_3q_{23} = C_4q_{l_1+1\,l_1+2}$. This is possible only in the case $l_1 = 1$. We get

$$r_1^{C_1} \equiv f_2^{C_2} \mod \gamma_2(F)$$
$$r_2^{C_1} \equiv f_1^{C_2} \mod \gamma_2(F).$$

Since any free abelian group is torsion free, we can assume that C_1 and C_2 are of coprime order. Then we have

$$w^{C_2} \equiv [r_1, \, f_1^{C_2}][r_2, \, f_2^{C_2}] \equiv [r_1, \, r_2^{C_1}][r_2, \, r_1^{C_1}] \equiv 1 \mod [R, \, F, \, F].$$

From the fact that C_1 and C_2 are of coprime order it follows that $w \in \gamma_j(F) \setminus (\gamma_j(F) \cap [R, \, F, \, F])$ for some $j \geq 3$, if and only if $w^{C_1} \in \gamma_j(F) \setminus (\gamma_j(F) \cap [R, \, F, \, F])$. Then

$$e := [f_1^{C_1}, \, f_2][f_2^{C_1}, \, f_1] \in \gamma_j(F).$$

We have

$$e \equiv [f_1, \, f_2, \, f_1]^k [f_2, \, f_1, \, f_2]^k \mod \gamma_4(F),$$

for some $k \neq 0$. Then

$$e \equiv [[f_1, \, f_2], \, f_1^k f_2^{-k}] \mod \gamma_4(F). \tag{1.88}$$

Since $f_1 \in F_2$, $f_1 \in F_1$, $[f_1, \, f_2] \notin \gamma_3(F)$, Lemma 1.90 (case (i)) implies that $f_1^k f_2^{-k} \in \gamma_2(F)$. But this is possible only in the case $f_i \in \gamma_2(F)$. Thus, in the case $r_i \notin \gamma_2(F)$, $i = 1, 2$, (1.84) implies that $m \leq 3$. The case $n = 2$ is thus proved and the following holds:

If $r_1 \in F_1$, $r_2 \in F_2$ and there exists m, which depends on r_1 and r_2, such that $[r_1, \, f_1][r_2, \, f_2] \in \gamma_m(F)$ for some $f_1, \, f_2 \in F$, then $[r_1, \, f_1][r_2, \, f_2] \in [R, \, F, \, F]$.

Now let $n \geq 3$. We shall prove the statement by induction on n, assuming that for $n - 1$ the statement is proved. Suppose that for a given m, there exist $f_1, \ldots, f_n$, which satisfy (1.83). We can assume that $f_j \notin \gamma_2(F)$. Then, for $f_1 \notin \langle F_j \rangle^F$ for almost all $j = 1, \ldots, n$, excepted maybe one of them, i.e. for all $j_1 \neq j_2$:

$$f_1 \notin \langle F_{j_1} \rangle^F \cap \langle F_{j_2} \rangle^F.$$

It follows from the fact that $\langle F_{j_1} \rangle^F \cap \langle F_{j_2} \rangle^F \subseteq \gamma_2(F)$. Then there exists $j \neq 1$, such that $f_i \notin \langle F_j \rangle^F$. Considering the projection

$$\hat{d}_j : F \to F_1 * \cdots * F_{j-1} * F_{j+1} \ldots F_n,$$

we get that $\hat{d}_j([r_1, \, f_1] \ldots [r_n, \, f_n])$ is an element from $\hat{d}_j(R)$, where $\hat{d}_j(R)$ is the normal closure of elements $r_1, \ldots, r_{j-1}, r_{j+1}, r_n$ in $\hat{d}_j(F)$. It is easy to see that (1.83) implies

$$d_j([r_1, \, f_1] \ldots [r_n, \, f_n]) \in \gamma_m(\hat{d}_j(F)) \setminus (\gamma_m(\hat{d}_j(F)) \cap [\hat{d}_j(R), \hat{d}_j(F), \hat{d}_j(F)]). \tag{1.89}$$

This reduces the problem to the case $n - 1$, which is proved by inductive assumption, i.e., there exists m, depending on $r_1, \ldots, r_n$, such that (1.89) impossible. Induction is complete and thus the statement is proved. $\square$

As a consequence of the foregoing result, we have the following generalization of Theorem 1.95.

Theorem 1.97 *Let G be a free product of one-relator groups and*

$$1 \to N \to \widetilde{G} \xrightarrow{\pi} G \to 1$$

a k-central extension ($k = 1, 2$). Then G is residually nilpotent if and only if $\widetilde{G}$ is residually nilpotent.

Proof. Suppose the group G is a free product of n one-relator groups with $n \geq 1$. The case $n = 1$ is already covered by Theorem 1.95. Suppose $n > 1$. If G is a residually nilpotent, then the residual nilpotence of $\widetilde{G}$ follows from Theorem 1.96 and Theorem 1.78.

Next suppose $\widetilde{G}$ is a residually nilpotent group and $x \in \gamma_\omega(G)$, $\pi(g) = x, g \in \widetilde{G}$. Then, by Lemma 1.93, $[g, {}_k\widetilde{G}] = 1$; hence $[x, {}_kG] = 1$. It follows that $[x, {}_kG] = 1$. Since any nontrivial free product has trivial centre it follows that $x = 1$ and hence G is residually nilpotent. $\square$

In view of the preceding results it is natural to raise the following

Problem 1.98 *Generalize the preceding results to the case of k-central extensions with $k \geq 3$.*

In the case ($k \geq 3$) the analysis of the generalized Magnus embedding looks much more complicated and we leave this problem open.

Para-free Groups

Let F be a free group. Recall that a group G is called F-*para-free* (or simply *para-free*) if it is residually nilpotent and there exists a homomorphism $F \to G$ which induces an isomorphism $F/\gamma_n(F) \to G/\gamma_n(G)$ for every natural number n. Such groups have been studied by G. Baumslag ([Bau67], [Bau69]). Call a group G to be *weakly para-free* if $G/\gamma_\omega(G)$ is para-free. Weakly para-free groups can be easily described in terms of the Dwyer filtration.

Proposition 1.99 *Let G be a group with $G/\gamma_2(G)$ free abelian. Then G is weakly para-free if and only if $H_2(G) = \bigcap_{m \geq 2} \varphi_m^{(1)}(G)$.*

Proof. Choose a homomorphism $f : F \to G$, with F free, which induces an isomorphism on abelianizations. Then the assertion follows from Theorem 1.76. $\square$

Example 1.100 (Baumslag [Bau67]).
Let G be the group given by the following presentation:

$$G = \langle a, b, c \mid a = [c^{-1}, a][c, b] \rangle.$$

Let F be a free group with generators x_1, x_2. Consider the homomorphism $f : F \to G$, defined by $f : x_1 \mapsto b$, $x_2 \mapsto c$. Clearly f induces an isomorphism $F/\gamma_2(F) \to G/\gamma_2(G)$. Since $H_2(G) = 0$, f induces isomorphisms

$$F/\gamma_n(F) \simeq G/\gamma_n(G), \ n \geq 2,$$

by Theorem 1.76. Hence G is a weakly para-free group.

In general, it is clear from Proposition 1.99 that if G is a residually nilpotent group with $G/\gamma_2(G)$ free abelian and $H_2(G) = 0$, then G is para-free. The converse statement, namely that $H_2(G) = 0$ for a para-free group G is false in general (see Theorem 1.126). Whether the converse holds for finitely generated groups is an open problem (see[Coc98]).

Problem 1.101 (Para-free Conjecture) *If G is a finitely generated para-free group, then $H_2(G) = 0$.*

The following result provides an equivalent formulation of the Parafree Conjecture.

Proposition 1.102 *Let G be a para-free group. Then $H_2(G) = 0$ if and only if G is absolutely residually nilpotent.*

Proof. First suppose that $H_2(G) = 0$. Then $M^{(k)}(G) = 0$ for all $n \geq 1$ by Theorem 5.14. Hence G is absolutely residually nilpotent by Corollary 1.80. Conversely, let G be an absolutely residually nilpotent para-free group. Let F/R be a free presentation of G. Then $\gamma_\omega(F/[R, F]) \simeq \bigcap_{m \geq 2} \varphi_m^{(1)}(G) = 0$ by Proposition 1.83. Therefore $H_2(G) = 0$ by Proposition 1.99. $\square$

Theorem 1.103 *Let G be a para-free group and $k \geq 1$. Then $M^{(k)}(G) = 0$ if $M^{(k+1)}(G) = 0$.*

Proof. Let $f : F \to G$ be a homomorphism which induces isomorphisms $f_n : F/\gamma_n(F) \to G/\gamma_n(G)$, for all $n \geq 1$. Suppose that $M^{(k+1)}(G) = 0$. Since f induces isomorphisms on lower central quotients, it also induces isomorphisms

$$M^{(c)}(F)/\varphi_k^{(c)}(F) \to M^{(c)}(G)/\varphi_k^{(c)}(G),$$

for all $c \geq 1$, $k \geq c+1$ by Proposition 1.76. Hence $M^{(c)}(G) = \varphi_k^{(c)}(G)$ for all c. Now the assertion follows from the epimorphism $M^{(k+1)}(G) \to \varphi_{k+1}^{(k)}(G) = M^{(k)}(G)$ (see Proposition 1.75). $\square$

Milnor's $\bar{\mu}$-invariants

Let L be a link in the three-dimensional sphere S^3, i.e. L is an embedding of certain number, say n, of circles:

$$f : S^1 \sqcup \cdots \sqcup S^1 \to S^3.$$

Let $T(L)$ be the tubular neighborhood of $\mathrm{im}(f)$ in S^3. The group $G(L) = \pi_1(S^3 \setminus T(L))$ is called *link group* of L. Given i, $1 \le i \le n$, let $T(L_i)$ be the tubular neighborhood of the ith component of $\mathrm{im}(f)$. For any point $x_0 \in S^3 \setminus T(L)$, connect it with an arbitrary point in $\partial T(L_i)$ by a path p, then transverse a closed loop in $T(L_i)$, which has linking number 1 with the ith component of $\mathrm{im}(f)$ and return to x_0 by p^{-1}. Such a loop then defines an element in the fundamental group $G(L)$ with the base point x_0, which is called (the ith) *meridian* of L. Clearly this element depends on the path p, but all meridians are conjugates in $G(L)$. The ith longitude l_i is the element of $G(L)$ determined by the the same path p and the loop in $T(L_i)$ whose linking number with ith component of $\mathrm{im}(f)$ is zero; any such pair (meridian, longitude) constitutes a set of homological generators of $H_1(\partial T(L_i))$.

There is a natural homomorphism, called a *meridional homomorphism* $f : F \to G(L)$, where F is a free group of rank n (with generator set $\{y_1, \ldots, y_n\}$). It maps the generators to the corresponding meridians of L. The collection of meridians determines a presentation of the quotient group $G(L)/\gamma_k(G(L))$ [Mil57]:

$$G(L)/\gamma_k(G(L)) \simeq$$
$$\langle x_1, \ldots, x_n \mid [x_i, l_i] = 1, \ i = 1, \ldots, n, \ \gamma_k(F(x_1, \ldots, x_n)) \rangle, \quad (1.90)$$

where $x_i = f(y_i)$, l_i is the element in $F(x_1, \ldots, x_n)$ representing the image of the ith longitude in $G(L)/\gamma_k(G(L))$.

J. Milnor introduced link isotopy invariants, the so called $\bar{\mu}$-invariants, which detect whether a meridional map induces an isomorphism

$$f_n : F/\gamma_k(F) \to G(L)/\gamma_k(G(L)).$$

Suppose the longitudes of L lie in the kth lower central series term of $G(L)$. Consider the Magnus embedding

$$\mu : \mathbb{Z}[F(x_1, \ldots, x_n)] \to \mathbb{Z}[[a_1, \ldots, a_n]],$$

with non-commutative variables $a_1, \ldots, a_n$, defined by

$$\mu : x_i \mapsto 1 + a_i, \ 1 \le i \le n,$$
$$\mu : x_i^{-1} \mapsto 1 - a_i + a_i^2 - a_i^3 + \ldots.$$

Then $\bar{\mu}(i_1, \ldots, i_k, j)$-*invariant is the coefficient in Magnus embedding of the jth longitude:*

$$\mu(l_j) = 1 + \sum_{(i_1, \ldots, i_k) \in S_k} \bar{\mu}(i_1, \ldots, i_k, j) a_{i_1} \ldots a_{i_k} + \ldots$$

The $\bar{\mu}$-invariants are not only isotopy invariants, but are also concordance invariants. Let L_1 and L_2 be two n-component links in S^3. Recall that a *link concordance* between L_1 and L_2 is an embedding:

$$H : (\bigsqcup_{i=1}^{n} S^1) \times I \to S^3 \times I,$$

such that $H(x,0)$ represents the first link L_1, $H(x,1)$ represents the second link L_2. In such a situation links L_1 and L_2 are called *concordant*. Let L_1 and L_2 be concordant links and H a concordance between them. Then the map $\pi_1(S^3 - L_1) \to \pi_1(S^3 \times I \backslash \mathrm{im}(H))$ induces isomorphisms of first homology H_1 and epimorphisms of second homology H_2. Hence, there are isomorphisms of lower central quotients of fundamental groups of L_1 and L_2.

The link L has all $\bar{\mu}$-invariants of length $\leq k$ trivial if and only if the meridional map induces isomorphisms $f_n : F/\gamma_{k+1}(F) \to G(L)/\gamma_{k+1}(G(L))$ [Mil57]. This is an interesting class of links which is still not fully understood. For example, all *slice links*, i.e. links which are concordant to trivial links have trivial $\bar{\mu}$-invariants. It is clear that for any such link L, *the group $G(L)$ is weakly para-free.*

A natural question that arises is the transfinite extension of $\bar{\mu}$-invariants [Mil57]. One of the versions of this question can be formulated as follows [Coc98]:

Problem 1.104 *Is it true that for any link L with trivial $\bar{\mu}$-invariants, $\gamma_\omega(G(L)) = \gamma_{\omega+1}(G(L))$?*

It follows directly from the para-free property of links with trivial $\bar{\mu}$-invariants that Para-free Conjecture, together with the exact sequence (1.40), implies a positive answer to the above problem, and thus provides a topological application of this conjecture.

1.5 Nilpotent Completion

For any group G, the *nilpotent completion* of G is defined to be the inverse limit of the tower of natural group epimorphisms

$$\cdots \to G/\gamma_n(G) \to \cdots \to G/\gamma_3(G) \to G/\gamma_2(G),$$

We denote it by $Z_\infty(G)$:

$$Z_\infty(G) = \varprojlim_{n \geq 2} G/\gamma_n(G).$$

We have the natural map

$$h : G \to Z_\infty(G), \qquad\qquad (1.91)$$

defined by setting

$$h : g \mapsto (g\gamma_2(G),\, g\gamma_3(G),\, \dots),\ \ g \in G,$$

which clearly is a monomorphism if and only if G is residually nilpotent.

The following result is due to Baumslag and Stammbach [Bau77].

Proposition 1.105 *Given a group G, let P be a subgroup of $Z_\infty(G)$ such that $h(G) \subseteq P$. Then the following statements are equivalent:*

(i) $h : G \to P$ induces isomorphisms

$$h_i : G/\gamma_i(G) \to P/\gamma_i(P),\ i \geq 2.$$

(ii) $h : G \to P$ induces an isomorphism

$$h_2 : G/\gamma_2(G) \to P/\gamma_2(P).$$

(iii) The map $P \to Z_\infty(G) \to G/\gamma_2(G)$ induces a monomorphism

$$P/\gamma_2(P) \to G/\gamma_2(G).$$

The above Proposition is a direct consequence of the fact that a homomorphism induces epimorphism of lower central quotients if and only if it induces epimorphism on the corresponding abelianizations.

Let F be a free group of infinite countable rank. Let X be a free basis of F. Enumerate X as follows

$$X = \{x_{1,1},\, x_{2,1},\, x_{2,2},\, \dots,\, x_{k,1},\, \dots,\, x_{k,k},\, x_{k+1,1},\, \dots\}.$$

Consider the element

$$\lambda = (\lambda_1\gamma_2(F),\, \lambda_2\gamma_3(F),\, \dots) \in Z_\infty(F),$$

where

$$\lambda_1 = 1,\ \lambda_k = [x_{2,1},\, x_{2,2}][x_{3,1},\, x_{3,2},\, x_{3,3}]\dots[x_{k,1},\, \dots,\, x_{k,k}],\ \ k \geq 2.$$

It is proved in (Baumslag-Stammbach [Bau77], Bousfield-Kan [Bou72]) that the element $\lambda \notin \gamma_2(Z_\infty(F))$. Hence we have

Theorem 1.106 *For the nilpotent completion map*

$$h : F \to Z_\infty(F) \tag{1.92}$$

the induced map

$$F/\gamma_2(F) \to Z_\infty(F)/\gamma_2(Z_\infty(F))$$

is, in general, not an epimorphism.

Orr's Link Invariants

Let L be an m-component link in S^3 such that all the $\bar{\mu}$-invariants of L are zero. [Vanishing of $\bar{\mu}$-invariants of L is equivalent to the fact that for the group $G = \pi_1(S^3 \setminus L)$ the meridian homomorphism $F_m \to G$ induces isomorphisms of lower central quotients, i.e., the group $G/\gamma_\omega(G)$ is para-free.] Then there exists a map $\rho : S^3 \setminus L \to K(Z_\infty(G), 1) = K(Z_\infty(F_m), 1)$. Consider the commutative diagram

$$
\begin{array}{ccc}
\sqcup_{i=1}^m S^1 \times S^1 & \xrightarrow{\;p\;} & \vee_{i=1}^m S^1 \\
{\scriptstyle r}\downarrow & & {\scriptstyle s}\downarrow \\
S^3 \setminus L & \xrightarrow{\;\rho\;} & K(Z_\infty(F), 1),
\end{array}
\tag{1.93}
$$

where p is the projection of torus to the wedge of circles which maps meridians to circles, r is an embedding of torus as boundaries of the tubular neighbourhood $\partial T(L)$, s is induced by the homomorphism $F_m \to Z_\infty(F_m)$. The cone of the map r is the 3-sphere S^3. Denote by K_ω the cone of the map s. Then the diagram (1.93) implies the existence of the map

$$f : S^3 \to K_\omega,$$

whose homotopy class $\theta(L) \in \pi_3(K_\omega)$ is, by definition, the *Orr's invariant* of L. This invariant vanishes for homology boundary links (see Cochran [Coc91]).

Problem 1.107 *Is it true that $\theta(L) = 0$ for any link L with zero $\bar{\mu}$-invariants?*

Problem 1.108 *Is it true that $\pi_i(K_\omega)$ is uncountable for $i > 2$?*

We shall show below (Theorem 1.126) that the group $H_2(Z_\infty(F))$ is uncountable for any free group F of rank ≥ 2 (Bousfield's theorem [Bou77]); thus the group $\pi_2(K_\omega)$ is also uncountable.

Some Subgroups of Nilpotent Completion

Theorem 1.109 (Baumslag-Stammbach [Bau77], Bousfield [Bou77]). *For every group G, there exists a subgroup $\bar{G} \subseteq Z_\infty(G)$ such that, under the map (1.92), $h(G) \subseteq \bar{G}$ and h induces isomorphisms*

$$h_i : G/\gamma_i(G) \to \bar{G}/\gamma_i(\bar{G}), \ i \geq 2,$$

with the following universal property:

For any group homomorphism $f : G \to H$, with H residually nilpotent, which induces isomorphisms $f_i : G/\gamma_i(G) \to H/\gamma_i(H), i \geq 2$, there exists a unique map $\bar{f} : H \to \bar{G}$ such that $\bar{f} \circ f = h : G \to \bar{G}$.

Proof. Let $\mathcal{U}$ be the set of subgroups U of $Z_\infty(G)$ which contain $h(G)$ and are such that the induced maps

$$h_i : G/\gamma_i(G) \to U/\gamma_i(U), \ i \geq 2$$

are isomorphisms. Let $U_1, U_2 \in \mathcal{U}$, and let W be the subgroup of $Z_\infty(G)$, generated by U_1 and U_2. Clearly $h(G) \subseteq W$. Consider the amalgamated free product $W_1 = U_1 *_{h(G)} U_2$. Then we have an exact sequence of homology groups

$$H_1(h(G)) \to H_1(U_1) \oplus H_1(U_2) \to H_1(W_1) \to 0,$$

which enables us to conclude that the natural map $\widetilde{h} : G \to W_1$ induces epimorphism of abelianizations:

$$\widetilde{h}_2 : G/\gamma_2(G) \to W_1/\gamma_2(W_1).$$

Since the map h_2 can be viewed as the composition of $\widetilde{h}_2$ with the natural epimorphism $W_1/\gamma_2(W_1) \to W/\gamma_2(W)$, we conclude that $h : G \to W$ induces epimorphism $h_2 : G/\gamma_2(G) \to W/\gamma_2(W)$ and hence $W \in \mathcal{U}$. Hence the set $\mathcal{U}$ is directed and we can define

$$\bar{G} = \varinjlim_{U \in \mathcal{U}} U.$$

We assert that the group $\bar{G}$ has the desired property. First observe that $\bar{G} \in \mathcal{U}$. Let H be a residually nilpotent group with a homomorphism $f : G \to H$ which induces isomorphisms $f_i : G/\gamma_i(G) \to H/\gamma_i(H)$, $i \geq 2$. Then H can be viewed as a subgroup of $Z_\infty(G)$ which lies in $\mathcal{U}$, and the obvious map $\bar{f} : H \to \bar{G}$ meets the requirement of the Theorem. $\square$

Theorem 1.110 ([Bou77]; see also [Bau77]). *If G is a finitely generated group, then $\bar{G} = Z_\infty(G)$.*

In particular, if F is a free group of finite rank, then its nilpotent completion $Z_\infty(F)$ is para-free.

Proof. By Proposition 1.105, it is enough to show that the map

$$\sigma : Z_\infty(G)/\gamma_2(Z_\infty(G)) \to G/\gamma_2(G)$$

is a monomorphism; i.e., we have to show that any element

$$\lambda = (\lambda_1\gamma_2(G),\ \lambda_2\gamma_3(G),\ \dots) \in Z_\infty(G),\ \lambda_i \in \gamma_2(G)$$

lies in $\gamma_2(Z_\infty(G))$.

Let $\{x_1, \dots, x_n\}$ be a set of generators of G. We define by induction a series of elements $u_i^{(k)} \in G$, $i \geq 1$, $k \geq 1$ such that

$$u_{i+1}^{(k)} \equiv u_i^{(k)} \mod \gamma_{i+1}(G),\ 1 \leq k \leq n,\ i \geq 1;$$

i.e., an element

$$u^{(k)} = (u_1^{(k)}\gamma_2(G),\ u_2^{(k)}\gamma_3(G),\ \dots)$$

of $Z_\infty(G)$ with the property:

$$\lambda = [u^{(1)}, h(x_1)] \dots [u^{(n)}, h(x_n)],$$

where h is the natural map (1.91). Observe that element λ_2 modulo $\gamma_3(G)$ can be written as

$$\lambda_2 \equiv [u_1^{(1)}, x_1] \dots [u_1^{(n)}, x_n] \mod \gamma_3(G)$$

for some elements $u_1^{(k)} \in G$, $1 \leq k \leq n$. Suppose we defined elements $u_i^{(k)}$, $1 \leq k \leq n$, $1 \leq i \leq t$, such that

$$u_{i+1}^{(k)} \equiv u_i^{(k)} \mod \gamma_{i+1}(G),\ 1 \leq k \leq n,\ 1 \leq i \leq t-1,$$
$$\lambda_i = [u_i^{(1)}, x_1] \dots [u_i^{(n)}, x_n] \mod \gamma_{i+1}(G),\ 1 \leq i \leq t,$$
$$\lambda_{t+1} = [u_t^{(1)}, x_1] \dots [u_t^{(n)}, x_n] \mod \gamma_{t+2}(G).$$

Observe then that we can find elements $v_1, \dots, v_n \in G$ such that

$$\lambda_{t+2} \equiv [u_t^{(1)}, x_1] \dots [u_t^{(n)}, x_n][v_1, x_1] \dots [v_n, x_n] \mod \gamma_{t+3}(G). \qquad (1.94)$$

We set $u_{t+1}^{(k)} = u_t^{(k)} v_k$, $1 \leq k \leq n$. Then (1.94) implies

$$\lambda_{t+1} \equiv [u_{t+1}^{(1)}, x_1] \dots [u_{t+1}^{(n)}, x_n] \mod \gamma_{t+2}(G),$$

which completes the construction of the elements $u^{(i)}$ ($i = 1, \dots, n$), thereby establishing the Theorem. $\square$

For a free group F with basis X of infinite cardinality, one more interesting subgroup $\widetilde{F}$ of the nilpotent completion $Z_\infty(F)$ was introduced in [Bau77]. It is defined as the direct limit

$$\widetilde{F} := \varinjlim_{(Y \subset X, \ Y \text{ finite})} Z_\infty(F(Y)).$$

It is easy to see that the group $\widetilde{F}$ is para-free, hence there are the following inclusions:

$$F \subseteq \widetilde{F} \subseteq \bar{F} \subseteq Z_\infty(F).$$

It was shown in [Bau77] that the inclusion $\widetilde{F} \subset \bar{F}$ is proper.

We next recall a construction which comes from topology and has properties similar to the one considered in the present section.

Vogel Localization

For a finite CW-complex X, the *Vogel localization* [Le 88] EX has the following properties:

(i) EX is the inductive limit of finite subcomplexes

$$X = X_0 \subset X_1 \subset X_2 \subset \dots,$$

such that X_n/X is contractible, i.e., the inclusion $X \subset X_n$ is a homological equivalence and $\pi_1(X_n)$ is the normal closure of the image of $\pi_1(X)$;

(ii) for any pair (K, L) of finite subcomplexes with K/L contractible, any map $L \to EX$ can be extended uniquely (up to homotopy) to a map $K \to EX$.

The theory of Vogel localization was developed by LeDimet for applications in high-dimensional link theory [Le 88] (see also [Coc05]). Consider the following group:

$$\hat{F} \simeq \pi_1(EK(F, 1))/\gamma_\omega(\pi_1(EK(F, 1))).$$

Clearly, the property (i) of Vogel localization implies that the map $F \to \hat{F}$ induces isomorphisms of lower central quotients; hence $\hat{F}$ is F-para-free and there is a chain of inclusions

$$F \subseteq \hat{F} \subseteq Z_\infty(F).$$

1.6 Bousfield-Kan Completion

The first derived functors of the lower central quotient functors are the same as Baer invariants (see Appendix, Example A.16). The question of studying the derived functors of the nilpotent completion functor thus arises naturally.

For a commutative ring R and a simplicial set X, Bousfield and Kan defined the simplicial set $R_\infty X$, now called the *Bousfield-Kan completion*, having natural R-*localization* properties [Bou72]. For example, in the case of a simply connected simplicial set X and $R = \mathbb{Z}_p$, the simplicial set $R_\infty X$ is homotopically equivalent to the pro-p-completion of X in the sense of Sullivan [Sul71]. One of the main properties of the Bousfield-Kan completion

$$R_\infty : X \to R_\infty X,$$

is the following result:

If $f : X \to Y$ is a map of two simplicial sets, then f induces isomorphisms $\tilde{H}_*(X, R) \to \tilde{H}_*(Y, R)$ of reduced homology if and only if it induces the weak homotopy equivalence $R_\infty X \simeq R_\infty Y$.

We consider only the case of $R = \mathbb{Z}$, i.e., $\mathbb{Z}$-completion. One of the major results in the theory of $\mathbb{Z}$-localized spaces is the "group-theoretical" construction of the functor $\mathbb{Z}_\infty$. It can be defined with the help of R-nilpotent completion in the following way.

Theorem 1.111 (Bousfield - Kan[Bou72]). *For a given simplicial set X, define the functor*

$$Z_\infty : X \mapsto \overline{W}(Z_\infty(GX)).$$

Then we have a weak homotopy equivalence

$$\mathbb{Z}_\infty X \simeq Z_\infty(X).$$

[For the definition of the operator $\overline{W}$ see Appendix, Section A.8.]

The properties of Bousfield-Kan completion make it possible to identify the classical K-theory functors with the derived functors of the nilpotent completion of general linear groups. To show this, let us first recall the basic definitions.

For a connected CW-complex X with a base point x_0, one may obtain a new CW-complex X^+ by attaching 2-cells and 3-cells to X, so that the following properties hold:

(i) The map $X \to X^+$ is a homology equivalence.

(ii) The homomorphism $\pi_1(X) \to \pi_1(X^+)$ is the quotient homomorphism $\pi_1(X) \to \pi_1(X)/\mathcal{P}(\pi_1(X))$, where $\mathcal{P}(\pi_1(X))$ is the perfect radical, i.e., the maximal perfect subgroup, of $\pi_1(X)$.

Such a space X^+ is unique up to homotopy. Given a ring R, K-theoretical functors $K_i(R)$, $i \geq 1$ can be defined as

$$K_i(R) := \pi_i(BGL(R)^+), \ i \geq 1.$$

Recall that a space X with a base point x_0 is called an *H-space* if there is a "multiplication" map $\mu : X \times X \to X$ with $\mu(x_0, x_0) = x_0$, such that the maps $x \mapsto \mu(x, x_0)$ and $x \mapsto \mu(x_0, x)$ are homotopic to identity. We next recall the following natural properties of Bousfield-Kan completion and Quillen's plus-construction:

(i) For an H-space X, the integral completion is the weak homotopy equivalent to X:
$$X \simeq Z_\infty X.$$

(ii) For any ring R, the plus-construction of the classifying space of its general linear group $BGL(R)^+$ is an H-space.

Theorem 1.112 (Keune [Keu]). *Let R be a ring with identity. Then*
$$\mathcal{L}_n Z_\infty(GL(R)) = K_{n+1}(R), \ n \geq 1,$$
where K_* is the classical K-theory of R.

[For the definition of the left derived functors $\mathcal{L}_n$ see Appendix, Section A.13.]

Proof. Consider the following commutative diagram

$$\begin{array}{ccc} BGL(R) & \longrightarrow & BGL(R)^+ \\ \downarrow & & \downarrow \\ Z_\infty BGL(R) & \xrightarrow{g} & Z_\infty BGL(R)^+, \end{array}$$

where g is the Bousfield-Kan completion of the plus-construction map. Since plus-construction is a homology equivalence, g is a weak homotopy equivalence. The fact that $BGL(R)^+$ is an H-space implies that the right vertical map is also the weak homotopy equivalence. Therefore,

$$\mathcal{L}_n Z_\infty(GL(R)) = \pi_n(\overline{W}Z_\infty(G\overline{W}(GL(R)))) =$$
$$\pi_{n+1}(Z_\infty BGL(R)) = \pi_{n+1}(BGL(R)^+) = K_{n+1}(R).$$

That is, the derived functors of the integral nilpotent completion functor define K-theory of rings. $\square$

The derived functors of the nilpotent completion functor, besides their K-theoretical meaning, play important role in topology. The celebrated theorem of Barratt-Kahn-Priddy-Quillen [Pri] says that there exists a weak homotopy equivalence:
$$K(\Sigma_\infty, 1)^+ \simeq QS^0,$$

where QS^0 is a space defined as a limit of loop spaces $\Sigma^n S^n$ such that its homotopy groups $\pi_*(QS^0)$ are naturally isomorphic to the stable homotopy groups of spheres, i.e.,
$$\pi_n(QS^0) = \pi_n^S, \ n \geq 1.$$

Proposition 1.113 ([Bou72], p. 207). *The derived functors $\mathcal{L}_n Z_\infty(\Sigma_\infty)$ of the nilpotent completion for the infinite symmetric group Σ_∞ are the stable homotopy groups π_{n+1}^S of spheres.*

Proof. The proof repeats the proof of Theorem 1.112. Consider the following commutative diagram

$$
\begin{array}{ccc}
K(\Sigma_\infty, 1) & \longrightarrow & K(\Sigma_\infty, 1)^+ \\
\downarrow & & \downarrow \\
Z_\infty K(\Sigma_\infty, 1) & \xrightarrow{\ g\ } & Z_\infty K(\Sigma_\infty, 1)^+,
\end{array}
$$

where g is the Bousfield-Kan completion of the plus-construction map. Since plus-construction is a homology equivalence, g is a weak homotopy equivalence. The fact that $K(\Sigma_\infty, 1)^+ \simeq QS^0$ is an H-space implies that the right vertical map is also a weak homotopy equivalence. Therefore,

$$
\mathcal{L}_n Z_\infty(\Sigma_\infty) = \pi_n(\overline{W} Z_\infty(G\overline{W}(\Sigma_\infty))) =
$$
$$
\pi_{n+1}(Z_\infty K(\Sigma_\infty, 1)) = \pi_{n+1}(K(\Sigma_\infty, 1)^+) = \pi_{n+1}^S.
$$

That is, the derived functors of the integral nilpotent completion functor define stable homotopy groups of spheres. $\square$

The natural inclusion $f : \Sigma_\infty \to GL(\mathbb{Z})$ induces a map between derived functors of nilpotent completions, which is identical to the well-known map

$$
\pi_*^S \to K_*(\mathbb{Z})
$$

between stable homotopy groups of spheres and K-theory of integers (see [Qui]).

We conclude with a mention of the derived functors of the p-adic completion functor (p-adic completion means the inverse limit over all finite p-quotients). Let p be a prime and

$$
Z_\infty^p : G \to \hat{G}_p
$$

be the p-adic completion functor. What can one say about its derived functors? This question was considered in [Bou92]. One can find the following beautiful and surprising properties of these derived functors in [Bou92]:

1. Let Σ_3 be the symmetric group of degree 3. Then there is a short exact sequence:

$$
0 \to \pi_{n+1} S^3 \otimes \mathbb{Z}_3 \to \mathcal{L}_n Z_\infty^3(\Sigma_3) \to \mathrm{Tor}(\pi_n(S^3), \mathbb{Z}_3) \to 0.
$$

2. For the infinite symmetric group Σ_∞, $\mathcal{L}_n Z_\infty^p(\Sigma_\infty)$ is the p-torsion part of the $(n+1)$th stable homotopy group π_{n+1}^S.

Problem 1.114 *What can one say about the derived functors of the functors* $G \mapsto \varprojlim G/\delta_i(G)$ *(prosolvable completion) and* $G \mapsto \hat{G}$ *(Vogel localization)?*

1.7 Homological Localization

General Idea of Localization

Let $\mathfrak{C}$ be a category and $\mathfrak{R}$ a set of morphisms in $\mathfrak{C}$. Recall that an object $X \in \mathfrak{C}$ is called $\mathfrak{R}$-*local* if, for every morphism $f : A \to B$ from $\mathfrak{R}$, the induced map

$$\mathrm{Mor}(B, X) \to \mathrm{Mor}(A, X)$$

is a bijection. Denote by $loc(\mathfrak{R})$ the class of $\mathfrak{R}$-local objects in $\mathfrak{C}$. For an object $Y \in \mathfrak{C}$, by an $\mathfrak{R}$-*localization* of Y we mean a morphism $Y \to Z$ in $\mathfrak{R}$ such that Z is $\mathfrak{R}$-local.

The concept of *colocalization* can be defined analogously. An object $X \in \mathfrak{C}$ is called $\mathfrak{R}$-*colocal* if, for every morphism $f : A \to B$ from $\mathfrak{R}$, the induced map $\mathrm{Mor}(X, A) \to \mathrm{Mor}(X, B)$ is a bijection. We denote by $coloc(\mathfrak{R})$ the class of $\mathfrak{R}$-colocal objects. For an object $Y \in \mathfrak{C}$, by an $\mathfrak{R}$-*colocalization* of Y, we mean a morphism $Z \to Y$ in $\mathfrak{R}$ such that Z in $\mathfrak{R}$-colocal.

The (co)localization concept plays an important role in modern algebra and topology. A lot of important theories, such as algebraic K-theory, or motivic homotopy theory, can be defined via localizations. A well-written interesting survey about (co)localizations is [Dwy04]. We present here two illustrative examples of applications of localizations.

Example 1.115
Let $\mathfrak{C}$ be a category of pointed CW-complexes. For a given $W \in \mathfrak{C}$, consider the trivialization map $W \to pt$. The localization in $\mathfrak{C}$ with respect to the map $W \to pt$ is called a W-*nullification*. For a given CW-complex X, n-dimensional sphere S^n and the localization map with respect to the map $S^n \to pt$, i.e., S^n-nullification, is homotopically equivalent to the $(n-1)$th stage of Postnikov tower for X.

Example 1.116
(Berrick-Dwyer [Ber00]) Let W be a CW-complex. Then, for any space X, the W-nullification is homotopically equivalent to the plus-construction X^+ if and only if W is acyclic and there is a nontrivial homomorphism $\pi_1(W) \to GL(\mathbb{Z})$. In [Ber00] such complexes W are called *spaces that define algebraic K-theory*.

Example 1.117
As an application of the cohomological approach to the dimension subgroup theory, in Chapter 2, we shall describe the quasi-variety of groups with trivial nth dimension subgroup as a suitable set of local objects.

HZ-tower

We will now consider the transfinite version of nilpotent completion, the so-called *HZ-localization*, due to Bousfield. This is a localization in the category of groups with respect to homomorphisms which induce isomorphism on abelianizations and epimorphism on the second homologies. We construct a tower of group homomorphisms, which have certain special properties, connected with transfinite lower central series and homology.

Let G be a group. Define the initial terms of the tower as the trivialization and abelianization:

$$\eta_1 : G \to T_1 G, \ T_1 G = 1,$$
$$\eta_2 : G \to T_2 G, \ T_2 G = G/\gamma_2(G).$$

Suppose for a given ordinal number α we have constructed the homomorphism $\eta_\alpha : G \to T_\alpha G$ such that the induced map

$$H_1(\eta_\alpha) : H_1(G) \to H_1(T_\alpha G)$$

is an isomorphism. The homomorphism η_α induces a map of classifying spaces:

$$\bar{\eta}_\alpha : K(G, 1) \to K(T_\alpha G, 1). \qquad (1.95)$$

Consider the cylinder $C_{\bar{\eta}_\alpha}$ of the map $\bar{\eta}_\alpha$ and the homology $H_2(\eta_\alpha)$, which is defined as the relative homology

$$H_2(\eta_\alpha) := H_2(C_{\bar{\eta}_\alpha}, K(G, 1)).$$

One has the following long exact sequence of homology groups of the pair $(C_{\bar{\eta}_\alpha}, K(G, 1))$:

$$\cdots \to H_2(G) \to H_2(T_\alpha G) \to H_2(\eta_\alpha) \to H_1(G) \to H_1(T_\alpha G) \to \cdots$$

Since the map $H_1(\eta_\alpha)$ is an isomorphism, therefore

$$H_2(\eta_\alpha) = \operatorname{coker}\{H_2(G) \to H_2(T_\alpha G)\}.$$

The cohomology $H^2(\eta_\alpha)$ can be defined in analogy with (1.95) as relative cohomology. Denote by k_α the fundamental class in $\operatorname{Hom}(H_2(\eta_\alpha), H_2(\eta_\alpha)) = H^2(\eta_\alpha, H_2(\eta_\alpha))$. The natural map

$$\beta : H^2(\eta_\alpha, H_2(\eta_\alpha)) \to H^2(T_\alpha G, H_2(\eta_\alpha))$$

determines the element $\beta(k_\alpha) \in H^2(T_\alpha G, H_2(\eta_\alpha))$. Identifying cohomology classes with homotopy classes of maps between Eilenberg-MacLain spaces:

$$H^2(T_\alpha G, H_2(\eta_\alpha)) = [K(T_\alpha G, 1), K(H_2(\eta_\alpha), 2)],$$

we have a map $\bar{\beta}(k_\alpha) : K(T_\alpha G, 1) \to K(H_2(\eta_\alpha), 2)$, uniquely defined by η_α up to homotopy, such that the following diagram is commutative:

$$
\begin{array}{ccc}
K(G, 1) & \longrightarrow & PK(H_2(\eta_\alpha), 2) \\
\downarrow{\scriptstyle \bar{\eta}_\alpha} & & \downarrow{\scriptstyle p} \\
K(T_\alpha G, 1) & \xrightarrow{\ \bar{\beta}(k_\alpha)\ } & K(H_2(\eta_\alpha), 2),
\end{array}
\qquad (1.96)
$$

where p is the path fibration with the fibre the loop space $\Omega K(H_2(\eta_\alpha), 2)$. Taking the induced principal fibration over $K(T_\alpha G, 1)$ and applying the fundamental group functor, we obtain the required diagram:

$$
\begin{array}{ccccccccc}
& & & & G & \longrightarrow & G & & \\
& & & & \downarrow{\scriptstyle \eta_{\alpha+1}} & & \downarrow{\scriptstyle \eta_\alpha} & & \\
1 & \longrightarrow & H_2(\eta_\alpha) & \longrightarrow & T_{\alpha+1}G & \xrightarrow{t_\alpha} & T_\alpha G & \longrightarrow & 1,
\end{array}
$$

where the bottom row is a central extension.

In the case of a limit ordinal number α, suppose we have already defined homomorphisms $\eta_\tau : G \to T_\tau G$ for all ordinals $\tau < \alpha$. Consider the limit homomorphism

$$
h_\alpha : G \to \varprojlim_{\tau < \alpha} T_\tau G
$$

and define $T_\alpha G$ to be the maximal subgroup of $\varprojlim_{\tau < \alpha} T_\tau G$, which contains $\mathrm{im}(h_\alpha)$ and is such that the map $h_\alpha : G \to T_\alpha G$ induces an isomorphism $H_1(h_\alpha) : H_1(G) \to H_1(T_\alpha G)$. This is the required homomorphism

$$
\eta_\alpha : G \to T_\alpha G \quad (\subseteq \varprojlim_{\tau < \alpha} T_\tau G)
$$

Thus we get the following transfinite tower of group homomorphisms:

$$
\begin{array}{ccccccccc}
G & \xleftarrow{\ id\ } & G & \xleftarrow{\ id\ } & \cdots \longleftarrow & G & \xleftarrow{\ id\ } & G & \longleftarrow \cdots \\
\downarrow{\scriptstyle \eta_1} & & \downarrow{\scriptstyle \eta_2} & & & \downarrow{\scriptstyle \eta_\alpha} & & \downarrow{\scriptstyle \eta_{\alpha+1}} & \\
T_1 G & \xleftarrow{\ t_1\ } & T_2 G & \xleftarrow{\ t_2\ } & \cdots \longleftarrow & T_\alpha G & \xleftarrow{\ t_\alpha\ } & T_{\alpha+1}G & \longleftarrow \cdots
\end{array}
$$

which is called the *HZ-tower* of the group G. The *HZ-localization* of G is the inverse limit of this HZ-tower:

$$
L : G \to L(G) := \varprojlim_\alpha T_\alpha G. \qquad (1.97)
$$

The functor L, constructed above, has the following properties:

(i) L induces isomorphism

$$
H_1(L) : H_1(G) \to H_1(L(G))
$$

and epimorphism

$$H_2(L) : II_2(G) \to H_2(L(G));$$

therefore, it induces isomorphisms

$$G/\gamma_n(G) \simeq L(G)/\gamma_n(L(G))$$

for all finite $n \geq 1$.

(ii) $L(G)$ is transfinitely nilpotent for any group G.

(iii) There exists a canonical homomorphism

$$L(G) \to Z_\infty(G),$$

where $Z_\infty(G)$ is the free nilpotent completion of G, which is an epimorphism with kernel $\gamma_\omega(L(G))$ in the case of a finitely generated group G.

(iv) For any ordinal number α, there is a natural isomorphisms

$$T_\alpha G = L(G)/\gamma_\alpha(L(G)). \tag{1.98}$$

Problem 1.118 *Is it true that for every group G, the kernel $\mathrm{ker}(L)$ is equal to the intersection of the transfinite lower central series of G?*

This problem is related to some other questions of localization theory (see [Rod04]).

The class of *HZ-local* groups is the smallest class, containing the class of abelian groups, which is closed under central extensions and inverse limits [Bou77]. The *HZ*-localization $L : G \to L(G)$ of a given group G can be completely described as a map, which induces an isomorphism $H_1(L) : H_1(G) \to H_1(L(G))$, and an epimorphism $H_2(G) \to H_2(L(G))$ with the target group $L(G)$ *HZ*-local. In particular, for any *HZ*-local group, the *HZ*-localization is the identity homomorphism.

We next present an illustrative example of a method for constructing *HZ*-local groups with transfinitely long lower central series or, equivalently, with transfinitely long *HZ*-tower.

For a given ordinal number τ, denote by $\mathfrak{N}_\tau$ the class of groups G with $\gamma_\tau(G) = 1$. For elements a, b in a given group G, the left-normed Engel elements $[a, {}_ib]$ are defined inductively by setting

$$[a, {}_0b] = a, \text{ and } [a, {}_ib] = [[a, {}_{(i-1)}b], b] \text{ for } i \geq 1.$$

Example 1.119
Let $F = \langle a, b \mid \emptyset \rangle$ be the free group of rank 2, $H_i = F/\gamma_i(F)$, $i \geq 2$, and $H = \prod_{i>2} H_i$, the unrestricted direct product of the groups H_i. Let $x_i = [a, {}_{(i-2)}b] \in F$, $y_i = x_i\gamma_i(F)$ and h_i the element in H with its ith entry y_i, and identity everywhere else. Consider the group $G = H/K$, where K is the subgroup generated by the elements $h_ih_j^{-1}$, $i, j \geq 2$. We claim that:

$$G \text{ is an } HZ\text{-local group in } \mathfrak{N}_{\omega+1} \text{ with } \gamma_\omega(G) = \prod_{i\geq 2}\langle h_i\rangle \neq 1.$$

Observe that $h_i K \in \gamma_\omega(G)$, $i \geq 2$. Let $N = \prod_{i \geq 2} \langle h_i \rangle$ and

$$h = (y_1^{k_1}, y_2^{k_2}, \ldots, y_m^{k_m}, \ldots), \quad k_i \in \mathbb{Z},$$

an arbitrary element of N. Let $n \geq 1$. Since $h_i K \in \gamma_n(G)$, $2 \leq i \leq n$, therefore

$$hK = (1, \ldots, 1, y_{n+1}^{k_{n+1}}, \ldots)K \quad \mathrm{mod}\ \gamma_n(G).$$

Note that $y_i^{k_i} = [y_{i-1}, b^{k_i}]\gamma_i(F)$ for all i; therefore, modulo $\gamma_n(H)$ we have:

$$(1, \ldots, 1, y_{n+1}^{k_{n+1}}, \ldots) =$$

$$[[(1, \ldots, 1, y_2, y_3, \ldots),\ _{(n-2)}(1, \ldots, 1, b, b, \ldots)], (1, \ldots, 1, b^{k_{n+1}}, \ldots)] = 1.$$

Hence $hK \in \gamma_n(G)$ for all n and so $N/K \subseteq \gamma_\omega(G)$. Clearly $H/N \simeq \prod_{i \geq 2} H_i/\langle y_i \rangle$. Since H_i, $i \geq 2$, is nilpotent it follows that H/N is residually nilpotent and HZ-local. Therefore $\gamma_\omega(G) = N/K$ which is non-identity (in fact, it is uncountable), and furthermore, since N is central in H, it follows that $\gamma_{\omega+1}(G) = 1$ and G is HZ-local.

Recall that the HZ-local groups have the following properties:

Limit property: *If τ is a limit ordinal and $f : G \to H$ is a homomorphism between HZ-local groups which induces an isomorphism $f_\alpha : G/\gamma_\alpha(G) \to H/\gamma_\alpha(H)$ for every $\alpha < \tau$, then f induces an isomorphism $f_\tau : G/\gamma_\tau(G) \to H/\gamma_\tau(H)$* ([Bou77], §3.16).

HZ-closure property: *If τ is a limit ordinal and G is an HZ-local group, then $G/\gamma_\tau(G)$ is an HZ-closed subgroup of $\varprojlim_{\alpha < \tau} G/\gamma_\alpha(G)$* ([Bou77], Theorem 3.11).
[A subgroup $H \subseteq G$ is called *HZ-closed* in G if whenever we have subgroups $H \subseteq W \subseteq G$ such that the induced map $H_1(H) \to H_1(W)$ is an epimorphism, then $H = W$.]
   We observe next that the subgroup $\bar{G}$ of the nilpotent completion (see Theorem 1.109) can be described as a quotient of the HZ-localization of G.

Theorem 1.120 *For every group G, $L(G)/\gamma_\omega(L(G)) \simeq \bar{G}$.*

Proof. Let p be the natural projection $L(G) \to L(G)/\gamma_\omega(L(G))$ and $q = p \circ L$. It is clear that q induces an isomorphism $G/\gamma_n(G) \simeq L(G)/\gamma_n(L(G))$ for all $n \geq 1$. Therefore by the property (ii) of $h : G \to \bar{G}$ (Proposition 1.105), there exists a monomorphism $\theta : L(G)/\gamma_\omega(L(G)) \to \bar{G}$ such that $\theta \circ q = h$. Clearly θ must induce an isomorphism

$$H_1(L(G)/\gamma_\omega(L(G))) \to H_1(\bar{G}).$$

From the HZ-closure property of the group $L(G)$ it follows that the group $L(G)/\gamma_\omega(L(G))$ is HZ-closed in $\varprojlim_n G/\gamma_n(G)$. Therefore θ is an isomorphism.

□

Homology of Nilpotent Completion

Let G be a simplicial group. Then one can define two natural objects, connected with nilpotent completion: $\pi_0(Z_\infty(G))$ and $Z_\infty(\pi_0(G))$. Also one can observe that there is a natural map from the first object to the second one. The fibration exact sequences together with properties of lower central series of groups give the following result.

Proposition 1.121 ([Bou77], Lemma 5.4). *For every simplicial group G, there is a natural short exact sequence of groups*

$$1 \to \varprojlim{}_n^1 \pi_1(G/\gamma_n(G)) \to \pi_0(Z_\infty(G)) \to Z_\infty(\pi_0(G)) \to 1. \qquad (1.99)$$

Note that for a free simplicial resolution $F \to G$ of a group G, the groups $\pi_1(F/\gamma_n(F))$ are Baer invariants. Therefore we get the natural example, when the $\varprojlim^1$-term in (1.99) vanishes.

Example 1.122
Let G be a finitely-presented group with $H_2(G)$ finite. Then

$$\varprojlim{}_n^1 M^{(n)}(G) = 0,$$

by Theorem 1.74, hence, for any free simplicial resolution $F \to G$ there is the following natural isomorphism:

$$\pi_0(Z_\infty(F)) \simeq Z_\infty(G).$$

Let G be a group and $F \to G$ a free simplicial resolution of G. Then for a given ordinal number α, the map η_α induces the map

$$\eta_\alpha : F \to L(F)/\gamma_\alpha(L(F))$$

of simplicial groups. This map induces the natural homomorphism

$$\pi_0(\eta_\alpha) : G \to \pi_0(L(F))/\gamma_\alpha(L(F)).$$

First consider the induced map of homology groups

$$H_2(\pi_0(\eta_\alpha)) : H_2(G) \to H_2(\pi_0(L(F)/\gamma_\alpha(L(F)))).$$

Lemma 1.123 *For any ordinal number α, there exists the natural epimorphism*

$$\gamma_\alpha(L(F_0))/\gamma_{\alpha+1}(L(F_0)) \rightarrow coker(H_2(\pi_0(\eta_\alpha))).$$

Proof. Consider the first quadrant spectral sequence associated with the simplicial group $L(F)/\gamma_\alpha(L(F))$ (see A.18). It has the initial terms $E^1_{p,q}$:

$$E^1_{p,0} = \mathbb{Z}, \ E^1_{p,1} = H_1(F_p), E^1_{p,2} = H_2(L(F_p)/\gamma_\alpha(L(F_p)))$$

At the next step, we get the following terms $E^2_{p,q}$:

$$E^2_{p,0} = 0, \ p > 0, \ E^2_{p,1} = H_{p+1}(G).$$

Since $E^1_{1,1} = E^\infty_{1,1}$, we get the following exact sequence

$$0 \rightarrow H_2(G) \rightarrow H_2(\overline{W}(L(F)/\gamma_\alpha(F))) \rightarrow E^\infty_{0,2} \rightarrow 0. \tag{1.100}$$

The natural epimorphism

$$y_\alpha : H_2(\overline{W}(L(F)/\gamma_\alpha(F))) \rightarrow H_2(\pi_0(L(F)/\gamma_\alpha(F)))$$

can be viewed as a part of the following commutative diagram:

$$
\begin{array}{ccccc}
H_2(G) & \rightarrowtail & H_2(\overline{W}(\frac{L(F)}{\gamma_\alpha(L(F))})) & \twoheadrightarrow & E^\infty_{0,2} \\
\downarrow & & \downarrow & & \\
H_2(\pi_0(\frac{L(F)}{\gamma_\alpha(L(F))})) & = & H_2(\pi_0(\frac{L(F)}{\gamma_\alpha(L(F))})) & &
\end{array}
\tag{1.101}
$$

The snake lemma applied to the diagram (1.101) implies the chain of natural epimorphisms

$$E^2_{0,2} \rightarrow E^\infty_{0,2} \rightarrow coker(H_2(\pi_0(\eta_\alpha))).$$

Now observe that

$$E^2_{0,2} = H_2(L(F_0)/\gamma_\alpha(L(F_0))) = \gamma_\alpha(L(F_0))/\gamma_{\alpha+1}(L(F_0)). \ \square$$

Corollary 1.124 *Let G be a finitely presented group with $H_2(G) = 0$ and $f : F_0 \rightarrow G$ an epimorphism with F_0 a free group of finite rank. Then the induced map*

$$H_2(f_*) : H_2(Z_\infty(F_0)) \rightarrow H_2(Z_\infty(G))$$

is an epimorphism of the second homologies of nilpotent completions.

Proof. Consider free simplicial resolution $F \to G$ with the zero-th term F_0 equal to the free group from the statement of Corollary. Lemma 1.123 implies that there is the following epimorphism:

$$H_2(Z_\infty(F_0)) \simeq \gamma_\omega(L(F_0))/\gamma_{\omega+1}(L(F_0)) \to$$
$$\operatorname{coker}\{H_2(G) \to H_2(\pi_0(Z_\infty(F)))\} = H_2(\pi_0(Z_\infty(F))).$$

It follows from the previous example that $\pi_0(Z_\infty(F))$ is naturally isomorphic to the nilpotent completion $Z_\infty(G)$ and the assertion follows. $\square$

Proposition 1.125 ([Bou77], Proposition 4.3). *Let G be the fundamental group of the Klein bottle:*

$$G = \langle a, b \mid aba^{-1}b = 1 \rangle.$$

Then $H_2(Z_\infty(G))$ is uncountable.

Proof. The group G is a metabelian group and, for any $k \geq 2$, there exists the following commutative diagram with exact horizontal maps:

$$
\begin{array}{ccccccccc}
1 & \longrightarrow & \mathbb{Z}_{2^{k-1}} & \longrightarrow & G/\gamma_k(G) & \longrightarrow & \mathbb{Z} & \longrightarrow & 1 \\
& & \uparrow & & \uparrow & & \| & & \\
1 & \longrightarrow & \mathbb{Z}_{2^k} & \longrightarrow & G/\gamma_{k+1}(G) & \longrightarrow & \mathbb{Z} & \longrightarrow & 1,
\end{array}
\tag{1.102}
$$

where the subgroup $\mathbb{Z}_{2^{i-1}}$ is generated by element $b\gamma_i(G)$ $i = k, k+1$. Consider the inverse limit over k. Using the fact that $\varprojlim_k^1 \mathbb{Z}_{2^k} = 0$, we get the following exact sequence:

$$1 \to \mathbb{Z}_{(2)} \to Z_\infty(G) \to \mathbb{Z} \to 1, \tag{1.103}$$

where $\mathbb{Z}_{(2)}$ is a group of 2-adic integers. The Hochschild-Serre spectral sequence for extension (1.103) gives the following short exact sequence for all $k \geq 1$:

$$0 \to H_0(\mathbb{Z}, H_k(\mathbb{Z}_{(2)})) \to H_k(Z_\infty(G)) \to H_1(\mathbb{Z}, H_{k-1}(\mathbb{Z}_{(2)})) \to 0. \tag{1.104}$$

Since $\mathbb{Z}_{(2)}$ is torsion-free uncountable, tensoring it with $\mathbb{Q}$ we get the uncountable $\mathbb{Q}$-vector space with uncountable second homology. Hence $H_0(\mathbb{Z}, H_2(\mathbb{Z}_{(2)}))$ is uncountable and the needed statement follows.

Furthermore, one can easily show that $H_2(Z_\infty(G)) = H_0(\mathbb{Z}, H_2(\mathbb{Z}_{(2)}))$. The action of the infinite cyclic group Z on $\mathbb{Z}_{(2)}$ is given by:

$$z \circ (x_1, x_2, \ldots) \mapsto (x_1^{-1}, x_2^{-1}, \ldots), \quad x_i \in \mathbb{Z}_{2^i},$$

where z denotes the generator of Z. For the group Z the augmentation extension

$$0 \to \mathfrak{z} \to \mathbb{Z}[Z] \to \mathbb{Z} \to 0$$

is a free resolution over $\mathbb{Z}$. By definition, we have

$$H_1(Z, \mathbb{Z}_{(2)}) = \ker\{\mathfrak{z} \otimes_{\mathbb{Z}[Z]} \mathbb{Z}_{(2)} \xrightarrow{q} \mathbb{Z}_{(2)}\},$$

$$q : \alpha \otimes w \mapsto \alpha \circ w, \ \alpha \in \mathfrak{z}, \ w \in \mathbb{Z}_{(2)}.$$

The augmentation ideal $\mathfrak{z}$ is a principal ideal in $\mathbb{Z}[Z]$, generated by the element $(1 - z)$; thus every element from $\mathfrak{z} \otimes_{\mathbb{Z}[Z]} \mathbb{Z}_{(2)}$ can be presented uniquely as $(1 - z) \otimes w$, $w \in \mathbb{Z}_{(2)}$. Then $q : (1 - z) \otimes w \mapsto w^2$ in $\mathbb{Z}_{(2)}$. Since $\mathbb{Z}_{(2)}$ is torsion free, therefore the kernel of q is trivial and $H_2(Z_\infty(G)) = H_0(Z, H_2(\mathbb{Z}_{(2)}))$. □

Theorem 1.126 (see [Bou77], Proposition 4.4). *Let F be a finitely generated non-cyclic free group. Then $H_2(Z_\infty(F))$ is uncountable.*

Proof. Construct epimorphism $F \to G$, where G is the fundamental group of Klein bottle and apply Proposition 1.125 and Corollary 1.124. □

Theorem 1.126 shows that the transfinite lower central length of the *HZ*-localization of a free group of finite rank is greater than the first limit ordinal, i.e.

$$\gamma_\omega(L(F)) \neq \gamma_{\omega+1}(L(F)).$$

The following problem thus arises naturally:

Problem 1.127 *What is the transfinite lower central length of $L(F)$ for a free group F?*

Transfinitely Para-free Groups

As a generalization of the notion of para-free groups, we define, for a given ordinal number τ, a group G to be τ-*para-free* if there exists a homomorphism $F \to G$, where F is a free group, which induces an isomorphism

$$L(F)/\gamma_\tau L(F) \simeq L(G)/\gamma_\tau L(G);$$

here $L : G \mapsto L(G)$ is the HZ-localization functor [see 1.97]. Note that, in view of the limit property of HZ-local groups, the following holds:

A group G is ω-para-free if and only if it is weakly para-free.

There exist groups which are ω-para-free but not $(\omega + 1)$-para-free.

Example 1.128
If F is the free group of finite rank, then its nilpotent completion $Z_\infty(F)$ is ω-para-free but not $(\omega + 1)$-para-free. The fact that $Z_\infty(F)$ is para-free has already been mentioned (Theorem 1.110). Suppose $Z_\infty(F)$ is $(\omega+1)$-para-free, i.e., there exists a homomorphism $f : \mathcal{F} \to Z_\infty(F)$, where $\mathcal{F}$ is a free group,

which induces an isomorphism $L(\mathcal{F})/\gamma_{\omega+1}L(\mathcal{F}) \simeq L(Z_\infty(F))/\gamma_{\omega+1}L(Z_\infty(F))$. Observe that $Z_\infty(F)$, being inverse limit of nilpotent groups, is residually nilpotent and HZ-local and so $L(Z_\infty(F)) = Z_\infty(F)$. Thus

$$L(Z_\infty(F))/\gamma_{\omega+1}L(Z_\infty(F))) = Z_\infty(F)/\gamma_{\omega+1}(Z_\infty(F)) = Z_\infty(F).$$

Therefore $L(\mathcal{F})/\gamma_{\omega+1}L(\mathcal{F})$ is residually nilpotent and so $\gamma_\omega L(\mathcal{F}) = \gamma_{\omega+1}L(\mathcal{F})$. Since $L(\mathcal{F})$ is transfinitely nilpotent, it follows that

$$\gamma_\omega L(\mathcal{F}) = 1 \quad \text{and} \quad L(\mathcal{F}) \simeq Z_\infty(F).$$

This, however, is not possible since $H_2(L(\mathcal{F})) = 0$ and $H_2(Z_\infty(F))$ is uncountable by Theorem 1.126. Hence $Z_\infty(F)$ is not $(\omega + 1)$-para-free.

It follows directly from the construction of HZ-localization that:

If G is a group such that $H_1(G)$ is free abelian and $H_2(G) = 0$, then G is τ-para-free for every ordinal number τ.

We call a group G to be a *globally para-free group* if it is τ-para-free for every ordinal number τ. The following problem, which may be compared with Para-free Conjecture, then arises naturally:

Problem 1.129 *Is it true that if G is a globally para-free group, then $H_2(G) = 0$?*

1.8 Crossed Modules and Cat^1-Groups

The category $\mathcal{H}o_n$ of homotopy n-types consists of connected CW-complexes X whose homotopy groups $\pi_i(X)$ are trivial in dimension $\geq n + 1$. There is a natural map $p_n : \mathsf{Top} \to \mathcal{H}o_n$, where Top is the category of topological spaces and $p_n(X)$ is the nth stage of the Postnikov tower for $X \in \mathsf{Top}$. It is well-known that the category $\mathcal{H}o_1$ is equivalent to the category Gr of groups. The corresponding equivalence map $p_1 : \mathsf{Top} \to \mathcal{H}o_1$ is the usual classifying space functor

$$p_1 : X \to K(\pi_1(X), 1), \quad X \in \mathsf{Top}.$$

In the case of the category of homotopy 2-types, there are a lot of algebraic models. The following categories are equivalent and present the algebraic models of the category $\mathcal{H}o_2$ [Lod82]:

- Category $\mathcal{CM}$ of crossed modules;
- Category $\mathcal{C}at^1$ of cat^1-groups ;
- Category $\mathcal{SG}r(1)$ of simplicial groups with Moore complex of length 1;
- Category $\mathcal{C}at(Gr)$ of internal categories in the category of groups;

Here we consider the equivalence between categories $\mathcal{CM}, \mathcal{C}at^1$ and $\mathcal{SG}r(1)$.

A *crossed module* is a triple (M, ∂, G), where G and M are groups, G acts on M (we denote the action as $g \circ m$ for $g \in G$, $m \in M$), and $\partial : M \to G$ is a group homomorphism such that the following conditions are satisfied:

$$CM1: \quad \partial(g \circ m) = g\partial(m)g^{-1} \ (g \in G, \ m \in M);$$
$$CM2: \quad mnm^{-1} = \partial(m) \circ n \ (m, \ n \in M).$$

Sometimes a crossed module (M, ∂, G) is also called a *G-crossed module*.

A morphism $(f, h) : (M, \partial, G) \to (M', \partial', G')$ between crossed modules consists of a pair of group homomorphisms $f : M \to M'$, $h : G \to G'$, such that the following diagram is commutative:

$$
\begin{array}{ccc}
M & \xrightarrow{\ f\ } & M' \\
\partial \downarrow & & \downarrow \partial' \\
G & \xrightarrow{\ h\ } & G',
\end{array}
$$

and $f(g \circ m) = h(g) \circ f(m)$, $g \in G, m \in M$. Thus, we have a category, which we denote by $\mathcal{CM}$ (the category of crossed modules). Analogically we can define morphisms in the category of G-crossed modules (for a fixed acting group G), which we denote $\mathcal{CM}_G$.

Example 1.130
(i) Let G be a group, and H its normal subgroup. Then the natural inclusion $i : H \to G$ defines the structure of a crossed module (H, i, G), where the action of G on H is by conjugation.
(ii) Let G be a group and $i : G \to \text{Aut}(G)$ the group homomorphism, which maps $g \in G$ to the inner automorphism $x \mapsto gxg^{-1}$, $x \in G$, of G. Then the triple $(G, i, \text{Aut}(G))$ is a crossed module.
(iii) **(Quillen).** For any fibration

$$F \xrightarrow{i} E \xrightarrow{p} X, \tag{1.105}$$

the induced map on fundamental groups

$$i_* : \pi_1(F) \to \pi_1(E)$$

defines the structure of the crossed module $(\pi_1(F), i_*, \pi_1(E))$. This crossed module is called the *fundamental crossed module* of the fibration (1.105).
(iv) **(Baues-Conduche [Bau97], see Lemma 3.2).** Let G be a simplicial group with face maps $\partial_i : G_n \to G_{n-1}$, $0 \leq i \leq n$, and $(N_n(G), \bar{\partial}_n)$ its Moore complex. Then the homomorphism $\bar{\partial}_n(G)$ induces an exact sequence of groups

$$0 \to \pi_n(G) \to \text{coker}(\bar{\partial}_{n+1}) \xrightarrow{\bar{\partial}_n^*} \ker(\bar{\partial}_{n-1}) \to \pi_{n-1}(G) \to 0.$$

The triple $(\operatorname{coker}(\bar{\partial}_{n+1}),\ \bar{\partial}_n^*,\ \ker(\bar{\partial}_{n-1}))$ has a natural structure of a crossed module.

For more examples, see the recent paper of Martins [Mar].

Free and projective objects can be naturally introduced in the categories $\mathcal{CM}$ and $\mathcal{CM}_G$. A crossed module $(M,\ \partial,\ G)$ is called a *free crossed module* with a basis $\{m_\alpha\} \subseteq M$, if the following condition holds:

> For any crossed module $\partial' : M' \to G'$, any subset $\{m'_\alpha\} \subseteq M$ and a homomorphism $f : G \to G'$, such that $f(\partial(m_\alpha)) = \partial'(m'_\alpha)$, there exists a unique homomorphism $h : M \to M'$, such that $h(m_\alpha) = m'_\alpha$ and the pair $(h,\ f)$ is a morphism of crossed modules.

A *free G-crossed module*, i.e., a free object in the category $\mathcal{CM}_G$, is defined in a similar manner.

A G-crossed module $(M,\ \partial,\ G)$ is called *G-projective*, if for any epimorphism

$$f := (f,\ \mathrm{id}) : (M_1,\ \partial_1,\ G) \to (M_2,\ \partial_2,\ G)$$

and any G-homomorphism

$$h := (h,\ \mathrm{id}) : (M,\ \partial,\ G) \to (M_2,\ \partial_2,\ G),$$

there exists a morphism

$$q := (q,\ \mathrm{id}) : (M,\ \partial,\ G) \to (M_1,\ \partial_1,\ G),$$

such that $fq = h$ (product of morphisms means composition).

The characterization of free and projective G-crossed modules was given by Ratcliffe [Rat80] (Ratcliffe's results and many other results from the crossed module theory can be found in [Dye93]). We will use the following equivalence (see, for example [Dye93], Theorem 2.3). For a G-crossed module $(M,\ \partial,\ G)$ the following conditions are equivalent:

(i) $(M,\ \partial,\ G)$ is a projective crossed G-module;

(ii) the abelianization $M_{ab} = H_1(M)$ is a projective $\mathbb{Z}[\operatorname{coker}(\partial)]$-module and the homomorphism $H_2(M) \to H_2(\operatorname{im}(\partial))$ is trivial.

A crossed module $(M,\ \partial,\ G)$ is called *aspherical*, if ∂ is injective. If furthermore, ∂ is an isomorphism then the crossed module is called *contractible*.

Let $(X,\ Y)$ be a pair of topological spaces. Then the boundary map

$$\partial : \pi_2(X,\ Y) \to \pi_1(Y) \qquad\qquad (1.106)$$

defines the structure of a crossed module. J. H. C. Whitehead [Whi41] proved that any free crossed module can be realized topologically as (1.106) for the case when X is obtained from Y by adding only 2-cells.

In the case of topological pairs $(X,\ X^{(1)})$, where $X^{(1)}$ is the 1-skeleton of a space X, we have $X^{(1)}$ homotopically equivalent to the wedge of circles and $\pi_1(X^{(1)})$ a free group. In this case there is the following exact sequence:

$$0 \to \pi_2(X) \to \pi_2(X, X^{(1)}) \to \pi_1(X^{(1)}) \to \pi_1(X) \to 1. \qquad (1.107)$$

The motivation for the definition of aspherical and contractible crossed modules directly follows from this sequence. The natural functor

$$\mathcal{H}o_2 \to \mathcal{CM}$$

which provides an algebraic model for $\mathcal{H}o_2$ is given by

$$X \mapsto (\pi_2(X, X^{(1)}), \partial, \pi_1(X^{(1)})). \qquad (1.108)$$

For a given space X, the crossed module defined by (1.108) is called the *fundamental crossed module* of X. The functor

$$B : \mathcal{CM} \to \mathsf{Top},$$

which can be viewed as a *classifying space* functor, was defined by Loday [Lod82]. For a given crossed module (M, ∂, G), the space $B(M, \partial, G)$ has the following homotopy groups

$$\pi_1 B(M, \partial, G) = \mathrm{coker}(\partial),$$
$$\pi_2 B(M, \partial, G) = \ker(\partial),$$
$$\pi_i B(M, \partial, G) = 0, \ \ i \geq 3,$$

and has the property that the second stage of Postnikov tower $p_2 B(M, \partial, G)$ is weakly equivalent to (M, ∂, G) in $\mathcal{CM}$. It is shown in [Lod82] that such a functor can be constructed by taking the geometric realization of the diagonal of the bisimplicial set $\mathcal{NNL}(M, \partial, G)$.

There is the equivalent description of the category $\mathcal{CM}$ in terms of the so-called *cat^1-groups*. Recall that a *cat^1-group* consists of a group G with two endomorphisms $s, t : G \to G$, satisfying

$$ss = s, \ st = t, \ ts = s, \ [\ker(s), \ker(t)] = 1.$$

Example 1.131
In analogy with the concept of the fundamental crossed complex for a given fibration, the concept of the fundamental cat^1-group can be naturally defined. For a given fibration (1.105), consider its pullback

$$E \times_X E = \{(e_1, e_2) \in E \times E \mid p(e_1) = p(e_2)\}.$$

It is easy to show that there is a natural isomorphism of groups

$$\pi_1(E \times_X E) \simeq \pi_1(E) \rtimes \pi_1(F).$$

Let s and t be the a compositions of the projection of $E \times_X E$ on the first and second coordinates respectively with the diagonal embedding. Then

$$(E \times_X E, s, t) = (\pi_1(E), i_*, \pi_1(F))$$

is a cat^1-group, called the fundamental cat^1-group for the fibration (1.105).

Theorem 1.132 (Loday [Lod82]). *The categories $\mathcal{CM}$, $\mathcal{C}at^1$ and $\mathcal{SG}r(1)$ are equivalent.*

We will define the equivalence maps between these categories omitting the details of the proof. For a detailed proof the reader can refer, for example, to [Lod82].

Equivalence map $F : \mathcal{CM} \to \mathcal{C}at^1$ can be constructed by setting

$$F : (M, \partial, P) \mapsto (M \rtimes P, s : (m, p) \mapsto p, t : (m, p) \mapsto \partial(m)p). \quad (1.109)$$

Its inverse $F^{-1} : \mathcal{C}at^1 \to \mathcal{CM}$ is constructed by setting

$$F^{-1} : (E, s, t) \mapsto (\ker(s), t|_{\ker(s)}, \mathrm{im}(s)),$$

where the action of $\mathrm{im}(s)$ on $\ker(s)$ is by conjugation.

For a space X, the cat^1-group defined as

$$X \mapsto (\pi_2(X, X^{(1)}) \rtimes \pi_1(X^{(1)}),$$
$$s : (m, p) \mapsto p, \ t : (m, p) \mapsto \partial(m)p, \ m \in \pi_2(X, X^{(1)}), \ p \in \pi_1(X^{(1)}))$$
$$(1.110)$$

is called the *fundamental cat^1-group* of X. We will denote it by $\mathcal{L}(X)$.

The equivalence map $P : \mathcal{SG}r(1) \to \mathcal{CM}$ can be constructed as follows:

$$P : G \mapsto (NG_1, d_1|_{NG_1}, N_0G), \ G \in \mathcal{SG}r(1).$$

For the construction of its converse $P^{-1} : \mathcal{CM} \to \mathcal{SG}r(1)$, we define the notion of the nerve of a given crossed module. Given a crossed module $\mathcal{M} = (M, \partial, G)$, its *nerve* $E(\mathcal{M})$ is the simplicial group

$$E_n(\mathcal{M}) = M \rtimes (\cdots \rtimes (M \rtimes G)\dots) \ \ (n \text{ semi-direct products}),$$

$$d_0(m_1, \dots, m_n, g) = (m_2, \dots, m_n, g),$$

$$d_i(m_1, \dots, m_n, g) = (m_1, \dots, m_i m_{i+1}, \dots, m_n, g), \ 0 < i < n,$$

$$d_n(m_1, \dots, m_n, g) = (m_1, \dots, m_{n-1}, \partial(m_n)g),$$

$$s_i(m_1, \dots, m_n, g) = (m_1, \dots, m_i, 1, m_{i+1}, m_n, g), \ 0 \le i \le n,$$

for $m_i \in M$, $g \in G$. It can be directly checked that the Moore complex of $E(\mathcal{M})$ is exactly the crossed sequence $\ker(\partial) \to M \to G$, thus the nerve can be viewed as a functor

$$E : \mathcal{CM} \to \mathcal{SG}r(2),$$

where $\mathcal{S}Gr(2)$ is the category of simplicial groups with Moore complex of length ≤ 2. The required map can be defined as

$$P^{-1} : \mathcal{M} \mapsto \mathrm{sk}^1(E(\mathcal{M})).$$

Peiffer Central Series for Pre-crossed Modules

If M and G are groups such that G acts on M:

$$g : m \mapsto g \circ m, \ g \in G, m \in M, \tag{1.111}$$

$\partial : M \to G$ is a group homomorphism which satisfies

$$\partial(g \circ m) = g\partial(m)g^{-1}, \ g \in G, m \in M,$$

then the triple (M, ∂, G) is called a *pre-crossed module*. Clearly any pre-crossed module (M, ∂, G) which satisfies the additional condition:

$$mnm^{-1} = \partial(m) \circ n, \ m, n \in M,$$

has the structure of a crossed module.

Example 1.133
Let G and H be groups such that G acts on H. Then the natural epimorphism

$$f : H \rtimes G \to G$$

defines the pre-crossed module $(H \rtimes G, f, G)$, which is not a crossed module in the case when G is non-abelian.

For elements $m, n \in M$, define the *Peiffer commutator*

$$\langle m, n \rangle = mnm^{-1}\partial(m) \circ n^{-1}.$$

Then a pre-crossed module (M, ∂, G) is a crossed module if and only if every Peiffer commutator is trivial.

For two subgroups M_1, M_2 of M, let $\langle M_1, M_2 \rangle$ denote the subgroup of M, generated by all commutators $\langle m_1, m_2 \rangle$, $m_1 \in M_1, m_2 \in M_2$. The *Peiffer central series* [Bau90] $P_n := P_n(M, \partial, G)$, $n \geq 1$, of the pre-crossed module (M, ∂, G) is defined inductively by setting

$$P_1 = M, \ P_n = \prod_{i+j=n} \langle P_i, P_j \rangle, \ n \geq 2.$$

It is easy to see that

$$P_n = \langle P_{n-1}, M \rangle . \langle M, P_{n-1} \rangle, \ n \geq 2.$$

The pre-crossed modules with P_3 trivial play a central role in the theory of homotopy 3-types (see Baues [Bau91]).

For $n \geq 2$, the quotients P_n/P_{n+1} are abelian groups. The action (1.111) defines a $\mathbb{Z}[\mathrm{coker}(\partial)]$-module structure on the abelian group P_n/P_{n+1}, $n \geq 2$.

Another interesting series $\{Q_n\}_{n\geq 1}$ in the the pre-crossed module (M, ∂, G) is defined as follows:

$$Q_n = P_n \cdot \gamma_n(M).$$

The action (1.111) again defines a $\mathbb{Z}[\mathrm{coker}(\partial)]$-module structure on the abelian group Q_n/Q_{n+1}.

Remark. In analogy with the concept of crossed module, the *crossed module of Lie algebras* can be defined. This is a triple (M, ∂, L), where M, L are Lie algebras, $\partial : M \to L$ is a homomorphism of Lie algebras, L acts on M and the following conditions are satisfied:

$$(i)\ \partial(g \circ m) = [g, \partial(m)],\ g \in L,\ m \in M,$$
$$(ii)\ \partial(b) \circ c = [b, c],\ b, c \in M.$$

The concept of cat^1-Lie algebra can be naturally introduced. The same type of constructions as in the case of groups show that the categories of crossed modules of Lie algebras, cat^1-Lie algebras and simplicial Lie algebras with Moore complex of length one are equivalent [Ell92].

First k-invariant of a Crossed Module

For a given space X, the pair $(\pi_1(X), \pi_2(X))$ is *not* a complete set of information to determine the corresponding 2-type. That is, it is possible to construct two spaces X and Y, such that there is an isomorphism of groups $\pi_1(X) \simeq \pi_1(Y)$ and of $\mathbb{Z}[\pi_1]$-modules $\pi_2(X) \simeq \pi_2(Y)$, but the corresponding 2-types $p_2(X)$ and $p_2(Y)$ are not the same. To complete the set of algebraic invariants which determine the 2-type, define the cohomological class $k(X) \in H^3(\pi_1(X), \pi_2(X))$, called *the first k-invariant of X* (or *the first Postnikov invariant of X*).

Let (M, ∂, F) be a crossed module with F a free group. Choose a set-theoretical section $s : \mathrm{coker}(\partial) \to F$ of the map $h : F \to \mathrm{coker}(\partial)$, i.e., $h \circ s = \mathrm{id}_{\mathrm{coker}(\partial)}$. Consider the corresponding 2-cocycle

$$W(q_1, q_2) = s(q_1 q_2)^{-1} s(q_1) s(q_2) \in \partial(M),\ q_1, q_2 \in \mathrm{coker}(\partial).$$

Since F is free, we can choice the homomorphic section $t : \partial(M) \to M$, such that $\partial \circ t = \mathrm{id}_{\partial(M)}$. Let $v(q_1, q_2) = t(W(q_1, q_2)) \in M$. Now define the 3-cocycle

$$w(q_1, q_2, q_3) = v(q_2, q_3)^{-1} v(q_1, q_2 q_3)^{-1} v(q_1 q_2, q_3) v(q_1, q_2)^{u(q_3)} \in \ker(\partial),$$

with $q_1, q_2, q_3 \in \mathrm{coker}(\partial)$. It can be shown that the image of $w(q_1, q_2, q_3)$ in $H^3(\mathrm{coker}(\partial), \ker(\partial))$ is independent of the choice of sections s and t. The corresponding element $k(M, \partial, F) \in H^3(\mathrm{coker}(\partial), \ker(\partial))$ is called *the k-invariant* of (M, ∂, F).

Now, for a given space X, the first k-invariant can be defined as a k-invariant of the fundamental crossed module of X:

$$k(X) = k(\pi_2(X, X^{(1)}), \partial, \pi_1(X^{(1)})).$$

Now the set $(\pi_1(X), \pi_2(X), k(X))$ completely determines the 2-type of X. That is, two spaces X and Y have the same 2-types $p_2(X) \simeq p_2(Y)$ if and only if there is an isomorphism of groups $\pi_1(X) \simeq \pi_1(Y)$, an isomorphism of $\mathbb{Z}[\pi_1]$-modules $\pi_2(X) \simeq \pi_2(Y)$ and the corresponding k-invariants coincide $k(X) = k(Y)$.

In the case of the trivial first k-invariant, the corresponding 2-type of a space is equivalent to the product of Eilenberg-MacLain spaces: $p_2(X) \simeq K(\pi_1(X), 1) \times K(\pi_2(X), 2)$.

Example 1.134

(Arlettaz, Baues-Conduche, [Bau97]). Let X be a space and $\Omega^2(X)$ the path component of the double loop space. Then $k(\Omega^2(X)) = 0$ and hence there is a homotopy equivalence:

$$p_2(\Omega^2(X)) \simeq K(\pi_3(X), 1) \times K(\pi_4(X), 2).$$

It is known that any abstract 2-type, i.e. a set (π_1, π_2, k), where π_1 is a group, π_2 a $\mathbb{Z}[\pi_1]$-module, $k \in H^3(\pi_1, \pi_2)$, can be realized by 3-dimensional complex [Mac50]. That is, there exists a 3-dimensional connected complex X with $\pi_1(X) = \pi_1$, $\pi_2(X) = \pi_2$, as modules over $\pi_1(X)$ and $k(X) = k \in H^3(\pi_1(X), \pi_2(X))$. However, not every 2-type can be realized by a connected 2-dimensional complex. The set of necessary and sufficient conditions for the triple (π_1, π_2, k) to be realizable by a 2-dimensional complex is given in [Dye75].

Homology and Lower Central Series of Crossed Modules

There are different ways to define (co)homologies in the above categories. G. Ellis defined homologies of the given crossed module (M, ∂, G) as $H_* B(M, \partial, G)$ [Ell92]. It directly follows from the definition of the classifying space of a crossed module that, for any crossed module (M, ∂, G), there exists a fibration

$$K(\ker(\partial), 2) \to B(M, \partial, G) \to K(\operatorname{coker}(\partial), 1)$$

and hence, the spectral sequence

$$H_p(\operatorname{coker}(\partial), H_q(K(\ker(\partial), 2))) \Rightarrow H_{p+q}B(M, \partial, G). \qquad (1.112)$$

For the computation of the homology groups $H_q(K \ker(\partial), 2)$ one can use the functorial method given by the Dold-Puppe spectral sequence (A.32) (see also [Bre99]).

Clearly a similar spectral sequence exists with coefficients in a given $\mathbb{Z}[\operatorname{coker}(\partial)]$-module N.

Proposition 1.135 *If* (M, ∂, F) *is a projective crossed module,* F *a free group,* N *a* $\mathbb{Z}[\operatorname{coker}(\partial)]$-*module, then the map*

$$d_3 = -\cap k : H_n(\operatorname{coker}(\partial), N) \to H_{n-3}(\operatorname{coker}(\partial), \ker(\partial) \otimes N)$$

is an isomorphism for all $n \geq 4$.

Proof. Consider the following central extension of the free group $\operatorname{im}(\partial)$:

$$1 \to \ker(\partial) \to M \to \operatorname{im}(\partial) \to 1. \qquad (1.113)$$

Since $H_2(\operatorname{im}(\partial)) = 0$, we have $\ker(\partial) \cap \gamma_2(M) = 1$; therefore we have the abelianization of (1.113):

$$0 \to \ker(\partial) \to M_{ab} \to \operatorname{im}(\partial)_{ab} \to 0, \qquad (1.114)$$

which can be viewed as a short exact sequence of $\mathbb{Z}[\operatorname{coker}(\partial)]$-modules. The module $\operatorname{im}(\partial)_{ab}$ is exactly the relation module of the group $\operatorname{coker}(\partial)$, given by presentation $F/im(\partial)$. Therefore, the Magnus embedding gives the following exact sequence of $\mathbb{Z}[\operatorname{coker}(\partial)]$-modules:

$$0 \to \ker(\partial) \to M_{ab} \to W \to \mathbb{Z}[\operatorname{coker}(\partial)] \to \mathbb{Z} \to 0, \qquad (1.115)$$

where W is a free $\mathbb{Z}[\operatorname{coker}(\partial)]$-module with a basis in 1-1 correspondence with a basis of F. Then the map $-\cap k$ can be viewed through the functor $H_*(\operatorname{coker}(\partial), -\otimes N)$, applied to the sequence (1.115). The required isomorphism follows from the dimension shifting principle. $\square$

Corollary 1.136 *Let* (M, ∂, F) *be a projective crossed module with* F *free. Then* $H_3B(M, \partial, F) = 0$.

Proof. The assertion follows from the spectral sequence (1.112), Proposition 1.135 and the fact that $H_1(K(A, 2)) = 0$ for any abelian group A. $\square$

In the work [Car02] the authors show that $\mathcal{CM}$ is a tripleable category over sets and define homology of crossed modules as cotriple homology in the sense of Barr-Beck [Bar69]. Homologies in the sense of [Car02] are functors $H_* : \mathcal{CM} \to \mathcal{Ab}(\mathcal{CM})$ from the category $\mathcal{CM}$ to the category $\mathcal{Ab}(\mathcal{CM})$ of abelian crossed modules. That is, for any crossed module (M, ∂, G), we have

$$H_*(M, \partial, G) = (\xi H_*(M, \partial, G), h_*, kH_*(M, \partial, G)),$$

where $\xi H_*(M, \partial, G)$, $kH_*(M, \partial, G)$ are abelian groups with trivial action of $kH_*(M, \partial, G)$ on $\xi H_*(M, \partial, G)$. We shall use the following properties of these homologies:

1. [Gra00]. For any crossed module (M, ∂, G) there exists a natural long exact sequence

$$\cdots \to H_{n+1}B(M, \partial, G) \to \xi H_n(M, \partial, G) \to H_n(G) \to H_n B(M, \partial, G) \to \cdots$$
$$(1.116)$$

2. **5-term sequence for crossed modules [Car02].**

Let
$$(N, \mu, R) \to (Q, \mu, F) \to (T, \partial, G)$$
be a short exact sequence of crossed modules. Then there exists a long exact sequence of abelian crossed modules:

$$H_2(Q, \mu, F) \to H_2(T, \partial, G) \to (\frac{N}{[F, N][R, Q]}, \mu, R/[F, R]) \to$$
$$H_1(Q, \mu, F) \to H_1(T, \partial, G) \to 0 \quad (1.117)$$

Define the lower central series $\{\gamma_i(G, M)\}_{i \geq 1}$ for a crossed module $\partial : M \to G$ by induction: $\gamma_1(G, M) := M$ and $\gamma_{i+1}(G, M)$ is the subgroup of M, generated by elements

$$[g, m] := (g \circ m)m^{-1}, \quad m \in \gamma_i(G, M), g \in G. \quad (1.118)$$

Denote the intersection of the series $\{\gamma_i(G, M)\}_{i \geq 1}$ by $\gamma_\omega(G, M)$. A crossed module (M, ∂, G) is called *residually nilpotent* if $\gamma_\omega(G, M) = \{1\}$. We will use the standard notation $\gamma_i(G)$, $1 \leq i \leq \omega$, for the lower central series of a group G.

Define *Baer invariants* of a crossed module (M, ∂, G) as

$$B^{(k)}(M, \partial, F) = \ker\{\partial_* : M/\gamma_{k+1}(G, M) \to F/\gamma_{k+1}(F)\}.$$

For any crossed module (M, ∂, G) there exists the following exact sequence:

$$H_2(G) \to H_2(B(M, \partial, G)) \to M/\gamma_2(G, M) \to G/\gamma_2(G) \to H_1(\mathrm{coker}(\partial)) \to 0.$$

In particular, if $H_2(G) = 0$, for example if G is free, then $B^{(1)}(M, \partial, G) = H_2(B(M, \partial, G))$.

Theorem 1.137 (Mikhailov [Mik07b]). *Let (M, ∂, F) be a crossed module with F free, $H_1(\mathrm{coker}(\partial))$ torsion free and $H_2(B(M, \partial, F)) = 0$. Then for all $n \geq 1$,*

$$B^{(n)}(M, \partial, F) = 0.$$

Proof. The proof is by induction on k. Suppose $B^{(k)}(M, \partial, F) = 0$ for some k. Consider the following short exact sequence of crossed modules:

$$(\gamma_k(F, M), \partial_k, \gamma_k(F)) \to (M, \partial, F) \to (M/\gamma_k(F, M), \partial_k^*, F/\gamma_k(F)),$$

where $\partial_k^* : M/\gamma_k(F, M) \to F/\gamma_k(F)$ is induced by $\partial : M \to F$. It induces the following epimorphism, coming from the 5-lemma analog for crossed modules (1.117):

$$H_2(M/\gamma_k(F, M), \partial_k^*, F/\gamma_k(F)) \to (\gamma_k(F, M)/\gamma_{k+1}(F, M), \partial_k^*, \gamma_n(F)/\gamma_{n+1}(F)).$$

That is, we have the following commutative diagram:

$$
\begin{array}{ccccccc}
0 & \longrightarrow & \ker(h_2) & \xrightarrow{\ h_2\ } & \xi H_2(M/\gamma_k(F, M), \partial_k^*, F/\gamma_k(F)) & \longrightarrow & H_2(F/\gamma_k(F)) \\
 & & \downarrow{\scriptstyle q} & & \downarrow{\scriptstyle s} & & \| \\
0 & \longrightarrow & \ker(\partial_k^*) & \longrightarrow & \gamma_k(F, M)/\gamma_{k+1}(F, M) & \longrightarrow & \gamma_k(F)/\gamma_{k+1}(F),
\end{array}
$$
$$(1.119)$$

where s is an epimorphism. Hence q is also an epimorphism.

Since we have the following exact sequence

$$0 \to \ker(\partial_k^*) \to B^{(k+1)}(M, \partial, F) \to B^{(k)}(M, \partial, F),$$

and commutative diagram (1.119), for the inductive step it is enough to prove that $\ker(h_2) = 0$. In view of sequence (1.116), $\ker(h_2)$ can be presented as

$$\ker(h_2) = \mathrm{coker}\{H_3(F/\gamma_n(F)) \to H_3(B(M/\gamma_k(F, M), \partial_k, F/\gamma_k(F))\}.$$

Note that the assumption of induction implies that $\ker(\partial_k) = B^{(k)}(M, \partial, F) = 0$, hence

$$B(M/\gamma_k(F, M), \partial_k, F/\gamma_k(F)) = K(\mathrm{coker}(\partial)/\gamma_k(\mathrm{coker}(\partial)), 1)$$

and

$$\ker(h_2) = \mathrm{coker}\{H_3(F/\gamma_k(F)) \to H_3(\mathrm{coker}(\partial)/\gamma_k(\mathrm{coker}(\partial)))\}. \quad (1.120)$$

Since $H_1(\mathrm{coker}(\partial))$ is free abelian, there exists a subgroup H in F, such that the restriction of the homomorphism ∂:

$$\partial|_H : H \to \mathrm{coker}(\partial)$$

induces isomorphism of abelianizations

$$H/\gamma_2(H) \simeq \mathrm{coker}(\partial)/\gamma_2(\mathrm{coker}(\partial)).$$

Observe that $H_2(\mathrm{coker}(\partial)) = 0$, since $H_2(B(M, \partial, F))$ maps epimorphically onto $H_2(\mathrm{coker}(\partial))$. Therefore, $\partial|_H$ induces isomorphisms of quotients

$$H/\gamma_n(H) \simeq \mathrm{coker}(\partial)/\gamma_n(\mathrm{coker}(\partial))$$

for all $n \geq 1$. Hence we have the following commutative diagram

$$
\begin{array}{ccc}
H_*(H/\gamma_n(H)) & \longrightarrow & H_*(\mathrm{coker}(\partial)/\gamma_n(\partial)) \\
\downarrow & & \uparrow \\
H_*(H/H \cap \gamma_n(F)) & \longrightarrow & H_*(F/\gamma_n(F)),
\end{array}
$$

where the upper horizontal map is an isomorphism. It implies that the natural map $H_*(F/\gamma_n(F)) \to H_*(G/\gamma_n(G))$ is an epimorphism for all $n \geq 1$ and therefore, $\ker(h_2) = 0$ by (1.120). $\square$

Theorem 1.138 *If (M, ∂, F) is a crossed module with F free, $H_1(\mathrm{coker}(\partial))$ free abelian and $H_2(B(M, \partial, F)) = 0$, then $\ker(\partial) = \gamma_\omega(F, M)$.*

Proof. It is clear that

$$\frac{\ker(\partial)}{\ker(\partial) \cap \gamma_n(F, M)} \subseteq \ker\{M/\gamma_n(F, M) \to F/\gamma_n(F)\}, \quad n \geq 1. \qquad (1.121)$$

It follows from Theorem 1.137 that $B^{(n)}(M, \partial, F) = 0$, i.e. the kernel (1.121) is trivial and $\ker(\partial) \in \gamma_n(F, M)$ for all $n \geq 1$.

If $m \in \gamma_2(F, M)$, then m is presented as the product of the elements of the form $(g \circ s)s^{-1}$, $s \in M$, $g \in F$. But $\partial((g \circ s)s^{-1}) = [g, \partial(s)]$, due to CM1. Therefore $\partial(m) \in \gamma_2(F)$. By induction we conclude that $\partial(m) \in \gamma_n(F)$, $m \in \gamma_n(F, M)$ for all $n \geq 1$. The residual nilpotence of F then implies that $m \in \ker(\partial)$ for $m \in \gamma_\omega(F, M)$, and the result is thus proved. $\square$

Proposition 1.139 (Mikhailov [Mik07b]). *Let (M, ∂, F) be a crossed module with F free group and $H_2B(M, \partial, F) \otimes \mathbb{Q} = 0$, then $B^{(k)}(M, \partial, F) \otimes \mathbb{Q} = 0$ for all $k \geq 1$.*

Proof. The proof is by induction on k. Suppose $B^{(k)}(M, \partial, F) \otimes \mathbb{Q} = 0$, for some $k \geq 1$. The diagram (1.119) and arguments used in the proof of Theorem 1.137 imply that the completion of the inductive step follows from the fact that the group

$$\ker(h_2) = \mathrm{coker}\{H_3(F/\gamma_n(F)) \to H_3B(M/\gamma_k(F, M), \partial_k^*, F/\gamma_k(F))\}$$

consists of torsion elements.

The fibration

$$K(B^{(k)}(M, \partial, F), 2) \to B(M/\gamma_k(F, M), \partial_k^*, F/\gamma_k(F))$$
$$\to K(\mathrm{coker}(\partial)/\gamma_k(\mathrm{coker}(\partial)), 1)$$

defines the following spectral sequence

$$E_{p,q}^2 = H_p(\mathrm{coker}(\partial)/\gamma_k(\mathrm{coker}(\partial)), H_q K(B^{(k)}(M, \partial, F), 2)) \Rightarrow$$
$$H_{p+q}B(M/\gamma_k(F, M), \partial_k^*, F/\gamma_k(F)) \quad (1.122)$$

It is well-known that for any abelian group A,

$$H_1 K(A, 2) = 0, \quad H_2 K(A, 2) = A, \quad H_3 K(A, 2) = 0;$$

therefore the spectral sequence (1.122) implies the following exact sequence:

$$H_1(\mathrm{coker}(\partial)/\gamma_k(\mathrm{coker}(\partial)), B^{(k)}(M, \partial, F)) \to$$
$$H_3 B(M/\gamma_k(F, M), \partial_k^*, F/\gamma_k(F)) \to H_3(\mathrm{coker}(\partial)/\gamma_k(\mathrm{coker}(\partial)))$$
$$\to H_0(\mathrm{coker}(\partial)/\gamma_k(\mathrm{coker}(\partial)), B^{(k)}(M, \partial, F)) \quad (1.123)$$

We have $B^{(k)}(M, \partial, F) \otimes \mathbb{Q} = 0$ by induction hypothesis. Therefore, tensoring the sequence (1.123) by $\mathbb{Q}$, we get the natural isomorphism

$$H_3 B(M/\gamma_k(F, M), \partial_k^*, F/\gamma_k(F)) \otimes \mathbb{Q} \simeq H_3(\mathrm{coker}(\partial)/\gamma_k(\mathrm{coker}(\partial))) \otimes \mathbb{Q}.$$

Thus we have

$$\ker(h_2) \otimes \mathbb{Q} = \ker\{H_3(F/\gamma_3(F)) \to H_3(\mathrm{coker}(\partial)/\gamma_k(\mathrm{coker}(\partial)))\}.$$

The next step is the same as in the proof of Theorem 1.137: we choose a subgroup H in F such that there is an isomorphism of $\mathbb{Q}$-modules:

$$f|_H : H \otimes \mathbb{Q} \to \mathrm{coker}(\partial) \otimes \mathbb{Q}.$$

The $\mathbb{Q}$-version of Stallings Theorem then implies that $f|_H$ induces isomorphisms of $\mathbb{Q}$-localizations

$$H/\gamma_k(H) \otimes \mathbb{Q} \simeq \mathrm{coker}(\partial)/\gamma_k(\mathrm{coker}(\partial)) \otimes \mathbb{Q}.$$

Now the same argument as in the proof of Theorem 1.137 implies the required statement, namely $\ker(h_2) \otimes \mathbb{Q} = 0$. $\square$

   The definition of the lower central series for a crossed module (M, ∂, G) can be extended to transfinite ordinal terms; more precisely, for an arbitrary ordinal number α, by transfinite induction we define $\gamma_{\alpha+1}(G, M)$ to be a subgroup in $\gamma_\alpha(G, M)$, generated by elements $[g, m]$, $m \in \gamma_\alpha(G, M)$, $g \in G$. For a limit ordinal τ, we define $\gamma_\tau(G, M)$ to be the intersection $\bigcap_{\alpha<\tau} \gamma_\alpha(G, M)$. A crossed module (M, ∂, G) is called *transfinitely nilpotent*, if $\gamma_\alpha(G, M) = \{1\}$ for some ordinal number α.

The exact sequences (1.116) and (1.117) imply the following commutative diagram:

$$
\begin{array}{ccccc}
H_3B(M, \partial, G) & \longrightarrow & \xi H_2(M, \partial, G) & \longrightarrow & H_2(G) \\
\downarrow & & \downarrow & & \downarrow \\
H_3B(\frac{M}{\gamma_\omega(G, M)}, \bar\partial, G/\gamma_\omega(G)) & \longrightarrow & \xi H_2(\frac{M}{\gamma_\omega(G, M)}, \bar\partial, G/\gamma_\omega(G)) & \longrightarrow & H_2(G/\gamma_\omega(G)) \\
\downarrow & & \downarrow & & \downarrow \\
\ker(\partial_\omega) & \longrightarrow & \gamma_\omega(G, M)/\gamma_{\omega+1}(G, M) & \xrightarrow{\partial_\omega} & \gamma_\omega(G)/\gamma_{\omega+1}(G) \\
& & \downarrow & & \downarrow \\
& & 0 & & 0
\end{array}
\tag{1.124}
$$

In the case of a free group $G = F$, we get the following exact sequence:

$$H_3B(M, \partial, F) \to H_3B(M/\gamma_\omega(F, M), \bar\partial, F) \to \gamma_\omega(F, M)/\gamma_{\omega+1}(F, M) \to 0, \tag{1.125}$$

which implies the following

Proposition 1.140 *If (M, ∂, F) is a projective crossed module, F a free group, then there is the following isomorphism*

$$H_3B(M/\gamma_\omega(F, M), \bar\partial, F) \simeq \gamma_\omega(F, M)/\gamma_{\omega+1}(F, M). \tag{1.126}$$

Furthermore, if $H_2B(M, \partial, F) = 0$ and $H_1(\mathrm{coker}(\partial))$ is free abelian, then

$$H_3(\mathrm{coker}(\partial)) \simeq \gamma_\omega(F, M)/\gamma_{\omega+1}(F, M). \tag{1.127}$$

Proof. By Proposition 1.135, $H_3B(M, \partial, F) = 0$. The isomorphism (1.126) follows from the exact sequence (1.125). Now suppose $H_2B(M, \partial, F) = 0$ and $H_1(\mathrm{coker}(\partial))$ is free abelian. Theorem 1.138 then implies that the crossed module $(M/\gamma_\omega(F, M), \bar\partial, F)$ is aspherical and $\mathrm{coker}(\partial) = \mathrm{coker}(\bar\partial)$; consequently, $B(M/\gamma_\omega(F, M), \bar\partial, F)$ is the classifying space $K(\mathrm{coker}(\partial), 1)$ and the required isomorphism follows from (1.126).

Obviously, one can get (1.127) immediately from Theorem 1.138 and Hopf's sequence (the sequence obtained by abelianization of the crossed module together with an application of the functor $H_0(\mathrm{coker}(\partial), -)$. $\square$

Faithfulness and Residual Nilpotence

It follows from CM2 that $\ker(\partial)$ is a central subgroup in M, therefore $\ker(\partial)$ has the structure of a $\mathbb{Z}[\operatorname{coker}(\partial)]$-module.

Theorem 1.141 (Mikhailov, [Mik06a]) *Let F be a free group and (M, ∂, F) a non-aspherical projective F-crossed module. Then $\operatorname{coker}(\partial)$ acts on $\ker(\partial)$ faithfully.*

Proof. As noted in the proof of Proposition 1.135, $\ker(\partial) \cap \gamma_2(M) = 1$; therefore we have an exact sequence (1.115) of $\mathbb{Z}[\operatorname{coker}(\partial)]$-modules.

Suppose $g \in \operatorname{coker}(\partial)$ is such that $g \circ x = x$ for all $x \in \ker(\partial)$. Denote by E the cyclic subgroup in $\operatorname{coker}(\partial)$ generated by g. By hypothesis, the crossed module (M, ∂, F) is projective; therefore, using Ratcliffe's characterization [Rat80] of the projective modules, we conclude that M_{ab} is a projective $\mathbb{Z}[\operatorname{coker}(\partial)]$-module, i.e. there exists a $\mathbb{Z}[\operatorname{coker}(\partial)]$-module N such that $M_{ab} \oplus N$ is a free $\mathbb{Z}[\operatorname{coker}(\partial)]$-module. Since $\ker(\partial)$ is embedded into the free $\mathbb{Z}[\operatorname{coker}(\partial)]$-module $M_{ab} \oplus N$, we conclude that $g - 1$ has a nontrivial annihilator in $\mathbb{Z}[\operatorname{coker}(\partial)]$ and, therefore, g is of finite order in $\operatorname{coker}(\partial)$.

Applying the homology functor $H_*(E, -)$ to the sequence (1.115), considered as a short exact sequence of $\mathbb{Z}[\operatorname{coker}(\partial)]$-modules, and using the fact that

$$H_i(E, M_{ab}) = 0, \; H_i(E, W) = 0, \; i \geq 1,$$

we get

$$H_i(E, \ker(\partial)) \simeq H_{i+3}(E, \mathbb{Z}), \; i \geq 1. \tag{1.128}$$

Let $i = 2j$, then (1.128) implies that

$$H_{2j}(E, \ker(\partial)) \simeq E$$

On the other hand, by assumption, $\ker(\partial)$ is a trivial $\mathbb{Z}[E]$-module, and so we get

$$H_{2j}(E, \ker(\partial)) = 0, \; j \geq 1, \tag{1.129}$$

due to the fact that $\ker(\partial)$ is free as an abelian group (it is embedded in $M_{ab} \oplus N$). Therefore E is a trivial group, so $g = 1$ in $\operatorname{coker}(\partial)$. Hence the action of the group $\operatorname{coker}(\partial)$ on $\ker(\partial)$ is faithful. $\square$

Corollary 1.142 *Let X be a nonaspherical two-dimensional complex. Then $\pi_1(X)$ acts faithfully on $\pi_2(X)$.*

Remark. It may however be noted that for a two-dimensional complex X the module $\pi_2(X)$ is, in general, not faithful. For example, let X be the projective plane $\mathbb{P}^2$. Then $\pi_2(\mathbb{P}^2) = \mathbb{Z}$ is annihilated by the element $1 + x \in \mathbb{Z}[\pi_1(\mathbb{P}^2)]$, where x is the nontrivial element of $\pi_1(\mathbb{P}^2) = \mathbb{Z}_2$.

We now give an application of Corollary 1.142 to the study of residual properties of some groups.

Proposition 1.143 *Let F_1 be a free group with generating set $\{x_i, y_j \mid i \in I, j \in J\}$, $\{r_j\}_{j \in J}$ a subset of $F = F(x_i \mid i \in I)$ and let R the normal closure of this subset in F. Suppose*

$$\langle y_j \mid j \in J \rangle^{F_1} \cap \langle y_j r_j^{-1} \mid j \in J \rangle^{F_1} \neq [\langle y_j \mid j \in J \rangle^{F_1}, \langle y_j r_j^{-1} \mid j \in J \rangle^{F_1}], \quad (1.130)$$

and $f \in F$ is such that

$$[f, x] \in [\langle y_j \mid j \in J \rangle^{F_1}, \langle y_j r_j^{-1} \mid j \in J \rangle^{F_1}],$$
$$\text{for all } x \in \langle y_j \mid j \in J \rangle^{F_1} \cap \langle y_j r_j^{-1} \mid j \in J \rangle^{F_1}.$$

Then $f \in R$.

Condition (1.130) always holds in case group F/R has cohomological dimension greater than two.

Let $H = F_1/[\langle y_j \mid j \in J \rangle^{F_1}, \langle y_j r_j^{-1} \mid j \in J \rangle^{F_1}]$. If, furthermore, F/R is not residually solvable, then

$$\delta_\omega(H) \neq \delta_{\omega+1}(H).$$

Proof. Consider a free presentation of a group F/R:

$$\langle x_i, \ i \in I \mid r_j, \ j \in J \rangle. \qquad (1.131)$$

The explicit structure (1.13) shows that the condition (1.130) is equivalent to the nontriviality of $\pi_1(\mathrm{sk}^1 S(X, \mathcal{R}))$. Identifying $\pi_1(\mathrm{sk}^1 S(X, \mathcal{R}))$ with the second homotopy module of the standard two-complex, constructed for the presentation (1.131), we get the fact that the action of G on $\pi_1(\mathrm{sk}^1 S(X, \mathcal{R}))$ is faithful, by Corollary 1.142, and the first statement follows.

Now let $g \in \delta_\omega(F/R)$. Then we conclude that

$$[g, x] \in \delta_\omega(H), \ x \in \langle y_j \mid j \in J \rangle^{F_1} \cap \langle y_j r_j^{-1} \mid j \in J \rangle^{F_1}.$$

Hence, the faithfulness implies that $\delta_\omega(H) \neq 1$. On the other hand, we have the following exact sequence

$$1 \to \pi_1(\mathrm{sk}^1 S(X, \mathcal{R})) \to H \to F \times F;$$

therefore, $\delta_{\omega+1}(H) = 1$. $\square$

Example 1.144
Let F be a free group with generator set x, y, z, t. Define its normal subgroups:

$$R = \langle z, t \rangle^F, \quad S = \langle zy^2, tx[x^y, x] \rangle^F.$$

Then

$$\delta_\omega(F/[R, S]) \neq \delta_{\omega+1}(F/[R, S]).$$

We next observe that simply connected simplicial groups provide examples of residually nilpotent groups.

Proposition 1.145 *Let G be a free simplicial group with $\pi_0(G) = 1$. Then $G_n/B_n(G)$ is residually nilpotent for $n \geq 1$.*

Proof. By Theorem 5.27, we have

$$\pi_j(\gamma_s(G)) = 0, \ s \geq 2^n, \ j \leq n.$$

Therefore, the simplicial map $G \to G/\gamma_s(G)$ implies isomorphism of homotopy groups:

$$\pi_n(G) \to \pi_n(G/\gamma_s(G)), \ s \geq 2^n.$$

Hence, $Z_n(G) \cap \gamma_s(G_n) \subseteq B_n(G)$. Therefore,

$$\gamma_i(G_n/B_n(G)) = \gamma_i(G_n/Z_n(G)), \ i \geq s, \tag{1.132}$$

The group $G_n/Z_n(G)$ is a subgroup of the direct product of free groups $G_n/\ker(\partial_t) \subseteq G_{n-1}$, $0 \leq t \leq n$; therefore, $G_n/Z_n(G)$ is residually nilpotent. The equality (1.132) implies that $G_n/B_n(G)$ is residually nilpotent. $\square$

Chapter 2
Dimension Subgroups

Let $\mathbb{Z}[G]$ be the integral group ring of a group G and let $\mathfrak{g}$ be its augmentation ideal. For each natural number $n \geq 1$, $D_n(G) = G \cap (1 + \mathfrak{g}^n)$ is a normal subgroup of G called the nth *integral dimension subgroup* of G. It is easy to see that the decreasing series

$$G = D_1(G) \supseteq D_2(G) \supseteq \ldots \supseteq D_n(G) \supseteq \ldots$$

is a central series in G, i.e., $[G, D_n(G)] \subseteq D_{n+1}(G)$ for all $n \geq 1$. Therefore, $\gamma_n(G) \subseteq D_n(G)$ for all $n \geq 1$, where $\gamma_n(G)$ is the nth term in the lower central series of G. The identification of dimension subgroups, and, in particular, whether $\gamma_n(G) = D_n(G)$, has been a subject of intensive investigation for the last over fifty years. It is now known that, whereas $D_n(G) = \gamma_n(G)$ for $n = 1$, 2, 3 for every group G (see [Pas79]), there exist groups G whose series $\{D_n(G)\}_{n \geq 1}$ of dimension subgroups differs from the lower central series $\{\gamma_n(G)\}_{n \geq 1}$ ([Rip72], [Tah77b], [Tah78b], [Gup90]). The various developments in this area have been reported in [Pas79] and [Gup87c]. In the present exposition, we will primarily concentrate on the results that have appeared since the publication of [Gup87c]. We particularly focus attention on the fourth and the fifth dimension subgroups. We recall the description of the fifth dimension subgroup due to Tahara (Theorem 2.29) and give a proof of one of his theorems which states that, for every group G, $D_5(G)^6 \subseteq \gamma_5(G)$ (see Theorem 2.27). The proof here is, hopefully, shorter than the original one.

2.1 Groups Without Dimension Property

Given a group G, we call the quotient $D_n(G)/\gamma_n(G)$, $n \geq 1$, the nth *dimension quotient* of G. In case all dimension quotients are trivial, we say that the group G has the *dimension property*. We begin with examples of groups which do *not* have the dimension property.

Example 2.1 (Rips [Rip72]).

The first example of a group without dimension property was given by E. Rips. Following the notation from [Rip72], consider the group G with generators

R. Mikhailov, I.B.S. Passi, *Lower Central and Dimension Series of Groups*,
Lecture Notes in Mathematics 1952,
© Springer-Verlag Berlin Heidelberg 2009

$$a_0, \ a_1, \ a_2, \ a_3, \ b_1, \ b_2, \ b_3, \ c$$

and defining relations

$b_1^{64} = b_2^{16} = b_3^4 = c^{256} = 1,$

$[b_2, \, b_1] = [b_3, \, b_1] = [b_3, \, b_2] = [c, \, b_1] = [c, \, b_2] = [c, \, b_3] = 1,$

$[a_0^{64} = b_1^{32}, \ a_1^{64} = b_2^{-4}b_3^{-2}, \ a_2^{16} = b_1^4 b_3^{-1}, \ a_3^4 = b_1^2 b_2,$

$[a_1, \, a_0] = b_1 c^2, \ [a_2, \, a_0] = b_2 c^8, \ [a_3, \, a_0] = b_3 c^{32},$

$[a_2, \, a_1] = c, \ [a_3, \, a_1] = c^2, \ [a_3, \, a_2] = c^4,$

$[b_1, \, a_1] = c^4, \ [b_2, \, a_2] = c^{16}, \ [b_3, \, a_3] = c^{64},$

$[b_i, \, a_j] = 1 \text{ if } i \neq j, \ [c, \, a_i] = 1 \text{ for } i = 0, \ 1, \ 2, \ 3.$

Then $\gamma_4(G) = 1$ while the element

$$[a_1, \, a_2]^{128}[a_1, \, a_3]^{64}[a_2, \, a_3]^{32} = c^{128}$$

is a non-identity element in $D_4(G)$.

Example 2.2 (Tahara [Tah78a]).

The above example was generalized by Ken-Ichi Tahara as follows:

Let $G_{k,l}$ $(k \geq 2, \ l \geq 0)$ be a group with generators x_1, x_2, x_3, x_4 and defining relations

$x_1^{2^{k+l+4}} = [x_2, \, x_3]^{-2^{k+l+4}}[x_2, \, x_1]^{-2^{k+l+3}},$

$x_2^{2^{k+4}} = [x_1, \, x_3]^{2^k}[x_1, \, x_4]^{-2^{k-1}}[x_2, \, x_3]^{3 \cdot 2^{k+3}},$

$x_3^{2^{k+2}} = [x_1, \, x_2]^{2^k}[x_1, \, x_4]^{2^{k-2}}[x_3, \, x_2]^{5 \cdot 2^{k+1}},$

$x_4^{2^k} = [x_1, \, x_2]^{2^{k-1}}[x_2, \, x_3]^{2^k}[x_1, \, x_3]^{2^{k-2}},$

$[x_3, \, x_2]^4 = [x_1, \, x_2, \, x_2], \ [x_2, \, x_3]^{16} = [x_1, \, x_3, \, x_3], \ [x_3, \, x_2]^{64} = [x_1, \, x_4, \, x_4],$

$[x_2, \, x_3, \, x_1] = [x_2, \, x_3, \, x_2] = [x_2, \, x_3, \, x_3] = [x_2, \, x_3, \, x_4] = 1,$

$[x_1, \, x_2, \, x_1] = [x_1, \, x_2, \, x_3] = [x_1, \, x_2, \, x_4] = 1,$

$[x_1, \, x_3, \, x_1] = [x_1, \, x_3, \, x_2] = [x_1, \, x_3, \, x_4] = 1,$

$[x_1, \, x_4, \, x_1] = [x_1, \, x_4, \, x_2] = [x_1, \, x_4, \, x_3] = 1.$

Then

$$w = [x_2, \, x_3]^{2^{k+5}} \in D_4(G_{k,l}) \setminus \gamma_4(G_{k,l}).$$

The case $k = 2, \ l = 0$ is exactly the example due to Rips.

We continue the above constructions of groups without dimension property by constructing a 4-generator and 3-relator example of a group G with $D_4(G) \neq \gamma_4(G)$ and, for each $n \geq 5$, a 5-generator 5-relator example of a group G with $D_n(G) \neq \gamma_n(G)$. Our motivation for constructing these examples is to develop a closer understanding of groups without dimension property and also to look for simpler, and in a sense minimal, examples of such groups.

Example 2.3 *Let G be the group defined by the presentation*

$$\langle x_1,\ x_2,\ x_3,\ x_4 \mid x_1^4[x_4,\ x_3]^2[x_4,\ x_2] = 1,$$
$$x_2^{16}[x_4,\ x_3]^4[x_4,\ x_1]^{-1} = 1,\ x_3^{64}[x_4,\ x_2]^{-4}[x_4,\ x_1]^{-2} = 1\rangle. \quad (2.1)$$

Then $w = [x_1,\ x_2^{32}][x_1,\ x_3^{64}][x_2,\ x_3^{128}] \in D_4(G) \setminus \gamma_4(G)$.

To prove the above statement, we need the following lemma:

Lemma 2.4 *Let Π be a group. If $x_1,\ x_2,\ x_3 \in \Pi$ and there exist $\xi_j \in \gamma_2(\Pi)$, $j = 1,\ \ldots,\ 6$ and $\eta_i \in \gamma_3(\Pi)$, such that*

$$x_1^4 = \xi_1,\ x_2^{16} = \xi_2,\ x_2^{32}x_3^{64} = \xi_4^4\eta_1,\ x_1^{-32}x_3^{128} = \xi_5^{16}\eta_2,\ x_1^{-64}x_2^{-128} = \xi_6^{64}\eta_3$$

then

$$w = [x_1,\ x_2^{32}][x_1,\ x_3^{64}][x_2,\ x_3^{128}] \in D_4(\Pi).$$

Proof. Since $\gamma_2(\Pi) \subseteq 1 + \Delta^2(\Pi)$, we have

$$1 - w \equiv \alpha_1 + \alpha_2 + \alpha_3 \quad \mod \Delta^4(\Pi),$$

where $\alpha_1 = (1 - [x_1,\ x_2^{32}])$, $\alpha_2 = (1 - [x_1,\ x_3^{64}])$, $\alpha_3 = (1 - [x_2,\ x_3^{128}])$. Now, working modulo $\Delta^4(\Pi)$, we have

$$\alpha_1 \equiv (1 - x_2^{32})(1 - x_1) - (1 - x_1)(1 - x_2^{32})$$
$$\equiv (1 - x_2^{32})(1 - x_1) - 32(1 - x_1)(1 - x_2) + \binom{32}{2}(1 - x_1)(1 - x_2)^2$$
$$\equiv (1 - x_2^{32})(1 - x_1) - (1 - x_1^{32})(1 - x_2) \quad (\text{since } x_1^4 \in \gamma_2(\Pi)).$$

Similarly, we have

$$\alpha_2 \equiv (1 - x_3^{64})(1 - x_1) - (1 - x_1^{64})(1 - x_3),$$
$$\alpha_3 \equiv (1 - x_3^{128})(1 - x_2) - (1 - x_2^{128})(1 - x_3).$$

Therefore,

$$\alpha_1 + \alpha_2 + \alpha_3 \equiv (2 - x_2^{32} - x_3^{64})(1 - x_1) + (x_1^{32} - x_3^{128})(1 - x_2)+$$

$$(x_1^{64} + x_2^{128} - 2)(1 - x_3) \equiv (1 - \xi_4^4\eta_1)(1 - x_1) + (1 - \xi_5^{16}\eta_2)(1 - x_2) + (1 - \xi_6^{64}\eta_2)(1 - x_3) \equiv$$

$$(1 - \eta_1)(1 - x_1) + (1 - \eta_2)(1 - x_2) + (1 - \eta_3)(1 - x_3)+$$
$$(1 - \xi_4)(1 - \xi_1) + (1 - \xi_5)(1 - \xi_2) + (1 - \xi_6)(1 - \xi_2^{-2}\xi_4^4) \equiv 0,$$

and hence $w \in D_4(\Pi)$. $\square$

Proof of Example 2.3. Modulo $\gamma_4(G)$, we have

$$x_2^{32} x_3^{64} = [x_4,\, x_1]^2 [x_3, x_4]^8 [x_4, x_1]^2 [x_4,\, x_2]^4 = [x_3,\, x_4]^8 [x_4,\, x_1]^4 [x_4,\, x_2]^4 = \xi_4^4 \eta_1,$$

with $\xi_4 = [x_3,\, x_4]^2 [x_4,\, x_2] \in \gamma_2(G)$, $\eta_1 = [x_4,\, x_1]^4 \in \gamma_3(G)$;

$$x_1^{-32} x_3^{128} = [x_4,\, x_3]^{16} [x_4,\, x_2]^8 [x_4,\, x_1]^4 [x_4,\, x_2]^8 =$$
$$[x_4,\, x_3]^{16} [x_4,\, x_2]^{16} [x_4,\, x_1]^4 = \xi_5^{16} \eta_2,$$

with $\xi_5 = [x_4,\, x_3] \in \gamma_2(G)$, $\eta_2 = [x_4,\, x_2]^{16} [x_4,\, x_1]^4 \in \gamma_3(G)$;

$$x_1^{-64} x_2^{-128} = [x_4,\, x_2]^{16} [x_4,\, x_3]^{32} [x_4,\, x_1]^{-8} [x_4,\, x_3]^{32} =$$
$$[x_4,\, x_2]^{16} [x_4,\, x_3]^{64} [x_4,\, x_1]^{-8} = \eta_3 \in \gamma_3(G).$$

By Lemma 2.4, $w \in D_4(G)$. It remains to show that $w \notin \gamma_4(G)$. We shall construct a nilpotent group H of class 3, which is an epimorphic image of G with nontrivial image of w.

The construction of H is a slight simplification of the construction of Passi and Gupta (see [Gup87c], Example 2.1, p. 76). Let F be a free group with generators $\langle x_1,\, x_2,\, x_3,\, x_4 \rangle$. Define R_1 to be the fourth term of the lower central series of F, i.e., $R_1 = \gamma_4(F)$. Define

$$R_2 = \langle R_1,\, [x_i,\, x_j,\, x_k] \notin \langle \alpha,\, \beta,\, \gamma \rangle R_1 \text{ for all } i,\, j,\, k,\; \alpha^4 \beta^{-1},\, \beta^4 \gamma^{-1},\, \gamma^4 \rangle,$$

where $\alpha = [x_4,\, x_3,\, x_3]$, $\beta = [x_4,\, x_2,\, x_2]$, $\gamma = [x_4,\, x_1,\, x_1]$;

$$R_3 = \langle R_2, [x_4,\, x_3]^{64} \alpha^{32},\, [x_4,\, x_2]^{16} \beta^8,\, [x_4,\, x_1]^4 \gamma^2,$$
$$[x_3,\, x_2]^{16} \beta^{-1},\, [x_3,\, x_1]^4 \alpha^{-2},\, [x_2,\, x_1]^4 \beta^{-1} \rangle,$$

$$R_4 = \langle R_3,\, c_1,\, c_2,\, c_3 \rangle,$$

where

$$c_1 = x_1^4 [x_4,\, x_3]^2 [x_4,\, x_2],$$
$$c_2 = x_2^{16} [x_4,\, x_3]^4 [x_4,\, x_1]^{-1},$$
$$c_3 = x_3^{64} [x_4,\, x_2]^{-4} [x_4,\, x_1]^{-2}.$$

We set $H = F/R_4$.

Clearly, the group H is a natural epimorphic image of G. Hence it remains to show that the element

$$w_0 = [x_1,\, x_2^{32}][x_1,\, x_3^{64}][x_2,\, x_3^{128}]$$

is nontrivial in H.

We claim that $[R_{i+1}, F] \subseteq R_i$, $i = 1, 2, 3$. This is obvious for $i = 1, 2$. We show it for $i = 3$. Working modulo R_3, we have:

$[c_1, x_1] = 1$,

$[c_1, x_2] = [x_1, x_2]^4 [x_4, x_2, x_2] = [x_1, x_2]^4 \beta = 1$,

$[c_1, x_3] = [x_1, x_3]^4 [x_4, x_3, x_3]^2 = [x_1, x_3]^4 \alpha^2 = 1$,

$[c_1, x_4] = [x_1, x_4]^4 [x_1, x_4, x_1]^2 = [x_1, x_4]^4 \gamma^{-2} = 1$,

$[c_2, x_1] = [x_2, x_1]^{16} [x_4, x_1, x_1]^{-1} = \beta^4 \gamma^{-1} = 1$,

$[c_2, x_2] = 1$,

$[c_2, x_3] = [x_2, x_3]^{16} [x_4, x_3, x_3]^4 = \beta^{-1} \alpha^4 = 1$,

$[c_2, x_4] = [x_2, x_4]^{16} [x_2, x_4, x_2]^8 = [x_2, x_4]^{16} \beta^{-8} = 1$,

$[c_3, x_1] = [x_3, x_1]^{64} [x_4, x_1, x_1]^{-2} = \alpha^{32} \gamma^{-2} = 1$,

$[c_3, x_2] = [x_3, x_2]^{64} [x_4, x_2, x_2]^{-4} = [x_3, x_2]^{64} \beta^{-4} = 1$,

$[c_3, x_3] = 1$,

$[c_3, x_4] = [x_3, x_4]^{64} [x_3, x_4, x_3]^{32} = [x_3, x_4]^{64} \alpha^{-32} = 1$.

Clearly, R_2/R_1 is cyclic of order 64, generated by the element α. We claim that the element α has order exactly 64 in H. Suppose $\alpha^s \in R_4$, $s > 0$ and s is not divisible by 64. Then R_4/R_2 has a torsion element α^s, since $\alpha^{64} \in R_2$. We have the following group extension:

$$1 \to R_3/R_2 \to R_4/R_2 \to R_4/R_3 \to 1.$$

Hence at least one of two groups: R_3/R_2 or R_4/R_3 has a torsion. Since $[R_4, F] \subseteq R_3$, every element of R_4/R_3 can be written as $c_1^{h_1} c_2^{h_2} c_3^{h_3}$ for some integers h_1, h_2, h_3. Clearly it is a free abelian group of rank 3, since $R_4/R_4 \cap \gamma_2(F)$ is free abelian, which is an epimorphic image of R_4/R_3. The same argument works for the quotient R_3/R_2, since all commutators which we added to R_2 to get R_3 are of the form $[x_i, x_j]^{h_{ij}} q_{ij}$, $q_{ij} \in \gamma_3(F)$, $h_{ij} \in \mathbb{Z}$, but commutators $[x_i, x_j]$ are basic commutators in F, i.e., they are linearly independent modulo $\gamma_3(F)$. Hence both R_4/R_3 and R_3/R_2 are free abelian and the element α has order exactly 64 in H.

Finally, note that $w_0 = \alpha^{32}$ is nontrivial in H; hence the element w does not lie in $\gamma_4(G)$. $\square$

Example 2.5 *Let G be the group defined by the presentation*

$$\langle x_1, x_2, x_3, x_4 \mid x_1^4 = \xi_1, \ x_2^{16} = \xi_2,$$
$$x_2^{32} x_3^{64} = \xi_4^4, \ x_1^{-32} x_3^{128} = \xi_5^{16}, \xi_1^{16} \xi_2^8 = 1 \rangle, \quad (2.2)$$

where

$$\xi_1 = [x_2, x_4][x_3, x_4]^2, \ \xi_2 = [x_4, x_1][x_3, x_4]^4,$$
$$\xi_4 = [x_3, x_4]^2 [x_4, x_2][x_4, x_1], \ \xi_5 = [x_4, x_3].$$

Then $w = [x_1, x_2^{32}][x_1, x_3^{64}][x_2, x_3^{128}] \in D_4(G) \setminus \gamma_4(G)$.

Proof. By Lemma 2.4, $w \in D_4(G)$. Note that the group H occurring in the proof of Example 2.3 is a natural epimorphic image of G. Indeed, the first two relations of G are also among the defining relations of H (due to relators c_1, c_2), and therefore we only need to check the other three. In H,

$$x_2^{32} x_3^{64} = [x_4, x_1]^2 [x_3, x_4]^8 [x_4, x_1]^2 [x_4, x_2]^4 = [x_3, x_4]^8 [x_4, x_1]^4 [x_4, x_2]^4 = \xi_4^4;$$

$$x_1^{-32} x_3^{128} = [x_4, x_3]^{16} [x_4, x_2]^8 [x_4, x_1]^4 [x_4, x_2]^8 =$$

$$[x_4, x_3]^{16} [x_4, x_2, x_2]^{-8} [x_4, x_1, x_1]^{-2} = [x_4, x_3]^{16} [x_4, x_1, x_1]^{-4} =$$

$$[x_4, x_3]^{16} = \xi_5^{16};$$

$$\xi_1^{16} \xi_2^8 = [x_2, x_4]^{16} [x_3, x_4]^{32} [x_4, x_1]^8 [x_3, x_4]^{32} =$$

$$[x_4, x_2, x_2]^8 [x_4, x_1, x_1]^4 [x_4, x_3, x_3]^{32} = [x_4, x_3, x_3]^{64} = 1.$$

Thus we have an epimorphism $\theta : G \to H$, $x_i \mapsto x_i$, $1 \leq i \leq 4$. It is shown in the proof of Example 2.3, that $w_0 = \theta(w)$ is nontrivial in H, which is nilpotent of class 3. Hence $w \notin \gamma_4(G)$. $\square$

Example 2.6 *Let*

$$\Pi = \langle x_1, x_2, x_3, y_1 \ldots, y_{26} \mid x_1^4 = \prod_{i=0}^{2} [y_{2i+1}, y_{2i+2}], x_2^{16} = \prod_{i=3}^{7} [y_{2i+1}, y_{2i+2}],$$

$$x_2^{32} x_3^{64} = \left(\prod_{i=8}^{11} [y_{2i+1}, y_{2i+2}] \right)^4, x_1^{-32} x_3^{128} = [y_{25}, y_{26}]^{16}, x_1^{-64} x_2^{-128} = 1 \rangle. \quad (2.3)$$

Then $w = [x_1, x_2^{32}][x_1, x_3^{64}][x_2, x_3^{128}] \in D_4(\Pi) \setminus \gamma_4(\Pi)$.

Proof. By Lemma 2.4, $w \in D_4(\Pi)$. Consider the group $\Pi' = \Pi * Z$, where Z is an infinite cyclic group with generator x, say. It is easy to see that there exists an epimorphism $\theta : \Pi' \to G$, where G is the group considered in Example 2.5 and θ maps $x_i \mapsto x_i$, $i = 1, 2, 3, x \mapsto x_4$. Clearly, for such an epimorphism θ, $\theta(w) \notin \gamma_4(G)$ by Example 2.5, and therefore $w \notin \gamma_4(\Pi)$. $\square$

Example 2.7 *For $n \geq 5$, let*

$$G(n) = \langle x_1, x_2, x_3, y_1, \ldots, y_{10n} \mid x_1^4 = \xi_{1,(n)}, x_2^{16} = \xi_{2,(n)},$$

$$x_2^{32} x_3^{64} = \xi_{3,(n)}^4, x_1^{-32} x_3^{128} = \xi_{4,(n)}^{16}, x_1^{-64} x_2^{-128} = \xi_{5,(n)}^{64} \rangle, \quad (2.4)$$

where

$$\xi_{i,(n)} = [y_{(2i-2)n+1}, y_{(2i-2)n+2}] \cdots [y_{2in-1}, y_{2in}], \quad 1 \le i \le 5.$$

Then $w = [x_1, x_2^{32}][x_1, x_3^{64}][x_2, x_3^{128}] \in D_4(G(n)) \setminus \gamma_4(G(n))$.

Proof. Observe that there exists an epimorphism $G(n) \to \Pi$, Π being the group considered in Example 2.6, which maps $x_i \mapsto x_i$, $i = 1, 2, 3$. The assertion thus follows from Lemma 2.4. □

The same principle can be used to construct more examples of groups without dimension property. The following example is a base for a later construction in Theorem 2.14.

Example 2.8 *Let* $k \ge 9$, *and* G *the group given by the following presentation:*

$$\langle x_1, x_2, x_3, x_4 \mid x_1^8 [x_4, x_3]^4 [x_4, x_2],$$
$$x_2^{64} [x_4, x_3]^{-16} [x_4, x_1]^{-1}, x_3^{2^k} [x_4, x_2]^{16} [x_4, x_1]^{-4} \rangle. \quad (2.5)$$

Then $[x_1, x_2^{256}][x_1, x_3^{2^k}][x_2, x_3^{2^{k+1}}] \in D_4(G) \setminus \gamma_4(G)$.

Example 2.9 *Let*

$$r \ge t \ge 2, \ k \ge q + r,$$
$$s \ge l + 3, \ q \ge s + r + 2$$

and G *the group with generators* x_1, x_2, x_3, x_4, x_5 *and relators*

$$x_1^{2^l} = [x_4, x_2]^{-2^{l-r-2}} [x_4, x_3]^{2^{l-1}} [x_4, x_5]^{-2^{l-t}};$$
$$x_2^{2^s} = [x_4, x_1]^{2^{l-r-2}} [x_4, x_3]^{-2^{s-3}} [x_4, x_5]^{2^{s-t-1}};$$
$$x_3^{2^k} = [x_4, x_1]^{-2^{l-1}} [x_4, x_2]^{2^{s-3}} [x_4, x_5]^{2^{q-5}};$$
$$x_5^{2^q} = [x_4, x_1]^{2^{l-t}} [x_4, x_2]^{-2^{s-t-1}} [x_4, x_3]^{-2^{q-5}}.$$

Then, for

$$w = [x_1, x_2^{2^{s+r}}][x_1, x_3^{2^k}][x_1, x_5^{2^{q+t-2}}][x_2, x_3^{2^{k+1}}][x_2, x_5^{2^{q+t}}][x_5, x_3^{2^{k+3}}],$$

we have

$$w \in D_4(G) \setminus \gamma_4(G).$$

Proof. Since $x_1^{2^l}, x_2^{2^s}, x_3^{2^k}, x_5^{2^q} \in \gamma_3(G)$, we have

$$1 - w \equiv \sum_{i=1}^{6} (1 - \alpha_i) \quad \mod \mathfrak{g}^6,$$

where

$$\alpha_1 = [x_1, x_2^{2^{s+r}}], \ \alpha_2 = [x_1, x_3^{2^k}], \ \alpha_3 = [x_1, x_5^{2^{q+t-2}}],$$
$$\alpha_4 = [x_2, x_3^{2^{k+1}}], \ \alpha_5 = [x_2, x_5^{2^{q+t}}], \ \alpha_6 = [x_5, x_3^{2^{k+3}}].$$

Clearly, we have

$$1 - \alpha_1 \equiv (1 - x_2^{2^{s+r}})(1 - x_1) - (1 - x_1)(1 - x_2^{s+r})$$
$$\equiv (1 - x_2^{2^{s+r}})(1-x_1) - 2^{s+r}(1-x_1)(1-x_2) + \binom{2^{s+r}}{2}(1-x_1)(1-x_2)^2$$
$$\equiv (1 - x_2^{2^{s+r}})(1 - x_1) - (1 - x_1^{2^{s+r}})(1 - x_2) \mod \mathfrak{g}^4;$$
$$1 - \alpha_2 \equiv (1 - x_3^{2^k})(1 - x_1) - (1 - x_1^{2^k})(1 - x_3) \mod \mathfrak{g}^4;$$
$$1 - \alpha_3 \equiv (1 - x_5^{2^{q+t-2}})(1 - x_1) - (1 - x_1^{2^{q+t-2}})(1 - x_5) \mod \mathfrak{g}^4;$$
$$1 - \alpha_4 \equiv (1 - x_3^{2^{k+1}})(1 - x_2) - (1 - x_2^{2^{k+1}})(1 - x_3) \mod \mathfrak{g}^4;$$
$$1 - \alpha_5 \equiv (1 - x_5^{2^{q+t}})(1 - x_2) - (1 - x_2^{2^{q+t}})(1 - x_5) \mod \mathfrak{g}^4;$$
$$1 - \alpha_6 \equiv (1 - x_3^{2^{k+3}})(1 - x_5) - (1 - x_5^{2^{k+3}})(1 - x_3) \mod \mathfrak{g}^4.$$

Hence

$$1 - w \equiv (1 - x_2^{2^{s+r}} x_3^{2^k} x_5^{2^{q+t-2}})(1 - x_1) + (1 - x_1^{-2^{s+r}} x_3^{2^{k+1}} x_5^{2^{q+t}})(1 - x_2) +$$
$$(1 - x_1^{-2^k} x_2^{-2^{k+1}} x_5^{-2^{k+3}})(1 - x_3) + (1 - x_1^{-2^{q+t-2}} x_2^{-2^{q+t}} x_3^{2^{k+3}})(1 - x_5) \mod \mathfrak{g}^4$$

In the group G, we have:

$$x_2^{2^{s+r}} x_3^{2^k} x_5^{2^{q+t-2}} = [x_4, x_1]^{2^{l-2}} [x_4, x_3]^{-2^{s+r-3}} [x_4, x_5]^{2^{s-t-1+r}}$$
$$[x_4, x_1]^{-2^{l-1}} [x_4, x_2]^{2^{s-3}} [x_4, x_5]^{2^{q-5}} [x_4, x_1]^{2^{l-2}} [x_4, x_2]^{-2^{s-3}} [x_4, x_3]^{-2^{q+t-7}} =$$
$$[x_4, x_2]^{2^{s-t-1+r}} [x_4, x_2]^{-2^{s-3}} [x_4, x_3]^{-2^{s+r-3}} [x_4, x_3]^{-2^{q+t-7}} [x_4, x_5]^{2^{q-5}} [x_4, x_5]^{2^{s-t-1+r}}.$$

$$x_1^{-2^{s+r}} x_3^{2^{k+1}} x_5^{2^{q+t}} = [x_4, x_2]^{2^{s-2}} [x_4, x_3]^{-2^{s+r-1}} [x_4, x_5]^{2^{s+r-t}}$$
$$[x_4, x_1]^{-2^l} [x_4, x_2]^{2^{s-2}} [x_4, x_5]^{2^{q-4}} [x_4, x_1]^{2^l} [x_4, x_2]^{-2^{s-1}} [x_4, x_3]^{-2^{q-5+t}} =$$
$$[x_4, x_3]^{-2^{s+r-1}} [x_4, x_3]^{-2^{q+t-5}} [x_4, x_5]^{2^{s+r-t}} [x_4, x_5]^{2^{q-4}}.$$

$$x_1^{-2^k} x_2^{-2^{k+1}} x_5^{-2^{k+3}} = [x_4, x_2]^{2^{k-r-2}} [x_4, x_3]^{-2^{k-1}} [x_4, x_5]^{2^{k-t}} [x_4, x_1]^{-2^{l-r-1+k-s}}$$
$$[x_4, x_3]^{2^{k-2}} [x_4, x_5]^{-2^{k-t}} [x_4, x_1]^{-2^{l-t+k-q+3}} [x_4, x_2]^{2^{s-t+k-q+2}} [x_4, x_3]^{2^{k-2}} =$$
$$[x_4, x_1]^{-2^{l-r+k-s}} [x_4, x_1]^{-2^{l-t+k-q+3}} [x_4, x_2]^{2^{k-r-2}} [x_4, x_2]^{2^{s-t+k-q+2}}.$$

$$x_1^{-2^{q+t-2}} x_2^{-2^{q+t}} x_3^{2^{k+3}} = [x_4, x_2]^{2^{q+t-r-4}} [x_4, x_3]^{-2^{q+t-3}} [x_4, x_5]^{2^{q-2}}$$

$$[x_4,x_1]^{-2^{l-r-2+q+t-s}} [x_4,x_3]^{2^{q+t-3}} [x_4, x_5]^{-2^{q-1}} [x_4,x_1]^{2^{l+2}} [x_4,x_2]^{2^{s}} [x_4,x_5]^{2^{q-2}} =$$

$$[x_4, x_2]^{2^{q+t-r-4}} [x_4, x_2]^{2^{s}} [x_4, x_1]^{q+l+t-s-r-2} [x_4, x_1]^{2^{l+2}}.$$

Hence,

$$1 - w \equiv (1 - \eta_1^{2^l})(1 - x_1) + (1 - \eta_2^{2^s})(1 - x_2) +$$
$$(1 - \eta_3)(1 - x_3) + (1 - \eta_5)(1 - x_5) \quad \mathrm{mod}\ \mathfrak{g}^4,$$

where

$$\eta_1 = [x_4, x_2]^{2^{s-t-1+r-l}} [x_4, x_2]^{-2^{s-3-l}} [x_4, x_3]^{-2^{s+r-l-3}} [x_4, x_3]^{-2^{q+t-l-7}}$$
$$[x_4, x_5]^{2^{q-l-5}} [x_4, x_5]^{2^{s-t-1+r-l}},$$

$$\eta_2 = [x_4, x_3]^{-2^{r-1}} [x_4, x_3]^{-2^{q+t-s-5}} [x_4, x_5]^{2^{r-t}} [x_4, x_5]^{2^{q-s-4}},$$

$$\eta_3 = [x_4, x_1]^{-2^{l-r+k-s}} [x_4, x_1]^{-2^{l-t+k-q+3}} [x_4, x_2]^{2^{k-r-2}} [x_4, x_2]^{2^{s-t+k-q+2}} \in \gamma_3(G),$$

$$\eta_4 = [x_4, x_2]^{2^{q+t-r-4}} [x_4, x_2]^{2^{s}} [x_4, x_1]^{q+l+t-s-r-2} [x_4, x_1]^{2^{l+2}} \in \gamma_3(G).$$

Therefore,

$$1 - w \equiv (1 - \eta_1)(1 - x_1^{2^l}) + (1 - \eta_2)(1 - x_2^{2^s}) \equiv 0 \quad \mathrm{mod}\ \mathfrak{g}^4.$$

The proof that $w \notin \gamma_4(G)$ is by the same principle as that in the proof of Example 2.3.

Let F be a free group with generators x_1, x_2, x_3, x_4, x_5. Define R_1 to be the fourth term of the lower central series of F, i.e., $R_1 = \gamma_4(F)$. Define

$$R_2 = \langle R_1, [x_i, x_j, x_k] \notin \langle \alpha, \beta, \gamma, \delta \rangle R_1 \text{ for all } i, j, k,$$
$$\alpha^{2^{k-q}} \beta^{-1},\ \beta^{2^{q-s}} \gamma^{-1},\ \gamma^{2^{s-l}} \delta^{-1},\ \delta^{2^l} \rangle,$$

where $\alpha = [x_4, x_3, x_3], \beta = [x_4, x_5, x_5], \gamma = [x_4, x_2, x_2], \delta = [x_4, x_1, x_1]$;

$$R_3 = \langle R_2, [x_4, x_3]^{2^k} \alpha^{2^{k-1}}, [x_4, x_5]^{2^q} \beta^{2^q}, [x_4, x_2]^{2^s} \gamma^{2^{s-1}},$$
$$[x_4, x_1]^{2^l} \delta^{2^{l-1}}, [x_3, x_5]^{-2^q} \alpha^{2^{q-5}}, [x_3, x_2]^{-2^s} \alpha^{2^{s-3}},$$
$$[x_3, x_1]^{2^l} \alpha^{2^{l-1}}, [x_5, x_2]^{2^s} \beta^{2^{s-t-1}}, [x_5, x_1]^{-2^l} \beta^{-2^{l-t}}, [x_2, x_1]^{-2^l} \gamma^{2^{l-r-2}} \rangle,$$

$$R_4 = \langle R_3, c_1, c_2, c_3, c_4 \rangle,$$

where

$$c_1 = x_1^{-2^l}[x_4, x_2]^{-2^{l-r-2}}[x_4, x_3]^{2^{l-1}}[x_4, x_5]^{-2^{l-t}};$$

$$c_2 = x_2^{-2^s}[x_4, x_1]^{2^{l-r-2}}[x_4, x_3]^{-2^{s-3}}[x_4, x_5]^{2^{s-t-1}};$$

$$c_3 = x_3^{-2^k}[x_4, x_1]^{-2^{l-1}}[x_4, x_2]^{2^{s-3}}[x_4, x_5]^{2^{q-5}};$$

$$c_4 = x_5^{-2^q}[x_4, x_1]^{2^{l-t}}[x_4, x_2]^{-2^{s-t-1}}[x_4, x_3]^{-2^{q-5}}.$$

We set $H = F/R_4$.

Clearly, the group H is a natural epimorphic image of G. Hence it remains to show that the element

$$w_0 = [x_1, x_2^{2^{s+r}}][x_1, x_3^{2^k}][x_1, x_5^{2^{q+t-2}}][x_2, x_3^{2^{k+1}}][x_2, x_5^{2^{q+t}}][x_5, x_3^{2^{k+3}}]$$

is nontrivial in H. We claim that $[R_{i+1}, F] \subseteq R_i$, $i = 1, 2, 3$. The proof is straightforward. In analogy with Example 2.3, one can show that $\gamma_3(G)$ is cyclic of order 2^k with generator α, but $w_0 = \alpha^{2^{k-1}}$. Therefore, $w_0 \neq 1$ and $w \notin \gamma_4(G)$. $\square$

We next discuss examples of groups without dimension property in arbitrary dimension. First examples of groups without dimension property in higher dimensions were constructed by N. Gupta [Gup90].

Example 2.10 (Gupta [Gup90], [Gup91a]). *Let $n \geq 4$ be fixed and let F be a free group of rank 4 with basis $\{r, a, b, c\}$. Set $x_0 = y_0 = z_0 = r$, and define commutators $x_i = [x_{i-1}, a]$, $y_i = [y_{i-1}, b]$, $z_i = [z_{i-1}, c]$, $i = 1, 2, \ldots$. Let $\mathfrak{G}_n$ be the quotient of F with the following defining relations:*

(i) $r^{2^{2n-1}} = 1$, $a^{2^{n+2}} = y_{n-3}^4 z_{n-3}^2$, $b^{2^n} = x_{n-3}^{-4} z_{n-3}$, $c^{2^{n-2}} = x_{n-3}^{-2} y_{n-3}^{-1}$;

(ii) $z_{n-2} = y_{n-2}^4$, $y_{n-2} = x_{n-2}^4$;

(iii) $x_{n-1} = 1$, $y_{n-1} = 1$, $z_{n-1} = 1$;

(iv) $[a, b, g] = 1$, $[b, c, g] = 1$, $[a, c, g] = 1$, *for all* $g \in F$;

(v) $[x_i, b] = 1$, $[x_i, c] = 1$, $[y_i, a] = 1$, $[y_i, c] = 1$, $[z_i, a] = 1$, $[z_i, b] = 1$, $i \geq 1$;

(vi) $[x_i, x_j] = 1$, $[x_i, y_j] = 1$, $[x_i, z_j] = 1$, $[y_i, y_j] = 1$, $[y_i, z_j] = 1$, $[z_i, z_j] = 1$, $i, j \geq 0$.

Let

$$g = [a, b]^{2^{2n-1}}[a, c]^{2^{2n-2}}[b, c]^{2^{2n-3}}.$$

Then $g \in D_n(\mathfrak{G}_n) \setminus \gamma_n(\mathfrak{G}_n)$.

Example 2.11 *For every integer $n \geq 0$, there exist integers $k > l$, such that for the group $\mathfrak{G}_n$ defined by the presentation*

$$\langle x_1, x_2, x_3, x_4, x_5 \mid x_1^4 \xi_1 = 1, \ x_2^{2^l} \xi_2 = 1, \ x_3^{2^k} \xi_3 = 1,$$
$$[[x_5, {}_n x_4], x_1]^4 [[x_5, {}_n x_4], x_3, x_3]^{2^{k-1}} = 1, \ \xi_1^{2^{k-2}} \xi_2^{2^{k-l+1}} = 1 \rangle,$$

where

$$\xi_1 = [[x_5, {}_n x_4], x_3]^2 [[x_5, {}_n x_4], x_2][x_5, {}_{n+1} x_4]^2,$$
$$\xi_2 = [[x_5, {}_n x_4], x_3]^{2^{l-2}} [[x_5, {}_n x_4], x_1]^{-1} [x_5, {}_{n+1} x_4]^2,$$
$$\xi_3 = [[x_5, {}_n x_4], x_2]^{-2^{l-2}} [[x_5, {}_n x_4], x_1]^{-2},$$

the element $w_n = [x_1, x_2^{2^{l+1}}][x_1, x_3^{2^k}][x_2, x_3^{2^{k+1}}] \in D_{4+n}(\mathfrak{G}_n) \setminus \gamma_{4+n}(\mathfrak{G}_n)$.

To prove the above assertion we need some technical lemmas. The following lemma is a generalization of Lemma 2.4.

Lemma 2.12 *Let Π be a group and $n \geq 4$ an integer. If $x_1, x_2, x_3 \in \Pi$ are such that there exist $\xi_i \in \gamma_{n-2}(\Pi)$, $i = 1, \ldots, 4$, satisfying*

$$x_1^4 = \xi_1, \ x_2^{2^l} = \xi_2, \ x_2^{2^{l+1}} x_3^{2^k} = \xi_3^4, \ x_1^{-2^{l+1}} x_3^{2^{k+1}} = \xi_4^{2^l}, \ x_1^{2^k} x_2^{2^{k+1}} = 1$$

then

$$w = [x_1, x_2^{2^{l+1}}][x_1, x_3^{2^k}][x_2, x_3^{2^{k+1}}] \in D_n(\Pi),$$

provided k, l are sufficiently large integers.

Proof. Since $1 - x \in \Delta^{n-2}(\Pi)$ for $x \in \gamma_{n-2}(\Pi)$, we have

$$1 - w \equiv \alpha_1 + \alpha_2 + \alpha_3 \mod \Delta^n(\Pi),$$

where $\alpha_1 = (1 - [x_1, x_2^{2^{l+1}}])$, $\alpha_2 = (1 - [x_1, x_3^{2^k}])$, $\alpha_3 = (1 - [x_2, x_3^{2^{k+1}}])$. Now, working modulo $\Delta^n(\Pi)$, we have

$$\alpha_1 \equiv (1 - x_2^{2^{l+1}})(1 - x_1) - (1 - x_1)(1 - x_2^{2^{l+1}})$$
$$\equiv (1 - x_2^{2^{l+1}})(1 - x_1) - 2^{l+1}(1 - x_1)(1 - x_2) +$$
$$\sum_{i=2}^{n-1} (-1)^i \binom{2^{l+1}}{i} (1 - x_1)(1 - x_2)^i.$$

Note that, for sufficiently large l and $i \leq n$, the integer $\binom{2^{l+1}}{i}$ is divisible by 4^n. Hence, for such an integer l, we have

$$\sum_{i=2}^{n-1}(-1)^i\binom{2^{l+1}}{i}(1-x_1)(1-x_2)^i \in \Delta^n(\Pi),$$

and

$$2^{l+1}(1-x_1)(1-x_2) \equiv (1-x_1^{2^{l+1}})(1-x_2) \mod \Delta^n(\Pi).$$

Therefore,

$$\alpha_1 \equiv (1-x_2^{2^{l+1}})(1-x_1) - (1-x_1^{2^{l+1}})(1-x_2) \mod \Delta^n(\Pi).$$

Assuming k to be large enough so that $\binom{2^{k+1}}{i}$ is divisible by 2^{ln} for $i \leq n$, we have

$$\alpha_2 \equiv (1-x_3^{2^k})(1-x_1) - (1-x_1^{2^k})(1-x_3),$$
$$\alpha_3 \equiv (1-x_3^{2^{k+1}})(1-x_2) - (1-x_2^{2^{k+1}})(1-x_3).$$

Therefore, $\mod \Delta^n(\Pi)$,
$\alpha_1 + \alpha_2 + \alpha_3$

$$\equiv (2 - x_2^{2^{l+1}} - x_3^{2^k})(1-x_1) + (x_1^{2^{l+1}} - x_3^{2^{k+1}})(1-x_2) +$$
$$(x_1^{2^k} + x_2^{2^{k+1}} - 2)(1-x_3)$$

$$\equiv (1-\xi_3^4)(1-x_1) + (1-\xi_4^{2^l})(1-x_2) + (1-x_1^{2^k}x_2^{2^{k+1}})(1-x_3)$$

$$\equiv (1-\xi_3)(1-\xi_1) + (1-\xi_4)(1-\xi_2)$$

$$\equiv 0,$$

and hence $w \in D_n(\Pi)$. $\square$

Lemma 2.13 *Let $k \geq l+2$, $l \geq 4$, be integers and G the group defined by the presentation*

$$\langle x_1, x_2, x_3, x_4, x_5 \mid x_1^4\xi_1 = 1, \ x_2^{2^l}\xi_2 = 1, \ x_3^{2^k}\xi_3 = 1$$
$$[x_4, x_1]^4[x_4, x_3, x_3]^{2^{k-1}} = 1, \ \xi_1^{2^{k-2}}\xi_2^{2^{k-l+1}} = 1,$$
$$[x_4, x_i, x_4] = 1, \ i = 1, \ldots, 4\rangle,$$

where

$$\xi_1 = [x_4, x_3]^2[x_4, x_2][x_4, x_5]^2,$$
$$\xi_2 = [x_4, x_3]^{2^{l-2}}[x_4, x_1]^{-1}[x_4, x_5]^2,$$
$$\xi_3 = [x_4, x_2]^{-2^{l-2}}[x_4, x_1]^{-2}.$$

Then the element $w = [x_1, x_2^{2^{l+1}}][x_1, x_3^{2^k}][x_2, x_3^{2^{k+1}}]$ does not lie in $\gamma_4(G)$.

Proof. We shall construct a nilpotent group of class 3 which is an epimorphic image of the given group G and is such that the image of the element w is nontrivial.

Let F be a free group with basis $\{x_1, \ldots, x_5\}$. Consider the following four types of relations:

$$R_1 = \gamma_4(F),$$

$R_2 = \langle R_1 \cup \{[x_i, x_j, x_k] : (i, j, k) \neq (4, 1, 1), (4, 2, 2), (4, 3, 3)\}, \alpha^{2^{k-l}}\beta^{-1}, \beta^{2^{l-2}}\gamma^{-1}, \gamma^4 \rangle$, where $\alpha = [x_4, x_3, x_3]$, $\beta = [x_4, x_2, x_2]$, $\gamma = [x_4, x_1, x_1]$. Now define R_3 to be the product of R_2 and the normal closure in F of the following words:

$$[x_4, x_3]^{2^k}\alpha^{2^{k-1}}, \quad [x_4, x_2]^{2^l}\beta^{2^{l-1}},$$
$$[x_4, x_1]^4\gamma^2, \quad [x_3, x_2]^{2^l}\alpha^{-2^{l-2}},$$
$$[x_3, x_1]^4\alpha^{-2}, \quad [x_2, x_1]^4\beta^{-1},$$
$$[x_4, x_5]^{2^{k-l+2}}\alpha^{-2^{k-1}}$$
$$[x_5, x_i], \quad i \neq 1.$$

Finally, let R_4 be the product of R_3 and the normal closure in F of the following words:

$$c_1 = x_1^4[x_4, x_3]^2[x_4, x_2][x_4, x_5]^2,$$
$$c_2 = x_2^{2^l}[x_4, x_3]^{2^{l-2}}[x_4, x_1]^{-1}[x_4, x_5]^2,$$
$$c_3 = x_3^{2^k}[x_4, x_2]^{-2^{l-2}}[x_4, x_1]^{-2}.$$

We claim that
$$[R_{i+1}, F] \subseteq R_i, \text{ for } i = 1, 2, 3.$$

This is obvious for $i = 1$, and 2 and it remains only to check for $i = 3$.
We note that, modulo R_3, we have:

$[c_1, x_1] = 1$,
$[c_1, x_2] = [x_1, x_2]^4[x_4, x_3, x_3] = [x_1, x_2]^4\beta = 1$,
$[c_1, x_3] = [x_1, x_3]^4[x_4, x_3, x_3]^2 = [x_1, x_3]^4\alpha^2 = 1$,
$[c_1, x_4] = [x_1, x_4]^4[x_1, x_4, x_1]^2 = [x_1, x_4]^4\gamma^2 = 1$,
$[c_1, x_5] = 1$,
$[c_2, x_1] = [x_2, x_1]^{2^l}[x_4, x_1, x_1]^{-1} = \beta^{2^{l-2}}\gamma^{-1} = 1$,
$[c_2, x_2] = 1$,
$[c_2, x_3] = [x_2, x_3]^{2^l}[x_4, x_3, x_3]^{2^{l-2}} = [x_2, x_3]^{2^l}\alpha^{2^{l-2}} = 1$,
$[c_2, x_4] = [x_2, x_4]^{2^l}[x_2, x_4, x_2]^{2^{l-1}} = [x_2, x_4]^{2^l}\beta^{-2^{l-1}} = 1$,
$[c_2, x_5] = 1$,
$[c_3, x_1] = [x_3, x_1]^{2^k}[x_4, x_1, x_1]^{-2} = \alpha^{2^{k-1}}\gamma^{-2} = 1$,

$$[c_3, x_2] = [x_3, x_2]^{2^k}[x_4, x_2, x_2]^{-2^{l-2}} = \alpha^{2^{k-2}}\beta^{-2^{l-2}} = 1,$$
$$[c_3, x_3] = 1,$$
$$[c_3, x_4] = [x_3, x_4]^{2^k}[x_3, x_4, x_3]^{2^{k-1}} = [x_3, x_4]^{2^k}\alpha^{2^{k-1}} = 1,$$
$$[c_3, x_5] = 1.$$

Clearly, $\gamma_3(F)/R_2$ is a cyclic group of order 2^k generated by the element α. To see that α has order exactly 2^k in the group F/R_4, as in the case of the proof of Theorem 2.5, we note that the groups R_3/R_2 and R_4/R_3 are free abelian. Hence, the relation $\alpha^s \in R_4$, $s > 0$ implies that s is divisible by 2^k. As a consequence we get that α has order exactly 2^k in F/R_4. Hence, modulo R_4, the word $w = [x_1, x_2^{2^{l+1}}][x_1, x_3^{2^k}][x_2, x_3^{2^{k+1}}] \equiv \alpha^{2^{k-1}} \equiv \beta^8 \equiv \gamma^2 \not\equiv 1$.

We claim that F/R_4 is a natural epimorphic image of the given group G. The first three relations of G hold in F/R_4 by construction. The relation $[x_4, x_1]^4[x_4, x_3, x_3]^{2^{k-1}} = [x_4, x_1]^4\gamma^2$ holds modulo R_3. Now, modulo R_4, we have

$$\xi_1^{2^{k-2}}\xi_2^{2^{k-l+1}}$$

$$= [x_4, x_3]^{2^{k-1}}[x_4, x_2]^{2^{k-2}}[x_4, x_5]^{2^{k-1}}[x_4, x_3]^{2^{k-1}}[x_4, x_1]^{-2^{k-l+1}}[x_4, x_5]^{2^{k-l+2}}$$
$$= [x_4, x_3]^{2^k}\alpha^{2^{k-1}}\alpha^{2^{k+l-4}}[x_4, x_1]^{-2^{k-l+1}}[x_4, x_2]^{2^{k-2}}$$
$$= \alpha^{2^{k+l-4}}\beta^{-2^{k-3}}\gamma^{2^{k-l-2}} = 1.$$

The relations $[x_4, x_i, x_4]$, $i \in \{1, 2, 3, 4\}$, clearly lie in R_2. Hence F/R_4 is a natural epimorphic image of G and the image of w is nontrivial in F/R_4. Therefore, $w \notin \gamma_4(G)$. $\square$

Proof of Example 2.11. The case $n = 0$ is exactly Lemma 2.13. Assume that the result holds for some $n \geq 0$, i.e., $w_n \notin \gamma_{4+n}(\mathfrak{G}_n)$. We shall prove it for $n + 1$, i.e., that $w_{n+1} \notin \gamma_{5+n}(\mathfrak{G}_{n+1})$.

Consider the quotient $\mathfrak{G}'_n = \mathfrak{G}_n/\gamma_{4+n}(\mathfrak{G}_n)N_n$, where N_n is the normal subgroup in $\mathfrak{G}_n$, generated by all left-normed commutators $[y_1, \ldots, y_s]$, $s \geq 3$, such that there are at least two entries with $y_i = x_4$. The automorphism of the free group of rank 5, given by

$$x_1 \mapsto x_1,$$
$$x_2 \mapsto x_2,$$
$$x_3 \mapsto x_3,$$
$$x_4 \mapsto x_4,$$
$$x_5 \mapsto x_5x_4$$

can be extended to an automorphism of $\mathfrak{G}'_n$; this follows from the fact that this automorphism preserves all relations. This automorphism defines a semidirect product $H_n = \mathfrak{G}'_n \rtimes \langle x \rangle$, where x acts as the described automorphism. Clearly, we have $[x, x_i] = 1$, $i = 1, 2, 3, 4$ and $[x_5, x] = x_4$ in H_n and H_n is nilpotent:

$\gamma_{5+n}(H_n) = 1$. Evidently the natural map $f : \mathfrak{G}'_n \to H_n$ is a monomorphism. However, it is easy to see that H_n is an epimorphic image of $\mathfrak{G}_{n+1}$, which sends w_{n+1} to $f(w_n)$. Hence, w_{n+1} can not lie in $\gamma_{5+n}(\mathfrak{G}_{n+1})$. $\square$

We next make somewhat more complicated constructions, working on the same principles as above, and show that there exists a nilpotent group of class 4 with nontrivial sixth dimension subgroup.

Theorem 2.14 *There exists a nilpotent group G of class 3 with*

$$G \cap (1 + \Delta(\gamma_2(G))^2 \mathbb{Z}[G] + \mathfrak{g}^5) \neq 1.$$

Proof. Let F be a free group with basis $\{x_1, x_2, x_3, x_4, x_5\}$. Let $R_1 := \gamma_4(F)$. Define

$$R_2 = \langle R_1, [x_i, x_j, x_k] \notin \langle \alpha, \beta, \gamma, \delta \rangle R_1 \text{ for all } i, j, k, \delta^{16}\beta, \alpha^{2^{k-10}}\delta, \beta^8\gamma^{-1}, \gamma^8 \rangle,$$

where $\delta = [x_4, x_5, x_5]$, $\alpha = [x_4, x_3, x_3]$, $\beta = [x_4, x_2, x_2]$, $\gamma = [x_4, x_1, x_1]$; Let R_3 be R_2 together with the following set of words:

$$[x_1, x_2]^8[x_4, x_2, x_2],$$
$$[x_1, x_3]^8[x_4, x_3, x_3]^4,$$
$$[x_2, x_3]^{64}[x_4, x_3, x_3]^{-16},$$
$$[x_1, x_4]^8[x_4, x_1, x_1]^4,$$
$$[x_2, x_4]^{64}[x_4, x_2, x_2]^{32},$$
$$[x_3, x_4]^{2^k}[x_4, x_3, x_3]^{2^{k-1}},$$
$$[x_2, x_5]^{64}[x_4, x_5, x_5]^{16},$$
$$[x_5, x_4]^{1024}[x_4, x_5, x_5]^{512},$$
$$[x_1, x_5]^8, \ [x_3, x_5]^{1024}.$$

Let R_4 be R_3 together with the following set of words:

$$c_1 = x_1^8[x_4, x_3]^4[x_4, x_2];$$
$$c_2 = x_2^{64}[x_4, x_3]^{-16}[x_4, x_1]^{-1}[x_4, x_5]^{16};$$
$$c_3 = x_3^{2^k}[x_4, x_2]^{16}[x_4, x_1]^{-4}$$
$$c_4 = x_5^{1024}[x_4, x_2]^{16};$$

For any $i = 1, 2, 3$, $[F, R_{i+1}] \subseteq R_i$ and $k \geq 12$. The case $i = 1$ is obvious. The case $i = 2$ easily can be checked. We shall consider the most difficult case $i = 3$. Working modulo R_3, we shall show that $[c_i, x_j] = 1$ for all i, j:

$[c_1, x_1] = 1$;
$[c_1, x_2] = [x_1, x_2]^8[x_4, x_2, x_2] = 1$;
$[c_1, x_3] = [x_1, x_3]^8[x_4, x_3, x_3]^4 = 1$;

$[c_1, x_4] = [x_1, x_4]^8 [x_4, x_1, x_1]^4 - 1$, since $\gamma^8 \in R_3$;

$[c_1, x_5] = [x_1, x_5]^8 = 1$;

$[c_2, x_1] = [x_2, x_1]^{64} [x_4, x_1, x_1]^{-1} = [x_4, x_2, x_2]^8 [x_4, x_1, x_1]^{-1} = 1$;

$[c_2, x_2] = 1$;

$[c_2, x_3] = [x_2, x_3]^{64} [x_4, x_3, x_3]^{-16}$;

$[c_2, x_4] = [x_2, x_4]^{64} [x_4, x_2, x_2]^{32} = 1$;

$[c_2, x_5] = [x_2, x_5]^{64} [x_4, x_5, x_5]^{16} = 1$;

$[c_3, x_1] = [x_3, x_1]^{2^k} [x_4, x_1, x_1]^{-4} = [x_4, x_3, x_3]^{2^{k-1}} [x_4, x_1, x_1]^{-1} = 1$;

$[c_3, x_2] = [x_3, x_2]^{2^k} [x_4, x_2, x_2]^{16} = [x_4, x_3, x_3]^{-2^{k-2}} [x_4, x_2, x_2]^{16} = 1$;

$[c_3, x_3] = 1$;

$[c_3, x_4] = [x_3, x_4]^{2^k} [x_4, x_3, x_3]^{2^{k-1}} = 1$;

$[c_3, x_5] = [x_3, x_5]^{2^k} = 1$, since $k > 10$;

$[c_4, x_1] = [x_5, x_1]^{1024} = 1$;

$[c_4, x_2] = [x_5, x_2]^{1024} [x_4, x_2, x_2]^{16} = [x_4, x_5, x_5]^{256} [x_4, x_2, x_2]^{16} = 1$;

$[c_4, x_3] = [x_5, x_3]^{1024} = 1$;

$[c_4, x_4] = [x_5, x_4]^{1024} [x_4, x_5, x_5]^{512} = 1$;

$[c_4, x_5] = 1$;

Clearly, $\gamma_3(F/R_2)$ is a cyclic group of order 2^k generated by element α. To see that α has order exactly 2^k in the group F/R_4, as in the case of the proof of Theorem 2.3, we note that the groups R_3/R_2 and R_4/R_3 are free abelian. Hence, the relation $\alpha^s \in R_4$, $s > 0$ implies that s divides 2^k. As a consequence we get the fact that α has order exactly 2^k in F/R_4. And therefore, our element

$$w = [x_1, x_2^{256}][x_1, x_3^{2^k}][x_2, x_3^{2^{k+1}}]$$

is equal to $\alpha^{2^{k-1}} = \delta^{512} = \beta^{32} = \gamma^4 \neq 1$.

Since $x_1^8, x_2^{64}, x_3^{2^k} \in \gamma_2(G)$, modulo $\mathfrak{g}^6$, we have the following equivalences:

$$1 - w \equiv (1 - [x_1, x_2^{256}]) + (1 - [x_1, x_3^{2^k}]) + (1 - [x_2, x_3^{2^{k+1}}]).$$

Since $64(1 - x_1)^2$, $2^k(1 - x_1)^2$, $2^{k+1}(1 - x_2)^2 \in \mathfrak{g}^4$, $64(1 - x_2) \in \mathfrak{g}^2$, modulo $\mathfrak{g}^5$ we have

$$1 - [x_1, x_2^{256}]$$
$$\equiv (1 - x_2^{256})(1 - x_1) - (1 - x_1)(1 - x_2^{256})$$
$$\equiv (1 - x_2^{256})(1 - x_1) - (1 - x_1^{256})(1 - x_2) + \binom{256}{2}(1 - x_1)(1 - x_2)^2$$
$$\equiv (1 - x_2^{256})(1 - x_1) - (1 - x_1^{256})(1 - x_2) + (1 - x_1^{128})(1 - x_2)^2.$$

Note that modulo $\mathfrak{g}^5$:

$$
\begin{aligned}
&(1 - x_1^{128})(1 - x_2)^2 \\
&\equiv (1 - [x_4,\, x_3]^{-64}[x_4,\, x_2]^{-16})(1 - x_2)^2 \\
&\equiv (1 - x_2) + (1 - x_5^{1024})(1 - x_2)^2 \\
&\equiv 1024(1 - x_5)(1 - x_2)^2 \\
&\equiv (1 - x_5)(1 - x_2)(1 - x_2^{1024}) \\
&\equiv (1 - x_5)(1 - x_2)(1 - [x_4,\, x_3]^{256}[x_4,\, x_1]^{16}[x_4,\, x_5]^{-256}) \\
&\equiv 0,
\end{aligned}
$$

therefore, modulo $\mathfrak{g}^5$,

$$
1 - [x_1,\, x_2^{256}] \equiv (1 - x_2^{256})(1 - x_1) - (1 - x_1^{256})(1 - x_2). \tag{2.6}
$$

Analogically, it is easy to show that modulo $\mathfrak{g}^5$,

$$
1 - [x_1,\, x_3^{2^k}] \equiv (1 - x_3^{2^k})(1 - x_1) - (1 - x_1^{2^k})(1 - x_3) + (1 - x_1^{2^{k-1}})(1 - x_3)^2, \tag{2.7}
$$

$$
1 - [x_2,\, x_3^{2^{k+1}}] \equiv (1 - x_3^{2^{k+1}})(1 - x_2) - (1 - x_2^{2^{k+1}})(1 - x_3) + (1 - x_2^{2^k})(1 - x_3)^2. \tag{2.8}
$$

Note that

$$
x_1^{2^{k-1}} x_2^{2^k} = [x_4,\, x_3]^{-2^{k-2}}[x_4,\, x_2]^{-2^{k-4}}[x_4,\, x_3]^{2^{k-2}}[x_4,\, x_1]^{2^{k-6}}[x_4,\, x_5]^{-2^{k-2}} =
$$
$$
[x_4,\, x_2]^{-2^{k-4}}[x_4,\, x_1]^{2^{k-6}}[x_4,\, x_5]^{-2^{k-2}}.
$$

Hence, for $k \geq 13$, we have $x_1^{2^{k-1}} x_2^{2^k} = 1$; therefore, modulo $\mathfrak{g}^5$, we have

$$
\begin{aligned}
(1 - x_1^{2^k})(1 - x_3) + (1 - x_1^{2^{k-1}})(1 - x_3)^2 + (1 - x_2^{2^{k+1}})(1 - x_3) + (1 - x_2^{2^k})(1 - x_3)^2 \equiv \\
(1 - x_1^{2^k} x_2^{2^{k+1}})(1 - x_3) + (1 - x_1^{2^{k-1}} x_2^{2^k})(1 - x_3)^2 \equiv 0. \tag{2.9}
\end{aligned}
$$

Equivalences (2.6) - (2.9) imply that, modulo $\mathfrak{g}^5$,

$$
\begin{aligned}
1 - w &\equiv (1 - x_2^{256})(1 - x_1) - (1 - x_1^{256})(1 - x_2) + \\
&\quad (1 - x_3^{2^{k+1}})(1 - x_2) + (1 - x_3^{2^k})(1 - x_1) \\
&= (1 - x_2^{256} x_3^{2^k})(1 - x_1) + (1 - x_1^{-256} x_3^{2^{k+1}})(1 - x_2) \\
&\equiv (1 - \zeta_1^{16})(1 - x_1) + (1 - \zeta_2^{128})(1 - x_2),
\end{aligned}
$$

where

$$
\begin{aligned}
\zeta_1 &= [x_4,\, x_3]^4 [x_4,\, x_5]^{-4} [x_4,\, x_2]^{-1} [x_4,\, x_2,\, x_2]^2, \\
\zeta_2 &= [x_4,\, x_3][x_4,\, x_5,\, x_5]^{-4}.
\end{aligned}
$$

Hence, modulo $\mathfrak{g}^5$,

$$1 - w \equiv 16(1 - \zeta_1)(1 - x_1) + 128(1 - \zeta_2)(1 - x_2)$$
$$\equiv (1 - \zeta_1)(1 - x_1^{16}) - 8(1 - \zeta_1)(1 - x_1)^2 + (1 - \zeta_2)(1 - x_2^{128})$$
$$- 64(1 - \zeta_2)(1 - x_2)^2$$
$$\equiv (1 - \zeta_1)(1 - x_1^{16}) + (1 - \zeta_2)(1 - x_2^{128}).$$

Since x_1^8, $x_2^{64} \in \gamma_2(G)$, we conclude

$$1 - w \in \Delta(\gamma_2(G))^2 \mathbb{Z}[G] + \mathfrak{g}^5.$$

Furthermore, the detailed analysis of the above construction shows that

$$1 - w \in \Delta([\langle x_4 \rangle^G, G])^2 \mathbb{Z}[G] + \Delta([\langle x_4 \rangle^G, {}_4 G]) \mathbb{Z}[G], \qquad (2.10)$$

since all commutators used in the words c_i, $i = 1, \ldots, 4$, have a nontrivial entry of the generator x_4. $\square$

Theorem 2.15 *There exists a nilpotent group Π of class 4 with $D_6(\Pi) \neq 1$.*

Proof. Consider the 5-generated group G of Theorem 2.14 which is nilpotent of class 3. Let $G_1 = G * \langle t \rangle / \gamma_4(G * \langle t \rangle)$, the quotient of the free product of G with infinite cyclic group with generator t modulo its fourth lower central subgroup. Clearly (2.10) implies that, for the image in G_1 of the element w (we retain the notation of elements of G when naturally viewed as elements of G_1), we have

$$1 - w \in \Delta([\langle x_4 \rangle^{G_1}, G_1])^2 \mathbb{Z}[G_1] + \Delta([\langle x_4 \rangle^{G_1}, {}_4 G_1]) \mathbb{Z}[G_1]. \qquad (2.11)$$

Clearly, $w \notin \langle t \rangle^{G_1}$. Define the quotient $G_2 = G_1 / \langle [x_4, t, x_4] \rangle^{G_1}$. Let f be an automorphism of the free group with basis $\{x_1, x_2, x_3, x_4, x_5, t\}$ defined by

$$x_i \mapsto x_i, \ i = 1, \ldots, 5$$
$$t \mapsto t x_4.$$

It is easy to see that f can be extended to an automorphism of the group G_2. Thus we can consider the semi-direct product $\Pi = G_2 \rtimes \langle x \rangle$. We have the following relations in the group Π:

$$[x_i, x] = 1, \ i = 1, \ldots, 5, \ [t, x] = x_4.$$

Since, in G_2, we have the relations $[x_4, x_i, x_4] = [x_4, t, x_4] = 1$ for all i, the group Π is nilpotent of class 4. The natural map $G_2 \to \Pi$ is a monomorphism; hence the image of the element

$$w = [x_1, x_2^{256}][x_1, x_3^{2^k}][x_2, x_3^{2^{k+1}}]$$

is nontrivial in Π. However, (2.11) implies that

$$1 - w \in \Delta([\langle [t, x] \rangle^\Pi, \Pi])^2 \mathbb{Z}[\Pi] + \Delta([\langle [t, x] \rangle^\Pi, {}_4\Pi]) \mathbb{Z}[\Pi] \subseteq \Delta^6(\Pi).$$

Therefore, $1 \neq w \in D_6(\Pi)$. $\square$

Example 2.16

The reader can check that the constructions given in the proofs of Theorems 2.14 and 2.15 show that for the group Γ given by the following presentation:

$$\langle x_1, x_2, x_3, x_4, x_5, x_6 \mid x_1^8 [x_4, x_6, x_3]^4 [x_4, x_6, x_2],$$
$$x_2^{64} [x_4, x_6, x_3]^{-16} [x_4, x_6, x_1]^{-1} [x_4, x_6, x_5]^{16},$$
$$x_3^{2^k} [x_4, x_7, x_2]^{16} [x_4, x_6, x_1]^{-4}, x_5^{1024} [x_4, x_6, x_2]^{16},$$
$$[x_4, x_6, x_5]^{2048}, [x_4, x_6, x_1]^{16}, [x_4, x_6, x_2]^{128},$$
$$[x_4, x_6, x_1, x_1][x_4, x_6, x_2, x_2]^{-8},$$
$$[x_4, x_6, x_2, x_2]^{-8} [x_4, x_6, x_3, x_3]^{2^{k-3}} \rangle,$$

for $k \geq 13$,

$$D_6(\Gamma) \not\subseteq \gamma_5(\Gamma).$$

The arguments from the proof of Theorem 2.14 imply that the relations of Γ are enough for the element

$$w = [x_1, x_2^{256}][x_1, x_3^{2^k}][x_2, x_3^{2^{k+1}}]$$

to lie in $D_6(\Gamma)$. However, the group Π, constructed in Theorem 2.15 is the natural epimorphic image of Γ, and consequently $w \notin \gamma_5(\Gamma)$.

2.2 Sjögren's Theorem

For every natural number k, let

$$b(k) = \text{ the least common multiple of } 1, 2, \ldots, k,$$

and let

$$c(1) = c(2) = 1, \ c(n) = b(1)^{\binom{n-2}{1}} \ldots b(n-2)^{\binom{n-2}{n-2}}, \ n \geq 3.$$

The most general result known about dimension quotients is the following:

Theorem 2.17 (Sjögren [Sjo79]). *For every group G,*

$$D_n(G)^{c(n)} \subseteq \gamma_n(G), \ n \geq 1.$$

Alternate proofs of Sjögren's theorem have been given by Gupta [Gup87c] and Cliff-Hartley [Cli87]. In case G is a metabelian group, Gupta [Gup87d] has given the following sharper bound for the exponents of dimension quotients:

Theorem 2.18 (Gupta [Gup87d]). *If G is a metabelian group, then*

$$D_n(G)^{2b(1)\cdots b(n-2)} \subseteq \gamma_n(G), \ n \geq 3.$$

Let F be a free group and R a normal subgroup of F. For $k \geq 1$, let

$$R(k) = [\ldots [[R, \underbrace{F, F], \ldots, F]}_{k-1},$$

and

$$\mathfrak{r}(k) = \sum \mathbb{Z}[F]\mathfrak{r}_1\mathfrak{r}_2\ldots\mathfrak{r}_k,$$

where $R_i \in \{R, F\}$ and exactly one $R_i = R$.

The following two lemmas are the key results in the proof of Sjögren's theorem.

Lemma 2.19 *Let $w \in \gamma_n(F)$, $n \geq 2$, be such that $w - 1 \in \mathfrak{f}^{n+1} + \mathfrak{r}(k)$ for some k, $1 \leq k \leq n$. Then $w^{b(k)} - 1 \equiv f_k - 1 \mod \mathfrak{f}^{n+1} + \mathfrak{r}(k+1)$ for some $f_i \in R(k)$.*

Lemma 2.20 *For $n \geq 1$, $F \cap (1 + \mathfrak{f}^{n+1} + \mathfrak{r}(n)) = \gamma_{n+1}(F)R(n)$.*

From Lemmas 2.19 and 2.20 Sjögren's theorem follows by using a process of descent:

Let $H_1 \supseteq H_2 \supseteq \ldots$ and $K_1 \supseteq K_2 \supseteq \ldots$ be two series, and $\{N_{m,l} : 1 \leq m \leq l\}$ a family of normal subgroups of a group G satisfying

$$\left.\begin{array}{l} N_{m,m+1} = H_m K_{m+1}, \\ H_m K_l \subseteq N_{m,l}, \\ N_{m,l+1} \subseteq N_{m,l} \text{ for all } m < l. \end{array}\right\} \tag{2.12}$$

Lemma 2.21 ([Gup87c], [Har82a]). *If n is a positive integer and there exist positive integers $a(l)$ such that*

$$(K_{l+m} \cap N_{l,l+m+1})^{a(l)} \subseteq N_{l+1,l+m+1}H_l, \ l+m \leq n+1,$$

then

$$N_{1,n+2}^{a(1,n+2)} \subseteq H_1 K_{n+2},$$

where

$$a(1, n+2) = \prod_{i=1}^{n} a(i)^{\binom{n}{i}}.$$

2.3 Fourth Dimension Subgroup

An identification of the fourth dimension subgroup is known.

Theorem 2.22 (see [Gup87c], [Tah77b]).
Let G be a nilpotent group of class 3 given by its pre-abelian presentation:

$$\langle x_1, \ldots, x_m \mid x_1^{d(1)}\xi_1, \ldots, x_k^{d(k)}\xi_k, \xi_{k+1}, \ldots, \gamma_4(\langle x_1, \ldots, x_m\rangle)\rangle$$

with $k \leq m$, $d(i) > 0$, $d(k)|\ldots|d(2)|d(1)$ and $\xi_i \in \gamma_2(\langle x_1, \ldots, x_m\rangle)$. Then, the group $D_4(G)$ consists of all elements

$$w = \prod_{1\leq i<j\leq k} [x_i^{d(i)}, x_j]^{a_{ij}}, \, a_{ij} \in \mathbb{Z}, \qquad (2.13)$$

such that

$$d(j)|\binom{d(i)}{2}a_{ij} \, (1 \leq i < j \leq m), \qquad (2.14)$$

and

$$y_l = \prod_{1\leq i<l} x_i^{-d(i)a_{il}} \prod_{l<j\leq k} x_j^{d(l)a_{lj}} \in \gamma_2(G)^{d(l)}\gamma_3(G) \text{ for } 1 \leq l \leq k. \qquad (2.15)$$

Theorem 2.23 (Losey [Los74], Tahara [Tah77a], Sjögren [Sjo79], Passi ([Pas68a], [Pas79])). *For any group G, $D_4(G)/\gamma_4(G)$ has exponent 2.*

In may be noted that every 3-generator group G has the property that $D_4(G) = \gamma_4(G)$ (see [Gup87c]). In Example 2.3 we have a 4-generator group G with 3 relators such that $D_4(G) \neq \gamma_4(G)$. We now show that every 2-relator group G has the property that $D_4(G) = \gamma_4(G)$. Thus, in a sense, Example 2.3 is a minimal example of a group G with $D_4(G) \neq \gamma_4(G)$.

Theorem 2.24 *Let $G = \langle X \mid r_1, r_2\rangle$ be a 2-relator group. Then $D_4(G) = \gamma_4(G)$.*

Proof. Observe that G has a pre-abelian presentation of the form

$$G = \langle x_1, \ldots, x_n, \ldots \mid x_1^{d(1)}\xi_1, x_2^{d(2)}\xi_2, \xi_3, \ldots\rangle$$

with $\xi_i \in \gamma_2\langle x_1, \ldots\rangle$ and $d(2)|d(1)$. Then, modulo $\gamma_4(G)$, the group $D_4(G)$ consists of the elements

$$w = [x_1^{d(1)}, x_2]^{a_{12}},$$

such that

$$d(2)\Big|\binom{d(1)}{2}a_{12},$$

and

$$y_2 = x_1^{-d(1)a_{12}} \in \gamma_2(G)^{d(2)}\gamma_3(G).$$

Therefore, modulo $\gamma_4(G)$, for some $z \in \gamma_2(G)$, we have

$$w = [x_1^{d(1)a_{12}},\, x_2] = [y_2,\, x_2] = [z^{d(2)},\, x_2] = [z,\, x_2^{d(2)}] = 1.$$

$\square$

Theorem 2.25 [Gup92] *For any group G, $[D_4(G), G] = \gamma_5(G)$.*

Proof. In view of Theorem 2.23, it suffices to prove the statement for finite 2-groups. Let G be a finite 2-group, generated by elements $x_1, \ldots, x_k$ such that $x_i^{d(i)} \in \gamma_2(G)$ for some $d(i) = 2^{\alpha_i}$, with ordering $\alpha_1 \geq \alpha_2 \geq \cdots \geq \alpha_k \geq 1$. Let $w \in D_4(G)$. Theorem 2.22 implies that modulo $\gamma_4(G)$, w can be expressed in the form (2.13), such that the conditions (2.14) and (2.15) are satisfied. Let h be arbitrary element of G. Then we have the following equivalences modulo $\gamma_5(G)$:

$$[w,\, h] = \Big[\prod_{1 \leq i < j \leq k}[x_i^{d(i)},\, x_j]^{a_{ij}},\, h\Big] \equiv \prod_{1 \leq i < j \leq k}[x_i^{d(i)},\, x_j,\, h]^{a_{ij}} \equiv$$

$$\prod_{1 \leq i < j \leq k}[x_i^{d(i)},\, h,\, x_j]^{a_{ij}} \prod_{1 \leq i < j \leq k}[x_j,\, h,\, x_i^{d(i)}]^{-a_{ij}} \quad \mathrm{mod}\ \gamma_5(G). \quad (2.16)$$

Condition (2.14) implies that

$$\prod_{1 \leq i < j \leq k}[x_i^{d(i)},\, h,\, x_j]^{a_{ij}} \equiv \prod_{1 \leq i < j \leq k}[x_i,\, h,\, x_j^{d(i)}]^{a_{ij}} \equiv \prod_{1 \leq i < j \leq k}[x_i,\, h,\, x_j^{d(i)a_{ij}}] \quad \mathrm{mod}\ \gamma_5(G);$$

$$\prod_{1 \leq i < j \leq k}[x_j,\, h,\, x_i^{d(i)}]^{-a_{ij}} \equiv \prod_{1 \leq i < j \leq k}[x_j,\, h,\, x_i^{-d(i)a_{ij}}] \quad \mathrm{mod}\ \gamma_5(G).$$

Therefore, by condition (2.15), we have

$$[w,\, h] \equiv \prod_{1 \leq i < j \leq k}[x_i,\, h,\, x_j^{d(i)a_{ij}}] \prod_{1 \leq i < j \leq k}[x_j,\, h,\, x_i^{-d(i)a_{ij}}] \equiv$$

$$\prod_{1 \leq t \leq k}\Big[x_t,\, h,\, \prod_{1 \leq s < t}x_s^{-d(s)a_{st}} \prod_{t < r \leq k}x_r^{-d(t)a_{tr}}\Big] \equiv$$

$$\prod_{1 \leq t \leq k}[x_t,\, h,\, y_t] \equiv 1 \quad \mathrm{mod}\ \gamma_5(G). \quad \square$$

An extensive analysis of the counter-examples to the equality of the fourth dimension subgroup with the fourth lower central subgroup has been carried out by M. Hartl [Har].

Theorem 2.26 (Hartl [Har]; see also [Har98, Theorem 7.2.6, p. 72]). *Let $A = \mathbb{Z}_{2^{\beta_1}} \oplus \mathbb{Z}_{2^{\beta_2}} \oplus \mathbb{Z}_{2^{\beta_3}} \oplus \mathbb{Z}_{2^{\beta_4}}$ with $\beta_1 \leq \beta_2 \leq \beta_3 \leq \beta_4$ and $n \geq 1$. Then there exists a finite nilpotent group G of class 3 with $G_{ab} \simeq A$, such that $D_4(G) \neq 1$ and $[v, x] = 1$ for every $v \in \gamma_2(G)$, $x \in G$, such that $x\gamma_2(G)$ is a generator of the summand $\mathbb{Z}_{2^{\beta_4}}$ in G_{ab} if and only if the following conditions hold:*

$$(i) \ \beta_1, \beta_2 - \beta_1, \beta_3 - \beta_2 \geq 2,$$
$$(ii) \ \beta_3 > n > max\{\beta_2, \beta_3 - \beta_1\}.$$

Moreover, under conditions (i) and (ii), the group G can be chosen to be of order $2^{4\beta_1 + 3\beta_2 + 2\beta_3 + \beta_4 + n + 1}$.

2.4 Fifth Dimension Subgroup

The structure of the fifth dimension subgroup has been described by Tahara [Tah81], and it has been further shown that $D_5^6(G) \subseteq \gamma_5(G)$:

Theorem 2.27 (Tahara [Tah81]). *For every group G, $D_5(G)^6 \subseteq \gamma_5(G)$.*

Analysis of Tahara's description of the fifth dimension subgroup leads us to the following result.

Theorem 2.28 *For every group G, $D_5(G)^2 \subseteq \delta_2(G)\gamma_5(G)$.*

Let G be a finite group of class 4. Choose the elements

$$\{x_{1i} \in G \setminus \gamma_2(G)\}_{i=1,\ldots,s},$$
$$\{x_{2i} \in \gamma_2(G) \setminus \gamma_3(G)\}_{i=1,\ldots,t},$$
$$\{x_{3i} \in \gamma_3(G) \setminus \gamma_4(G)\}_{i=1,\ldots,u},$$
$$\{x_{4i} \in \gamma_4(G)\}_{i=1,\ldots,v}$$

to be such that $\{x_{li}\gamma_{l+1}(G)\}$ forms a basis of $\gamma_l(G)/\gamma_{l+1}(G)$. Let $d(i)$ be the order of $x_{1i}\gamma_2(G)$ in $G/\gamma_2(G)$, $e(i)$ the order of $x_{2i}\gamma_3(G)$ in $\gamma_2(G)/\gamma_3(G)$, $f(i)$ the order of $x_{3i}\gamma_4(G)$ in $\gamma_3(G)/\gamma_4(G)$. We then have

$$x_{1i}^{d(i)} = \prod_{1 \leq j \leq t} x_{2j}^{b_{ij}} \prod_{1 \leq j \leq u} x_{3j}^{c_{ij}} y_{4i}, \ y_{4i} \in \gamma_4(G), \ 1 \leq i \leq s;$$

$$x_{2i}^{e(i)} = \prod_{1 \leq j \leq t} x_{3i}^{d_{ij}} y'_{4i}, \ y'_{4i} \in \gamma_4(G), \ 1 \leq i \leq t;$$

$$x_{3i}^{f(i)} = \prod_{1 \leq j \leq v} x_{4j}^{f_{ij}}, \ 1 \leq i \leq u;$$

$$[x_{1i}^{d(i)}, x_{1j}] = \prod_{1 \leq k \leq u} x_{3k}^{\alpha_k^{(ij)}} y''_{ij}, \ y''_{ij} \in \gamma_4(G), \ 1 \leq i < j \leq s.$$

We choose the element x_{ij} in such a way that $d(i)|d(i+1)$, $e(i)|e(i+1)$, $f(i)|f(i+1)$.

Theorem 2.29 (Tahara [Tah81]). *With the above notations, the subgroup $D_5(G)$ is equal to the subgroup generated by the elements*

$$\prod_{1\le i\le j\le s} [x_{1i}^{u_{ij}d(j)}, x_{1j}] \prod_{1\le i\le s,\ 1\le k\le t k<l} \prod [x_{2l}, x_{2k}]^{b_{il}v_{ik}} \prod_{1\le i\le j\le k\le s} [x_{1i}^{d(i)}, x_{1j}, x_{1k}]^{w_{ijk}},$$

$$(2.17)$$

where

$$u_{ij},\ 1\le i<j\le s,$$
$$v_{ik},\ 1\le i\le s,\ 1\le k\le t,$$
$$v'_{ik},\ 1\le i\le s, 1\le k\le t,$$
$$w_{ijk},\ 1\le i\le j\le k\le s,$$
$$w'_{ijk},\ 1\le i<j\le k\le s,$$
$$w''_{ijk},\ 1\le i\le j<k\le s,$$

are integers satisfying the following conditions:

$$w_{iii}=0,\ 1\le i\le s; \tag{2.18}$$

$$u_{ij}\frac{d(j)}{d(i)}\binom{d(i)}{2}+w_{iij}d(i)+w''_{iij}d(j)=0,\ 1\le i<j\le s; \tag{2.19}$$

$$-u_{ij}\binom{d(j)}{2}+w_{ijj}d(i)+w'_{ijj}d(j)=0,\ 1\le i<j\le s; \tag{2.20}$$

$$w_{ijk}d(i)+w'_{ijk}d(j)+w''_{ijk}d(k)=0,\ 1\le i<j<k\le s; \tag{2.21}$$

$$\sum_{i<h}u_{ih}b_{hk}-\sum_{h<i}u_{hi}\frac{d(i)}{d(h)}b_{hk}+v_{ik}d(i)+v'_{ik}e(k)=0,\ 1\le i\le s,\ 1\le k\le t; \tag{2.22}$$

$$u_{ij}\frac{d(j)}{d(i)}\binom{d(i)}{3}+w_{iij}\binom{d(i)}{2}\equiv 0 \mod d(i),\ 1\le i<j\le s; \tag{2.23}$$

$$w_{iij}\binom{d(i)}{2}+w''_{iij}\binom{d(j)}{2}\equiv 0 \mod d(i),\ 1\le i<j\le s; \tag{2.24}$$

$$-u_{ij}\binom{d(j)}{3}+w'_{ijj}\binom{d(j)}{2}\equiv 0 \mod d(i),\ 1\le i<j\le s; \tag{2.25}$$

$$w_{ijk}\binom{d(i)}{2},\ w'_{ijk}\binom{d(j)}{2},\ w''_{ijk}\binom{d(k)}{2}\equiv 0 \mod d(i),\ 1\le i<j<k\le t; \tag{2.26}$$

$$v_{ik}\binom{d(i)}{2} - \sum_{h \le i} w_{hii} b_{hk} - \sum_{i < h} w''_{iih} b_{hk} \equiv 0 \quad \mod (d(i),\, e(k)),$$

$$1 \le i \le s,\ 1 \le k \le t; \quad (2.27)$$

$$\sum_{h \le i} w_{hij} b_{hk} + \sum_{i < h \le j} w'_{ihj} b_{hk} + \sum_{j < h} w''_{ijh} b_{hk} \equiv 0 \quad \mod (d(i),\, e(k)),$$

$$1 \le i < j \le s,\ 1 \le k \le t; \quad (2.28)$$

$$-\sum_{h < i} u_{hi} \frac{d(i)}{d(h)} \alpha_l^{(hi)} + \sum_{i < h} u_{ih} c_{hl} - \sum_{h < i} u_{hi} \frac{d(i)}{d(h)} c_{hl} - \sum_{k} v'_{ik} d_{kl} -$$

$$\sum_{g \le i \le h} w_{gih} \alpha_l^{(gh)} - \sum_{g \le h \le i} w_{ghi} \alpha_l^{(gh)} - \sum_{i < g \le h} w'_{igh} \alpha_l^{(gh)} \equiv 0 \quad \mod (d(i),\, f(l)),$$

$$1 \le i \le s,\ 1 \le l \le s; \quad (2.29)$$

$$\sum_{i} v_{ik} b_{ik} \equiv 0 \quad \mod e(k),\ 1 \le k \le t; \quad (2.30)$$

$$\sum_{i} v_{ik} b_{il} + \sum_{i} v_{il} b_{ik} \equiv 0 \quad \mod e(k),\ 1 \le k < l \le t. \quad (2.31)$$

Proof of Theorem 2.28. Standard reduction argument shows that it is enough to consider finite groups. Commutator identities (see Chapter 1, 1.1) and condition (2.25) imply

$$[x_{1i}^{u_{ij}d(j)},\, x_{1j}] = [x_{1i},\, x_{1j}^{u_{ij}d(j)}][x_{1i},\, x_{1j},\, x_{1j}]^{-u_{ij}\binom{d(j)}{2}}[x_{1j},\, x_{1i},\, x_{1i}]^{-u_{ij}\binom{d(j)}{2}}.$$

$$[x_{1i},\, x_{1j},\, x_{1j},\, x_{1j}]^{-u_{ij}\binom{d(j)}{3}}[x_{1j},\, x_{1i},\, x_{1i},\, x_{1i}]^{-u_{ij}\binom{d(j)}{3}} =$$

$$[x_{1i},\, x_{1j}^{u_{ij}d(j)}][x_{1i},\, x_{1j},\, x_{1j}]^{-u_{ij}\binom{d(j)}{2}}[x_{1j},\, x_{1i},\, x_{1i}]^{-u_{ij}\binom{d(j)}{2}}.$$

$$[x_{1i},\, x_{1j},\, x_{1j},\, x_{1j}]^{-w'_{ijj}\binom{d(j)}{2}}[x_{1j},\, x_{1i},\, x_{1i},\, x_{1i}]^{-w'_{ijj}\binom{d(j)}{2}}. \quad (2.32)$$

Observe that

$$[x_{1j},\, x_{1i},\, x_{1i},\, x_{1i}]^{-2w'_{ijj}\binom{d(j)}{2}} = 1,$$

$$[x_{1j},\, x_{1i},\, x_{1i},\, x_{1i}]^{-3w'_{ijj}\binom{d(j)}{2}} = [x_{1j},\, x_{1i},\, x_{1i},\, x_{1i}]^{-3u_{ij}\binom{d(j)}{3}}$$

$$= [x_{1j},\, x_{1i},\, x_{1i},\, x_{1i}]^{-u_{ij}d(j)\frac{(d(j)-1)(d(j)-2)}{2}} = 1$$

Therefore,

$$[x_{1j},\, x_{1i},\, x_{1i},\, x_{1i}]^{w'_{ijj}\binom{d(j)}{2}} = 1.$$

Analogically,

$$[x_{1i},\, x_{1j},\, x_{1j},\, x_{1j}]^{w'_{ijj}\binom{d(j)}{2}} = 1.$$

$$\prod_{i<j}[x_{1i}^{u_{ij}d(j)},\, x_{1j}]^2 = \prod_j(\prod_{i<j}[x_{1i}^{u_{ij}d(j)},\, x_{1j}]\prod_{j<k}[x_{1j}^{u_{jk}d(k)},\, x_{1k}]) =$$

$$\prod_j([\prod_{i<j}x_{1i}^{u_{ij}d(j)}\prod_{j<k}x_{1k}^{-u_{jk}d(k)},\, x_{1j}])\cdot B, \quad (2.33)$$

where

$$B := \prod_{j<k}([x_{1j},\, x_{1k},\, x_{1k}]^{-u_{jk}\binom{d(k)}{2}}[x_{1k},\, x_{1j},\, x_{1j}]^{-u_{jk}\binom{d(k)}{2}}).$$

Also we have

$$[\prod_{i<j}x_{1i}^{u_{ij}d(j)}\prod_{j<k}x_{1k}^{-u_{jk}d(k)},\, x_{1j}] =$$

$$[\prod_l x_{2l}^{\sum_{i<j}u_{ij}\frac{d(j)}{d(i)}b_{il}-\sum_{j<k}u_{jk}b_{kl}}\prod_q x_{3q}^{\sum_{i<j}u_{ij}\frac{d(j)}{d(i)}c_{iq}-\sum_{j<k}u_{jk}c_{kq}},\, x_{1j}] =$$

$$[\prod_l x_{2l}^{v_{jl}d(j)+v'_{jl}e(l)},\, x_{1j}][\prod_q x_{3q}^{A_{qj}},\, x_{1j}] \qquad\qquad (2.34)$$

$$\prod_l[x_{2l},\, x_{1j},\, x_{1j}]^{-v_{jl}\binom{d(j)}{2}}[\prod_l x_{2l}^{v_{jl}},\, x_{1j}^{d(j)}][\prod_l x_{2l}^{v'_{jl}e(l)},\, x_{1j}][\prod_q x_{3q}^{A_{qj}},\, x_{1j}],$$

where $A_{qj} = \sum_{i<j}u_{ij}\frac{d(j)}{d(i)}c_{iq} - \sum_{j<k}u_{jk}c_{kq}$.
   The condition (2.27) implies that

$$\prod_l[x_{2l},\, x_{1j},\, x_{1j}]^{-v_{jl}\binom{d(j)}{2}} =$$

$$\prod_l[x_{2l},\, x_{1j},\, x_{1j}]^{-\sum_{i\le j}w_{ijj}b_{il}-\sum_{j\le k}w''_{jjk}b_{kl}} = \qquad (2.35)$$

$$\prod_{i\le j}[x_{1i},\, x_{1j},\, x_{1j}]^{-w_{ijj}d(i)}[x_{1j},\, x_{1i},\, x_{1i}]^{-w''_{iij}d(j)}$$

The condition (2.31) implies that

$$C := \prod_j[\prod_l x_{2l}^{v_{jl}},\, x_{1j}^{d(j)}] =$$

$$\prod_j(\prod_{l'>l}[x_{2l},\, x_{2l'}]^{v_{jl}b_{jl'}}\prod_{l'<l}[x_{2l},\, x_{2l'}]^{v_{jl}b_{jl'}}) = \qquad (2.36)$$

$$\prod_{l'>l}[x_{2l},\, x_{2l'}]^{\sum_j v_{jl}b_{jl'}}\prod_{l>l'}[x_{2l},\, x_{2l'}]^{\sum_j v_{jl}b_{jl'}} =$$

$$\prod_{l'>l}[x_{2l},\, x_{2l'}]^{\sum_j v_{jl}b_{jl'}}\prod_{l>l'}[x_{2l},\, x_{2l'}]^{-\sum_j v_{jl}b_{jk}} = \prod_{l'>l}[x_{2l},\, x_{2l'}]^{2\sum_j v_{jl}b_{jl'}}.$$

The condition (2.29) implies that

$$D := \prod_j [\prod_l x_{2l}^{v'_{jl}e(l)}, x_{1j}][\prod_q x_{3q}^{A_{qj}}, x_{1j}] = \prod_{1 \le q \le u, 1 \le j \le s} [x_{3q}, x_{1j}]^{\sum_l v'_{jl} d_{lq} + A_{qj}} =$$

$$\prod_{\substack{1 \le j \le s, \\ 1 \le q \le u}} [x_{3q}, x_{1j}]^{-\sum_{h<j} u_{hj} \frac{d(j)}{d(h)} \alpha_q^{(hj)} - \sum_{g \le j \le h} w_{gjh} \alpha_q^{(gh)} - \sum_{g \le h \le j} w_{ghj} \alpha_q^{(gh)} - \sum_{j<g \le h} w'_{jgh} \alpha_q^{(gh)}} =$$

$$\prod_{h<j} [x_{1h}^{d(j)}, x_{1j}, x_{1j}]^{-u_{hj}} \prod_{g \le j \le h} [x_{1g}^{d(g)}, x_{1h}, x_{1j}]^{-w_{gjh}}. \tag{2.37}$$

$$\prod_{g<h \le j} [x_{1g}^{d(g)}, x_{1h}, x_{1j}]^{-w_{ghj}} \prod_{j<g \le h} [x_{1g}^{d(g)}, x_{1h}, x_{1j}]^{-w'_{jgh}}.$$

Since $[x_{1i}^{d(i)}, x_{1k}, x_{1j}][x_{1k}, x_{1j}, x_{1i}^{d(i)}][x_{1j}, x_{1i}^{d(i)}, x_{1k}] = 1$, we have

$$\prod_{i \le j \le k} [x_{1i}^{d(i)}, x_{1k}, x_{1j}]^{-w_{ijk}} = \prod_{i \le j \le k} [x_{1k}, x_{1j}, x_{1i}^{d(i)}]^{w_{ijk}} \prod_{i \le j \le k} [x_{1i}^{d(i)}, x_{1j}, x_{1k}]^{-w_{ijk}}.$$

We change the subscripts g, h in (2.37) by appropriate subscripts i, j, k. The conditions (2.20) and (2.26) then imply

$$D = \prod_{i<j} [x_{1i}, x_{1j}, x_{1j}]^{-u_{ij}d(j)} \prod_{i \le j \le k} [x_{1k}, x_{1j}, x_{1i}^{d(i)}]^{w_{ijk}}.$$

$$\prod_{i \le j \le k} [x_{1i}^{d(i)}, x_{1j}, x_{1k}]^{-2w_{ijk}} \prod_{j<i \le k} [x_{1i}^{d(i)}, x_{1k}, x_{1j}]^{-w'_{jik}} = \tag{2.38}$$

$$\prod_{i<j} [x_{1i}, x_{1j}, x_{1j}^{d(j)}]^{-u_{ij}} \prod_{i \le j \le k} [x_{1k}, x_{1j}, x_{1i}^{d(i)}]^{w_{ijk}}.$$

$$\prod_{i \le j \le k} [x_{1i}^{d(i)}, x_{1j}, x_{1k}]^{-2w_{ijk}} \prod_{j<i \le k} [x_{1k}, x_{1i}, x_{1j}^{d(i)}]^{-w'_{jik}}.$$

Hence

$$g^2 = B \cdot C \cdot D \cdot \prod_{i \le j} [x_{1i}, x_{1j}, x_{1j}]^{-w_{ijj}d(i)} [x_{1j}, x_{1i}, x_{1i}]^{-w''_{iij}d(j)}.$$

$$\prod_{1 \le i \le s, \, 1 \le k \le t} \prod_{k<l} [x_{2l}, x_{2k}]^{2b_{il}v_{ik}} \prod_{1 \le i \le j \le k \le s} [x_{1i}^{d(i)}, x_{1j}, x_{1k}]^{2w_{ijk}}$$

$$= \prod_{j<k} ([x_{1j}, x_{1k}, x_{1k}]^{-u_{jk}\binom{d(k)}{2}} [x_{1k}, x_{1j}, x_{1j}]^{-u_{jk}\binom{d(k)}{2}}).$$

$$\prod_{i \le j} [x_{1i}, x_{1j}, x_{1j}]^{-w_{ijj}d(i)} [x_{1j}, x_{1i}, x_{1i}]^{-w''_{iij}d(j)}.$$

$$\prod_{i<j} [x_{1i}, x_{1j}, x_{1j}^{d(j)}]^{-u_{ij}} \prod_{i \le j \le k} [x_{1k}, x_{1j}, x_{1i}^{d(i)}]^{w_{ijk}} \prod_{i<j \le k} [x_{1k}, x_{1j}, x_{1i}^{d(j)}]^{w'_{ijk}}$$

$$\tag{2.39}$$

$$= \prod_{j<k}([x_{1j}, x_{1k}, x_{1k}]^{w'_{jkk}d(k)}[x_{1k}, x_{1j}, x_{1j}]^{w_{jjk}d(j)}).$$

$$\prod_{i\leq j\leq k}[x_{1k}, x_{1j}, x_{1i}^{d(i)}]^{w_{ijk}} \prod_{i<j\leq k}[x_{1k}, x_{1j}, x_{1i}^{d(j)}]^{w'_{ijk}}$$

$$= \prod_{j<k}([x_{1j}, x_{1k}, x_{1k}^{d(k)}]^{w'_{jkk}}[x_{1k}, x_{1j}, x_{1j}^{d(j)}]^{w_{jjk}}[x_{1k}, x_{1j}, x_{1j}, x_{1j}]^{-w_{jjk}\binom{d(j)}{2}}).$$

$$\prod_{i\leq j\leq k}[x_{1k}, x_{1j}, x_{1i}^{d(i)}]^{w_{ijk}} \prod_{i<j\leq k}[x_{1k}, x_{1j}, x_{1i}^{d(j)}]^{w'_{ijk}}$$

$$= \prod_{j<k}([x_{1j}, x_{1k}, x_{1k}^{d(k)}]^{w'_{jkk}}[x_{1k}, x_{1j}, x_{1j}^{d(j)}]^{w_{jjk}}) \prod_{i\leq j\leq k}[x_{1k}, x_{1j}, x_{1i}^{d(i)}]^{w_{ijk}}.$$

$$\prod_{i<j\leq k}[x_{1k}, x_{1j}, x_{1i}^{d(j)}]^{w'_{ijk}},$$

since $[x_{1k}, x_{1j}, x_{1j}, x_{1j}]^{2w_{jjk}\binom{d(j)}{2}} = 1$, and

$$[x_{1k}, x_{1j}, x_{1j}, x_{1j}]^{3w_{jjk}\binom{d(j)}{2}} = [x_{1k}, x_{1j}, x_{1j}, x_{1j}]^{-3u_{jk}\frac{d(k)}{d(j)}\binom{d(j)}{3}} = 1$$

by (2.23). Consequently, $g^2 \in \delta_2(G)$, and the proof is complete. $\square$

Proof of Theorem 2.27. Multiplying $[x_{1i}, x_{1j}, x_{2k}]$ by left hand side of (2.28) and taking the product over all $i < j$ and k, we obtain the following:

$$1= \prod_{i\leq j<k}[x_{1j}, x_{1k}, x_{1i}^{d(i)}]^{w_{ijk}} \prod_{i<j\leq k}[x_{1i}, x_{1k}, x_{1j}^{d(j)}]^{w'_{ijk}} \prod_{i<j<k}[x_{1i}, x_{1j}, x_{1k}^{d(k)}]^{w''_{ijk}}$$

$$= (\prod_{i<j}[x_{1i}, x_{1j}, x_{1i}]^{w_{iij}d(i)}[x_{1i}, x_{1j}, x_{1j}]^{w'_{ijj}d(j)}).$$

$$\prod_{i<j<k}[x_{1j}, x_{1k}, x_{1i}]^{w_{ijk}d(i)} \prod_{i<j<k}[x_{1i}, x_{1k}, x_{1j}^{d(j)}]^{w'_{ijk}}.$$

$$\prod_{i<j<k}[x_{1j}, x_{1i}, x_{1k}]^{w'_{ijk}d(j)+w_{ijk}d(i)}$$

$$= (\prod_{i<j}[x_{1i}, x_{1j}, x_{1i}]^{w_{iij}d(i)}[x_{1i}, x_{1j}, x_{1j}]^{w'_{ijj}d(j)}).$$

$$\prod_{i<j<k}[x_{1j}, x_{1k}, x_{1i}]^{w_{ijk}d(i)} \prod_{i<j<k}[x_{1j}, x_{1i}, x_{1k}]^{w_{ijk}d(i)} \prod_{i<j<k}[x_{1j}, x_{1k}, x_{1i}]^{w'_{ijk}d(j)}.$$

Therefore,

$$\prod_{i<j<k} [x_{1k}, x_{1j}, x_{1i}]^{w'_{ijk}d(j)}$$

$$= (\prod_{i<j}[x_{1i}, x_{1j}, x_{1i}]^{w_{iij}d(i)}[x_{1i}, x_{1j}, x_{1j}]^{w'_{ijj}d(j)}) \prod_{i<j<k} [x_{1j}, x_{1k}, x_{1i}]^{w_{ijk}d(i)}.$$

$$\prod_{i<j<k} [x_{1j}, x_{1i}, x_{1k}]^{w_{ijk}d(i)}$$

Now consider the element g^2 given in (2.39):

$$g^2 = \prod_{j<k}([x_{1j}, x_{1k}, x_{1k}^{d(k)}]^{w'_{jkk}}[x_{1k}, x_{1j}, x_{1j}^{d(j)}]^{w_{jjk}}).$$

$$\prod_{i\leq j\leq k} [x_{1k}, x_{1j}, x_{1i}^{d(i)}]^{w_{ijk}} \prod_{i<j\leq k} [x_{1k}, x_{1j}, x_{1i}^{d(j)}]^{w'_{ijk}}$$

$$= \prod_{j<k}([x_{1j}, x_{1k}, x_{1k}^{d(k)}]^{w'_{jkk}}[x_{1k}, x_{1j}, x_{1j}^{d(j)}]^{w_{jjk}}) \prod_{i\leq j\leq k} [x_{1k}, x_{1j}, x_{1i}^{d(i)}]^{w_{ijk}}.$$

$$(\prod_{i<j}[x_{1i}, x_{1j}, x_{1i}]^{w_{iij}d(i)}[x_{1i}, x_{1j}, x_{1j}]^{w'_{ijj}d(j)}). \tag{2.40}$$

$$\prod_{i<j<k} [x_{1j}, x_{1k}, x_{1i}]^{w_{ijk}d(i)} \prod_{i<j<k} [x_{1j}, x_{1i}, x_{1k}]^{w_{ijk}d(i)}$$

$$= (\prod_{i<j}[x_{1j}, x_{1i}, x_{1i}]^{w_{iij}d(i)}[x_{1i}, x_{1j}, x_{1j}]^{2w'_{ijj}d(j)}) \prod_{i<j<k} [x_{1j}, x_{1i}, x_{1k}]^{w_{ijk}d(i)}$$

Analogously, multiplying $[x_{2k}, x_{1j}, x_{1i}]$ by left hand side of (2.28) and taking the product over all $i < j$ and k, we obtain the following:

$$1= \prod_{i\leq j<k} [x_{1i}^{d(i)}, x_{1k}, x_{1j}]^{w_{ijk}} \prod_{i<j<k} [x_{1j}^{d(j)}, x_{1k}, x_{1i}]^{w'_{ijk}} \prod_{i<j<k} [x_{1k}^{d(k)}, x_{1j}, x_{1i}]^{w''_{ijk}}$$

$$= \prod_{i\leq j<k} [x_{1i}^{d(i)}, x_{1k}, x_{1j}]^{w_{ijk}} \prod_{i<j<k} [x_{1j}, x_{1k}, x_{1i}]^{w'_{ijk}d(j)}.$$

$$\prod_{i<j<k} [x_{1j}, x_{1k}, x_{1i}]^{w_{ijk}d(i)+w'_{ijk}d(j)}$$

$$= \prod_{i<j}[x_{1i}^{d(i)}, x_{1j}, x_{1i}]^{w_{iij}} \prod_{i<j<k} [x_{1i}, x_{1k}, x_{1j}]^{w_{ijk}d(i)}.$$

$$\prod_{i<j<k} [x_{1j}, x_{1k}, x_{1i}]^{w_{ijk}d(i)} \prod_{i<j<k} [x_{1j}, x_{1k}, x_{1i}]^{2w'_{ijk}d(j)}$$

Hence

$$\prod_{i<j<k} [x_{1k}, x_{1j}, x_{1i}]^{2w'_{ijk}d(j)}$$
$$= \prod_{i<j}[x_{1i}^{d(i)}, x_{1j}, x_{1i}]^{w_{iij}} \prod_{i<j<k} [x_{1i}, x_{1k}, x_{1j}]^{w_{ijk}d(i)} \prod_{i<j<k} [x_{1j}, x_{1k}, x_{1i}]^{w_{ijk}d(i)}.$$

Now consider the element g^4 obtained by squaring the element given in (2.39):

$$g^4 = \prod_{j<k}([x_{1j}, x_{1k}, x_{1k}^{d(k)}]^{2w'_{jkk}}[x_{1k}, x_{1j}, x_{1j}^{d(j)}]^{2w_{jjk}}).$$
$$\prod_{i\le j\le k} [x_{1k}, x_{1j}, x_{1i}^{d(i)}]^{2w_{ijk}} \prod_{i<j<k} [x_{1k}, x_{1j}, x_{1i}^{d(j)}]^{2w'_{ijk}}$$
$$= \prod_{j<k}([x_{1j}, x_{1k}, x_{1k}^{d(k)}]^{2w'_{jkk}}[x_{1k}, x_{1j}, x_{1j}^{d(j)}]^{4w_{jjk}}) \prod_{i<j<k} [x_{1k}, x_{1j}, x_{1i}^{d(i)}]^{2w_{ijk}}.$$
$$\prod_{i<j}[x_{1i}^{d(i)}, x_{1j}, x_{1i}]^{w_{iij}} \prod_{i<j<k} [x_{1i}, x_{1k}, x_{1j}]^{w_{ijk}d(i)} \prod_{i<j<k} [x_{1j}, x_{1k}, x_{1i}]^{w_{ijk}d(i)}$$
$$= \prod_{i<j}([x_{1i}, x_{1j}, x_{1j}^{d(j)}]^{2w'_{ijj}}[x_{1j}, x_{1i}, x_{1i}^{d(i)}]^{3w_{iij}}) \prod_{i<j<k} [x_{1i}, x_{1j}, x_{1k}]^{w_{ijk}d(i)}.$$

$$(2.41)$$

Multiplying (2.40) and (2.41), we obtain

$$g^6 = \prod_{i<j}([x_{1i}, x_{1j}, x_{1j}^{d(j)}]^{4w'_{ijj}}[x_{1j}, x_{1i}, x_{1i}^{d(i)}]^{4w_{iij}}).$$

The condition (2.27) implies that

$$\prod_{i<j}[x_{1i}^{d(i)}, x_{1j}, x_{1j}]^{w_{ijj}}[x_{1j}^{d(j)}, x_{1i}, x_{1i}]^{w''_{iij}} = \prod_{i,k} v_{ik}\binom{d(i)}{2}[x_{2k}, x_{1i}, x_{1i}].$$

Conditions (2.19), (2.20), (2.22) imply that

$$1 = \prod_{i<j}[x_{1i}, x_{1j}, x_{1j}]^{2w_{ijj}d(i)}[x_{1j}, x_{1i}, x_{1i}]^{2w''_{iij}d(j)}$$
$$= \prod_{i<j}[x_{1i}, x_{1j}, x_{1j}]^{-2w'_{ijj}d(j)-u_{ij}d(j)}[x_{1j}, x_{1i}, x_{1i}]^{-2w_{iij}d(i)+u_{ij}d(j)}$$
$$= (\prod_{i<j}[x_{1i}, x_{1j}, x_{1j}]^{-2w'_{ijj}d(j)}[x_{1j}, x_{1i}, x_{1i}]^{-2w_{iij}d(i)}).$$
$$\prod_{i<j}[x_{1i}, x_{1j}, x_{1j}]^{-u_{ij}d(j)}[x_{1i}, x_{1j}, x_{1i}]^{-u_{ij}d(j)}$$
$$= \prod_{i<j}[x_{1i}, x_{1j}, x_{1j}]^{-2w'_{ijj}d(j)}[x_{1j}, x_{1i}, x_{1i}]^{-2w_{iij}d(i)}.$$

Hence $g^6 = 1$. $\square$

Problem 2.30 *If G is a nilpotent group of class three, then must $D_5(G)$ be trivial?*

We illustrate the complexity of the above problem by verifying it for a group, without dimension property, considered by Gupta-Passi ([Gup87c], p. 76). Let us recall the construction of this group.

Let F be the free group with basis x_1, x_2, x_3, x_4 and let R be the normal subgroup generated by

$$r_1 = x_4^{64}[x_4, x_3]^{32}, \ r_2 = x_3^{64}[x_4, x_2]^{-4}[x_4, x_1]^{-2}, \ r_3 = x_2^{16}[x_4, x_3]^4[x_4, x_1]^{-1},$$

$$r_4 = x_1^4[x_4, x_3]^2[x_4, x_2], \ r_5 = [x_4, x_3]^{64}[x_4, x_3, x_3]^{32},$$

$$r_6 = [x_4, x_2]^{16}[x_4, x_2, x_2]^8, \ r_7 = [x_4, x_1]^4[x_4, x_1, x_1]^2,$$

$$r_8 = [x_3, x_2]^{16}[x_4, x_2, x_2]^{-1}, \ r_9 = [x_3, x_1]^4[x_4, x_3, x_3]^{-2},$$

$$r_{10} = [x_2, x_1]^4[x_4, x_2, x_2]^{-1}, \ r_{11} = [x_4, x_3, x_3]^4[x_4, x_2, x_2]^{-1},$$

$$r_{12} = [x_4, x_2, x_2]^4[x_4, x_1, x_1]^{-1}, \ r_{13} = [x_4, x_1, x_1]^4,$$

$\gamma_4(F)$, and all commutators $[x_i, x_j, x_k](1 \le i, j, k \le 4)$ which do not belong to

$$\langle [x_4, x_1, x_1], [x_4, x_2, x_2], [x_4, x_3, x_3]\rangle\gamma_4(F).$$

Then the group

$$G := F/R \tag{2.42}$$

is a finite 2-group of class 3 with the non-identity element

$$w_0 = [x_3^{64}, x_2]^2[x_3^{64}, x_1][x_2^{16}, x_1]^2 R$$

in $D_4(G)$.

With the notations of Theorem 2.29, we choose

$$x_{11} = x_1, \ x_{12} = x_2, \ x_{13} = x_3, \ x_{14} = x_4,$$

$$x_{21} = [x_1, x_2], \ x_{22} = [x_1, x_3], \ x_{23} = [x_1, x_4],$$

$$x_{24} = [x_2, x_3], \ x_{25} = [x_2, x_4], \ x_{26} = [x_3, x_4],$$

$$x_{31} = [x_4, x_3, x_3].$$

For this group we have the following constants:

$$d(1) = 4, \ d(2) = 16, \ d(3) = 64, \ d(4) = 64,$$

$$e(1) = 4, \ e(2) = 4, \ e(3) = 4, \ e(4) = 16, \ e(5) = 16, \ e(6) = 64,$$

$$b_{15} = 1, \ b_{16} = 2, \ b_{23} = -1, \ b_{26} = 4,$$

$$b_{33} = 2, \ b_{35} = -4, \ b_{46} = 32, \text{ all other } b_{ij} \text{ are zero},$$

$$d_{11} = -4, \ d_{21} = -2, \ d_{31} = 32, \ d_{41} = -4, \ d_{51} = 32, \ d_{61} = 32,$$
all other d_{ij} are zero,

$$\alpha_1^{(12)} = -4, \ \alpha_1^{(23)} = -4, \ \alpha_1^{(13)} = -2, \text{ all other } \alpha_1^{(ij)} \text{ are zero},$$

$$f(1) = 64.$$

Theorem 2.31 *For the group G defined by the presentation* (2.42),

$$D_5(G) = 1.$$

Proof. With the constants $d(i)$, $e(i)$, $f(i)$, d_{ij}, described above, let

$$u_{ij}, \ v_{ik}, \ v'_{ik}, \ w_{ijk}, \ w'_{ijk}, \ w''_{ijk},$$

be constants satisfying the conditions (2.18)-(2.31), and let g be the corresponding element, defined by (2.17). Since the group G is nilpotent of class 3, the element g can be written as

$$g = \prod_{1 \le i \le j \le s} [x_{1i}^{u_{ij}d(j)}, \ x_{1j}];$$

by Theorem 2.29, the fifth dimension subgroup $D_5(G)$ is generated by such elements. From the defining relations of the group G, it follows that $[x_i^{d(i)}, x_4] = 1$, $i = 1, 2, 3$; therefore,

$$g = [x_1, \ x_2]^{16u_{12}}[x_1, \ x_3]^{64u_{13}}[x_2, \ x_3]^{64u_{23}}.$$

Consider the condition (2.22) for the case $i = 1$, $k = 6$:

$$u_{12}b_{26} + u_{14}b_{46} + v_{16}d(1) + v'_{16}e(6) = 4u_{12} + 32u_{14} + 4v_{16} + 64v'_{16} = 0.$$

It follows that
$$u_{12} + v_{16} \equiv 0 \pmod 4. \tag{2.43}$$

Next, consider the condition (2.30) for the case $k = 6$, we have:
$$2v_{16} + 4v_{26} + 32v_{46} \equiv 0 \pmod{64},$$

and we have
$$v_{16} + 2v_{26} \equiv 0 \pmod{16}. \tag{2.44}$$

From the condition (2.31) for the case $k = 3$, $l = 6$, we have:
$$2v_{13} + 4v_{23} + 32v_{43} - v_{26} + 2v_{36} \equiv 0 \pmod 4,$$

and thus we conclude that
$$v_{26} \equiv 0 \pmod 2. \tag{2.45}$$

The conditions (2.43), (2.44), (2.45) then imply that
$$u_{12} \equiv 0 \pmod 4. \tag{2.46}$$

It is clear from the defining relations of the group G that
$$[x_1, x_2]^{64} = [x_4, x_3, x_3]^{64} = 1.$$

Therefore,
$$g = [x_1, x_3]^{64u_{13}}[x_2, x_3]^{64u_{23}} = [x_1^{64u_{13}} x_2^{64u_{23}}, x_3] =$$
$$[x_4, x_3, x_3]^{-32u_{13}-16u_{23}} = x_{31}^{-32u_{13}-16u_{23}}.$$

Now consider the condition (2.22) for the case $i = 3$, $k = 6$. We have
$$32u_{34} - 32u_{13} - 16u_{23} + 64v_{36} + 64v_{36}' = 0.$$

Hence,
$$32u_{34} - 32u_{13} - 16u_{23} \equiv 0 \pmod{64}. \tag{2.47}$$

Note that the condition (2.20) for the case $i = 3$, $j = 4$, implies that
$$u_{34} \binom{64}{2} \equiv 0 \pmod{64};$$

hence
$$u_{34} \equiv 0 \pmod 2. \tag{2.48}$$

Congruences (2.47) and (2.48) imply that
$$32u_{13} + 16u_{23} \equiv 0 \pmod{64}.$$

Therefore, we have

$$g = [x_4, \, x_3, \, x_3]^{-32u_{13}-16u_{23}} = 1. \quad \Box$$

Problem 2.32 *Is it true that* $[D_5(G), G, G] = \gamma_7(G)$ *for every group* G*?*

2.5 Quasi-varieties of Groups

Our discussion in this and the next section follows [Mik06c].

Recall that a variety $\mathcal{V}$ of groups is a class of groups defined by a set of identities. Let $\mathcal{D}_n$ $(n \geq 2)$ denote the class of groups with trivial nth dimension subgroup. The existence of groups without dimension property shows that $\mathcal{D}_n$ is not a variety of groups for $n \geq 4$, since a variety of groups is always quotient closed. The classes $\mathcal{D}_n$, however, are quasi-varieties (Theorem 2.35). We recall in this section some of the basic notions about quasi-varieties.

Let F_∞ be a free group of countable rank with basis $\{x_1, x_2, \dots\}$ and $w_1, \dots, w_k, v$ some words in F_∞. A *quasi-identity* is a formal implication:

$$(w_1 = 1 \, \& \, \dots \, \& \, w_n = 1) \Longrightarrow (v = 1). \qquad (2.49)$$

A quasi-identity (2.49) is said to hold in a given group G if it is a true implication for every substitution $x_i = g_i$, $g_i \in G$.

A *quasi-variety* $\mathcal{V}_S$ is a class of groups defined by a set S of quasi-identities, i.e., $\mathcal{V}_S$ is the class of all groups in which every quasi-identity from S holds.

Example 2.33

The class $\mathcal{T}_0$ of all torsion-free groups is a quasi-variety; it is defined by the infinite set of quasi-identities

$$x^p = 1 \Longrightarrow x = 1,$$

where p runs over the set of all primes. Trivially, $\mathcal{T}_0$ is not a variety.

Recall that a non-empty class $\mathcal{F}$ of subsets of a given set I is called a *filter* on I if the following conditions are satisfied:

(i) $\emptyset \notin \mathcal{F}$;
(ii) $A \in \mathcal{F}, B \in \mathcal{F} \Longrightarrow A \cap B \in \mathcal{F}$;
(iii) $A \in \mathcal{F}, A \subseteq B \Longrightarrow B \in \mathcal{F}$.

Let $\{A_i\}_{i \in I}$ be a family of groups indexed by the elements of a set I, and $\mathcal{F}$ a filter on I. Let A be the Cartesian product

$$A = \prod_{i \in I} A_i.$$

For a given $a \in A$, denote by a_i the ith component of a in A. Consider the relation $\sim_{\mathcal{F}}$ on A defined by setting

$$a \sim_{\mathcal{F}} b \text{ if and only if } \{i \mid a_i = b_i\} \in \mathcal{F}, \ a, b \in A.$$

It follows directly from the properties of a filter that this relation is, in fact, an equivalence relation. The *filtered product of the family* $\{A_i\}_{i \in I}$ *of groups, with respect to the filter* $\mathcal{F}$, is, by definition, the quotient group

$$\prod_{\mathcal{F}} A_i := A/\sim_{\mathcal{F}}.$$

The following result of A. I. Mal'cev gives a characterization of quasi-varieties of groups.

Theorem 2.34 (Mal'cev [Mal70]). *A class $\mathcal{X}$ of groups is a quasi-variety if and only if it contains the trivial group and is closed under subgroups and filtered products.*

Recall that $\mathcal{D}_n$ $(n \geq 2)$ denotes the class of groups with trivial nth dimension subgroup. For $n = 2$, and 3, the class $\mathcal{D}_n$ coincides with the variety $\mathfrak{N}_n$ of nilpotent groups of nilpotency class $\leq n$. On the other hand, for all $n \geq 4$, as already mentioned, the existence of groups without dimension property shows that the class $\mathcal{D}_n$ is not a variety of groups. However, there is the following result:

Theorem 2.35 (Plotkin [Plo71]). *For all $n \geq 1$, the class $\mathcal{D}_n$ is a quasi-variety of groups.*

Proof. The fact that the class $\mathcal{D}_n$, $n \geq 1$, is nonempty and closed under subgroups is obvious.

Let $\{A_i\}_{i \in I}$ be a family of groups in the class $\mathcal{D}_n$, and let $\mathcal{F}$ be a filter on I. Consider the Cartesian product $A = \prod_{i \in I} A_i$. Let N be the normal subgroup of A consisting of elements $(g_i)_{i \in I}$ with $J := \{i \in I \mid g_i = 1\} \in \mathcal{F}$. If $\prod_{\mathcal{F}} A_i \notin \mathcal{D}_n$, then there exists an element $g \in A$ such that

$$g - 1 \in \mathfrak{a}^n + \sum_{s \in S}(y_s - 1)\alpha_s, \qquad (2.50)$$

where the sum is finite, $\alpha_s \in \mathbb{Z}[A]$, and $y_s \in N$. Define

$$J_s := \{i \in I \mid \text{the } i\text{th component of } y_s \text{ is } 1\}.$$

By definition, $J_s \in \mathcal{F}$. Since the set S in the sum (2.50) is finite, we have

$$\bar{J} = \bigcap_{s \in S} J_s \in \mathcal{F}.$$

For $j \in \bar{J}$, projecting g to the j-th component, we get $g_j \in D_n(A_j)$ and hence $g_j = 1$, $j \in J$. Consider the set

$$K := \{i \in I \mid g_i = 1\}.$$

Since $\bar{J} \subseteq K$, we conclude that $K \in \mathcal{F}$. Hence $g \in N$ and therefore, $\prod_{\mathcal{F}} A_i \in \mathcal{D}_n$. Consequently, the class $\mathcal{D}_n$ is closed under filtered products. Hence, by Mal'cev's criterion (Theorem 2.34), the class $\mathcal{D}_n$ is a quasi-variety. $\square$

In view of Theorem 2.22 the quasi-variety $\mathcal{D}_4$ is defined by the following implications:

Given integers k, c_i, d_{ij} $(1 \le i, \ j \le k)$ and elements $g_1, \ \ldots \ , \ g_k$ of the group G, if the following conditions hold

(1) $2^{c_i} d_{ij} + 2^{c_j} d_{ji} = 0$ $(1 \le i, \ j \le k)$,
(2) if $c_i = c_j$, then d_{ij} is even,
(3) $g_i^{2^{c_i}} \in \gamma_2(G)$ $(1 \le i \le k)$,
(4) $\prod_{i=1}^{k} g_i^{2^{c_i} d_{ij}} \in \gamma_2(G)^{2^{c_j}} \gamma_3(G)$ $(1 \le j \le k)$,

then

$$\prod_{i=1}^{k} \prod_{j=i+1}^{k} [g_i, g_j]^{2^{c_i} d_{ij}} = 1.$$

Clearly, this set of implications is equivalent to a suitable set of quasi-identities.

A quasi-variety $\mathcal{Q}$ is said to be *finitely based* if it can be defined by a finite number of quasi-identities.

Let $\mathcal{Q}$ be a quasi-variety of groups. Then the rank $rk(\mathcal{Q})$ of $\mathcal{Q}$ is the minimal number n (which may be infinite) such that there exists a system of quasi-identities

$$(w_1^i = 1 \ \& \ \ldots \& \ w_{n_i}^i = 1) \Longrightarrow (v_i = 1), \ i = 1, 2, \ldots \quad (2.51)$$

such that all words w_i^j, v_i are from a free group F_n of rank n.

Example 2.36

(i) For the quasi-variety $\mathcal{T}_0$ of torsion-free groups, $rk(\mathcal{T}_0) = 1$.
(ii) The quasi-variety defined by the quasi-identity

$$([x, y]^2 = 1) \Longrightarrow ([x, y] = 1)$$

clearly has rank 2.

Proposition 2.37 *Let $\mathcal{Q}$ be a quasi-variety and G a group. Then $G \in \mathcal{Q}$ if and only if all $rk(\mathcal{Q})$-generated subgroups of G lie in $\mathcal{Q}$.*

Proof. One side is clear, due to the fact that quasi-varieties are closed under the operation of taking subgroups.

Suppose G is a group such that all its $rk(\mathcal{Q})$-generated subgroups lie in $\mathcal{Q}$. Consider the quasi-identity system (2.51) which defines $\mathcal{Q}$ and the total number of variables entering in (2.51) is $rk(\mathcal{Q})$, i.e., all words w_i^j, v_i in (2.51) are from a free group of rank $rk(\mathcal{Q})$. Then (2.51) holds for any choice of elements $g_1, \ldots, g_{rk(\mathcal{Q})}$ from G, since it holds for any elements from the subgroup in G generated by $g_1, \ldots, g_{rk(\mathcal{Q})}$ (which is at most $rk(\mathcal{Q})$-generated. Hence (2.51) holds for all possible substitutions of elements from G and $G \in \mathcal{Q}$ by definition. $\square$

The following observation is immediate:

Proposition 2.38 *If $\mathcal{Q}$ is finitely based, then $rk(\mathcal{Q})$ is finite.*

The next result provides a method for showing that a given quasi-variety is not finitely based.

Proposition 2.39 *Let $\mathcal{Q}$ be a quasi-variety such that there exists a sequence of finitely-generated groups G_i, $i = 1, 2, \ldots$, such that the following conditions are satisfied:*

    (i) $G_i \notin \mathcal{Q}$.
    (ii) For any i there exists $f(i)$ such that all $f(i)$-generated subgroups of G_i lie in $\mathcal{Q}$.
    (iii) The function $f(i)$ is not bounded, i.e., $f(i) \to \infty$ for $i \to \infty$.

Then $rk(\mathcal{Q}) = \infty$ and hence $\mathcal{Q}$ is not finitely based.

Proof. Suppose $rk(\mathcal{Q}) < \infty$. Then, by (iii), there exists an integer i that $f(i) > rk(\mathcal{Q})$. Since every $f(i)$-generated subgroup of G_i lies in $\mathcal{Q}$, every $rk(\mathcal{Q})$-generated subgroup also lies in $\mathcal{Q}$. Therefore, $G_i \in \mathcal{Q}$ by Proposition 2.37; but this contradicts (i). Hence $rk(\mathcal{Q}) = \infty$, and $\mathcal{Q}$ is not finitely based. $\square$

2.6 The Quasi-variety $\mathcal{D}_4$

For the study of the quasi-variety $\mathcal{D}_4$, recall that the precise structure of the fourth dimension subgroup for finitely generated nilpotent groups of class 3 is given by Theorem 2.22. It has been shown by Mikhailov-Passi [Mik06c] that the quasi-variety $\mathcal{D}_4$ is not finitely based, thus answering a problem of Plotkin ([Plo83], p. 144, Probelm 12.3.2). The proof requires a technical result about certain finite groups of class 2.

Lemma 2.40 *Let n, s be natural numbers,*

$$G = \langle x_1, \ldots, x_{2n} \mid x_i^s = 1 \ (1 \leq i \leq 2n) \rangle$$

and $\Pi = G/\gamma_3(G)$. If

$$[x_1, x_2]^k \ldots [x_{2n-1}, x_{2n}]^k = [h_1, h_2] \ldots [h_{2l-1}, h_{2l}], \tag{2.52}$$

with $0 < k < s$, $h_1, \ldots, h_{2l} \in \Pi$, then $l \geq n$.

In particular, if H be an m-generator subgroup of Π and

$$[x_1, x_2]^k \ldots [x_{2n-1}, x_{2n}]^k \in \gamma_2(H),$$

then $\binom{m}{2} \geq n$.

Proof. Suppose

$$h_i \equiv x_1^{a_{i,1}} \ldots x_{2n}^{a_{i,2n}} \mod \gamma_2(\Pi),$$

where $0 \leq a_{i,j} < s$, $1 \leq i \leq 2l$, $1 \leq j \leq 2n$. Substituting in the equation (2.52), we have the following equation in Π:

$$[x_1, x_2]^k \ldots [x_{2n-1}, x_{2n}]^k = \prod_{1 \leq i < j \leq 2n} [x_i, x_j]^{b_{ij}}, \tag{2.53}$$

where

$$b_{ij} = \sum_{r=1}^{l} (a_{2r-1,i} a_{2r,j} - a_{2r-1,j} a_{2r,i}).$$

Observe that $\gamma_2(\Pi) = \prod_{1 \leq i < j \leq 2n} \langle [x_i, x_j] \rangle$ and $\langle [x_i, x_j] \rangle$ is a cyclic group of order s. Therefore, from equation (2.53), comparing the exponents of the generators $[x_i, x_j]$, $1 \leq i < j \leq 2n$ of the summands, we have:

$$b_{2t-1,2t} \equiv k \mod s, \quad 1 \leq t \leq n, \tag{2.54}$$

$$b_{i,j} \equiv 0 \mod s, \quad 1 \leq i < j \leq 2n, \ (i,j) \neq (2t-1, 2t). \tag{2.55}$$

Let $M_{p,q}(\mathbb{Z}_s)$ denote the set of $p \times q$ matrices over the ring $\mathbb{Z}_s$ of integers mod s. Let $A = (a_{i,j})_{1 \leq i \leq 2l, 1 \leq j \leq 2n} \in M_{2l,2n}(\mathbb{Z}_s)$ and define a matrix $D \in M_{2n,2l}(\mathbb{Z}_s)$ as follows:

$$D = (D_{p,q})_{1 \leq p \leq n, 1 \leq q \leq l},$$

where

$$D_{p,q} = \begin{pmatrix} a_{2q,2p} & -a_{2q-1,2p} \\ -a_{2q,2p-1} & a_{2q-1,2p-1} \end{pmatrix} \in M_{2,2}(\mathbb{Z}_s).$$

A straightforward verification shows that

$$DA = kI_{2n,2n},$$

where $I_{2n,2n} \in M_{2n,2n}(\mathbb{Z}_s)$ is the identity matrix, and it follows that $l \geq n$.

Next let H be an m-generator subgroup of Π. It is easy to see that every element of $\gamma_2(H)$ can be expressed as a product of at most $\binom{m}{2}$ commutators of elements in H, since H is nilpotent of class 2. The second assertion in Lemma thus follows from the preceding result. $\square$

Theorem 2.41 *The quasi-variety $\mathcal{D}_4$ is not finitely based.*

Proof. For $n \geq 5$, let $\Pi = G(n)/\gamma_4(G(n))$ be the lower central quotient of the group considered in Example 2.7. We assert that every m-generator subgroup H of Π, with $\binom{m}{2} < n$, has the property that $D_4(H) = 1$. Clearly then $rk(\mathcal{D}_4) = \infty$ (by Proposition 2.39) and the assertion in Theorem 2.41 is an immediate consequence. We conitnue to denote by n x_i, y_i the set of generators of Π.

Let H be an m-generator subgroup of Π and $h_1, \ldots, h_m$ a set of generators of H. Assume that, modulo $\gamma_2(H)$, $h_1, \ldots, h_k$ ($k \leq m$) are of finite order and $h_{k+1}, \ldots, h_m$ are of infinite order.

For $g \in \Pi$, let $\bar{g}$ denote the image of g in $\Pi/\gamma_2(\Pi)$ under the natural projection. Observe from the structure of Π that the torsion subgroup of $\Pi/\gamma_2(\Pi)$ is equal to

$$\langle \bar{x}_1 \rangle \oplus \langle \bar{x}_2 \rangle \oplus \langle \bar{x}_3 \rangle \simeq \mathbb{Z}_4 \oplus \mathbb{Z}_{16} \oplus \mathbb{Z}_{64}.$$

By suitably replacing $h_1, \ldots, h_k$, if necessary, we can assume that

$$h_1 = x_1^{l_{1,1}} x_2^{l_{1,2}} x_3^{l_{1,3}} \lambda_1, \; h_2 = x_1^{l_{2,1}} x_2^{l_{2,2}} \lambda_2, \; h_3 = x_1^{l_{3,1}} \lambda_3,$$

$$h_j = \lambda_j \; (4 \leq j \leq k),$$

where $l_{i,j} \in \mathbb{Z}$, $\lambda_i \in H \cap \gamma_2(\Pi)$ $(1 \leq i \leq k)$.

Let $d(i)$ be the order of h_i modulo $\gamma_2(H)$. Then, in particular,

$$l_{3,1}d(3) \equiv 0 \mod 4, \tag{2.56}$$

$$l_{2,2}d(2) \equiv 0 \mod 16. \tag{2.57}$$

We can assume also that

$$d(k)|d(k-1)|\ldots|d(2)|d(1).$$

Therefore, by Theorem 2.22, the group $D_4(H)$ consists of the following elements:

$$w = \prod_{1 \leq i < j \leq k} [h_i^{d(i)}, h_j]^{a_{ij}},$$

where the integers a_{ij} satisfy the conditions (2.14) and (2.15).

We have, for $j \geq 4$, $[h_i^{d(i)}, h_j] = [h_i^{d(i)}, \lambda_j] = 1$; therefore,

$$w = [h_1^{d(1)}, h_2]^{a_{12}} [h_1^{d(1)}, h_3]^{a_{13}} [h_2^{d(2)}, h_3]^{a_{23}} =$$
$$[h_1, h_2^{d(1)a_{12}} h_3^{d(1)a_{13}}][h_2^{d(2)}, h_3]^{a_{23}}. \quad (2.58)$$

Since

$$y_1 = \prod_{1 < j \leq k} h_j^{d(1)a_{1j}} \in \gamma_2(\Pi)^{d(1)} \gamma_3(\Pi) \quad \text{by (2.15)},$$

we have,

$$w = [h_1, y_1][h_2^{d(2)}, h_3]^{a_{23}} = [h_2^{d(2)}, h_3]^{a_{23}}.$$

We claim that $[h_2^{d(2)}, h_3]^{a_{23}} = 1$.

Consider the element $h_3 = x_1^{l_{3,1}} \lambda_3$. We have

$$x_1^{l_{3,1}d(3)} \lambda_3^{d(3)} = \prod_{1 \leq i < j \leq m} [h_i, h_j]^{u_{ij}} \gamma, \quad (2.59)$$

for some $\gamma \in \gamma_3(\Pi)$ and $u_{ij} \in \mathbb{Z}$.

Let E be the normal subgroup in Π generated by x_2, x_3, Y_2, Y_3, $[x_1, Y_j]$ ($j \in \{1, 4, 5\}$), $[Y_i, Y_j]$ ($i, j \in \{1, 4, 5\}$, $i \neq j$) and $\gamma_3(\Pi)$, where

$$Y_i = \{y_{(2i-2)n+1}, \ldots, y_{2in}\}, \, i = 1, \ldots, 5.$$

Let

$$S = \langle x_1, Y_1, Y_4, Y_5 \mid x_1^4 = \xi_{1,(n)}, \, x_1^{-32} = \xi_{4,(n)}^{16}, \, x_1^{-64} = \xi_{5,(n)}^{64},$$
$$[x_1, Y_i] = 1 \, (i \in \{1, 4, 5\}), \, [Y_i, Y_j] = 1 \, (i, j \in \{1, 4, 5\}, \, i \neq j)\rangle, \quad (2.60)$$

We note that

$$\Pi/E \simeq S/\gamma_3(S).$$

Let $p : \Pi \to S/\gamma_3(S)$ be the composition of the projections $\Pi \to \Pi/E$ and $\Pi/E \to S/\gamma_3(S)$. Applying the projection p to the equation (2.59) in Π, we have the following equation in $S/\gamma_3(S)$:

$$x_1^{l_{3,1}d(3)} p(\lambda_1)^{d(3)} = \prod_{i < j} [p_1(h_i), p_1(h_j)]^{u_{ij}}, \quad (2.61)$$

Note that

$$S/\gamma_3(S) = (\langle x_1 \rangle \oplus \mathcal{Y}_1/\gamma_3(\mathcal{Y}_1) \oplus \mathcal{Y}_4/\gamma_3(\mathcal{Y}_4) \oplus \mathcal{Y}_5/\gamma_3(\mathcal{Y}_3)) / N,$$

where $\mathcal{Y}_i$, $1 \leq i \leq 5$ is a free group with basis Y_i,

$$N = \langle x_{1,(n)}^4 \xi_{1,(n)}^{-1}, \; \xi_{1,(n)}^8 \xi_{4,(n)}^{16}, \; \xi_{1,(n)}^{16} \xi_{5,(n)}^{64} \rangle.$$

Therefore (2.61) implies that in the direct product

$$\mathcal{Y} := \mathcal{Y}_1/\gamma_3(\mathcal{Y}_1) \oplus \mathcal{Y}_4/\gamma_3(\mathcal{Y}_3) \oplus \mathcal{Y}_5/\gamma_3(\mathcal{Y}_3).$$

We have

$$l_{3,1}d(3) \equiv 0 \mod 4, \tag{2.62}$$

and

$$\xi_{1,(n)}^{\frac{l_{3,1}d(3)}{4}} \mu_1^{d(3)} (\xi_{1,(n)}^8 \xi_{4,(n)}^{16})^{k_1} (\xi_{1,(n)}^{16} \xi_{5,(n)}^{64})^{k_2} = \prod_{1 \leq i < j \leq m} [z_i, z_j]^{u_{ij}}, \tag{2.63}$$

for some integers k_1, k_2 and elements $\mu_1 \in \gamma_2(\mathcal{Y})$, $z_i \in \mathcal{Y}$, $1 \leq i \leq m$. Projecting (2.63) to each of the three summands of $\mathcal{Y}$ we have the following three equations:

$$\xi_{1,(n)}^{d_1} \mu_{1,1}^{d(3)} = \prod_{1 \leq i < j \leq m} [z_{i,1}, z_{j,1}]^{u_{ij}}, \; \text{in } \mathcal{Y}_1/\gamma_3(\mathcal{Y}_1), \; d_1 = \frac{l_{3,1}d(3)}{4} + 8k_1 + 16k_2,$$

$$\tag{2.64}$$

$$\xi_{4,(n)}^{d_4} \mu_{1,4}^{d(3)} = \prod_{1 \leq i < j \leq m} [z_{i,4}, z_{j,4}]^{u_{ij}}, \; \text{in } \mathcal{Y}_4/\gamma_3(\mathcal{Y}_4), \; d_4 = 16k_1, \tag{2.65}$$

$$\xi_{5,(n)}^{d_5} \mu_{1,5}^{d(3)} = \prod_{1 \leq i < j \leq m} [z_{i,5}, z_{j,5}]^{u_{ij}}, \; \text{in } \mathcal{Y}_5/\gamma_3(\mathcal{Y}_5), \; d_5 = 64k_2, \tag{2.66}$$

for some $\mu_{1,i} \in \gamma_2(\mathcal{Y}_i)/\gamma_3(\mathcal{Y}_i)$, $z_{i,l} \in \mathcal{Y}_l/\gamma_3(\mathcal{Y}_l)$, $1 \leq i \leq m$, $l \in \{1, 4, 5\}$.

Case (a): $l_{3,1}$ *is odd.*
In view of (2.62), we have $d(3) = 4s$ for some integer s. Let

$$Z_i = \langle Y_i \mid y_i^{4s} = 1 \; (y_i \in Y_i), \; \gamma_3(\mathcal{Y}_1) \rangle,$$

and $p_i : \mathcal{Y}_i/\gamma_3(\mathcal{Y}_i) \rightarrow Z_i$ be the natural projection, $i \in \{1, 4, 5\}$. Projecting the equations (2.64), (2.65) and (2.66) into Z_1, Z_4, Z_5 respectively, we conclude, by an application of Lemma 2.40, that

$$d_i \equiv 0 \mod 4s \; (i \in \{1, 4, 5\}).$$

From (2.64) and (2.65), we therefore have

$$l_{3,1}s + 8k_1 + 16k_2 \equiv 0 \pmod{4s}, \tag{2.67}$$

$$16k_1 \equiv 0 \pmod{4s}. \tag{2.68}$$

It follows easily that $s \equiv 0 \pmod{16}$, and consequently,

$$d(3) \equiv 0 \pmod{64}.$$

Let $d(3) = 64f$, $f \in \mathbb{Z}$, and suppose $d(2) = d(3)c$ ($c \in \mathbb{Z}$) (recall that $d(3)|d(2)$). Then we have

$$w = [h_2^{d(2)}, h_3]^{a_{23}} = [h_2, h_3^{d(3)}]^{ca_{23}} =$$
$$[h_2, x_1^{64l_{3,1}f} \lambda_3^{64f}]^{ca_{23}} = [h_2^{16}, x_1^{4l_{3,1}f} \lambda_3^{4f}]^{ca_{23}}.$$

Since $h_2^{16} \in \gamma_2(\Pi)$, it follows that $w = 1$.

Case (b): $l_{3,1} = 2l$ *and* l *is odd.* We assert that in this case

$$d(3) \equiv 0 \pmod{16}. \tag{2.69}$$

Since $x_1^{2ld(3)} \in \gamma_2(\Pi)$, we have $d(3) = 2r$ for some $r > 0$. Projecting the square of the equation (2.64) to Z_1 under the map p_1, we conclude, by an application of Lemma 2.40, that $2d_1 \equiv 0 \pmod{4r}$.

Therefore we have

$$2d_1 = 2lr + 16k_1 + 32k_2 \equiv 0 \pmod{4r},$$

which implies that $r \equiv 0 \pmod{8}$, and consequently, we have (2.69).

Now consider the element $h_2 = x_1^{l_{2,1}} x_2^{l_{2,2}} \lambda_2$. We have

$$h_2^{d(2)} = (x_1^{l_{2,1}} x_2^{l_{2,2}})^{d(2)} \lambda_2^{d(2)} = \prod_{1 \le i < j \le m} [h_i, h_j]^{v_{ij}} \gamma, \tag{2.70}$$

for some $\gamma \in \gamma_3(\Pi)$ and $v_{ij} \in \mathbb{Z}$.

Let I be the normal subgroup in Π generated by x_1, x_3, Y_1, Y_4, $[x_2, Y_j]$, $j \in \{2, 3, 5\}$), $[Y_i, Y_j]$ ($i, j \in \{2, 3, 5\}$, $i \ne j$) and $\gamma_3(\Pi)$. Let

$$Q = \langle x_2, Y_2, Y_3, Y_5 \mid x_2^{16} = \xi_{2,(n)}, \xi_{2,(n)}^2 = \xi_{3,(n)}^4, \xi_{2,(n)}^8 = \xi_{5,(n)}^{64} \rangle.$$

Note that $\Pi/I \simeq Q/\gamma_3(Q)$ and

$$Q/\gamma_3(Q) \simeq (\langle x_2 \rangle \oplus \mathcal{Y}_2/\gamma_3(\mathcal{Y}_2) \oplus \mathcal{Y}_3/\gamma_3(\mathcal{Y}_3) \oplus \mathcal{Y}_5/\gamma_3(\mathcal{Y}_5))/M,$$

where $M = \langle x_2^{16}\xi_{2,(n)}^{-1}, \xi_{2,(n)}^2\xi_{3,(n)}^{-4}, \xi_{2,(n)}^8\xi_{5,(n)}^{64} \rangle$. Let $q : \Pi \to Q$ be the natural projection. Applying q to the equation (2.70), we have the following equation

$$q(x_2)^{l_{2,2}d(2)}q(\lambda_2)^{d(2)} = \prod_{1\leq i<j\leq m} [q(h_i),\, q(h_j)]^{v_{ij}} \qquad (2.71)$$

in the group $Q/\gamma_3(Q)$. This equation implies that, in the direct product

$$\mathcal{V} := \mathcal{Y}_2/\gamma_3(\mathcal{Y}_2) \oplus \mathcal{Y}_3/\gamma_3(\mathcal{Y}_3) \oplus \mathcal{Y}_5/\gamma_3(\mathcal{Y}_5),$$

we have (using (2.57))

$$\xi_{2,(n)}^{\frac{l_{2,2}d(2)}{16}}\mu_2^{d(2)}(\xi_{2,(n)}^2\xi_{3,(n)}^{-4})^{m_1}(\xi_{2,(n)}^8\xi_{5,(n)}^{64})^{m_2} = \prod_{1\leq i<j\leq m} [v_i,\, v_j]^{v_{ij}}, \qquad (2.72)$$

for some integers m_1, m_2 and the elements $\mu_2 \in \gamma_2(\mathcal{V})$, $v_i \in \mathcal{V}$, $1 \leq i \leq m$. Projecting (2.72) to the first summand of $\mathcal{V}$, we have the following equation:

$$\xi_{2,(n)}^{e_1}\mu_{2,1}^{d(2)} = \prod_{1\leq i<j\leq m} [v_{i,1},\, v_{j,1}],$$

where

$$e_1 = \frac{l_{2,2}d(2)}{16} + 2m_1 + 8m_2,$$

and $\mu_{2,1} \in \gamma_2(\mathcal{Y}_2)/\gamma_3(\mathcal{Y}_2)$, $v_{i,1} \in \mathcal{Y}_2/\gamma_3(\mathcal{Y}_2)$, $1 \leq i \leq m$. Since $d(3)|d(2)$, therefore $d(2) = 16t$ for some t. An application of Lemma 2.40 once again shows that $e_1 \equiv 0 \mod 16t$; consequently, $l_{2,2}t$ is even and so $l_{2,2}d(2) = 32f$ for some f. Hence

$$w = [h_2^{d(2)},\, h_3]^{a_{23}} = [(x_1^{l_{2,1}}x_2^{l_{2,2}}\lambda_2)^{d(2)},\, x_1^{2l}]^{a_{23}} = [x_1^{l_{2,1}}x_2^{l_{2,2}}\lambda_2,\, x_1^{l_{3,1}d(2)}]^{a_{23}}$$

$$= [x_2^{l_{2,2}},\, x_1^{l_{3,1}d(2)}]^{a_{23}} = [x_2,\, x_1^{l_{3,1}l_{2,2}d(2)}]^{a_{23}} = [x_2,\, x_1^{64lf}]^{a_{23}}$$

$$= [x_2,\, \xi_{1,(n)}^{16lf}]^{a_{23}} = [x_2^{16},\, \xi_{1,(n)}^{lf}]^{a_{23}} = 1.$$

Case (c): $l_{3,1} \equiv 0 \mod 4$. In this case $h_3 \in \gamma_2(\Pi)$, since $x_1^4 \in \gamma_2(\Pi)$; therefore, $w = 1$.

Thus, in all cases, $w = 1$, and consequently, $D_4(H) = 1$. This completes the proof. $\square$

2.7 Dimension Quotients

If G is a finite p-group, p odd, then $D_4(G) = \gamma_4(G)$ [Pas68a]. Refuting the long standing *dimension conjecture* that $D_n(G) = \gamma_n(G)$ always, Rips [Rip72] constructed a 2-group (Example 2.1) with $D_4(G) \neq \gamma_4(G) = 1$. Extending these results N. Gupta has shown that odd prime power groups have the dimension property [Gup02] and, for every $n \geq 4$, there exist 2-groups with $D_n(G) \neq \gamma_n(G)$ [Gup90]. For odd prime p, the dimension property was earlier shown to hold for metabelian p-groups by Gupta [Gup91b] and for centre-by-metabelian p-groups by Gupta-Gupta-Passi [Gup94]. The result for odd prime power groups is an immediate consequence of the following result.

Theorem 2.42 (N. Gupta [Gup02]). *The nth dimension quotient of a finite nilpotent group has exponent dividing 2^l, where l is the least natural number such that $2^l \geq n$.*

Let $n \geq 3$ be an arbitrary but *fixed* integer and let G be a finite nilpotent group with $\gamma_n(G) = 1$. Choose a non-cyclic free presentation (see [Mag66], Theorem 3.5, p. 140)

$$1 \to R \to F \to G \to 1,$$

where F is the free group with basis $\{x_1, \ldots, x_m\}$, $m \geq 2$, and R is the normal closure in F of the set of relators $\{x_1^{e(1)}\xi_1, \ldots, x_m^{e(m)}\xi_m\} \cup T$ such that $e(i) > 1$, $\xi_i \in [F, F]$ and T is a finite subset of $[F, F]$.

Let l be the least positive integer such that $2^l \geq n$. Let

$$G = \delta_0(G) \supseteq \delta_1(G) \ldots \supseteq \delta_{l-1}(G) \supseteq \delta_l(G) = 1$$

be the derived series of G. Then $\delta_k(G) \simeq \delta_k(F)R/R$, $0 \leq k \leq l-1$, and therefore we can have a presentation

$$1 \to R^{(k)} \to F^{(k)} \to \delta_k(G) \to 1$$

where $F^{(k)}$ is a free subgroup of the kth derived subgroup $\delta_k(F)$ of F with ordered basis $B(k) = \{x_{k,1}, \ldots, x_{k,m_k}\}$, $m_k \geq 2$, $R^{(k)}$ is the normal closure in $F^{(k)}$ of the set of relators $\{x_{k,1}^{e(k,1)}\xi_{k,1}, \ldots, x_{k,m_k}^{e(k,m_k)}\xi_{k,m_k}\} \cup T_k$ with $e(k,i) > 1$, $\xi_{k,i} \in [F^{(k)}, F^{(k)}]$ and $T_k \subset [F^{(k)}, F^{(k)}]$ a finite subset. Furthermore, it is possible to define a weight function and a weight-preserving order on the set $\cup_k B(k)$. To this end, we need the following basic results.

Lemma 2.43 *If S is a set of generators of a free group F which is linearly independent modulo $[F, F]$, then S is a basis of F.*

Proof. Let X be a set equinumerous with S and $\alpha : X \to S$ a bijection. Let $\mathfrak{F}$ be the free group on X. Then the map α extends to a homomorphism $\bar{\alpha} : \mathfrak{F} \to F$. Since S generates F and is linearly independent modulo $[F, F]$, the homomorphism $\bar{\alpha}$ is an epimorhism and the induced homomorphism $\mathfrak{F}/[\mathfrak{F}, \mathfrak{F}] \to F/[F, F]$ is an isomorphism. By Theorem 1.76, the induced homomorphisms $\mathfrak{F}/\gamma_m(\mathfrak{F}) \to F/\gamma_m(F)$, $m \geq 2$, are all isomorphisms, since, both F and $\mathfrak{F}$ being free, $H_2(\mathfrak{F}) = H_2(F) = 0$. Hence $\ker(\bar{\alpha}) \subseteq \gamma_\omega(\mathfrak{F}) = 1$. It thus follows that $\bar{\alpha}$ is an isomorphism, and so S is a free set of generators of F. $\square$

Lemma 2.44 *Let B be an ordered basis of a free group F. Then the basic commutators*

$$C(t) = [y_1, y_2, \ldots, y_t], \ y_i \in B, \ t \geq 2,$$

satisfying $y_1 > y_2 \leq y_3 \leq \ldots \leq y_t$ are linearly independent modulo $\delta_2(F)$. $\square$

Proof. Let $\mathfrak{a} = \mathbb{Z}[F]\Delta(\delta_1(F))$. Consider the Magnus embedding

$$\theta : \delta_1(F)/\delta_2(F) \rightarrow \mathfrak{f}/\mathfrak{f}\mathfrak{a}, \ x\delta_2(F) \mapsto (x-1) + \mathfrak{f}\mathfrak{a}, \ x \in \delta_1(F). \tag{2.73}$$

Suppose we have an inclusion

$$\prod_{i=1}^{m} \mathbf{y}_i^{n(\mathbf{y}_i)} \in \delta_2(F), \tag{2.74}$$

where $\mathbf{y_i}$, $i = 1, 2, \ldots, m$, are left-normed commutators $[y_{i1}, y_{i2}, \ldots, y_{it_i}]$ satisfying $y_{i1} > y_{i2} \leq \ldots \leq y_{it_i}$. On applying θ, we then have

$$\sum_{i=1}^{m}([y_{i1}, y_{i2}] - 1)(y_{i3} - 1)\ldots(y_{it_i} - 1) \equiv 0 \quad \mod \mathfrak{f}\mathfrak{a}. \tag{2.75}$$

Since $\mathfrak{f}$ is a free right $\mathbb{Z}[F]$-module with basis $B - 1$, it follows that

$$\sum n(\mathbf{y}_i)(y_{i2} - 1)\ldots(y_{im} - 1) \equiv 0 \quad \mod \mathfrak{a},$$

where the sum is taken over all i for which the first entry y_{i1} in $\mathbf{y}_i$ is the same. Since the elements $(y_1 - 1)(y_2 - 1)\ldots(y_r - 1)$, $y_1 \leq y_2 \leq \ldots y_r$ with y_i's in B are linearly independent modulo $\mathfrak{a}$, it follows that $n(\mathbf{y}_i) = 0$ for all $i = 1, 2, \ldots, m$. $\square$

The chain

$$F = F^{(0)} \supset F^{(1)} \supset \cdots \supset F^{(l)} = \{1\}, \tag{2.76}$$

can be constructed inductively as follows. Let the basis $\{x_1, \ldots, x_m\}$ of $F = F^{(0)}$ be renamed as $B(0) = \{x_{0,1}, \ldots, x_{0,m_0}\}$ by defining $m_0 = m$ and setting $x_{0,1} = x_1, \ldots, x_{0,m_0} = x_m$. To each basis element $x_{0,i}$ in $B(0)$, we assign weight 1:

$$\mathrm{wt}(x_{0,i}) = 1 \text{ for } i = 1, \ldots, m_0.$$

Having defined, for $k \geq 1$, the subgroup $F^{(k-1)}$ with an ordered basis

$$B(k-1) = \{x_{k-1,1}, \ldots, x_{k-1,m_{k-1}}\}$$

satisfying $x_{k-1,i} < x_{k-1,i+1}$ and $\mathrm{wt}(x_{k-1,i}) < n$ for $i = 1, \ldots, m_{k-1}$, to define the subgroup $F^{(k)}$ with a *weight preserving* ordered basis, list the finite set $B(k)$ of *all* left-normed basic commutators of the form

$$C(t) = [y_1, y_2, \ldots, y_t], y_i \in B(k-1), t \geq 2, \tag{2.77}$$

satisfying $y_1 > y_2 \leq \cdots \leq y_t$ and $\mathrm{wt}(y_1) + \cdots + \mathrm{wt}(y_t) < n$. Let $F^{(k)}$ be the subgroup generated by $B(k)$. By Lemmas 2.43 and 2.44 the commutators $C(t)$ constitute a free basis of $F^{(k)}$. Now define

$$\mathrm{wt}(C(t)) = \mathrm{wt}(y_1) + \cdots + \mathrm{wt}(y_t).$$

Define any weight-preserving order relation on the set $B(k)$ and relabel its elements following this order to obtain the basis

$$B(k) = \{x_{k,1}, \ldots, x_{k,m(k)}\} \tag{2.78}$$

of the subgroup $F^{(k)}$.

Let $k \in \{0, 1, \ldots, l-1\}$ be arbitrary but fixed. In the free group rings $\mathbb{Z}[F^{(k)}]$ set

$$\begin{aligned}
\mathbf{r}^{(k)} &= \mathbb{Z}[F^{(k)}](R^{(k)} - 1), \\
\mathbf{f}^{(n,k)} &= \mathbb{Z}\text{-span}\{(y_1^{\pm 1} - 1) \ldots (y_t^{\pm 1} - 1) \,|\, t \geq 2\} \tag{2.79} \\
&\text{with } y_i \in B(k) \text{ satisfying } \mathrm{wt}(y_1) + \cdots + \mathrm{wt}(y_t) \geq n.
\end{aligned}$$

Next, define the kth *partial dimension subgroup* by

$$D_{(n)}(R^{(k)}) = F^{(k)} \cap (1 + \mathbf{r}^{(k)} + \mathbf{f}^{(n,k)}) \tag{2.80}$$

and the kth *partial lower central subgroup* $\gamma_{(n)}(F^{(k)})$ to be the normal closure of the set

$$\{[y_1, \ldots, y_t], y_i \in B(k), t \geq 2, y_1 > y_2 \leq \cdots \leq y_t\},$$

where $\mathrm{wt}(y_1) + \cdots + \mathrm{wt}(y_t) \geq n$ and $\mathrm{wt}(y_1) + \cdots + \mathrm{wt}(y_{t-1}) < n$. We thus have the following subnormal chain of subgroups:

$$D_{(n)}(R^{(0)}) \supseteq D_{(n)}(R^{(1)}) \supseteq \cdots \supseteq D_{(n)}(R^{(l)}) = 1 \tag{2.81}$$

where clearly $R^{(k)}\gamma_{(n)}(F^{(k)}) \leq D_{(n)}(R^{(k)})$.

The main result in [Gup02] is the following

Theorem 2.45 *For each $k \in \{0, 1, \ldots, l-1\}$,*

$$D_{(n)}(R^{(k)})^2 \subseteq R^{(k)}\gamma_{(n)}(F^{(k)})D_{(n)}(R^{(k+1)}).$$

Theorem 2.42 is an immediate consequence of the above result. For, let $w \in F \cap (1 + \mathbf{r} + \mathbf{f}^n)$. Then $w - 1 \in \mathbf{r}^{(0)} + \mathbf{f}^{(n,0)}$ and $w \in D_{(n)}(R^{(0)})$.

Theorem 2.45 implies that there exist elements

$$g_0 \in R^{(0)}\gamma_{(n)}(F^{(0)}), \; g_1 \in R^{(1)}\gamma_{(n)}(F^{(1)}), \; \ldots, \; g_{l-1} \in R^{(l-1)}\gamma_{(n)}(F^{(l-1)})$$

such that

$$(\ldots((w^2 g_0)^2 g_1)^2 \ldots)^2 g_{l-1} = 1$$

and, since $R^{(k)}\gamma_{(n)}(F^{(k)}) \subseteq R\gamma_n(F)$ for each k, Theorem 2.42 follows.

If G is a group whose lower central factors $\gamma_n(G)/\gamma_{n+1}(G)$ are all torsion-free, then G has the dimension property (see [Pas79], p. 48). Thus, in particular, free nilpotent groups and the free poly-nilpotent groups have the dimension property.

Theorem 2.46 (Kuz'min [Kuz96]). *If G is an extension of a group whose lower central quotients are torsion-free by an abelian group, then G has the dimension property.*

It is known [Gup73] that the lower central factors of the free centre-by-metabelian group are, in general, not torsion-free. However, we have the following

Theorem 2.47 (Gupta-Levin [Gup86]). *Free centre-by-metabelian groups have the dimension property.*

Let $\mathfrak{f}$ be the augmentation ideal of the free group ring $\mathbb{Z}[F]$. For $c \geq 1$, let $\mathfrak{a}_c$ be the ideal $\mathbb{Z}[F](\gamma_c(F) - 1)$.

Theorem 2.48 (Gupta-Gupta-Levin [Gup87b]). *For all $n, \, c \geq 1$,*

$$F \cap (1 + \mathfrak{f}\mathfrak{a}_c + \mathfrak{f}^{n+1}) = [\gamma_c(F), \, \gamma_c(F)]\gamma_{n+1}(F).$$

In particular, the groups $F/[\gamma_c(F), \gamma_c(F)]$, $c \geq 1$, have the dimension property.

For $c = 2$, the above result was proved earlier by Gupta [Gup82].

Theorem 2.49 (Gupta-Kuz'min). *For any $n \geq 1$ and a group G, the subquotient group $D_n(G)/\gamma_{n+1}(G)$ is abelian.*

Proof. Let G be a nilpotent of class n. We have to show that $D_n(G)$ is abelian. Let A be a maximal abelian normal subgroup of G. It is easy to show that A coincides with its centralizer $C_G(A)$. We can view A as a G-module via conjugation. Then for any $k \geq 1$, we have

$$a \circ (g - 1) \subseteq \gamma_{k+1}(G), \; g \in D_k(G).$$

In particular, any $g \in D_n(G)$ lies in $C_G(A)$. Therefore $D_n(G) \subseteq C_G(A)$ and hence $D_n(G)$ is an abelian group. $\square$

2.8 Plotkin's Problems

The following problems have been raised and discussed by Plotkin in [Plo73] (see also Hartley [Har84]).

Problem 2.50 *For every group G, is it true that $D_\omega(G) = \gamma_\omega(G)$.*

Problem 2.51 *Is it true that for every nilpotent group G, there exists an integer $n(G)$ such that $D_{n(G)}(G) = 1$? In other words, does every nilpotent group have finite dimension series?*

Plotkin conjectures that problem 2.50 has an affirmative answer.

Theorem 2.52 (Hartley [Har82c]). *If G is a nilpotent group in which the torsion subgroup has finite dimension series, then G itself has finite dimension series.*

For a group G, let $s(G)$ denote the least natural number n, if it exists, such that $D_n(G) = 1$, and infinity otherwise. Let $\mathfrak{N}_c$ denote the variety of nilpotent groups of class $\leq c$. It is easy to see that *finitely generated nipotent groups and torsion-free nilpotent groups have finite dimension series.*

Let c be a natural number and suppose that every group in $\mathfrak{N}_c$ has finite dimension series. Then there exists a natural number $r = r(c)$ such that $D_r(G) = 1$ for every $G \in \mathfrak{N}_c$. For, if not, then we can find groups in $\mathfrak{N}_c$ having arbitrarily long dimension series. Choose groups G_1, G_2, ... in $\mathfrak{N}_c$ so that G_i has dimension series of length $\geq i$. Then the group $\Gamma = \oplus_{i=1}^{\infty} G_i$, is in $\mathfrak{N}_c$, but its dimension series does not terminate with identity in a finite number of steps. A standard reduction argument (see [Pas68a]) shows that if $s = s(c)$ is a number such that, for every finite p-group $G \in \mathfrak{N}_c$, $D_s(G) = 1$, then, for every group $\Gamma \in \mathfrak{N}_c$, $D_s(\Gamma) = 1$.

Lemma 2.53 *Let $H \triangleleft G$ and suppose that*

$$[H, {}_mG] := [\ldots[H, \underbrace{G], G].\ldots,], G}_{m \ terms}] = 1.$$

Let M be a right G-module such that $M.\mathfrak{g}^r \subseteq M.\mathfrak{h}$ for some integer $r \geq 1$. Then $M.\mathfrak{g}^{rn^m} \subseteq M.\mathfrak{h}^n$ for all $n \geq 1$.

Proof. We proceed by induction on $m \geq 1$. If $m = 1$, then H is a central subgroup. Therefore, repeated use of $M.\mathfrak{g}^r \subseteq M.\mathfrak{h}$ gives the required inclusion:

$$M.\mathfrak{g}^{rn} \subseteq M.\mathfrak{h}^n.$$

Now suppose $m > 1$ and the result holds for $m - 1$. Let $K = [H, {}_{m-1}G]$ and consider the groups $\bar{H} = H/K$ and $\bar{G} = G/K$. Note that $\bar{H} \triangleleft \bar{G}$ and

$[\bar{H}, {}_{m-1}\bar{G}] = 1$. The quotient $\bar{M} = M/M.\mathfrak{k}$ is a $\bar{G}$-module under the action induced by that of M as a G-module and

$$\bar{M}.\bar{\mathfrak{g}}^r \subseteq \bar{M}.\bar{\mathfrak{h}}.$$

Therefore, by induction hypothesis,

$$\bar{M}.\bar{\mathfrak{g}}^{r.n^{m-1}} \subseteq \bar{M}.\bar{\mathfrak{h}}^n,$$

for all $n \geq 1$. This implies that

$$M.\mathfrak{g}^{rn^{m-1}} \subseteq M.\mathfrak{h}^n + M.\mathfrak{k}.$$

Since K is a central subgroup of G, iteration gives

$$M.\mathfrak{g}^{rn^m} \subseteq M.\mathfrak{h}^n,$$

and the proof is complete. $\square$

Lemma 2.54 *Let G be a group, and suppose that $H \lhd G$, $G = HF$ for some finite p-group $F \subseteq G$, $[H, {}_mG] = 1$ for some integer $m \geq 1$. Then for every $r \geq 1$, there exists $u = u(r)$ such that*

$$\mathbb{Z}[H] \cap \mathfrak{g}^u \subseteq \mathfrak{h}^r.$$

Proof. Let $D = H \cap F$. Then D is a finite p-group. Let $r \geq 1$ be given. Choose $s \geq 1$ such that $p^s \mathfrak{d} \subseteq \mathfrak{d}^r \subseteq \mathfrak{h}^r$. Observe that $\mathfrak{h}^r \mathbb{Z}[G] + p^s \mathfrak{f} \mathbb{Z}[H]$ is a right ideal of $\mathbb{Z}[G]$. Consider the right G-module $M = \mathbb{Z}[G]/(\mathfrak{h}^r \mathbb{Z}[G] + p^s \mathfrak{h} \mathbb{Z}[H])$. Since G/H is a finite p-group, there exists $n \geq 1$ such that $\mathfrak{g}^n \subseteq \mathfrak{h}\mathbb{Z}[G] + p^s \mathfrak{f} \mathbb{Z}[H]$. Hence, by Lemma 2.53, we can conclude that there exists an integer $u = u(r) \geq 1$ such that $M.\mathfrak{g}^u \subseteq M.\mathfrak{h}^r$, i.e.,

$$\mathfrak{g}^u \subseteq \mathfrak{h}^r \mathbb{Z}[G] + p^s \mathfrak{f} \mathbb{Z}[H].$$

Intersecting with $\mathbb{Z}[H]$ we get

$$\mathbb{Z}[H] \cap \mathfrak{g}^u \subseteq \mathbb{Z}[H] \cap (\mathfrak{h}^r \mathbb{Z}[G] + p^s \mathfrak{f} \mathbb{Z}[H]). \tag{2.82}$$

If T is a transversal for D in F including 1, then by the choice of s, we have

$$\mathfrak{h}^r \mathbb{Z}[G] + p^s \mathfrak{f} \mathbb{Z}[H] = \mathfrak{h}^r \mathbb{Z}[G] + p^s \mathfrak{t} \mathbb{Z}[H],$$

where $\mathfrak{t}$ is the additive subgroup of $\mathbb{Z}[G]$ generated by $t - 1$, $t \in T$. Let $\theta : \mathbb{Z}[G] \to \mathbb{Z}[H]$ be the linear extension of the map $G \to H$ given by $g = th \mapsto h$ ($t \in T$, $h \in H$). Applying θ to the inclusion (2.82) we get

$$\mathbb{Z}[H] \cap (\mathfrak{h}^r \mathbb{Z}[G] + p^s \mathfrak{f} \mathbb{Z}[H]) = \mathfrak{h}^r.$$

Hence $\mathbb{Z}[H] \cap \mathfrak{g}^u \subseteq \mathfrak{h}^r$. $\square$

Theorem 2.55 (Kuskulei, see [Plo73]). *If G is a nilpotent group having a subgroup H of finite index whose dimension series is finite, then G has finite dimension series.*

Proof. It clearly suffices to consider the case when $H \lhd G$ and G/H is a cyclic group of prime order, p say. If the torsion subgroup T of G lies in H, then T has finite dimension series and therefore, by Theorem 2.52, G has finite dimension series. If $T \nsubseteq H$, then H has a supplement of p-power order in G, and Lemma 2.54 implies that G has finite dimension series. $\square$

Theorem 2.56 (Tokarenko and Rips [Plo73]). *If a semi-direct product $G = H \rtimes K$ is nilpotent and both H and K have finite dimension series, then G has finite dimension series and $s(G) \leq \max(s(H)^c, \ s(K))$.*

Proof. Regard $\mathbb{Z}[H]$ as a right G-module as follows. For $\alpha \in \mathbb{Z}[H]$, $g = hk \in G$, $h \in H$, $k \in K$, define

$$\alpha.g = \alpha^k h,$$

where α^k stands for the element of $\mathbb{Z}[H]$ obtained on conjugating by k each element in the support of α. Then, as can be seen by induction on the class of G,

$$\mathbb{Z}[H].\mathfrak{g}^{m^c} \subseteq \mathfrak{h}^m.$$

Since K has finite dimension series, $D_n(G) \subseteq H$ for $n \geq s(K)$. Let $n \geq \max(s(H)^c, \ s(K))$ and $x \in D_n(G)$. Then $x - 1 \in \mathbb{Z}[H] \cap \mathfrak{g}^{s(H)^c}$. Hence

$$1.(x - 1) \in \mathfrak{h}^{s(H)}.$$

However, under the G-module action we are considering, $1.(x - 1) = x - 1$. Therefore, it follows that $x - 1 \in \mathfrak{h}^{s(H)}$, and consequently $x = 1$, showing that G has finite dimension series with $s(G) \leq \max(s(H)^c, \ s(K))$. $\square$

Corollary 2.57 (Valenza [Val80]). *If G is a nilpotent group and $G = H \rtimes K$ with K abelian, then $s(G)$ is bounded by a function of $s(H)$ and the class of G.*

A group G is said to satisfy the minimal condition on subgroups if each nonempty collection of subgroups contains a minimal element; or, equivalently, each descending chain of subgroups stabilizes after a finite number of steps. A solvable group satisfies the minimum condition on subgroups if and only if it is an extension of a direct product of finitely many quasicyclic groups by a finite group (see [Rob95, p. 156]). Thus, in view of Theorem 2.55, we have:

Proposition 2.58 *Every nilpotent group which satisfies minimum condition on subgroups has finite dimension series.*

An A_3-group, in the notation of Mal'cev [Mal56], is an abelian group G whose periodic part P satisfies the minimum condition on subgroups and the quotient G/P has finite rank. A nilpotent A_3-group is a nilpotent group having a finite normal series in which the factor groups are A_3-groups. Clearly, the torsion subgroup of a nilpotent A_3-group satisfies the minimum condition on subgroups and therefore, by Proposition 2.58, the torsion part, and hence by Theorem 2.52, the group itself has finite dimension series:

Theorem 2.59 (Plotkin [Plo73]). *A nilpotent A_3-group has finite dimension series.*

2.9 Modular Dimension Subgroups

In contrast to the case of integral dimension subgroups, definitive answer for the identification of dimension subgroups over fields has long been known. To state the result we need the following definitions, given a group G and a prime p:

(i) Define the series $\{M_{n,p}(G)\}_{n\geq 1}$ by setting

$$M_{1,p}(G) = G, \ \ M_{2,p}(G) = \gamma_2(G), \ \ M_{n+1,p}(G) = [G, M_{n,p}(G)]M_{(\frac{n}{p}),p}^p(G) \tag{2.83}$$

for $n \geq 2$, where $(\frac{r}{s})$ denotes the least integer $\geq \frac{r}{s}$.

(ii) Define the series $\{G_{n,p}\}_{n\geq 1}$ by setting

$$G_{n,p} = \prod_{ip^j \geq n} \gamma_i(G)^{p^j}. \tag{2.84}$$

If H is a subset of a group G, we denote by $\sqrt{H}$ the radical of H:

$$\sqrt{H} = \{x \in G \,|\, x^m \in H, \text{ for some } m > 0\}. \tag{2.85}$$

Theorem 2.60 (Jennings, [Jen41], [Jen55]). *Let F be a field and G a group.*
(i) If $\mathrm{char}(F) = 0$, then $D_{n,F}(G) = \sqrt{\gamma_n(G)}$ for all $n \geq 1$.
(ii) If $\mathrm{char}(F) = p > 0$, then $D_{n,F}(G) = M_{(n),p}(G) = G_{n,p}$ for all $n \geq 1$.

Over general rings, it is known that the dimension subgroups of groups depend only on the ones over the rings $\mathbb{Z}_n$, $n \geq 0$ (see Passi [Pas79], p. 16 for details). We mention here a few results in low dimensions.

Theorem 2.61 (Moran [Mor70]; see also Tasić [Tas93]). *For every group G, prime p and integer $e \geq 1$,*

$$D_{n, \mathbb{Z}_{p^e}}(G) = G^{p^e} \gamma_n(G) \text{ for } 1 \leq n \leq p.$$

Let n be a non-negative integer; if n is even, let $n = 2qm$, where q is a power of 2 and m is odd. Let

$$K_n(G) = \begin{cases} G^n \gamma_3(G), & \text{if } n \text{ is odd or } 0, \\ (G^m \gamma_3(G)) \cap \langle x^{2q} \mid x^q \in G^{2q} \gamma_2(G) \rangle \gamma_3(G), & \text{if } n \text{ is even.} \end{cases}$$

Let $N/K_n(G)$ be the subgroup of the centre of $G/K_n(G)$ consisting of the elements of order dividing n.

Theorem 2.62 (Passi-Sharma [Pas74]).

(i) $G \cap (1 + \Delta_{\mathbb{Z}_n}^3(G) + \Delta_{\mathbb{Z}_n}(G) \Delta_{\mathbb{Z}_n}(N)) = K_n(G)$ *if n is odd or 0.*

(ii) $G \cap (1 + \Delta_{\mathbb{Z}_n}^3(G) + \Delta_{\mathbb{Z}_n}(G) \Delta_{\mathbb{Z}_n}(N)) = K_n(G) \langle x^n \mid x^{qm} \in N \rangle$ *if n is even.*

(iii) $G \cap (1 + \Delta_{\mathbb{Z}_n}^3(G)) = K_n(G)$ *for all n.*

2.10 Lie Dimension Subgroups

Given a multiplicative group G and a commutative ring R with identity, define ideals $\Delta_R^{(n)}(G)$, $n \geq 1$, inductively by setting $\Delta_R^{(1)}(G) = \Delta_R(G)$, the augmentation ideal of the group ring $R[G]$, and

$$\Delta_R^{(n)}(G) = [\Delta_R^{(n-1)}(G), \Delta_R(G)]R[G], \; n > 1, \tag{2.86}$$

the two-sided ideal of $R[G]$ generated by $[\alpha, \beta] = \alpha\beta - \beta\alpha$, $\alpha \in \Delta_R^{(n-1)}(G)$, $\beta \in \Delta_R(G)$. We then have a decreasing series

$$\Delta_R(G) = \Delta_R^{(1)}(G) \supseteq \Delta_R^{(2)}(G) \supseteq \cdots \supseteq \cdots \Delta_R^{(n)}(G) \supseteq \cdots$$

of two-sided ideals in $R[G]$; this series has the property that

$$\Delta_R^{(m)}(G).\Delta_R^{(n)}(G) \subseteq \Delta_R^{(n+m-1)}(G) \tag{2.87}$$

for all $m, n \geq 1$ (see [Pas79], Prop. 1.7 (iii), p.4). Let

$$D_{(n), R}(G) = G \cap (1 + \Delta_R^{(n)}(G)), \; n \geq 1.$$

We call $D_{(n),R}(G)$ the nth *upper Lie dimension subgroup* of G over R. In view of (2.87), $\{D_{(n),R}(G)\}_{n\geq 1}$ is a central series in G. When $R = \mathbb{Z}$, we drop the suffix and write simply $D_{(n)}(G)$ instead of $D_{(n),\mathbb{Z}}(G)$.

Let L be a Lie ring. For subsets H, K of L, we denote by $[H, K]$ the additive subgroup of L spanned by the commutators $[h, k] = hk - kh$, $h \in H$, $k \in K$. Recall that the *lower central series* $\{L_n\}_{n\geq 1}$ of L is defined inductively by setting $L_1 = L$, and $L_{n+1} = [L, L_n]$ for $n \geq 1$. The Lie ring L is said to be *nilpotent* if $L_n = 0$ for some $n \geq 1$.

Let A be an associative ring. We can view A as a Lie ring with the bracket operation defined by

$$[\alpha, \beta] = \alpha\beta - \beta\alpha, \ \alpha, \beta \in A.$$

Define a series of two-sided ideals $\{A^{[n]}\}_{n\geq 1}$ of A by setting $A^{[1]} = A$ and $A^{[n]}$, $n > 1$, to be the two-sided ideal of A generated by the nth term A_n in the lower central series of A viewed as a Lie ring. We say that A is *Lie nilpotent* (resp. *residually Lie nilpotent*) if $A^{[n]} = 0$ for some $n \geq 1$ (resp. $\cap A^{[n]} = 0$).

Theorem 2.63 (Gupta-Levin [Gup83]). *Let A be an associative ring with identity and let $\mathcal{U} = \mathcal{U}(A)$ be its group of units. Then*

$$A^{[m]}.A^{[n]} \subseteq A^{[m+n-2]} \text{ for all } m, \ n \geq 2.$$

Let G be a multiplicative group and R a commutative ring with identity. Consider the series $\{\Delta_R^{[n]}(G)\}_{n\geq 1}$ of two-sided ideals in $R[G]$. Clearly

$$\Delta_R^{[n]}(G) \subseteq \Delta^{(n)}(G) \text{ for all } n \geq 1,$$

and, by Theorem 2.63, we have

$$\Delta^{[n]}(G)\Delta^{[m]}(G) \subseteq \Delta^{[n+m-2]}(G)\mathbb{Z}[G] \tag{2.88}$$

for every group G. The filtration $\{\Delta_R^{[n]}(G)\}_{n\geq 1}$ of $\Delta_R(G)$ defines a series of normal subgroups $\{D_{[n],R}(G)\}_{n\geq 1}$ in G:

$$D_{[n],R}(G) = G \cap (1 + \Delta_R^{[n]}(G)). \tag{2.89}$$

We call $D_{[n],R}(G)$, $n \geq 1$, the nth *lower Lie dimension subgroup* of G over R. As usual, when the ring R is $\mathbb{Z}$, we drop the suffix R and write $D_{[n]}(G)$ instead of $D_{[n],\mathbb{Z}}(G)$.

From definitions, and in view of Theorems 1.6 and 2.63, it is then clear that for any group G and integer $n \geq 1$, we have the following inclusions:

$$\gamma_n(G) \subseteq D_{[n]}(G) \subseteq D_{(n)}(G) \subseteq D_n(G). \tag{2.90}$$

In general, not only the inclusion $\gamma_n(G) \subseteq D_n(G)$ can be strict, but even the inclusion $\gamma_n(G) \subseteq D_{[n]}(G)$ can be strict. To this end, we have

Theorem 2.64 *Let s be an arbitrary natural number. Then there exists a natural number n and a nilpotent group G of class n, such that $D_{[n+s]}(G) \neq 1$.*

We first prove two lemmas.

Lemma 2.65 *Let Π be a group, $k \gg l \gg 4$. If x_1, x_2, $x_3 \in \gamma_m(\Pi)$ and there exist $\xi_i \in \gamma_n(\Pi)$, $i = 1, \ldots, 5$, $n \geq 2m$, $m \geq 3$, such that*

$$x_1^4 = \xi_1,\ x_2^{2^l} = \xi_2,\ x_2^{2^{l+1}} x_3^{2^k} = \xi_3^4,\ x_1^{-2^{l+1}} x_3^{2^{k+1}} = \xi_4^{2^l},\ x_1^{2^k} x_2^{2^{k+1}} = \xi_5^{2^k},$$

then

$$w = [x_1, x_2^{2^{l+1}}][x_1, x_3^{2^k}][x_2, x_3^{2^{k+1}}] \in D_{[n+2m-6]}(\Pi).$$

Proof. Since $1 - x \in \Delta^{[n]}(\Pi)\mathbb{Z}[\Pi]$ for $x \in \gamma_n(\Pi)$, we have

$$1 - w \equiv \alpha_1 + \alpha_2 + \alpha_3 \mod \Delta^{[2n]}(\Pi)\mathbb{Z}[\Pi],$$

where $\alpha_1 = (1 - [x_1, x_2^{2^{l+1}}])$, $\alpha_2 = (1 - [x_1, x_3^{2^k}])$, $\alpha_3 = (1 - [x_2, x_3^{2^{k+1}}])$. Now, working modulo $\Delta^{[n+2m-6]}(\Pi)\mathbb{Z}[\Pi]$, we have

$$\alpha_1 \equiv (1 + (x_1^{-1} x_2^{-2^{l+1}} - 1))((1 - x_2^{2^{l+1}})(1 - x_1) - (1 - x_1)(1 - x_2^{2^{l+1}}))$$
$$\equiv (1 - x_2^{2^{l+1}})(1 - x_1) - (1 - x_1)(1 - x_2^{2^{l+1}}),$$

since $x_1^{-1} x_2^{-2^{l+1}} \in \gamma_m(\Pi)$ and

$$(x_1^{-1} x_2^{-2^{l+1}} - 1)((1 - x_2^{2^{l+1}})(1 - x_1) - (1 - x_1)(1 - x_2^{2^{l+1}})) \in \Delta^{[n+2m-6]}(\Pi)\mathbb{Z}[\Pi]$$

by (2.88). Modulo $\Delta^{[n+2m-6]}(\Pi)\mathbb{Z}[\Pi]$, we have:

$$\alpha_1 \equiv (1 - x_2^{2^{l+1}})(1 - x_1) - 2^{l+1}(1 - x_1)(1 - x_2)$$
$$+ \sum_{i=2}^{n-1} (-1)^i \binom{2^{l+1}}{i} (1 - x_1)(1 - x_2)^i$$

Note that $\binom{2^{l+1}}{i}$ is divisible by 4^n for sufficiently large l and $i \leq n$. Hence, for such a large l,

$$\sum_{i=2}^{n-1} (-1)^i \binom{2^{l+1}}{i} (1 - x_1)(1 - x_2)^i \in \Delta^{[n+2m-6]}(\Pi)\mathbb{Z}[\Pi].$$

By the same principle, we get

$$2^{l+1}(1 - x_1)(1 - x_2) \equiv (1 - x_1^{2^{l+1}})(1 - x_2) \mod \Delta^{[n+2m-6]}(\Pi)\mathbb{Z}[\Pi].$$

Therefore,

$$\alpha_1 \equiv (1 - x_2^{2^{l+1}})(1 - x_1) - (1 - x_1^{2^{l+1}})(1 - x_2) \quad \mathrm{mod}\ \Delta^{[n+2m-6]}(\Pi)\mathbb{Z}[\Pi].$$

Choosing k to be such that $\binom{2^{k+1}}{i}$ is divisible by 2^{ln} for any $i \le n$, we get

$$\alpha_2 \equiv (1 - x_3^{2^k})(1 - x_1) - (1 - x_1^{2^k})(1 - x_3) \quad \mathrm{mod}\ \Delta^{[n+2m-6]}(\Pi)\mathbb{Z}[\Pi],$$
$$\alpha_3 \equiv (1 - x_3^{2^{k+1}})(1 - x_2) - (1 - x_2^{2^{k+1}})(1 - x_3) \quad \mathrm{mod}\ \Delta^{[n+2m-6]}(\Pi)\mathbb{Z}[\Pi].$$

Therefore,

$$\alpha_1 + \alpha_2 + \alpha_3 \equiv (2 - x_2^{2^{l+1}} - x_3^{2^k})(1 - x_1) + (x_1^{2^{l+1}} - x_3^{2^{k+1}})(1 - x_2) +$$

$(x_1^{2^k} + x_2^{2^{k+1}} - 2)(1 - x_3) \equiv (1 - \xi_3^4)(1 - x_1) + (1 - \xi_4^{2^l})(1 - x_2) + (1 - \xi_5^{2^k})(1 - x_3) \equiv$

$(1 - \xi_3)(1 - \xi_1) + (1 - \xi_4)(1 - \xi_2) + (1 - \xi_5)(1 - \xi_3) \equiv 0 \quad \mathrm{mod}\ \Delta^{[n+2m-6]}(\Pi)\mathbb{Z}[\Pi],$

and hence $w \in D_{[n+2m-6]}(\Pi)$. $\square$

Lemma 2.66 *Let $W_{m,n}$ be the group given by the following presentation:*

$$\langle x_1, \ldots, x_{14} \mid [x_1,\ _m x_{11}]^4 \xi_1, [x_2,\ _m x_{12}]^{2^l} \xi_2, [x_3,\ _m x_{13}]^{2^k} \xi_3,$$
$$[x_4,\ _n x_{14}, x_{10}]^4 [x_4,\ _n x_{14}, x_3,\ _m x_{13}, x_7]^{2^{k-1}}, \xi_1^{2^{k-2}} \xi_2^{2^{k-l+1}} \rangle,$$

where

$$\xi_1 = [x_4,\ _n x_{14}, x_7]^2 [x_4,\ _n x_{14}, x_6][x_4,\ _n x_{14}, x_5]^2,$$
$$\xi_2 = [x_4,\ _n x_{14}, x_7]^{2^{l-2}} [x_4,\ _n x_{14}, x_{10}]^{-1} [x_4,\ _n x_{14}, x_5]^2,$$
$$\xi_3 = [x_4,\ _n x_{14}, x_6]^{-2^{l-2}} [x_4,\ _n x_{14}, x_{10}]^{-2}.$$

Then the element

$$w_{n,m} = [x_1,\ _m x_{11}, [x_2,\ _m x_{12}]^{2^{l+1}}][x_1,\ _m x_{11}, [x_3,\ _m x_{13}]^{2^k}]$$
$$[x_2,\ _m x_{12}, [x_3,\ _m x_{13}]^{2^{k+1}}]$$

does not lie in $\gamma_{n+m+4}(W_{n,m})$, $n \ge m \ge 0$.

Proof. Let F be a free group with basis $\{x_1, \ldots, x_{10}\}$. Consider four types of relations:

$$R_1 = \gamma_4(F),$$

$$R_2 = \langle R_1, [x_i, x_j, x_k] \notin \langle \alpha, \beta, \gamma, \delta, \epsilon, \theta \rangle R_1 \text{ for all } i, j, k,$$
$$\alpha^{2^{k-l}} \beta^{-1}, \beta^{2^{l-2}} \gamma^{-1}, \gamma^4, \beta\epsilon, \alpha\delta, \theta\gamma \rangle,$$

where

$$\alpha = [x_4, x_3, x_7], \ \beta = [x_4, x_2, x_6], \ \gamma = [x_4, x_{10}, x_1], \ \delta = [x_4, x_7, x_3],$$
$$\epsilon = [x_4, x_6, x_2], \ \theta = [x_4, x_1, x_{10}];$$

Now define R_3 to be the subgroup generated by R_2 together with the normal closure of the following words:

$$[x_3, x_4]^{2^k}, \ [x_2, x_4]^{2^l},$$

$$[x_4, x_1]^4, \ [x_2, x_3]^{2^l} \alpha^{-2^{l-2}},$$

$$[x_3, x_1]^4 \alpha^{-2}, \ [x_2, x_1]^4 \beta^{-1},$$

$$[x_4, x_5]^{2^{k-l+2}} \alpha^{-2^{k-1}}, \ [x_4, x_7]^{2^k} \gamma^2,$$

$$[x_4, x_6]^{2^{k-2}}, \ [x_4, x_{10}]^4 \gamma^2,$$

$$[x_5, x_i], \ i \neq 1, \ [x_1, x_i], \ [x_2, x_i], \ [x_3, x_i], \ i > 4.$$

Let R_4 be the subgroup generated by R_3 and the normal closure of words

$$c_1 = x_1^4 [x_4, x_7]^2 [x_4, x_6][x_4, x_5]^2,$$
$$c_2 = x_2^{2^l} [x_4, x_7]^{2^{l-2}} [x_4, x_{10}]^{-1} [x_4, x_5]^2,$$
$$c_3 = x_3^{2^k} [x_4, x_6]^{-2^{l-2}} [x_4, x_{10}]^{-2}.$$

Set $H = F/R_4$. We claim that $[R_{i+1}, F] \subseteq R_i$, $i = 1, 2, 3$. This is obvious for $i = 1, 2$. We show it for $i = 3$. Working modulo R_3, we have:

$[c_1, x_1] = 1,$

$[c_1, x_2] = [x_1, x_2]^4 [x_4, x_6, x_2] = [x_1, x_2]^4 \beta = 1,$

$[c_1, x_3] = [x_1, x_3]^4 [x_4, x_7, x_3]^2 = [x_1, x_3]^4 \alpha^2 = 1,$

$[c_1, x_4] = [x_1, x_4]^4 = 1,$

$[c_2, x_1] = [x_2, x_1]^{2^l} [x_4, x_{10}, x_1]^{-1} = \beta^{2^{l-2}} \gamma^{-1} = 1,$

$[c_2, x_2] = 1,$

$[c_2, x_3] = [x_2, x_3]^{2^l} [x_4, x_7, x_3]^{2^{l-2}} = [x_2, x_3]^{2^l} \alpha^{-2^{l-2}} = 1,$

$[c_2, x_4] = [x_2, x_4]^{2^l} = 1,$

$[c_3, x_1] = [x_3, x_1]^{2^k} [x_4, x_{10}, x_1]^{-2} = \alpha^{2^{k-1}} \gamma^{-2} = 1,$

$[c_3, x_2] = [x_3, x_2]^{2^k} [x_4, x_6, x_2]^{-2^{l-2}} = \alpha^{2^{k-2}} \beta^{-2^{l-2}} = 1,$

$[c_3, x_3] = 1,$

$[c_3, x_4] = [x_3, x_4]^{2^k} = 1.$

By standard arguments, one can show that the element

$$w = [x_1, x_2^{2^{l+1}}][x_1, x_3^{2^k}][x_2, x_3^{2^{k+1}}]$$

is nontrivial in H. Note that all brackets $[x_j, x_i, x_j]$ are trivial in H.

Let W be a group given by the following presentation:

$$\langle x_1,\ \ldots,\ x_{10}\ |\ x_1^4\xi_1,\ x_2^{2^l}\xi_2,\ x_3^{2^k}\xi_3,$$

$$[x_4,\ x_{10}]^4[x_4,\ x_3,\ x_7]^{2^{k-1}},\ \xi_1^{2^{k-2}}\xi_2^{2^{k-l+1}}\rangle,$$

where

$$\xi_1 = [x_4,\ x_7]^2[x_4,\ x_6][x_4,\ x_5]^2,$$

$$\xi_2 = [x_4,\ x_7]^{2^{l-2}}[x_4,\ x_{10}]^{-1}[x_4,\ x_5]^2,$$

$$\xi_3 = [x_4,\ x_6]^{-2^{l-2}}[x_4,\ x_{10}]^{-2}.$$

It is easy to see that the group $W_{0,0}$ is a free product of W with a free group of rank 5. The group W naturally maps onto H, and $W_{0,0}$ maps onto G_2. The image of $w_{0,0}$ is exactly the element w which is nontrivial, hence $w_{0,0} \notin \gamma_4(W_{0,0})$.

We shall prove first that $w_{m,m} \notin \gamma_{2m+4}(W_{m,m})$, i.e., the case $n = m$. For any m consider the quotient $W'_{m,m} = W_{m,m}/\gamma_{2m+4}(W_{m,m})N_m$, where N_m is the normal closure in $W_{m,m}$ of brackets $[y_1,\ \ldots,\ y_t]$, $t \geq 3$, such that there are at least two occurrences of $y_i = x_1$ or $y_i = x_2$, or $y_i = x_3$, or $y_i = x_4$ in this bracket, or at least three occurrences of elements from $\{x_1, x_2, x_3, x_4\}$ simultaneously. We see that all such brackets are trivial in H, hence $w_{0,0}$ is nontrivial in $W'_{0,0}$.

We assume that the element $w_{m,m}$ is nontrivial in $W'_{m,m}$ for a given m and we shall prove the statement for $m+1$.

Consider the following automorphism f of the free group of rank 14:

$$x_i \mapsto x_i,\ i \neq 11, 12, 13, 14,$$

$$x_{11} \mapsto x_{11}x_1,$$

$$x_{12} \mapsto x_{12}x_2,$$

$$x_{13} \mapsto x_{13}x_3,$$

$$x_{14} \mapsto x_{14}x_4.$$

Clearly, this automorphism can be extended to get an automorphism f' of a group $W'_{m,m}$. This automorphism defines the semi-direct product

$$W''_{m,m} = W'_{m,m} \rtimes \langle x \rangle,$$

where x acts as f'. Clearly, we have in $W''_{m,m}$:

$$[x,\ x_i] = 1,\ i \neq 11,\ 12,\ 13,\ 14,$$

$$[x_{11},\ x] = x_1,\ [x_{12},\ x] = x_2,\ [x_{13},\ x] = x_3,\ [x_{14},\ x] = x_4.$$

It is easy to see that $W''_{m,m}$ is nilpotent with $\gamma_{2m+6}(W''_{m,m}) = 1$. Note that $W''_{m,m}$ is an epimorphic image of $W_{m+1,m+1}$; thus the image of the element $w_{m+1,m+1}$ is nontrivial in $W''_{m,m}$, since it is the same as the element $w_{m,m}$ in $W''_{m,m}$. Thus we have proved that $w_{m+1,m+1} \notin \gamma_{2m+6}(W'_{m+1,m+1})$. The induction is thus complete and we have

$$w_{m,m} \notin \gamma_{2m+4}(W'_{m,m})$$

for any $m \geq 0$.

Now we shall prove the needed result for general case $n \geq m \geq 0$. We fix m and make an induction on $t = n - m$. For the case $t = 0$ we already proved the needed result. We consider the quotient $W'_{n,m} = W_{n,m}/\gamma_{n+m+4}(W_{n,m})N_m$. Consider the following automorphism f' of a free group on generators x_i:

$$x_i \mapsto x_i, \ i \neq 14,$$

$$x_{14} \mapsto x_{14}x_4.$$

Clearly, it extends to an automorphism of the group $W'_{n,m}$. Then the corresponding semi-direct product $W''_{n,m} = W_{n,m} \rtimes x$ is nilpotent with $\gamma_{n+m+5}(W''_{n,m}) = 1$. Observe that $W'_{n+1,m}$ naturally maps onto $W''_{n,m}$, sending non-trivially the element $w_{n+1,m}$. Hence $w_{n+1,m} \notin \gamma_{n+m+5}(W'_{n+1,m})$ and we have thus completed the induction. $\square$

Proof of Theorem 2.64. By Lemma 2.65, for $k \gg l \gg 4$, we have

$$w_{n,m} \in D_{[n+2m-2]}(W_{n,m}) \setminus \gamma_{n+m+4}(W_{n,m}).$$

Since the difference $(n+2m-2) - (n+m+4) = m-6$ can be taken arbitrarily, the statement of the Theorem 2.64 follows. $\square$

When R is a field, upper Lie dimension subgroups have been identified in [Pas75b] (see also [Pas79]). To state the result we need the following definitions, given a group G and a prime p:

(i) Define the series $\{M_{(n),p}(G)\}_{n\geq 1}$ by setting
$M_{(1),p}(G) = G$, $M_{(2),p}(G) = \gamma_2(G)$, $M_{(n+1),p}(G) = [G, M_{(n),p}(G)]M^p_{(\lceil \frac{n+p}{p} \rceil),p}(G)$
for $n \geq 2$, where $(\frac{r}{s})$ denotes the least integer $\geq \frac{r}{s}$.

(ii) Define the series $\{G_{(n),p}\}_{n\geq 1}$ by setting

$$G_{(n),p} = \prod_{(i-1)p^j \geq n} \gamma_i(G)^{p^j}.$$

Theorem 2.67 (Passi-Sehgal [Pas75b]). *Let G be a group and F a field. Then, for $n \geq 2$,*

$$D_{(n), F}(G) = \begin{cases} \sqrt{\gamma_n(G)} \cap \gamma_2(G), & \text{if } \operatorname{char}(F) = 0, \\ G_{(n-1), p} = M_{(n), p}(G), & \text{if } \operatorname{char}(F) = p > 0. \end{cases}$$

Some very interesting properties of lower central and dimension subgroups have been observed by A. Shalev [Sha90a]. To mention a sample, let us adopt the following

Notation. For integers $n \geq 1$, $k \geq 0$, write

$$D_{n, k}(G) = \prod_{ip^j \geq n} \gamma_{i+k}^{p^j}(G).$$

Proposition 2.68 (Shalev [Sha90a].) *For integers $n \geq 1$, $k \geq 0$,*

$$[D_{n, k}(G), G] = D_{n, k+1}(G).$$

We next consider the lower Lie dimension subgroups in characteristic $p > 0$. An identification of these subgroups is known when $p \neq 2$, 3. First note the following

Proposition 2.69 *The series $\{D_{[n], \mathbb{F}_p}(G)\}_{n \geq 1}$ is a central series of G satisfying*

(i) $[D_{[m], \mathbb{F}_p}(G), D_{[n], \mathbb{F}_p}(G)] \subseteq D_{[m+n-2], \mathbb{F}_p}(G)$, m, $n \geq 2$.

(ii) $(D_{[n], \mathbb{F}_p}(G))^p \subseteq D_{[p(n-2)+2], \mathbb{F}_p}(G)$, $n \geq 2$.

Theorem 2.70 (Bhandari-Passi [Bha92b]), Riley [Ril91]). *For every group G and field F with $\operatorname{char}(F) \neq 2$, 3,*

$$D_{[n], F}(G) = D_{(n), F}(G) \text{ for all } n \geq 1.$$

Theorem 2.71 (Bhandari-Passi [Bha92b]). *Let G be a group. Then for all $n \geq 0$*

(i) $D_{[2^n+2], \mathbb{F}_2}(G) = D_{(2^n+2), \mathbb{F}_2}(G)$;

(ii) $D_{[a3^n+2], \mathbb{F}_3}(G) = D_{(a3^n+2), \mathbb{F}_3}(G)$, $0 \leq a \leq 2$.

As a result of Theorems 2.70 and 2.71, we have

Corollary 2.72 *The following statements for a group algebra $F[G]$ are equivalent:*

(i) $F[G]$ *is residually Lie nilpotent.*

(ii) $\bigcap_{n \geq 1} D_{[n], F}(G) = 1$.

(iii) Either $\operatorname{char}(F)$ *is zero and G is residually "nilpotent with derived group torsion free", or $\operatorname{char}(F) = p > 0$ and G is residually "nilpotent with derived group a p-group of bounded exponent".*

We end this section with a review of the results on integral Lie dimension subgroups.

Theorem 2.73 $D_{(n)}(G) = \gamma_n(G)$ *for* $1 \leq n \leq 8$.

The above result for $1 \leq n \leq 6$ is due to Sandling [San72a] and the cases $n = 7, 8$ are due to Gupta-Tahara [Gup93].

Theorem 2.74 (Gupta-Srivastava [Gup91c]). *In general,*

$$D_{[n],\mathbb{Z}}(G) \neq \gamma_n(G) \text{ for } 9 \leq n \leq 13.$$

Theorem 2.75 (Hurley-Sehgal [Har91b]). *In general,*

$$D_{[n],\mathbb{Z}}(G) \neq \gamma_n(G) \text{ for } n \geq 14,$$

and

$$D_{(n),\mathbb{Z}}(G) \neq \gamma_n(G) \text{ for } n \geq 9.$$

2.11 Lie Nilpotency Indices

Theorem 2.76 (Passi, Passman and Sehgal [Pas73]). *The group algebra $F[G]$ of a group G over a field F is Lie nilpotent if and only if either the characteristic of F is zero and G is abelian, or the characteristic of F is a prime p, G is nilpotent and G', the derived subgroup of G, is a finite p-group.*

As a consequence of the above theorem, we have

Corollary 2.77 *The following two statements are equivalent:*

(i) $F(G)^{(m)} = 0$ *for some* $m \geq 1$.
(ii) $F(G)^{[n]} = 0$ *for some* $n \geq 1$.

For a Lie nilotent group algebra $F[G]$, define the *upper and lower Lie nilpoency indices* $t^L(F[G])$ and $t_L(F[G])$ as follows:

$$t^L(F[G]) = \min\{m \mid F[G]^{(m)} = 0\},$$

$$t_L(F[G]) = \min\{m \mid F[G]^{[m]} = 0\}.$$

Clearly $t_L(F[G]) \leq t^L(F[G])$, and by Theorem 2.63, the unit group $\mathcal{U}(F[G])$ is nilpotent of class c, say, with $c + 1 \leq t_L(F[G])$. In fact, in view of a result of Du [Du,92], $c + 1 = t_L(F[G])$ (see Theorems 2.79, 2.80 below).

Recall that a ring R is said to be a Jacobson radical ring if, for every $r \in R$, there exists $s \in R$ such that

$$r + s - rs = 0 = r + s - sr.$$

Let R be a Jacobson radical ring. Define a binary operation on R by setting

$$a \circ b = a + b - ab, \quad a, \, b \in R.$$

With this binary operation, R is a group, called the *adjoint group* of R; we denote this group by $(R, \circ)$.

Example 2.78

Let G be a finite p-group and F a field of characteristic p. Then the augmentation ideal $\Delta_F(G)$ is nilpotent; therefore $\Delta_F(G)$ is a Jacobson radical ring. Observe that the group $(\Delta_F(G), \circ)$ is isomorphic to the group $\mathcal{U}_1(F[G])$ of units of augmentation 1 under the map

$$\alpha \mapsto 1 - \alpha, \quad \alpha \in \Delta_F(G).$$

Theorem 2.79 *If G is a finite p-group, F a field of characteristic p, then*

$$nilpotency \ class \ of \ \mathcal{U}(F[G]) = t_L(F[G]).$$

This result is an immediate consequence of the following

Theorem 2.80 (Du [Du,92]). *The associated Lie ring of a Jacobson radical ring is nilpotent of class n if and only if its adjoint group is nilpotent of class n.*

Theorem 2.81 (Bhandari-Passi [Bha92a]). *Let F be a field of characteristic $p > 3$ and let G be a group such that $F[G]$ is Lie nilpoent. Then*

$$t_L(F[G]) = t^L(F[G]) = 2 + (p - 1) \sum_{m \geq 1} m d_{(m+1)},$$

where, for $m \geq 2$, $p^{d_{(m)}} = [D_{(m), F}(G) : D_{(m+1), F}(G)]$.

   The proof of the above theorem requires the following results of Sharma-Srivastava. Following their notation, let $L_n(R)$ denote the nth term in the lower central series of the ring R when viewed as a Lie ring under commutation.

Theorem 2.82 (Sharma-Srivastava [Sha90b], Theorem 2.8). *Let R be a ring in which both 2, 3 are invertible. If m and n are any two positive integers such that one of them is odd, then*

$$L_m(R)RL_n(R)R \subseteq L_{m+n-1}(R)R.$$

Lemma 2.83 (Sharma-Srivastava [Sha90b], Lemma 2.11). *Let R be a ring in which both 2, 3 are invertible. Then for any positive integers m and n and for all $g_1, g_2, \ldots, g_m \in \mathcal{U}(R)$,*

$$([g_1, g_2, \ldots, g_{m+1}] - 1)^n \in L_{mn+1}(R)R.$$

Proof of Theorem 2.81. Since $F[G]$ is Lie nilpotent, by Theorem 2.76, G is nilpotent and its derived subgroup G' is a finite p-group. If G is abelian, then the assertion is obviously true; thus we assume that $G' \neq 1$.

Let $H_i = D_{(i+1), F}(G)$, $i \geq 1$, and $p^{e_i} = [H_i : H_{i+1}]$ so that $e_i = d_{(i+1)}$. The series

$$G = H_1 \supset H_2 \supset \ldots \supset H_d \supset H_{d+1} = 1$$

is a restricted N-series in G', i.e.,

$$[H_i, H_j] \subseteq H_{i+j}, \quad H_i \subseteq H_{ip}, \text{ for all } i, \; j \geq 1.$$

By Theorem 2.67,

$$H_n = \prod_{(i-1)p^j \geq n} \gamma_i(G)^{p^j}.$$

Now observe that H_n is generated, modulo H_{n+1}, by the elements of the type x^{p^j}, where x is a left-normed group commutator of weight i and $(i-1)p^j = n$. Thus it is possible to choose a canonical basis (see [Pas79], p. 23) $\{x_{11}, x_{12}, \ldots, x_{1e_1}, x_{21}, x_{22}, \ldots, x_{2e_2}, \ldots, x_{d1}, x_{d2}, \ldots, x_{de_d}\}$ of G', where for $1 \leq r \leq d$, $1 \leq k \leq e_r$, x_{rk} is an element of of the type $\xi_i^{p^j}$, where ξ_i is a left-normed group commutator of weight i and $(i-1)p^j = r$. It then follows that the element

$$\alpha = (x_{11} - 1)^{(p-1)}(x_{12} - 1)^{(p-1)} \ldots (x_{1e_1} - 1)^{(p-1)} \ldots$$
$$(x_{d1} - 1)^{(p-1)} \ldots (x_{de_d} - 1)^{(p-1)} \quad (2.91)$$

is a non-zero element of $F[G]$. For $1 \leq r \leq d$, $1 \leq k \leq e_r$, $x_{rk} = \xi_i^{p^j}$, by Lemma 2.83, we have

$$(x_{rk} - 1)^{(p-1)} = (\xi_i - 1)^{p^j(p-1)} \in F[G]^{[(i-1)p^j(p-1)+1]} = F[G]^{[r(p-1)+1]}.$$

Moreover, by Theorem 2.82

$$(x_{r1} - 1)^{(p-1)}(x_{r2} - 1)^{(p-1)} \ldots (x_{re_r} - 1)^{(p-1)} \in F[G]^{[re_r(p-1)+1]},$$

which in turn yields that $\alpha \in F[G]^{[1+(p-1)\sum_{r=1}^d re_r]}$. Since $\alpha \neq 0$, it follows that

$$t_L(F[G]) \geq 2 + (p-1) \sum_{r=1}^d re_r = 2 + (p-1) \sum_{m \geq 1} md_{(m+1)},$$

as $e_r = d_{(r+1)}$ for $r \geq 1$.

Since, $t^L(F[G]) = 2 + (p-1) \sum_{r=1}^d re_r = 2 + (p-1) \sum_{m \geq 1} md_{(m+1)}$ (see [Pas79], p. 47) and $t^L(F[G]) \geq t_L(F[G])$ always, the proof is complete. $\square$

Let p be a prime and R a ring of characteristic p. Consider the Lie powers $R^{(m)}$, $m \geq 1$, of R defined inductively by setting $R^{(1)} = R$, and, for $m \geq 1$,

$R^{(m+1)}$ to be the two-sided ideal of R generated by the ring commutators $xy - yx$, $x \in R^{(m)}$, $y \in R$.

Theorem 2.84 (Shalev [Sha91]). *If $R^{(1+(p-1)p^{i-1})} = 0$ for some $i \geq 1$, then R satisfies the identity*

$$(X + Y)^{p^i} = X^{p^i} + Y^{p^i}. \tag{P_i}$$

Corollary 2.85 *Let G be a finite group of exponent p^e, and K a field of characteristic p. If $K[G]^{(1+(p-1)p^{e-1})} = 0$, then $\exp(\mathcal{U}_1(K[G])) = \exp(G)$, where $\mathcal{U}_1(K[G])$ is the group of units of $K[G]$ having augmentation 1.*

Proof. Since $G \subseteq \mathcal{U}_1(K[G])$, it only needs to be checked that $\exp(\mathcal{U}_1(K[G])) \leq \exp(G)$.

Let $u = \sum a_j g_j \in \mathcal{U}_1(K[G])$, $a_j \in K$, $g_j \in G$. Since, by Theorem 2.84, $K[G]$ satisfies the identity (P_i), we have

$$u^{p^e} = \sum (a_j g_j)^{p^e} = \sum a_j^{p^e}.1 = 1. \quad \square$$

In [Sha91] Shalev shows that the hypothesis of the above Corollary is satisfied if $p \geq 7$ and $\exp(G)^3 > |G|$, and thus we have

Theorem 2.86 (Shalev [Sha91]). *Let K be a field of prime characteristic p, and G a finite p-group. Then G and $\mathcal{U}_1(K[G])$ have the same exponent if $p \geq 7$ and $\exp(G)^3 > |G|$.*

2.12 Subgroups Dual to Dimension Subgroups

Let G be a group and R a commutative ring with identity. Define a series $\{Z_n(R[G])\}_{n \geq 0}$ of two-sided ideals in $R[G]$ inductively by setting

$$Z_0(R[G]) = 0$$

and

$$Z_{n+1}(R[G]) = \{\alpha \in R[G] \mid \alpha(g - 1) \in Z_n(R[G]), \ (g - 1)\alpha \in Z_n(R[G])\}$$

for $n \geq 0$. This ascending series $\{Z_n(R[G])\}_{n \geq 0}$ of two-sided ideals of $R[G]$ is the most rapidly ascending among all ascending series stabilized by G; it defines an ascending series $\{\mathfrak{z}_{n,R}(G)\}_{n \geq 0}$ of normal subgroups of G, by setting

$$\mathfrak{z}_{n,R}(G) := G \cap (1 + Z_n(R[G])).$$

The series $\{\mathfrak{z}_{n,R}(G)\}_{n \geq 0}$ has been investigated by R. Sandling [San72b]; it is in a sense dual to the dimension series $\{D_{n,R}(G)\}_{n \geq 1}$ defined by the series of augmentation powers $\Delta_R^n(G)$, $n \geq 1$, which is the most rapidly descending among the descending series stabilized by G. The subgroups $\mathfrak{z}_{n,R}(G)$, $n \geq 1$, are rarely non-trivial. More precisely, we have

Theorem 2.87 (Sandling [San72b]). *The normal subgroup $\mathfrak{z}_{n,R}(G)$ is non-trivial for some $n \geq 1$ if and only if the group G is a finite p-group and the ring R is of characteristic p^e for some $e \geq 1$.*

Proof. Suppose $N := \mathfrak{z}_{n,R}(G)$ has a non-trivial element g, say, for some $n \geq 1$. Then, by definition of $Z_n(R[G])$, $(g-1)\Delta_R^n(G) = 0$. This is not possible if G is infinite. Hence G must be finite. Now $\Delta_R^{n+1}(N) = 0$. Therefore, there exists a prime p such that N a finite p-group and R has characteristic p^e for some $e \geq 1$. If $1 \neq h \in G$ is a p'-element, then $(g-1).(h-1)^n = 0$. It then follows that $(g-1)(h-1) = 0$ and hence $h = 1$, a contradiction. Hence G is a finite p-group. $\square$

Let R be a commutative ring with identity and G a group. Let $W = R \wr G$, the standard wreath product of the abelian group R and the group G. Let

$$J_n(R, G) := \{\alpha \in R[G] \mid \alpha.\Delta_R^n(G) = 0\},$$

i.e., the left annihilator of $\Delta_R^n(G)$. The subgroups $\mathfrak{z}_n(R, G)$ are related to the upper central series $\{\zeta_n(W)\}_{n \geq 0}$ of W. More precisely, there is the following result which is easily proved by induction.

Theorem 2.88 (Sandling [San72b]). *For all natural numbers n,*

$$\zeta_n(W) = \mathfrak{z}_{n-1}(R, G)J_n(R, G).$$

Chapter 3
Derived Series

Closely related to the lower central series of a group is its derived series. We first give a method, due to Gupta-Passi [Gup07], for studying the derived series of free nilpotent groups. Our main aim in this Chapter, however, is the study of the trans-finite terms of the derived series of groups. We give an account of the homological approach of R. Strebel [Str74]. The impact on the asphericity conjecture of J. H. C. Whitehead [Whi41] is also explored.

3.1 Commutator Subgroups of Free Nilpotent Groups

Let F be a free group of finite rank. The structure of $\gamma_n(F), n \geq 2$, as a subgroup of the commutator subgroup $[F, F] (= \gamma_2(F))$ has been investigated by Gupta-Passi [Gup07] with the help of a filtration

$$F = F^{(0)} \supset F^{(1)} \supset \ldots \supset F^{(l-1)} \supset F^{(l)} = \{1\}$$

of F by finitely generated free groups $F^{(k)} \subseteq \delta_k(F)$, which we briefly discussed in Chapter 2, §2.7. This procedure, which is an interesting blend of the properties of lower central and derived series, provides a setting for recognizing, up to isomorphism, the derived subgroups $\delta_k(F/\gamma_n(F))$ of the free nilpotent group $F/\gamma_n(F)$ as $F^{(k)}/\gamma_{(n)}(F^{(k)})$, where $\gamma_{(n)}(F^{(k)}) = F^{(k)} \cap \gamma_n(F)$ is the normal closure of certain commutators of weight exceeding $n - 1$.

Let $F = \langle x_1, \ldots, x_m \,|\, \emptyset \rangle$ be the free group of rank $m \geq 2$ with basis $x_1, \ldots, x_m$ and let $n \geq 2$ be a *fixed* integer. Let l be the least natural number satisfying $2^l \geq n$. Set

$$F^{(0)} = F, \quad B(0) = \{x_{0,1}, \ldots, x_{0, m(0)}\},$$

where $m(0) = m$, $x_{0, i} = x_i$ for $1 \leq i \leq m$ and the elements of $B(0)$ are ordered as $x_{0,1} < \ldots < x_{0, m(0)}$. Define the *weight* of each $x_{0, i}$ $(1 \leq i \leq m(0))$ by $\mathbf{wt}(x_{0, i}) = 1$. Now assume that, for some $k \geq 1$, we have already defined the free subgroup $F^{(k-1)} \subseteq \delta_{k-1}(F)$ having ordered basis

R. Mikhailov, I.B.S. Passi, *Lower Central and Dimension Series of Groups*,
Lecture Notes in Mathematics 1952,
© Springer-Verlag Berlin Heidelberg 2009

$$B(k-1) = \{x_{k-1,1}, \, x_{k-1,2}, \, \ldots, \, x_{k-1,m(k-1)}\} \quad (x_{k-1,i} < x_{k-1,i+1})$$

with integral weights $\mathbf{wt}(x_{k-1,i})$ defined and satisfying

$$2^{k-1} \leq \mathbf{wt}(x_{k-1,i}) < n \quad \text{and} \quad \mathbf{wt}(x_{k-1,j}) \leq \mathbf{wt}(x_{k-1,j+1})$$

for $1 \leq i \leq m(k-1)$ and $1 \leq j \leq m(k-1) - 1$. Consider the set U_{k-1} of all *distinct* t-tuples $(z_1, \ldots, z_t)$, $t \geq 2$, of elements $z_j \in B(k-1)$ satisfying the following conditions:

(i) $z_1 \leq \ldots \leq z_t$; (ii) $z_1 < z_t$; (iii) $\mathbf{wt}(z_1) + \ldots + \mathbf{wt}(z_t) < n$.

For each t-tuple $\mathbf{z} = (z_1, \ldots, z_t) \in U_{k-1}$, consider the set $S(\mathbf{z})$ of all *distinct* left-normed commutators

$$\underline{\mathbf{z}} = [z_j, z_1, \cdots, z_{j-1}, z_{j+1}, \ldots, z_t] \quad (z_j > z_1, \, 2 \leq j \leq t)$$

and define, for each such $\underline{\mathbf{z}}$,

$$\mathbf{wt}(\underline{\mathbf{z}}) = \mathbf{wt}(z_1) + \cdots + \mathbf{wt}(z_t).$$

Let $B(k)$ be the subset

$$B(k) := \bigcup_{\mathbf{z} \in U_{k-1}} S(\mathbf{z}) \tag{3.1}$$

of $[F^{(k-1)}, F^{(k-1)}]$, and let

$$F^{(k)} := \langle B(k) \rangle. \tag{3.2}$$

In view of Lemmas 2.43 and 2.44, $B(k) \subseteq \delta_k(F)$ is a free basis of $F^{(k)}$. Now order the finite set $B(k)$ as follows:

For each pair of elements

$$\underline{\mathbf{z}} = [z_i, z_1, \ldots, z_{i-1}, z_{i+1}, \ldots, z_s], \quad \underline{\mathbf{z}}' = [z_j', z_1', \ldots, z_{j-1}', z_{j+1}', \ldots, z_t'],$$

in $B(k)$, define $\underline{\mathbf{z}} < \underline{\mathbf{z}}'$ if

(i) $\mathbf{wt}(\underline{\mathbf{z}}) < \mathbf{wt}(\underline{\mathbf{z}}')$, or
(ii) $\mathbf{wt}(\underline{\mathbf{z}}) = \mathbf{wt}(\underline{\mathbf{z}}')$ and $s < t$, or
(iii) $\mathbf{wt}(\underline{\mathbf{z}}) = \mathbf{wt}(\underline{\mathbf{z}}')$, $s = t$ and $(z_1, \ldots, z_s) < (z_1', \ldots, z_s')$ lexicographically relative to the ordering in $B(k-1)$, or
(iv) $(z_1, \ldots, z_s) = (z_1', \ldots, z_s')$ and $i < j$.

With this ordering, rename the elements of the set $B(k)$ as

$$x_{k,1}, \, \cdots, \, x_{k,m(k)}$$

with $x_{k,i} < x_{k,i+1}$ for $1 \leq i \leq m(k) - 1$, so that

$$F^{(k)} = \langle x_{k,1}, \ldots, x_{k,m(k)} \mid \emptyset \rangle.$$

Clearly $B(t) = \emptyset$ when $2^t \geq n$ and thus $F^{(l)} = \{1\}$.

Let $\gamma_{(n)}(F^{(k)})$ be the normal closure in $F^{(k)}$ of all commutators

$$[x_{k,i(1)}, x_{k,i(2)}, \ldots, x_{k,i(t)}], \ i(1) > i(2) \leq \ldots \leq i(t),$$

satisfying

$$\mathbf{wt}(x_{k,i(1)}) + \cdots + \mathbf{wt}(x_{k,i(t)}) \geq n,$$

and

$$\mathbf{wt}(x_{k,i(1)}) + \cdots + \mathbf{wt}(x_{k,i(t-1)}) < n.$$

Extend the order relation on the basis $B(k)$ to $B(k) \cup B(k)^{-1}$ be declaring for u, v in $B(k)$,

$$u < u^{-1} \text{ and } u^{-1} < v, \text{ if } u < v.$$

The commutators $[z_p^{\epsilon(p)}, z_1^{\epsilon(1)}, \ldots, z_{p-1}^{\epsilon(p-1)}, z_{p+1}^{\epsilon(p+1)}, \ldots, z_t^{\epsilon(t)}] \in F^{(k)}$, where $t \geq 2$, $z_q \in B(k)$ and $\epsilon(q) = \pm 1$ $(1 \leq q \leq t)$, $z_1^{\epsilon(1)} \leq \ldots \leq z_t^{\epsilon(t)}$, $z_p > z_1$, will be called *ordered commutators* of level k. An ordered commutator will be called *reduced*, if whenever $\epsilon(i) \neq \epsilon(i+1)$, then $z_i \neq z_{i+1}$. Define

$$\gamma_{<n>}(F^{(k)}) = \langle \{[z_p^{\epsilon(p)}, z_1^{\epsilon(1)}, \ldots, z_{p-1}^{\epsilon(p-1)}, z_{p+1}^{\epsilon(p+1)}, \ldots, z_t^{\epsilon(t)}]\} \rangle$$

to be the subgroup generated by all reduced commutators of level k and weight $\geq n$, i.e., satisfying $\sum_{1 \leq q \leq t} \mathbf{wt}(z_q) \geq n$. Further, let

$$\beta_{<n>}(F^{(k)}) = \langle \{[z_p^{\epsilon(p)}, z_1^{\epsilon(1)}, \ldots, z_{p-1}^{\epsilon(p-1)}, z_{p+1}^{\epsilon(p+1)}, \ldots, z_t^{\epsilon(t)}]\} \rangle$$

be the subgroup generated by all reduced commutators of level k, weight $< n$ and not lying in $B(k+1)$, i.e., satisfying the conditions that $\sum_{1 \leq q \leq t} \mathbf{wt}(z_q) < n$ and not all $\epsilon(i)$ are equal to 1.

From definitions and repeated use the commutator identities listed in Chapter 1, Section 1.1, we have

Lemma 3.1 *For* $0 \leq k < l$,

(i) $\gamma_{<n>}(F^{(k)}) \subseteq \gamma_{(n)}(F^{(k)})$;

(ii) $\beta_{<n>}(F^{(k)}) \subseteq F^{(k+1)} \gamma_{(n)}(F^{(k)})$;

(iii) $\gamma_{(n)}(F^{(k+1)}) \subseteq \gamma_{(n)}(F^{(k)})$.

Theorem 3.2 (Gupta-Passi [Gup07]). *The commutator subgroup* $[F^{(k)}, F^{(k)}]$ *of the free group* $F^{(k)}$ *is freely generated by the set of all reduced commutators of level* k, *namely commutators of the form*

$$\mathbf{z}(k, p, \underline{\epsilon}(t)) := [z_p^{\epsilon(p)}, z_1^{\epsilon(1)}, \ldots, z_{p-1}^{\epsilon(p-1)}, z_{p+1}^{\epsilon(p+1)}, \ldots, z_t^{\epsilon(t)}], \tag{3.3}$$

where $t \geq 2$, $z_q \in B(k)$ *and* $\epsilon(q) = \pm 1 (1 \leq q \leq t)$, $z_1^{\epsilon(1)} \leq \ldots \leq z_t^{\epsilon(t)}$, $z_p > z_1$, *and, furthermore, whenever* $\epsilon(i) \neq \epsilon(i+1)$, *then* $z_i \neq z_{i+1}$.

Proof. It is clear that $[F^{(k)}, F^{(k)}]$ is generated by the set of all commutators

$$[z_1^{\epsilon(1)}, z_2^{\epsilon(2)}, \ldots, z_t^{\epsilon(t)}], \ t \geq 2, \ z_q \in B(k), \ \epsilon(q) = \pm 1 \ (1 \leq q \leq t). \quad (3.4)$$

Modulo $\delta_2(F^{(k)})$, the commutators (3.4) can be decomposed, with the help of commutator identities (Ch. 1, 1.1 (v)), as products of ordered commutators of the form

$$[z_p^{\epsilon(p)}, z_1^{\epsilon(1)}, \ldots, z_{p-1}^{\epsilon(p-1)}, z_{p+1}^{\epsilon(p+1)}, \ldots, z_t^{\epsilon(t)}], \quad (3.5)$$

with $t \geq 2$, $\epsilon(i) = \pm 1$ and further satisfying $z_1 \leq \ldots \leq z_t$ and $z_p > z_1$. With repeated application of the standard congruence, modulo $\delta_2(F^{(k)})$,

$$[g_p, g_1, \ldots, g_i^{\epsilon}, g_i^{-\epsilon}, \ldots, g_t] \equiv$$
$$[g_p, g_1, \ldots, g_i^{\epsilon}, \ldots, g_t]^{-1}[g_p, g_1, \ldots, g_i^{-\epsilon}, \ldots, g_t]^{-1} \quad (g_i \in F^{(k)}),$$

we can write each ordered commutator of the form (3.5) as a product of reduced and ordered commutators of smaller weights. It follows that the set of all reduced commutators generates the commutator subgroup $[F^{(k)}, F^{(k)}]$. It suffices, therefore, to prove that the reduced commutators are linearly independent modulo $\delta_2(F^{(k)})$.

Consider the set of all *reduced* commutators $\mathbf{z}(k, p, \underline{\epsilon}(t))$ (3.3). To prove the linear independence of this set, we consider the Schumann-Magnus embedding

$$F^{(k)}/\delta_2(F^{(k)}) \longrightarrow \mathbb{Z}[F^{(k)}]/\mathfrak{f}^{(k)}\mathfrak{a}^{(k)}$$

(see, for instance, [Gup87c], p. 2), where $\mathbb{Z}[F^{(k)}]$ is the integral group ring of $F^{(k)}$, $\mathfrak{f}^{(k)}$ the augmentation ideal of $\mathbb{Z}[F^{(k)}]$ and the ideal

$$\mathfrak{a}^{(k)} := \mathbb{Z}[F^{(k)}]([F^{(k)}, F^{(k)}] - 1).$$

We then have, modulo $\mathfrak{f}^{(k)}\mathfrak{a}^{(k)}$,

$$\mathbf{z}(k, p, \underline{\epsilon}(t)) - 1 \equiv (z_p^{\epsilon(p)} - 1)(z_1^{\epsilon(1)} - 1) \ldots (z_{p-1}^{\epsilon(p-1)} - 1)(z_{p+1}^{\epsilon(p+1)} - 1)$$
$$\ldots (z_t^{\epsilon(t)} - 1) - (z_1^{\epsilon(1)} - 1)(z_2^{\epsilon(2)} - 1) \ldots (z_t^{\epsilon(t)} - 1).$$

Since $\mathfrak{f}^{(k)}$ is a free $\mathbb{Z}[F^{(k)}]$-module with basis $\{z - 1 : z \in B(k)\}$, the commutators $\mathbf{z}(k, p, \underline{\epsilon}(t))$ and $\mathbf{z}(k, p', \underline{\epsilon}(t))$ with $p \neq p'$ are linearly independent and so are the commutators $\mathbf{z}(k, p, \underline{\epsilon}(t))$ and $\mathbf{z}(k, p, \underline{\epsilon}(t'))$ with $t \neq t'$. Thus the linear independence of the commutators $\mathbf{z}(k, p, \underline{\epsilon}(t))$ reduces to proving only the linear independence of the following set of reduced commutators:

$$S = \{\mathbf{z}(k\,,\,p\,,\,\underline{\epsilon}(t)) : \underline{\epsilon}(t) = (\epsilon(1)\,,\,\epsilon(2)\,,\,\ldots\,,\,\epsilon(t))\} \tag{3.6}$$

with *fixed* $\mathbf{z} = (z_1\,,\,\ldots\,,\,z_t)$ and p with $z_p > z_1$, and the t-tuples $\underline{\epsilon}(t)$ ranging over distinct t-tuples $(\epsilon(1)\,,\,\epsilon(2)\,,\,\ldots\,,\,\epsilon(t))$ with $\epsilon(i) = \pm 1$ and $\epsilon(i) \neq \epsilon(i+1)$ only if $z_i \neq z_{i+1}$. Expansion of $\mathbf{z}(k\,,\,p\,,\,\underline{\epsilon}(t)) - 1$ modulo $\mathfrak{f}^{(k)}\mathfrak{a}^{(k)}$ yields ordered products of the form

$$(z_p^{\epsilon(p)} - 1)(z_1^{\epsilon(1)} - 1)\ldots(z_{p-1}^{\epsilon(p-1)} - 1)(z_{p+1}^{\epsilon(p+1)} - 1)\ldots(z_t^{\epsilon(t)} - 1) \equiv$$
$$(z_p - 1)(z_1 - 1)\ldots(z_{p-1} - 1)(z_{p+1} - 1)\ldots(z_t - 1) \prod_{1 \leq i \leq t} \epsilon(i) \prod_{1 \leq i \leq t} z_i^{(\epsilon(i)-1)/2}.$$

Since *distinct* $\underline{\epsilon}(t)$'s yield *distinct* elements $\prod_{1 \leq i \leq t} z_i^{(\epsilon(i)-1)/2}$, the linear independence of the set S, as defined in (3.6), follows from the fact that distinct group elements $\prod_{1 \leq i \leq t} \epsilon(i) \prod_{1 \leq i \leq t} z_i^{(\epsilon(i)-1)/2}$ associated with the distinct $\underline{\epsilon}(t)$'s are linearly independent modulo $[F^{(k)}\,,\,F^{(k)}]$, and this completes the proof of the theorem. $\square$

In particular, restricting to the case $k = 0$, we have the following description of the commutator subgroup of F :

Corollary 3.3 *The commutator subgroup* $[F,\,F]$ *of a free group* F *with ordered basis* B *is freely generated by the set of all reduced commutators of the form*

$$[z_p^{\epsilon(p)}\,,\,z_1^{\epsilon(1)}\,,\,\ldots\,,\,z_{p-1}^{\epsilon(p-1)}\,,\,z_{p+1}^{\epsilon(p+1)}\,,\,\ldots\,,\,z_t^{\epsilon(t)}]\,,\quad t \geq 2,$$

where $z_q \in B$ *and* $\epsilon(q) = \pm 1$ $(1 \leq q \leq t)$, $z_1^{\epsilon(1)} \leq \ldots \leq z_t^{\epsilon(t)}$, $z_p > z_1$, *and* $\epsilon(i) \neq \epsilon(i + 1)$ *only if* $z_i \neq z_{i+1}$.

Remark. Recall that the Cartesian group of a free product G is the kernel of the natural homomorphism of G onto the direct product of its free factors, and that the Cartesian group of a free product of abelian groups coincides with the derived group. The above result is, therefore, a special case of ([Gru57], Theorem 5.2) where Gruenberg gives a basis for the cartesian group of a free product of cyclic groups (see also Levi [Lev40]).

From the definitions of the various partial commutator subgroups we immediately have:

Theorem 3.4 (Gupta-Passi [Gup07]). *The commutator subgroup* $[F^{(k)}\,,\,F^{(k)}]$ *of the free group* $F^{(k)}$ *is the free product*

$$[F^{(k)}\,,\,F^{(k)}] = \beta_{<n>}(F^{(k)}) * \gamma_{<n>}(F^{(k)}) * F^{(k+1)}.$$

Fundamental Theorem of Free Group Rings

Theorem 3.5 (Magnus-Grün-Witt). *If F is a free group, then $D_n(F) = \gamma_n(F)$ for all $n \geq 1$.*

Proof (Gupta-Passi [Gup07]). It clearly suffices to consider the case when the free group F is of finite rank $m \geq 2$, and $n \geq 2$. In the free group ring $\mathbb{Z}[F^{(k)}]$, $k \geq 0$, let

$$\mathfrak{f}^{(k,n)} = \mathbb{Z}\text{-span of all products } (z_1^{\pm 1} - 1) \ldots (z_t^{\pm 1} - 1) \quad (t \geq 2),$$

where $z_j \in B(k)$ for $1 \leq j \leq t$, and $\mathbf{wt}(z_1) + \cdots + \mathbf{wt}(z_t) \geq n$. Note that $\gamma_{(n)}(F^{(k)}) \subseteq 1 + \mathfrak{f}^{(k,n)}$. We first prove the following result.

Lemma 3.6 *Let $u \in \mathfrak{f}^{(k,n)}$, $k \geq 0$. Then*

$$u = u_1 + u_2 + u_3 + \sum_i \pm (h_i - 1), \quad h_i \in \gamma_{(n)}(F^{(k)}),$$

where u_1 is a linear sum of ordered products of the form

$$(z_1^{\pm 1} - 1) \ldots (z_t^{\pm 1} - 1), \quad t \geq 2,$$

with $z_i \in B(k)$, $z_1 \leq \ldots \leq z_t$ and $\mathbf{wt}(z_1) + \cdots + \mathbf{wt}(z_t) \geq n$; u_2 is a linear sum of partly-ordered products of the form

$$(z_1^{\pm 1} - 1) \ldots (z_a^{\pm 1} - 1)(X_1 - 1) \ldots (X_b - 1), \quad a \geq 1, \ b \geq 1,$$

with $z_i \in B(k)$, $z_1 \leq \ldots \leq z_a, X_s = [z_{s(1)}^{\pm 1}, \ldots, z_{s(\ell)}^{\pm 1}], \ell \geq 2, z_{s(i)} \in B(k), \mathbf{wt}$ $(z_1) + \cdots + \mathbf{wt}(z_a) + \mathbf{wt}(X_1) + \cdots + \mathbf{wt}(X_b) \geq n, \mathbf{wt}(X_s) := \sum_i \mathbf{wt}(z_{s(i)}) < n;$ u_3 is a linear sum of products of the form

$$(X_1 - 1) \ldots (X_b - 1), \quad b \geq 2,$$

with X_s as above and satisfying $\mathbf{wt}(X_1) + \cdots + \mathbf{wt}(X_b) \geq n$.

Proof. By definition, $\mathbf{f}^{(k,n)}$ consists of finite linear sums of products of the form

$$(z_1^{\pm 1} - 1) \ldots (z_t^{\pm 1} - 1), \quad t \geq 2, \tag{3.7}$$

where $z_j \in B(k)$, and $\mathbf{wt}(z_1) + \cdots + \mathbf{wt}(z_t) \geq n$. By repeated application of the identity

$$(y - 1)(x - 1) = (x - 1)(y - 1) + ([y, x] - 1) + (x - 1)([y, x] - 1)$$
$$+ (y - 1)([y, x] - 1) + (x - 1)(y - 1)([y, x] - 1),$$

each product of the form (3.7), and hence every element $u \in \mathfrak{f}^{(k,n)}$, can be decomposed as claimed.

Proof of Theorem 3.5. Let $k \geq 0$ and let $f \in F^{(k)} \cap (1 + \mathfrak{f}^{(k,n)})$. Since $\mathfrak{f}^{(k,n)} \subseteq \Delta^2(F^{(k)})$, we have $f \in [F^{(k)}, F^{(k)}] = F^{(k+1)}\gamma_{(n)}(F^{(k)})$. Therefore, $f = gh$, $g \in F^{(k+1)}$, $h \in \gamma_{(n)}(F^{(k)})$. Since $\gamma_{(n)}(F^{(k)}) \subseteq 1 + \mathfrak{f}^{(k,n)}$, therefore $h - 1 \in \mathfrak{f}^{(k,n)}$, and consequently $g - 1 \in \mathfrak{f}^{(k,n)}$. With the notations of Lemma 5.2, we decompose $g - 1$ as

$$g - 1 = u_1 + u_2 + u_3 + \sum_i \pm(h_i - 1).$$

Observe that

$$u_1 \in \mathbb{Z}[F^{(k)}](F^{(k)} - 1) \setminus \{\mathbb{Z}[F^{(k)}]\left([F^{(k)}, F^{(k)}] - 1\right)\},$$

$$u_2 \in \mathbb{Z}[F^{(k)}]\left([F^{(k)}, F^{(k)}] - 1\right) \setminus \{\mathbb{Z}[[F^{(k)}, F^{(k)}]]\left([F^{(k)}, F^{(k)}] - 1\right)\},$$

$$u_3 \in \mathbb{Z}[[F^{(k)}, F^{(k)}]]\left([F^{(k)}, F^{(k)}] - 1\right)\left([F^{(k)}, F^{(k)}] - 1\right).$$

Since $g \in F^{(k+1)} \subseteq [F^{(k)}, F^{(k)}]$, we must have $u_1 = 0$, $u_2 = 0$, and consequently

$$g - 1 = u_3 + \sum_i \pm(h_i - 1). \tag{3.8}$$

Since $F^{(k+1)}$ is a free factor of $[F^{(k)}, F^{(k)}]$ (Theorem 3.4), projecting $\mathbb{Z}[[F^{(k)}, F^{(k)}]]$ onto $\mathbb{Z}[F^{(k+1)}]$ under the map which sends the basis elements of $[F^{(k)}, F^{(k)}]$ which lie in $F^{(k+1)}$ identically and the remaining basis elements into identity, the equation (3.8) yields that

$$g \in F^{(k+1)} \cap (1 + \mathfrak{f}^{(k+1,n)}).$$

Therefore

$$f = gh \in (F^{(k+1)} \cap (1 + \mathfrak{f}^{(k+1,n)}))\gamma_{(n)}(F^{(k)}).$$

Hence $F^{(k)} \cap (1 + \mathfrak{f}^{(k,n)}) \subseteq (F^{(k+1)} \cap (1 + \mathfrak{f}^{(k+1,n)}))\gamma_{(n)}(F^{(k)})$. Since $\mathfrak{f}^{(0,n)} = \mathfrak{f}^n$ and $\gamma_{(n)}(F^{(k)}) \subseteq \gamma_n(F)$, we conclude, by iteration, that

$$D_n(F) = F \cap (1 + \mathfrak{f}^n) \subseteq \gamma_n(F),$$

and the theorem is proved. $\square$

3.2 *E* and *D*-groups

In this section we give an exposition of Strebel's theory of *E*- and *D*-groups. Since this theory plays a key role in the theory of aspherical complexes (see next Section), we present this theory in some detail. All the material here is from Strebel's paper [Str74].

Let R be a nontrivial commutative ring with identity. A group G is called an $E(R)$-group if R, as a trivial $R[G]$-module, has an $R[G]$-projective resolution

$$\cdots \to P_2 \xrightarrow{\partial_2} P_1 \xrightarrow{\partial_1} P_0 \xrightarrow{\partial_0} R \to 0,$$

for which the map

$$\mathrm{id}_R \otimes \partial_2 : R \otimes_{R[G]} P_2 \to R \otimes_{R[G]} P_1$$

is injective. A group G is called an E-group if it is an $E(R)$-group for every ring R.

It follows directly from the definition, that $H_2(G, R) = 0$ for every $E(R)$-group G. One of the most important class of E-groups is that of knot groups.

A group G is called a $D(R)$-group if, for every homomorphism $\eta : A \to B$ of projective $R[G]$-modules, the injectivity of the induced map $\mathrm{id}_R \otimes_{R[G]} \eta : R \otimes_{R[G]} A \to R \otimes_{R[G]} B$ implies injectivity of η.

Example 3.7

The infinite cyclic group is a $D(R)$-group for any ring R. For, let C be an infinite cyclic group with generator $c \in C$ and $\eta : A \to B$ a map between projective $R[C]$-modules, such that $id_R \otimes_{R[C]} \eta$ is a monomorphism. The augmentation ideal of the group ring $R[C]$ is residually nilpotent, i.e. the intersection $\bigcap_n \Delta_R^n(C)$ is the zero ideal and for any projective $R[C]$-module A, the intersection $\bigcap_n \Delta_R^n(C)A$ is the trivial submodule in A. The augmentation quotients $\Delta_R^n(C)/\Delta_R^{n+1}(C)$ are naturally isomorphic to R as $R[G]$-modules. Therefore, the maps

$$\Delta_R^n(C)/\Delta_R^{n+1}(C) \otimes_{R[C]} A \to \Delta_R^n(C)/\Delta_R^{n+1}(C) \otimes_{R[C]} B,$$

induced by η, are monomorphisms for all $n \geq 0$. For every projective $R[C]$-module there is the natural isomorphism

$$\Delta_R^n(C)/\Delta_R^{n+1}(C) \otimes_{R[C]} A \simeq \Delta_R^n(C)A/\Delta_R^{n+1}(C)A,$$

hence η induces the monomorphisms

$$\Delta_R^n(C)A/\Delta_R^{n+1}(C)A \to \Delta_R^n(C)B/\Delta_R^{n+1}(C)B;$$

therefore, due to the fact that $\bigcap_n \Delta_R^n(C)A = 0$, the map η is a monomorphism. Hence C is a $D(R)$-group.

Let $\eta : A \to B$ be a map of projective $R[G]$-modules. Then there exist modules $\bar{A}$, $\bar{B}$ such that $A \oplus \bar{A}$ and $B \oplus \bar{B}$ are free $R[G]$-modules. Define the map

$$\widetilde{\eta} : A \oplus \bar{A} \to B \oplus \bar{B} \oplus A \oplus \bar{A}$$

by setting

$$\widetilde{\eta} : a \oplus \bar{a} \mapsto \eta(a) \oplus 0 \oplus 0 \oplus \bar{a}, \ a \in A, \ \bar{a} \in \bar{A}.$$

It is easy to see that η is injective if and only if $\widetilde{\eta}$ is, and $id_R \otimes_{R[G]} \eta$ is injective if and only if $id_R \otimes_{R[G]} \widetilde{\eta}$ is. Therefore, in the definition of the $D(R)$-group we can assume that modules A and B are free $R[G]$-modules.

The main result about E-groups is the following theorem, due to Strebel.

Theorem 3.8 *Let G be an E-group. Then*

(i) *Every term of the transfinite derived series $\delta_\alpha(G)$ is an E-group;*

(ii) *Every term $\gamma_\alpha(G)$, $1 \le \alpha \le \omega$, of the lower central series of G is an E-group;*

(iii) *The quotient $G/\mathcal{P}(G)$ is an E-group of cohomological dimension at most two, where $\mathcal{P}(G)$ is the perfect radical of G.*

We will need the following result which implies the statement (iii) of Theorem 3.8.

Theorem 3.9 *Let G be an E-group and*

$$\cdots \to P_2 \xrightarrow{\partial_2} P_1 \to P_0 \to \mathbb{Z} \to 0,$$

a $\mathbb{Z}[G]$-resolution of $\mathbb{Z}$ such that the induced map

$$id_{\mathbb{Z}} \otimes \partial_2 : \mathbb{Z} \otimes_{\mathbb{Z}[G]} P_2 \to \mathbb{Z} \otimes_{\mathbb{Z}[G]} P_1$$

is a monomorphism. Then, for every ordinal number α, the induced map

$$\widetilde{\partial}_2^\alpha := id_{\mathbb{Z}} \otimes \partial_2 : \mathbb{Z} \otimes_{\mathbb{Z}[\delta_\alpha(G)]} P_2 \to \mathbb{Z} \otimes_{\mathbb{Z}[\delta_\alpha(G)]} P_1$$

is a monomorphism.

Lemma 3.10 *Let G be a group with transfinite descending subnormal series*

$$G = G_0 \supset G_1 \supset G_2 \supset \cdots \supset G_\omega \supset G_{\omega+1} \supset \cdots \supset G_\alpha = 1$$

in which all the quotients $G_\beta/G_{\beta+1}$ are $D(R)$-groups. Then G is a $D(R)$-group.

Proof. Let A and B be $R[G]$-projective modules, $\eta : A \to B$ a map such that

$$id_R \otimes_{R[G]} \eta : R \otimes_{R[G]} A \to R \otimes_{R[G]} B$$

is a monomorphism. We use the transfinite induction on α to show that the maps

$$\mathrm{id}_R \otimes_{R[G_\beta]} \eta : R \otimes_{R[G_\beta]} A \to R \otimes_{R[G_\beta]} B$$

are monomorphisms. It will then follow that η is a monomorphism. For any ordinal number β, consider the group extension

$$1 \to G_{\beta+1} \to G_\beta \to G_\beta/G_{\beta+1} \to 1$$

and assume that $\mathrm{id}_R \otimes_{R[G_\beta]} \eta$ is a monomorphism. The map $\mathrm{id}_R \otimes_{R[G_{\beta+1}]} \eta$ can be viewed as a map between $R[G_\beta/G_{\beta+1}]$-projective modules. Then $\mathrm{id}_R \otimes_{R[G_\beta/G_{\beta+1}]} \otimes (\mathrm{id}_R \otimes_{R[G_{\beta+1}]} \eta)$ gets identified with $\mathrm{id}_R \otimes_{R[G_\beta]} \eta$ and is injective. Since $G_\beta/G_{\beta+1}$ is a $D(R)$-group, we conclude that the map $\mathrm{id}_R \otimes_{R[G_{\beta+1}]} \eta$ is a monomorphism.

Consider now the limit case. Let τ be a limit ordinal number and suppose that the maps $\mathrm{id}_R \otimes_{R[G_\beta]} \eta$ are monomorphisms for all $\beta < \tau$. For a given projective $R[G]$-module M, let

$$i_M : R \otimes_{R[G_\tau]} M \to \prod_{\beta<\tau} R \otimes_{R[G_\beta]} M$$

be the canonical map induced by inclusions $G_\tau \subset G_\beta$, $\beta < \tau$. We have the following commutative diagram:

$$
\begin{array}{ccc}
R \otimes_{R[G_\tau]} A & \xrightarrow{\ \mathrm{id}_R \otimes_{R[G_\tau]} \eta\ } & R \otimes_{R[G_\tau]} B \\
\Big\downarrow{i_A} & & \Big\downarrow{i_B} \\
\prod_{\beta<\tau} R \otimes_{R[G_\beta]} A & \xrightarrow{\ \prod_{\beta<\tau} \mathrm{id}_R \otimes_{R[G_\beta]} \eta\ } & \prod_{\beta<\tau} R \otimes_{R[G_\beta]} B,
\end{array}
$$

where the lower map is a monomorphism, since all maps $\mathrm{id}_R \otimes_{R[G_\beta]} \eta$ are. Observe that it suffices to show that the map i_A is a monomorphism.

Suppose first $A = R[G]$. Choose transversals $\{T_s\}_{s \in G/G_\tau}$ for G_τ in G. Suppose we have $x \in \ker(i_A)$ and $x \neq 0$. Then element $x \in R \otimes_{R[G_\tau]} R[G]$ has a unique expression of the form

$$x = \sum_s r_s(x) \otimes_{R[G_\tau]} T_s, \ r_s(x) \in R. \tag{3.9}$$

Since this sum is finite, there exists β such that for every pair of elements T_s, $T_{s'}$ from the sum (3.9) the element $T_s T_{s'}^{-1} \notin G_\beta$. Therefore, the element x has a nontrivial image in $R \otimes_{R[G_\beta]} A$ and hence $i_A(x) \neq 0$, which is a contradiction.

Now let $A = \oplus_k M_i$, where M_i are $R[G]$-modules. Then there exists the following natural commutative diagram:

$$\bigoplus_k R \otimes_{R[G_\tau]} M_k \xrightarrow{\ \oplus_k i_{M_k}\ } \bigoplus_k \prod_{\beta < \tau} R \otimes_{R[G_\beta]} M_k$$

$$\Big\| \qquad\qquad\qquad\qquad \sigma \Big\downarrow$$

$$\bigoplus_k R \otimes_{R[G_\tau]} M_k \xrightarrow{\ \oplus_k i_{M_k}\ } \prod_{\beta < \tau} \bigoplus_k R \otimes_{R[G_\beta]} M_k$$

$$\Big\| \qquad\qquad\qquad\qquad \Big\|$$

$$R \otimes_{R[G_\tau]} A \xrightarrow{\ i_A\ } \prod_{\beta < \tau} R \otimes_{R[G_\beta]} A$$

with monomorphism σ. The map i_A is injective for any free $R[G]$-module and furthermore for any direct summand of a free $R[G]$-module, i.e., for any projective $R[G]$-module. Hence the map

$$\mathrm{id}_R \otimes_{R[G_\tau]} \eta : R \otimes_{R[G_\tau]} A \to R \otimes_{R[G_\tau]} B$$

is a monomorphism. This completes the induction and so the proof of the Lemma is complete. $\square$

Lemma 3.11 *Let G be a torsion-free abelian group. Then G is a $D(R)$-group for any commutative ring R with identity.*

Proof. First observe that in the case of a finitely-generated group G, the statement follows from Lemma 3.10 and the fact that the infinite cyclic group is $D(R)$-group.

Next let G be arbitrary torsion-free abelian group. Then G is a direct limit of its finitely-generated subgroups:

$$G = \varinjlim_{i \in I} U_i,$$

where U_i, $i \in I$, are finitely-generated free abelian groups. Let $\eta : A \to B$ be a map of free $R[G]$-modules such that the map $\mathrm{id}_R \otimes_{R[G]} \eta$ is injective. Since A and B are unions of their finitely generated free direct summands, for testing the $D(R)$-property, we can assume that A and B are finitely generated. Choose $R[G]$-bases $\{a_1, \dots, a_m\}$ and $\{b_1, \dots, b_n\}$ of A and B respectively. Let H be the $m \times n$ matrix which defines the map η with respect to the above choice of bases of A, B. Since this matrix is finite, there exists $i \in I$ such that all entries of H are in U_i. It is easy to see that $\mathrm{id}_R \otimes_{R[U_i]} \eta$ is injective. Since U_i is a $D(R)$-group, the map η is also injective. $\square$

Lemma 3.12 *Let G be an E-group. Then for any ordinal number α, $\delta_\alpha(G)$ is an E-group and for any ring R, the group $G/\delta_\alpha(G)$ is a $D(R)$-group.*

Proof. The proof is by induction on α. Suppose that, for any $\beta < \alpha$, $\delta_\beta(G)$ is an E-group and $G/\delta_\beta(G)$ is a $D(R)$-group. Consider the group $G/\delta_\alpha(G)$. Since G_β are E-groups for $\beta < \alpha$, the quotients $G_\beta/G_{\beta+1}$ are torsion-free abelian groups. Hence $G/\delta_\alpha(G)$ has a descending series with torsion-free abelian quotients. Therefore $G/\delta_\alpha(G)$ is a $D(R)$-group by Lemmas 3.10 and 3.11.

Let

$$\cdots \to P_2 \overset{\partial_2}{\to} P_1 \to P_0 \to R \to 0,$$

be an $R[G]$-projective resolution of R such that $\mathrm{id}_R \otimes \partial_2$ is injective. Consider the $R[G/\delta_\alpha(G)]$-projective resolution

$$\cdots \to R \otimes_{R[\delta_\alpha(G)]} P_2 \overset{\mathrm{id}_R \otimes \partial_2}{\to} R \otimes_{R[\delta_\alpha(G)]} P_1 \to R \otimes_{R[\delta_\alpha(G)]} P_0 \to R \to 0.$$

For this resolution the map

$$\mathrm{id}_R \otimes_{R[G/\delta_\alpha(G)]} \left(\mathrm{id}_R \otimes_{R[\delta_\alpha(G)]} \partial_2\right) \simeq \mathrm{id}_R \otimes_{R[G]} \partial_2$$

is injective. But the group $G/\delta_\alpha(G)$ is a $D(R)$-group, hence $\mathrm{id}_R \otimes_{R[\delta_\alpha(G)]} \partial_2$ is injective and therefore $\delta_\alpha(G)$ is an $E(R)$-group for any R, i.e., it is an E-group. The induction, and therefore the proof, is thus complete. $\square$

Proof of Theorem 3.9. We have the following natural isomorphism of maps:

$$\mathrm{id}_\mathbb{Z} \otimes_{\mathbb{Z}[G/\delta_\alpha(G)]} \widetilde{\partial}_2^\alpha \simeq \mathrm{id}_\mathbb{Z} \otimes_{\mathbb{Z}[G]} \partial_2.$$

Therefore the map $\mathrm{id}_\mathbb{Z} \otimes_{\mathbb{Z}[G/\delta_\alpha(G)]} \widetilde{\partial}_2^\alpha$ is injective. By Lemma 3.12, the group $G/\delta_\alpha(G)$ is a $D(R)$-group; hence the map $\widetilde{\partial}_2^\alpha$ is also injective. $\square$

C. Gordon raised the following problem in [Gor81]:

Let G be a knot group with $\gamma_2(G)$ transfinitely nilpotent. Is it true that $\gamma_2(G)$ is residually nilpotent?

The above motivated the following more general

Problem 3.13 *Is it true that $\gamma_\omega(\gamma_2(G)) = \gamma_{\omega+1}(\gamma_2(G))$ for any finitely generated E-group G?*

3.3 Transfinite Derived Series

Recall that, for a group G, $\delta^+(G)$ denotes the subgroup generated by the torsion elements $g \in G$ which have only finitely many conjugates in G.

Theorem 3.14 (Mikhailov [Mik05a]). *Let F be a free group and $\{1\} \neq S \subseteq R$ its normal subgroups such that:*

(i) F/S *is residually solvable,*

(ii) $\delta^+(F/R) = 1$,

(iii) F/R *is not residually solvable.*

Then

$$\delta_\omega(F/[S,\,R]) \neq \delta_{\omega+1}(F/[S,\,R]). \tag{3.10}$$

Proof. By Theorem 1.60, the module $S/[S,\,R]$ is a faithful $\mathbb{Z}[F/R]$-module. Consider the following exact sequence of groups:

$$1 \to S/[S,\,R] \to F/[S,\,R] \to F/S \to 1.$$

By condition (i), $\delta_\omega(F/S) = \{1\}$, therefore $\delta_\omega(F/[S,\,R]) \subseteq S/[S,\,R]$. Since the subgroup $S/[S,\,R]$ is abelian, we have $\delta_{\omega+1}(F/[S,\,R]) = \{1\}$. Suppose that

$$\delta_\omega(F/[S,\,R]) = \delta_{\omega+1}(F/[S,\,R]) = \{1\}.$$

Let $1 \neq g \in \delta_\omega(F/R)$. Then

$$S/[S,\,R] \circ (1 - g) = 0,$$

which contradicts the faithfulness of $S/[S,\,R]$ as a $\mathbb{Z}[F/R]$-module. Hence $\delta_\omega(F/[S,\,R]) \neq 1$, and our assertion is proved. $\square$

Example 3.15
Let F be a free group of rank 3 with basis $\{a,\,b,\,c\}$. Let S be the normal closure of a in F, and R the normal closure of $\langle a,\,b[cbc^{-1},\,b]\rangle$ in F. Then the group F/S is free of rank 2, and is, therefore, residually solvable. The group F/R has the following presentation

$$\langle b,\,c \mid b[cbc^{-1},\,b] = 1\rangle.$$

It follows from properties of one-relator groups that:

   (i) the group F/R is torsion free and therefore $\delta^+(F/R) = 1$;
   (ii) the element b is nontrivial in F/R.

It is clear that the element $b \in \delta_\omega(F/R)$. Hence, by Theorem 3.14, F/R is not residually solvable and

$$\delta_\omega(F/[S,\,R]) \neq \delta_{\omega+1}(F/[S,\,R]) = \{1\}.$$

For a given group G, let

$$\delta_{(\omega)}(G) := \bigcap_n [\delta_\omega(G),\,\delta_n(G)].$$

Obviously, we have

$$\delta_\omega(G) \supseteq \delta_{(\omega)}(G) \supseteq \delta_{\omega+1}(G).$$

Lemma 3.16 *Let* $g \in \delta_{\omega+1}(F/RS)$, *then*

$$\frac{R \cap S}{[R,\,S]} \circ (g - 1) \subseteq \delta_{(\omega)}(F/[R,\,S]). \tag{3.11}$$

Proof. The element $g \in \delta_{\omega+1}(F/RS)$ can be written as

$$g = \prod_{j=1}^{k} [x_{2j-1}, x_{2j}],$$

for some k, where $x_j \in \delta_\omega(F/RS)$. Let $r[R, S] \in \frac{R \cap S}{[R, S]}$, $r \in R \cap S$. Then

$$r[R, S] \circ (g - 1) = [\prod_{j=1}^{k} [f_{2j-1}, f_{2j}], r^{-1}],$$

where the elements $f_l \in F$, $l = 1, \ldots, 2k$, are such that $f_l RS = x_l$. We thus have

$$r[R, S] \circ (g - 1) \in \langle [[f_j, r], f_l], \ j, l = 1, \ldots, 2k \rangle^F. \tag{3.12}$$

Since $f_l RS \in \delta_\omega(F/RS)$, it follows that $[f_j, r] \in \delta_\omega(F/[R, S])$. The assertion (3.11) then immediately follows because the elements $f_l RS$ in (3.12) lie in $\delta_\omega(F/RS)$. $\square$

Lemma 3.17 *Let F be a free group, R and S its normal subgroups such that F/RS is torsion free and*

(i) $\delta_\alpha(F/R) = \delta_\alpha(F/S) = 1$, $\alpha \geq \omega$,

(ii) $\delta_\omega \left(\frac{F}{[R \cap S, RS]} \right) = \delta_{\omega+1} \left(\frac{F}{[R \cap S, RS]} \right)$.

Then $\delta_\omega(F/RS) = 1$.

Proof. The module $\frac{R \cap S}{[R \cap S, RS]}$ is a faithful $\mathbb{Z}[F/RS]$-module (Theorem 1.60). Consider the following exact sequence of groups:

$$1 \to \frac{R \cap S}{[R \cap S, RS]} \to \frac{F}{[R \cap S, RS]} \to F/R \cap S \ (\subseteq F/R \times F/S). \tag{3.13}$$

From (3.13), and condition (i) it follows that $\delta_{\alpha+1} \left(\frac{F}{[R \cap S, RS]} \right) = 1$, therefore, due to (ii), we have

$$\delta_\omega \left(\frac{F}{[R \cap S, RS]} \right) = 1. \tag{3.14}$$

Now let $g \in \delta_\omega(F/RS)$. Then, using the fact that $\frac{R \cap S}{[R \cap S, RS]} \circ (g - 1) \subseteq \delta_\omega \left(\frac{F}{[R \cap S, RS]} \right)$ and (3.14) holds, we have

$$\frac{R \cap S}{[R \cap S, RS]} \circ (g - 1) = 0.$$

Since F/RS act faithfully on $\frac{R \cap S}{[R \cap S, RS]}$, it follows that $g = 1$. Hence $\delta_\omega(F/RS) = 1$. $\square$

By a similar argument, we have the following

Lemma 3.18 *Let F be a free group, R and S its normal subgroups such that the group F/RS is torsion free and*

(i) $\delta_\omega(F/R) = \delta_\omega(F/S) = 1$,

(ii) $\delta_{(\omega)}\left(\frac{F}{[R\cap S, RS]}\right) = \delta_{\omega+1}\left(\frac{F}{[R\cap S, RS]}\right)$,

then $\delta_{\omega+1}(F/RS) = 1$.

Proof. Observe that the group $F/R \cap S$ is a subgroup of the direct product $F/R \times F/S$; therefore the condition (i) implies that $F/R \cap S$ is residually solvable, i.e. $\delta_\omega(F/R \cap S) = 1$. The following short exact sequence

$$1 \to \frac{R\cap S}{[R\cap S, RS]} \to \frac{F}{[R\cap S, RS]} \to F/R\cap S \to 1$$

shows that $\delta_{\omega+1}\left(\frac{F}{[R\cap S, RS]}\right) = 1$, and in view of the condition (ii), we have

$$\delta_{(\omega)}\left(\frac{F}{[R\cap S, RS]}\right) = 1. \tag{3.15}$$

Recall (Theorem 1.60) that F/RS acts faithfully on $\frac{R\cap S}{[R\cap S, RS]}$. Let $g \in \delta_{\omega+1}(F/RS)$. Then, by (3.15) and Lemma 3.16, we have

$$\frac{R\cap S}{[R, S]} \circ (g - 1) = 0,$$

and therefore $g = 1$. Hence $\delta_{\omega+1}(F/RS) = 1$. $\square$

3.4 Applications to Asphericity

Let X be a two-dimensional CW-complex. Identifying $\pi_2(X)$ with the second homology group of the universal covering space $\widetilde{X}$ over X, we have an embedding of the second homotopy module $\pi_2(X)$ into a free $\mathbb{Z}[\pi_1(X)]$-module, namely:

$$\partial_2 : \pi_2(X) \to C_2(\widetilde{X}),$$

where $C_*(\widetilde{X})$ is the chain complex of $\widetilde{X}$. Define the *Fox ideal* $\mathsf{F}(\pi_2(\mathsf{X}))$ of $\pi_2(X)$ to be the two-sided ideal in $\mathbb{Z}[\pi_1(X)]$ generated by the coordinates of the embedding ∂_2. For a given subgroup $H \subseteq \pi_1(X)$, the complex X is called *H-Cockcroft* if

$$\mathsf{F}(\pi_2(\mathsf{X})) \subseteq \mathfrak{h}\mathbb{Z}[\pi_1(\mathsf{X})].$$

It is easy to see that a complex X is H-Cockcroft for a given subgroup $H \subseteq \pi_1(X)$ if and only if the Hurewicz homomorphism

$$h_H : \pi_2(X) \to H_2(X_H)$$

is the zero map, where X_H is a covering of X corresponding to the subgroup H.

Proposition 3.19 [Bra81]. *For a given two-dimensional complex X and a subgroup $H \subseteq \pi_1(X)$, the following conditions are equivalent:*

(i) *The induced map* $\mathrm{id}_{\mathbb{Z}} \otimes \partial_2 : \mathbb{Z} \otimes_{\mathbb{Z}[H]} C_2(\widetilde{X}) \to \mathbb{Z} \otimes_{\mathbb{Z}[H]} C_1(\widetilde{X})$ *is injective.*

(ii) *X is H-Cockcroft and $H_2(H) = 0$.*

(iii) *$H_2(X_H) = 0$.*

The proof follows directly from the structure of the chain complex of $\widetilde{X}$ on applying the functor $\mathbb{Z} \otimes_{\mathbb{Z}[H]} -$.

One of the most interesting and challenging problems in the theory of CW-complexes is the following:

Problem 3.20 (J. H. C. Whitehead). *Is any subcomplex of an aspherical two-dimensional complex itself aspherical?*

The above problem was first raised in [Whi41]; an affirmative answer is now conjectured and so it is often referred to as the *Whitehead asphericity conjecture*. The derived series techniques give conditions for a subcomplex of an aspherical complex which imply asphericity. We will see how Proposition 3.19 introduces the perfect radicals into the theory of the second homotopy modules.

Theorem 3.21 [Bra81]. *Let X be a two-dimensional complex such that $Y = X \bigcup_{i \in I} e_i$ is aspherical, where $\{e_i\}_{i \in I}$ is a collection of 2-cells. If $L = \ker\{\pi_1(X) \to \pi_1(Y)\}$, then X is $\mathcal{P}(L)$-Cockcroft and $H_2(\mathcal{P}(L)) = 0$.*

Proof. Observe that L_{ab} is isomorphic to $\mathbb{Z}[H]^{|I|}$ and, in particular, free abelian. It is easy to see that X is L-Cockcroft and $H_2(L) = 0$; therefore the map

$$\mathrm{id}_{\mathbb{Z}} \otimes_{\mathbb{Z}[L]} \partial_2 : \mathbb{Z} \otimes_{\mathbb{Z}[L]} C_2(\widetilde{X}) \to \mathbb{Z} \otimes_{\mathbb{Z}[L]} C_1(\widetilde{X})$$

is a monomorphism by Proposition 3.19. We conclude that L is an E-group in the sense of Strebel . By Theorem 3.9, the map

$$1 \otimes_{\mathbb{Z}[\mathcal{P}(L)]} \partial_2 : \mathbb{Z} \otimes_{\mathbb{Z}[\mathcal{P}(L)]} C_2(\widetilde{X}) \to \mathbb{Z} \otimes_{\mathbb{Z}[\mathcal{P}(L)]} C_1(\widetilde{X})$$

is a monomorphism. The assertion then follows from Proposition 3.19. $\square$

It is shown in [Gil] that for a given two-dimensional complex K, the following conditions are equivalent:

(i) The plus-construction K^+ is aspherical.

(ii) The covering space $K_{\mathcal{P}(\pi_1(K))}$ of the complex K with $\pi_1(K_{\mathcal{P}(\pi_1(K))}) = \mathcal{P}(\pi_1(K))$ is acyclic.

(iii) $H_2(\mathcal{P}(\pi_1(K))) = 0$ and K is $\mathcal{P}(\pi_1(K))$-Cockcroft.

Therefore, Theorem 3.21 implies that, for any subcomplex of a contractible two-dimensional complex, the plus-construction is aspherical. This is the case also for a subcomplex of an aspherical two-dimensional complex [Hau].

J. Howie [How83] proved that in case there exists a nonaspherical subcomplex of an aspherical two-dimensional complex, then there exists a connected two-dimensional complex L such that

either (1) L is finite and contractible and $L \setminus e$ is not aspherical for some open two-cell e of L;

or (2) L is the union of an infinite ascending chain of finite connected nonaspherical subcomplexes $K_0 \subset K_1 \subset \ldots$, where each inclusion $K_{i-1} \subset K_i$ is nullhomotopic.

Subsequently E. Luft [Luf96] has shown that in the case of a complex L with property (1), one can construct also a complex, satisfying property (2). Hence, Whitehead asphericity conjecture is the question of existence of a two-dimensional complex L, satisfying the condition (2).

Let F be a free group of finite rank n, say. A factorization $F = R_1 \ldots R_k$ of F into a product of normal subgroups R_i is called *efficient* if there exist natural numbers $r_i, 1 \leq i \leq k$, such that R_i is the normal closure in F of a set of words in F having cardinality r_i and $r_1 + \cdots + r_k = n$.

W. Bogley [Bog93] has proved that the following statements are equivalent:

(i) Every connected subcomplex of a finite contractible 2-complex is aspherical.

(ii) If R and S are distinct factors from an efficient normal factorization of a finitely generated free group, then $R \cap S = [R, S]$.

This equivalence follows from Theorem 1.45 and the following result of Cockcroft [Coc54], which solves Whitehead's asphericity conjecture in the simplest case.

Theorem 3.22 (Cockcroft [Coc54]). *If a connected subcomplex of an aspherical two-dimensional complex has just a single two-dimensional cell, then that subcomplex is aspherical.*

Bogley [Bog93] has also shown that if $F = RST$ is an efficient factorization, then

$$\frac{R \cap S}{[R, S]} \subseteq \gamma_\omega(F/[R, S])$$

and the lower central quotients of $F/[R, S]$ are free abelian. Hence the residual nilpotence of certain groups of the type $F/[R, S]$ implies asphericity of some subcomplexes of aspherical complexes. The following statement shows that one can replace the hypothesis of residual nilpotence with residual solubility.

Theorem 3.23 (Mikhailov-Passi [Mik05b]). *Let F be a free group with basis $x_1, \ldots, x_n$ $(n \geq 2)$, and*

$$P = \langle x_1, \ldots, x_n \mid r_1, \ldots, r_n \rangle$$

a balanced presentation of the trivial group. Let R_i, S_i $(i = 1, \ldots, n)$ be the normal closures in F of r_i and $\{r_1, \ldots, r_i\}$ respectively. Suppose that for some natural number $m < n$, $\delta_\omega(F/[R_{i+1}, S_i]) = 1$ $(i = 1, \ldots, m)$. Then the presentation

$$P_m = \langle x_1, \ldots, x_n \mid r_1, \ldots, r_m \rangle$$

is aspherical, i.e. $\pi_2(K_{P_m}) = 0$.

Proof. First note that K_P is contractible (see Example 2).

If $m = 1$, then K_{P_1} is a connected sub-complex of the aspherical 2-complex K_P and has only a single 2-cell; therefore, by Theorem 3.22, K_{P_1} is aspherical. Suppose now that $m > 1$ and $K_{P_{m-1}}$ is aspherical. Then, by the sequence (1.18), applied to the case $R = R_m$, $S = S_{m-1}$, we have

$$\pi_2(K_{P_m}) \simeq \frac{R_m \cap S_{m-1}}{[R_m, S_{m-1}]}. \tag{3.16}$$

By Corollary 1.142 (i), either $\frac{R_m \cap S_{m-1}}{[R_m, S_{m-1}]}$ is zero, or the action of F/S_m on this module is faithful. Suppose $\frac{R_m \cap S_{m-1}}{[R_m, S_{m-1}]}$ is not zero. Then, in view of (i), Theorem 1.36 (ii) (applied with $V = [R, S]$) implies that $\delta_\omega(F/S_m) = 1$. Therefore, by Theorem 3.21, P_m is aspherical, and so $\frac{R_m \cap S_{m-1}}{[R_m, S_{m-1}]} = 0$ by (3.16). This is a contradiction. Hence $\frac{R_m \cap S_{m-1}}{[R_m, S_{m-1}]} = 0$ and K_{P_m} is aspherical. $\square$

Results about crossed modules give some conditions on a subcomplex of an aspherical complex to be itself aspherical. First, let X be a subcomplex of a finite contractible complex. Then, clearly, $H_1(X)$ is free abelian and $H_2(X) = 0$. Therefore, we can apply Theorem 1.138 to get the following:

Proposition 3.24 (Conduché) [Con96] *Let X be a subcomplex of a finite contractible two-dimensional complex. Then $\pi_2(X) = \gamma_\omega(\pi_1(X^{(1)}), \pi_2(X, X^{(1)}))$.*

Corollary 1.142 gives the following *group-theoretical* criterion for the asphericity of a subcomplex of an aspherical complex.

Theorem 3.25 (Mikhailov [Mik07b]). *Let X be a subcomplex of an aspherical two-dimensional complex. Then the following conditions are equivalent:*

(i) X is aspherical.
(ii) The fundamental cat^1-group $\mathcal{L}(X)$ (see p. 88) is residually solvable.

Proof. The implication (i) $\Rightarrow$ (ii) is obvious. Indeed, let X be aspherical. Then $\pi_2(X, X^{(1)})$ is a normal subgroup of the free group $\pi_1(X^{(1)})$ by (1.107). Consequently, the group $\mathcal{L}(X)$ is residually nilpotent, and hence, residually solvable.

To see the implication (ii) $\Rightarrow$ (i), assume the contrary, i.e. that $\pi_2(X) \neq 0$ and that $\mathcal{L}(X)$ is residually solvable. Note that

$$m \circ (1 - f) = [m, f], \ f \in \pi_1(X^{(1)}), \ m \in \pi_2(X, X^{(1)}).$$

Let $g \in \pi_1(X)$, then g can be presented as a coset $g = f.\operatorname{im}(\partial), \ f \in \pi_1(X^{(1)})$. Then, for any $m \in \pi_2(X) = \ker(\partial)$, with these notations, we have

$$m \circ (1 - g) = [m, f]. \tag{3.17}$$

Suppose $g \in \delta_n(\pi_1(X))$, then $g = f.\operatorname{im}(\partial), \ f \in \delta_n(\pi_1(X^{(1)}))$, therefore, by

$$m \circ (1 - g) \in \delta_n(\mathcal{L}(X)), \ m \in \pi_2(X), \ g \in \delta_n(\pi_1(X)).$$

Hence,

$$m \circ (1 - g) \in \delta_\omega(\mathcal{L}(X)), \ m \in \pi_2(X), \ g \in \delta_\omega(\pi_1(X)).$$

The residual solubility of $\mathcal{L}(X)$ implies, therefore, the fact that any element from $\delta_\omega(\pi_1(X))$ annihilates all elements from $\pi_2(X)$. However, by Corollary 1.142, the action of $\pi_1(X)$ on $\pi_2(X)$ is faithful. Hence, $\delta_\omega(\pi_1(X)) = 1$. Therefore, $\pi_1(X)$ has a trivial perfect radical and $\pi_2(X) = 0$ by Theorem 3.21, a contradiction. Hence, (ii) implies (i). $\square$

Remark. It is easy to see that the preceding result can be proved for the more general case of a two-dimensional complex X with aspherical plus-construction X^+.

Note that for any crossed module (M, ∂, F) with F free, the group M is residually nilpotent, since it is a central extension of the free group $\operatorname{im}(\partial)$. Therefore, the group $M \rtimes F$ is transfinitely residually solvable, namely,

$$\delta_{2\omega}(M \rtimes F) = 1.$$

Hence, the condition (ii) from Theorem 3.25 is equivalent to the stabilization of the transfinite derived series:

$$\delta_\omega(\mathcal{L}(X)) = \delta_{\omega+1}(\mathcal{L}(X)).$$

This observation shows that the obstructions to the asphericity of certain 2-complexes lies in the difference between δ_ω and $\delta_{\omega+1}$ of fundamental cat^1-groups. Here we will show that for some important cases of the Whitehead asphericity conjecture, such kind of obstructions can be found in the difference between δ_ω and $\delta_{\omega+1}$ of certain finitely generated groups.

Proposition 3.26 *Let K be a subcomplex of a contractible 2-dimensional complex which is the union of its subcomplexes $K = K_1 \cup K_2$, such that $K_1 \cap K_2 = K^{(1)}$. Let $R = \ker\{\pi_1(K^{(1)}) \to \pi_1(K_1)\}$, $S = \ker\{\pi_1(K^{(1)}) \to \pi_1(K_2)\}$. Suppose that*

(i) $\delta_\omega(\pi_1(K_1)) = \delta_\omega(\pi_1(K_2)) = 1$,

(ii) $\delta_{(\omega)}(\pi_1(K^{(1)})/[R, S]) = \delta_{\omega+1}(\pi_1(K^{(1)})/[R, S])$.

Then $\pi_2(K) = 0$.

Proof. By Theorem 3.21, $\pi_2(K_1) = \pi_2(K_2) = 0$. Therefore,

$$\pi_2(K) \simeq \frac{R \cap S}{[R, S]}$$

by Theorem 1.45. Suppose that $\pi_2(K) \neq 0$. Then, by Corollary 1.142, the action of $\pi_1(K) \simeq \pi_1(K^{(1)})/RS$ on $\frac{R \cap S}{[R, S]}$ is faithful. By Lemma 2.69, we have

$$\frac{R \cap S}{[R, S]} \circ (g - 1) \subseteq \delta_{(\omega)}(\pi_1(K^{(1)})/[R, S]), \ g \in \delta_{\omega+1}(\pi_1(K)). \tag{3.18}$$

Due to the exact sequence

$$1 \to \frac{R \cap S}{[R, S]} \to \pi_1(K^{(1)})/[R, S] \to \pi_1(K^{(1)})/R \cap S \ (\subseteq \pi_1(K^{(1)})/R \times \pi_1(K^{(1)})/S),$$

we have $\delta_{\omega+1}(\pi_1(K^{(1)})/[R, S]) = 1$. By condition (ii) and (3.18), we have

$$\delta_{\omega+1}(\pi_1(K^{(1)})/RS) = 1$$

and complex K is aspherical by Theorem 3.21. $\square$

Proposition 3.27 *Let K be a subcomplex of a contractible 2-dimensional complex and $K = K_1 \cup \cdots \cup K_m$, $K_1 \cap \ldots \cap K_m = K^{(1)}$. Suppose*

$$\delta_\omega(K_i) = 1, \ 1 \leq i \leq m, \tag{3.19}$$

and for all $i = 1, \ldots, , m - 1$

$$\delta_\omega\left(\frac{\pi_1(K^{(1)})}{[Q_1 \ldots Q_i, Q_{i+1}]}\right) = \delta_{\omega+1}\left(\frac{\pi_1(K^{(1)})}{[Q_1 \ldots Q_i, Q_{i+1}]}\right), \tag{3.20}$$

$$\delta_\omega\left(\frac{\pi_1(K^{(1)})}{[Q_1 \ldots Q_i \cap Q_{i+1}, Q_1 \ldots Q_{i+1}]}\right) = \delta_{\omega+1}\left(\frac{\pi_1(K^{(1)})}{[Q_1 \ldots Q_i \cap Q_{i+1}, Q_1 \ldots Q_{i+1}]}\right), \tag{3.21}$$

where $Q_i = \ker\{\pi_1(K^{(1)}) \to \pi_1(K_i)\}$, $i = 1, \ldots, , m$. Then $\pi_2(K) = 0$.

Proof. The proof is by induction on m. The case $m = 1$ follows from Theorem 3.21. Suppose that the Theorem is proved for a given natural number $m - 1$, $m \geq 2$, i.e., a sub-complex of a contractible complex, which is the union of $m - 1$ sub-complexes, and is such that the corresponding conditions (3.19)-(3.27) are satisfied, is aspherical.

It is clear that, by induction hypothesis, conditions (3.19), (3.20) and (3.27) imply asphericity of $K_1 \cup \cdots \cup K_{m-1}$. Let

$$R = \ker\{\pi_1(K^{(1)}) \to \pi_1(K_1 \cup \ldots \cup K_{m-1})\}, \ S = Q_m,$$

and apply Proposition 3.26 to this case. It follows that the complex $K_1 \cup \ldots \cup K_m$ is aspherical and therefore its fundamental group is torsion-free. Now we can apply Lemma 3.17 (the condition (i) of Lemma 3.17 follows from (3.19) and the induction hypothesis, whereas the condition (ii) follows from (3.27)). We get $\delta_\omega(\pi_1(K^{(1)})/RS) = 1$ and the induction is complete. $\square$

Conditions (3.19) are satisfied in certain important cases of the Whitehead asphericity conjecture; for example, in the case of LOT (label oriented tree) presentations. Let $\mathcal{T}$ be a tree with vertices $\mathcal{X}$ and edges $\mathcal{E}$. Let $\varphi : \mathcal{E} \to \mathcal{X}$ be some function. For a given pair $(\mathcal{T}, \varphi)$ we have the following corresponding presentation, called *LOT-presentation*:

$$\langle \mathcal{X} \mid i(e)\varphi(e)t^{-1}(e)\varphi(e)^{-1}, \ e \in \mathcal{E} \rangle,$$

where $i(e)$ and $t(e)$ are respectively the initial and final vertices of a given edge e.

It is easy to see that the standard 2-complex, associated with a LOT-presentation is a subcomplex of a 2-dimensional contractible complex. Clearly, we have the condition (3.19) for a one-relator group, where the relator is taken from the set of relators of a given LOT-presentation. Hence the obstructions to the asphericity of such presentations lie in the difference between δ_ω and $\delta_{\omega+1}$ terms of the derived series of some finitely generated groups, associated with a given LOT-presentation. Note that modulo Andrews-Curtis conjecture about balanced presentations of the trivial group, the positive answer to the Whitehead asphericity conjecture for finite complexes is equivalent to the asphericity of LOT-presentations [How83].

Remark. The relation of homology with lower central series and derived series, with applications to topology, has been investigated by T. Cochran and S. Harvey ([Coc05], [Coc08], [Cocb], [Coca]).

Chapter 4
Augmentation Powers

Our main aim in this Chapter is to discuss certain (co)homological methods for the study of group rings, in particular the augmentation powers.

4.1 Augmentation Identities

Given a group G, let $\mathfrak{g}$ denote the augmentation ideal of its integral group ring $\mathbb{Z}[G]$. While studying augmentation powers $\mathfrak{g}^n$, certain identities play a crucial role. We list them here for the reader's convenience.

1. (Hartley [Har70], Prop. 2.1; [Har82c], Lemma 4.5) If x, y are elements of G and x has order q, then $(1 - y^{q^n})(1 - x) \in \mathfrak{g}^{n+2}$ for all $n \geq 0$.
2. If $y \in G$ is an element of infinite p-height, and $x \in G$ is a p-element, then $(1 - y)(1 - x) \in \mathfrak{g}^\omega$.
3. (Hartley [Har82c], Lemma 5.3) Let G be a nilpotent group. If $x \in G$ is an element of infinite p-height modulo the torsion subgroup T of G, and $y \in G$ is a p-element of infinite p-height, then $(1 - x)(1 - y) \in \mathfrak{g}^\omega$.

4.2 Integral Augmentation Powers

The structure of augmentation powers $\mathfrak{g}^n$ and augmentation quotients $P_n(G) := \mathbb{Z}[G]/\mathfrak{g}^{n+1}$, $Q_n(G) := \mathfrak{g}^n/\mathfrak{g}^{n+1}$ is of number-theoretic interest (for example, see Passi [Pas68b], Mazur-Tate [Maz87], [Dar92], Ki-Seng Tan [Tan95], Bak-Vavilov [Bak00]). A presentation of the abelian group $\mathfrak{g}^n$, $n \geq 1$, has been given by Bak and Tang [Bak04] in case G is a torsion-free or torsion abelian group; we begin with a brief account of this work.

The augmentation ideal $\mathfrak{g}$ is clearly a free abelian group generated by the elements $\theta(g) := g - 1$, $g \in G$, modulo the relation $\theta(1) = 0$. Therefore the ideal $\mathfrak{g}^n$, $n \geq 1$, is generated by the elements

R. Mikhailov, I.B.S. Passi, *Lower Central and Dimension Series of Groups.*
Lecture Notes in Mathematics 1952,
© Springer-Verlag Berlin Heidelberg 2009

$$\theta(g_1, \ldots, g_n) := \theta(g_1) \ldots \theta(g_n), \quad g_1, \, \ldots \, , \, g_n \in G.$$

This set of generators of $\mathfrak{g}^n$ are called the *standard generators*. Obviously there hold the following relations:

$N := \theta(g_1, \, \ldots \, , g_n) = 0$, *whenever some* $g_i = 1$ $(n \geq 1)$ (**normalizing relation**);

$R := \theta(g_1, \, \ldots \, , g_n)$ *is a 2-cocycle in* g_{i-1}, g_i, $(i \geq 2)$, *when the other variables are fixed* (**cocycle relation**);

$S := \theta(g_1, \, \ldots \, , g_n) = \theta(g_{\sigma(1)}, \, \ldots \, , g_{\sigma(n)})$ *for any permutation* σ *of* n *letters* $(n \geq 1)$ (**symmetric relation**).

If z is a rational number, let

$$\{z\}_{\geq 0} = \text{smallest nonnegative integer } \geq z.$$

If p is a prime number, let

$$v_p : \mathbb{Q} \to \mathbb{Z}, \, z \mapsto v_p(z)$$

denote the discrete p-adic valuation on the field $\mathbb{Q}$ of rational numbers. Let g, h be elements of a torsion abelian group G, $|g|$, $|h|$ their respective orders. Let $|h| = \prod_{\alpha=1}^{\kappa} p_\alpha^{t_\alpha}$ be a factorization of the order $|h|$ of h as a product of distinct prime powers. Let m be a natural number such that $3 \leq m \leq |g| + 1$. For each α $(1 \leq \alpha \leq \kappa)$ and each i $(1 \leq i \leq m-2)$, define

$$e_{\alpha, i}^{(m)}(h, \, g) := \left\{ \frac{m - (i+1)}{p_\alpha - 1} + t_\alpha - 1 - v_{p_\alpha}\binom{|g|}{i} \right\}_{\geq 0} \tag{4.1}$$

$$c_i^{(m)}(h, g) := \prod_\alpha p_\alpha^{e_{\alpha, i}^{(m)}(h, \, g)} \tag{4.2}$$

and let

$$c^{(m)}(h, \, g) := l.c.m.\{c_i^{(m)}(h, \, g) \,|\, 1 \leq i \leq m-2\}. \tag{4.3}$$

Consider the polynomial

$$F_{|g|}^{(m)}(X) := (X^{|g|-m+2} - 1)(X - 1)^{m-2} - X^{|g|} + 1. \tag{4.4}$$

Since the free $\mathbb{Z}$-module of all polynomials $f(X) \in \mathbb{Z}[X]$ such that degree $f(X) \leq |g| - 1$ and $f(1) = 0$ has $\mathbb{Z}$-basis consisting of the set

$$\{(X - 1)^i \,|\, 1 \leq i \leq m-2\} \cup \{(X^i - 1)(X - 1)^{m-2} \,|\, 1 \leq i \leq |g| - m + 1\},$$

the polynomial $F_{|g|}^{(m)}(X)$ can be written uniquely as

$$F_{|g|}^{(m)}(X) = \sum_{i=1}^{m-2} a_i^{(m)}(X - 1)^i + \sum_{i=1}^{|g|-m+1} b_i^{(m)}(X^i - 1)(X - 1)^{m-2}, \tag{4.5}$$

with $a_i^{(m)}$, $b_i^{(m)} \in \mathbb{Z}$. It can be checked that

$$a_i^{(m)} = -\binom{|g|}{i}. \tag{4.6}$$

Now there exist unique integers $a_{i,1}^{(m)}, \ldots, a_{i,|h|-1}^{(m)}$ $(1 \leq i \leq m - 2)$ such that

$$c^{(m)}(h, g)\binom{|g|}{i}(h-1) = \sum_{j=1}^{|h|-1} a_{i,j}^{(m)}(h^j - 1)(h-1)^{m-i-1}. \tag{4.7}$$

These considerations lead to the following relation in $\mathfrak{g}^m$:

$$T^m(g, h): \quad c^{(m)}(h, g)\theta(h, g^{|g|\cdot m+2}, \underbrace{g, \ldots, g}_{m-2}) =$$

$$\sum_{i=1}^{m-2}\sum_{j=1}^{|h|-1} a_{i,j}^{(m)}\theta(h^j, \underbrace{h, \ldots, h}_{m-i-1}, \underbrace{g, \ldots, g}_{i}) + \sum_{i=1}^{|g|-m+1} c^{(m)}(h, g)b_i^{(m)}\theta(h, g^i, \underbrace{g, \ldots, g}_{m-2}). \tag{4.8}$$

Set

$$T^n(G) := \{\theta(f_1, \ldots, f_k)T^m(h, g)\theta(f_{k+m+1}, \ldots, f_n) \mid n \geq m \geq 3, \; k \geq 0,$$
$$n \geq m + k, \; f_1, \; \ldots, \; f_k, \; h, g, \; f_{k+m+1}, \; \ldots, \; f_n \text{ ranges}$$
$$\text{over all sequences of elements in } G \text{ such that } |g| + 1 \geq m\}. \tag{4.9}$$

Let (I, g) denote an ordered basis for the torsion abelian group G, where I is an ordered set and $g : I \to G$ is a map, and let i_0 denote the smallest element of I, which might not exist. Let

$$M_I(G) = \{m \in \mathbb{N} \mid if 1 \leq k \leq \; \inf\text{imum } \{m - 2, \; |g_i| - 1 \mid i \in I\backslash\{i_0\}\}$$
$$\text{then } \forall i \in I\backslash\{i_0\}, \; \forall h \in G_{<i}\backslash\langle 1\rangle, \text{ and } \forall \text{ natural primes } p_\alpha$$

such that $p_\alpha \mid |h|$, $m \leq (k+1) + (p_\alpha - 1)v_{p_\alpha}\binom{|g|}{k} - v_{p_\alpha}(|h|) + 1)\}$, (4.10)

where

$$G_{<i} = \langle g_j \mid j < i\rangle.$$

Define

$$n_I(G) = \text{supremum } M_I(G). \tag{4.11}$$

For an arbitrary torsion abelian group G, define

$$n(G) = \text{supremum } \{n' \mid \text{given a finite subgroup} H \subseteq G, \exists \text{ a finite subgroup}$$
$$H' \supseteq H \text{and an ordered basis } (I', g) \text{ of } H' \text{ such that } n_{I'}(H') \geq n'\}. \tag{4.12}$$

Theorem 4.1 (Bak and Tang [Bak04]). *Let G be a torsion free or torsion abelian group. Then the following holds:*

- *N, R and S are a defining set of relations for $\mathfrak{g}^n$ when either $n = 2$ or $n \geq 2$ and G is torsion free or a direct limit of cyclic groups.*
- *N, R, S and T are a defining set of relations for $\mathfrak{g}^n$ when either G is p-elementary or G is torsion and $n \leq n(G)$.*

Bak and Tang [Bak04] define another set of relations, denoted by U, which together with the relations N, R, S and T provide a defining set of relations for $\mathfrak{g}^n$ when G is an arbitrary torsion abelian group.

As a consequence of the above work, there immediately follows a presentation for the augmentation quotient $Q_n(G) = \mathfrak{g}^n/\mathfrak{g}^{n+1}$ ($n \geq 1$), in the corresponding cases since $Q_n(G)$ is the quotient of $\mathfrak{g}^n$ modulo the relations

$$(B) : \theta(g_1, \ldots, g_k)(\theta(fg) - \theta(f) - \theta(g))\theta(g_{k+3}, \ldots, g_{n+1}) = 0.$$
$$\textbf{(bilinearizing relation)}$$

For some of the earlier work on augmentation powers, see [Pas77a], [Pas78], [Pas79], [Hal85], [Par01].

4.3 Intersection Theorems

Theorem 4.2 (Hartley [Har82c]). *Let G be a nilpotent group of class c having a normal subgroup K with G/K torsion-free. Let V be a $\mathbb{Z}[K]$-module satisfying $V\mathfrak{k}^b = 0$ for some integer $b \geq 1$. Let $W = V \otimes_{\mathbb{Z}[K]} \mathbb{Z}[G]$. Then*

$$W\mathfrak{g}^{(b-1)c(c+1)+1} \cap V = 0.$$

Theorem 4.3 (Hartley [Har82c]). *If there is no prime p such that G/K contains an element of infinite p-height and V contains an element of additive order p, then $\bigcap_{b=1}^{\infty} W\mathfrak{g}^b = 0$.*

Theorem 4.2 when applied to the module $V = \mathbb{Z}[K]/\mathfrak{k}^b$, immediately gives the following result.

Theorem 4.4 (Hartley [Har82c]). *If G is a nilpotent group of class c, and K is a normal subgroup of G with G/K torsion-free, then*

$$\mathfrak{g}^{(b-1)c(c+1)+1} \cap \mathbb{Z}[K] \subseteq \mathfrak{k}^b$$

for all integers $b \geq 1$.

Corollary 4.5 *If G is a nilpotent group, and T its torsion subgroup, then*

(i) $D_\omega(G) = D_\omega(T)$.
(ii) $[D_\omega(G), T] = 1$.

Theorem 4.6 (Hartley [Har82b]). *Let p be a given prime and let c, N be given natural numbers. Then there exists a function $\varphi(a) = \varphi_{p,c,N}(a)$ such that if $H = LG$ is a nilpotent group of class at most c, where $L \lhd H$ and G has finite exponent dividing p^N, then*

$$\mathfrak{h}^{\varphi(a)} \cap \mathbb{Z}[L] \subseteq \mathfrak{l}^a \quad \text{for all } a \geq 1.$$

4.4 Transfinite Augmentation Powers

Let G be a group and R a commutative ring with identity. The *transfinite augmentation powers* $\Delta_R^\alpha(G)$ are defined as follows:

$$\Delta_R^1(G) = \Delta_R(G), \quad \Delta_R^{\alpha+1}(G) = \Delta_R^\alpha(G)\Delta_R(G) + \Delta_R(G)\Delta_R^\alpha(G),$$

and

$$\Delta_R^\tau(G) = \bigcap_{\alpha < \tau} \Delta_R^\alpha(G)$$

if τ is a limit ordinal. Thus, for instance, if n is a natural number and ω denotes the least infinite limit ordinal, then

$$\Delta_R^{n\omega}(G) = \bigcap_m \Delta_R^{(n-1)\omega+m}(G),$$

where m runs over all natural numbers. In the case of the integral group ring $\mathbb{Z}[G]$ we drop the suffix R and write simply $\Delta^\alpha(G)$, or $\mathfrak{g}^\alpha$, in place of $\Delta_R^\alpha(G)$.

In analogy with the ordinary dimension subgroups, for a given ordinal number α, denote by $D_{\alpha,R}(G)$ the *transfinite dimension subgroup* $G \cap (1 + \Delta_R^\alpha(G))$ of G over R:

$$D_{\alpha,R}(G) := G \cap (1 + \Delta_R^\alpha(G)).$$

It is easy to see that $\gamma_\alpha(G) \subseteq D_{\alpha,R}(G)$ always, and $\gamma_{\alpha,\mathbb{Q}}(G) \subseteq D_{\alpha,R}(G)$ provided $\mathbb{Q} \subseteq R$.

Let G be a nilpotent group of class c containing a normal subgroup K with G/K torsion-free, and let T be the torsion subgroup of G. Then $T = \prod_p T_p$, where p runs over the rational primes and T_p is the p-torsion subgroup of T. Denote by $G(p)$ the set of elements of infinite p-height in G, by $G^*(p)/T$ the set of elements of infinite p-height in G/T and by $T_p(p)$ the set of elements of infinite p-height in T_p. With these notations, we have

Theorem 4.7 (Hartley [Har82a]). *If there is no prime p such that (i) G/K contains a non-trivial element of infinite p-height and (ii) T contains an element of order p, then*

$$\mathfrak{g}^\omega = \mathfrak{k}^\omega \mathbb{Z}[G].$$

Proof. Let π be the set of primes p for which T has an element of order p.

Trivially, $\mathfrak{k}^\omega \mathbb{Z}[G] \subseteq \mathfrak{g}^\omega$. Since G/T is a torsion-free nilpotent group, $\mathfrak{g}^\omega \subseteq \mathfrak{t}\mathbb{Z}[G]$. Observe that $\mathfrak{t} = \oplus_{p\in\pi} \mathfrak{t}_p \oplus \sum_{p\in\pi} \mathfrak{t}_p \mathfrak{t}_{p'}$ and $\mathfrak{t}_p \mathfrak{t}_{p'} \subseteq \mathfrak{t}^\omega$. Thus, to complete the proof, it suffices to show that $X = (\sum_{p\in\pi} \mathfrak{t}_p) \cap \mathfrak{g}^\omega \subseteq \mathfrak{t}^b\mathbb{Z}[G]$ for all $b \geq 1$. The $\mathbb{Z}[K]$-module $V/V_{\pi'}$ is such that the only additive torsion is π-torsion, whereas G/K has no element of infinite p-height. Therefore, an application of Theorem 4.3 to this $\mathbb{Z}[K]$-module shows that the image of X in V is contained in $V_{\pi'}\mathbb{Z}[G]$. Note that, modulo $\mathfrak{t}^b$, X is a π-torsion ablian group, it therefore follows that $X \subseteq \mathfrak{t}^b\mathbb{Z}[G]$ for all $b \geq 1$. $\square$

Theorem 4.8 (Hartley [Har82c]).

$$\mathfrak{g}^\omega = \sum_p (\mathfrak{g}(p)\mathfrak{t}_p + \mathfrak{g}^*(p)\mathfrak{t}_p(p) + \mathfrak{d}_\omega(T_p))\mathbb{Z}[G].$$

Proof. For the inclusion of the terms on the right hand side in $\mathfrak{g}^\omega$, see ([Pas79], Theorem 2.3, p. 97).

For the reverse inclusion, suppose first that T is a p-group.

If T has no element of infinite p-height, then $\mathfrak{t}\mathbb{Z}[G] \cap \mathfrak{g}(p)\mathbb{Z}[G] = \mathfrak{g}(p)\mathfrak{t}\mathbb{Z}[G]$. Since the augmentation ideals of $\mathbb{Z}[G/T]$ and $\mathbb{Z}[G(p)]$ are both residually nilpotent, it follows that $\mathfrak{g}^\omega \subseteq \mathfrak{t}\mathbb{Z}[G] \cap \mathfrak{g}(p)\mathbb{Z}[G]$. Hence $\mathfrak{g}^\omega = \mathfrak{g}(p)\mathfrak{t}\mathbb{Z}[G]$. Thus we have $\mathfrak{g}^\omega \subseteq \mathfrak{t}(p)\mathbb{Z}[G] + \mathfrak{g}(p)\mathfrak{t}\mathbb{Z}[G]$. Consequently

$$\mathfrak{g}^\omega = \mathfrak{g}(p)\mathfrak{t}\mathbb{Z}[G] + (\mathfrak{t}(p)\mathbb{Z}[G] \cap \mathfrak{g}^\omega).$$

Suppose every element of G is, modulo T, of infinite p-height. Then $\mathfrak{g}\mathfrak{t}(p) \subseteq \mathfrak{g}^\omega$. Therefore, in this case, every element of $\mathfrak{t}(p) \cap \mathfrak{g}^\omega$ can be expressed as $1 - x + \alpha$, with $x \in T(p)$, $\alpha \in \mathfrak{g}\mathfrak{t}(p)$. Since $1 - x$ then lies in $\mathfrak{g}^\omega$, we have $x \in D_\omega(G) \cap T$. Therefore, by Cor. 4.5, $x \in D_\omega(T)$. We have thus shown that if T is a p-group and $G = G^*(p)$, then

$$\mathfrak{g}^\omega = \mathfrak{g}\mathfrak{t}(p)\mathbb{Z}[G] + \mathfrak{g}\mathfrak{t}(p) + \mathfrak{d}_\omega(T).$$

Now suppose G is an arbitrary nilpotent group whose torsion subgroup is a p-group. By Theorem 4.7, we have $\mathfrak{g}^\omega = \mathfrak{k}^\omega\mathbb{Z}[G]$, where $K = G^*(p)$. Observe that the subgroup K is such that all its element are of infinite p-height within this subgroup. Therefore the preceding case is applicable to this group, and so we have

$$\mathfrak{g}^\omega = (\mathfrak{g}^*(p))^\omega\mathbb{Z}[G] = \mathfrak{g}(p)\mathfrak{t} + \mathfrak{g}^*(p)\mathfrak{t}(p) + \mathfrak{d}_\omega(T))\mathbb{Z}[G].$$

Finally, the general case can be handled by passing to the the quotients $G/T_{p'}$, with p varying over the primes for which G has p-torsion. $\square$

It may be mentioned that another definition of the transfinite augmentation powers was used in [Gru72], where it was defined inductively as follows:

$$\bar{\Delta}_R^1(G) = \Delta_R(G), \quad \bar{\Delta}_R^{\alpha+1}(G) = \bar{\Delta}_R^\alpha(G)\Delta_R(G),$$

and as intersection of the preceding terms in the case of a limit ordinal. We call these augmentation powers as the *right augmentation powers* [note that we have denoted them by $\bar{\Delta}_R^\alpha(G)$; as usual, we drop the suffix in case R is the ring $\mathbb{Z}$ of integers]. It is easy to see that $\bar{\Delta}_R^\alpha(G) \subseteq \Delta_R^\alpha(G)$ for any group G and ordinal number α, but the converse, in general, is not true (see Example 4.14).

For an ideal $\mathfrak{a}$ of $\mathbb{Z}[G]$, we define ideals $\mathfrak{a}^{(n)}$, $n \geq 0$, inductively by setting

$$\mathfrak{a}^{(0)} = \mathfrak{a}, \qquad \mathfrak{a}^{(n)} = \mathfrak{g}\mathfrak{a}^{(n-1)} + \mathfrak{a}^{(n-1)}\mathfrak{g} \text{ for } n \geq 1. \tag{4.13}$$

An example of a group with transfinite augmentation powers not stabilizing at the first limit ordinal can be found in the class of abelian groups.

Example 4.9 (Gruenberg-Roseblade [Gru72]).

Let p be a prime. If G be an abelian p-group with $\bigcap_{n\geq 1} G^{p^n}$ of exponent p, then

$$\bar{\Delta}^{\omega+k}(G) \neq \bar{\Delta}^{\omega+k+1}(G), \ k \geq 0;$$

for example, let G an abelian p-group with generators $x_1, x_2, \ldots$ and relations

$$x_n^{p^n} = 1, \ x_{n+1}^{p^n} = x_1, \ n = 1, 2, \ldots.$$

Example 4.10 (Gruenberg-Roseblade [Gru72]).

Let $d > 0$ be an integer, p an odd prime greater than d, and $\mathbb{F}_p$ the field with p elements. Let V_d be a vector space with basis $v_1, \ldots, v_d$ over the field $\mathbb{F}_p$. Consider the automorphism α of V_d defined by

$$\alpha : v_i \mapsto v_{i+1} - v_i, \ 1 \leq i < d,$$
$$\alpha : v_d \mapsto -v_d.$$

Define $G_d = V_d \rtimes \langle \alpha \rangle$. Then

$$\bar{\Delta}^{\omega+k}(G) \neq \bar{\Delta}^{\omega+k+1}(G), \ k < d, \ \bar{\Delta}^{\omega+d}(G) = \bar{\Delta}^{\omega+d+1}(G).$$

A simple result towards the description of the transfinite dimension subgroups is the following:

Theorem 4.11 (Mikhailov-Passi [Mik04]). *For any group G and natural number n*

$$D_{\omega+n,\mathbb{Q}}(G) = \gamma_{\omega+n,\mathbb{Q}}(G).$$

Proof. It suffices to prove that $D_{\omega+n,\mathbb{Q}}(G) \subseteq \gamma_{\omega+n,\mathbb{Q}}(G)$, the reverse inclusion being easy to see.

Recall that $\Delta_\mathbb{Q}^\omega(G) = \Delta(\gamma_{\omega,\mathbb{Q}}(G))\mathbb{Q}[G]$ (Jennings [Jen55]) and therefore $D_{\omega+n,\mathbb{Q}}(G) \subseteq \gamma_{\omega,\mathbb{Q}}(G)$. Let $g \in D_{\omega+n,\mathbb{Q}}(G)$. Then $g \in \gamma_{\omega,\mathbb{Q}}(G)$ and

$$g - 1 \in \Delta_\mathbb{Q}^{\omega+n}(G) \subseteq \Delta_\mathbb{Q}^n(G)\Delta_\mathbb{Q}(\gamma_{\omega,\mathbb{Q}}(G)) + \Delta_\mathbb{Q}(\gamma_{\omega,\mathbb{Q}}(G)))\Delta_\mathbb{Q}(G).$$

Therefore there exists an integer $m \geq 0$ such that

$$m(g - 1) \in \mathfrak{g}^n \Delta_\mathbb{Z}(\gamma_{\omega,\mathbb{Q}}(G)) + \Delta_\mathbb{Z}(\gamma_{\omega,\mathbb{Q}}(G))\mathfrak{g}.$$

Since $m(g - 1) = g^m - 1 \mod \Delta_\mathbb{Z}^2(\gamma_{\omega,\mathbb{Q}}(G))$, it follows that

$$g^m - 1 \in \mathfrak{g}^n \Delta_\mathbb{Z}(\gamma_{\omega,\mathbb{Q}}(G)) + \Delta_\mathbb{Z}(\gamma_{\omega,\mathbb{Q}}(G))\mathfrak{g}.$$

Hence, by Lemma 4.31, $g^m \in [\gamma_{\omega,\mathbb{Q}}(G), \gamma_{\omega,\mathbb{Q}}(G)] \cdot [\gamma_{\omega,\mathbb{Q}}(G), {}_nG]$. Now it is easy to see that $[\gamma_{\omega,\mathbb{Q}}(G), {}_nG]$ and $[\gamma_{\omega,\mathbb{Q}}(G), \gamma_{\omega,\mathbb{Q}}(G)]$ are both contained in $\gamma_{\omega+n,\mathbb{Q}}(G)$. Therefore $g \in \gamma_{\omega+n,\mathbb{Q}}(G)$ and the proof is complete. $\square$

Remark 4.12
The behavior of the rational augmentation powers is closely related to that of the rational lower central series; for instance, as is clear from the preceding theorem, or even otherwise, $\gamma_{\omega,\mathbb{Q}}(G) = \gamma_{\omega+1,\mathbb{Q}}(G)$ if and only if $\Delta_\mathbb{Q}^\omega(G) = \Delta_\mathbb{Q}^{\omega+1}(G)$.

The following Proposition can be proved by proceeding in analogy with the proof of Theorem 1.34.

Proposition 4.13 *Let* $1 \to N \to F \to G \to 1$ *be a non-cyclic free presentation of the group* G, *and* α *an ordinal number not less than* ω. *Then the following statements are equivalent:*

(i) $\bar{\Delta}_R^\alpha(G) = 0$;

(ii) $\gamma_{\alpha,R}(F/[N, N]) = 1$;

where R *is* $\mathbb{Z}$ *or* $\mathbb{Q}$ *and* $\{\bar{\Delta}_R^\alpha(G)\}$ *is the series of right augmentation powers of* G *over* R.

To conclude this section we give an example of a group G such that

$$\Delta_\mathbb{Q}^{\omega+1}(G) \neq \bar{\Delta}_\mathbb{Q}^{\omega+1}(G)$$

[recall that $\bar{\Delta}_\mathbb{Q}^{\omega+1}(G) = \Delta_\mathbb{Q}^\omega(G)\Delta_\mathbb{Q}(G)$], and consequently

$$\Delta_\mathbb{Q}^\omega(G)\Delta_\mathbb{Q}(G) \neq \Delta_\mathbb{Q}(G)\Delta_\mathbb{Q}^\omega(G).$$

Example 4.14
Consider the fundamental group $\mathcal{G}_K$ of the Klein bottle, namely

$$\mathcal{G}_K = \langle a, b \mid aba^{-1}b = 1 \rangle.$$

We claim that $\Delta_\mathbb{Q}^{\omega+1}(\mathcal{G}_K) \neq \bar{\Delta}_\mathbb{Q}^{\omega+1}(\mathcal{G}_K)$.

Observe that b is a generalized periodic element (i.e., b is periodic modulo every term of the lower central series of $\mathcal{G}_K$), the cyclic subgroup $\langle b \rangle$ is normal in $\mathcal{G}_K$ and $\mathcal{G}_K / \langle b \rangle$ is cyclic. Therefore it follows that

$$\gamma_{\omega, \mathbb{Q}}(\mathcal{G}_K) = \langle b \rangle,$$

and

$$\gamma_{\omega+1, \mathbb{Q}}(\mathcal{G}_K) = \sqrt{[\langle b \rangle, \mathcal{G}_K]} = \sqrt{\langle b^2 \rangle} = \sqrt{\langle b \rangle} = \langle b \rangle.$$

Therefore, by Remark 4.12, $\Delta_{\mathbb{Q}}^{\omega+1}(\mathcal{G}_K) = \Delta_{\mathbb{Q}}^{\omega}(\mathcal{G}_K) = \bar{\Delta}_{\mathbb{Q}}^{\omega}(\mathcal{G}_K)$, and hence

$$\Delta_{\mathbb{Q}}^{\alpha}(\mathcal{G}_K) = \Delta_{\mathbb{Q}}(\langle b \rangle)\mathbb{Q}[\mathcal{G}_K],$$

for all ordinals $\alpha \geq \omega$.

If $\Delta_{\mathbb{Q}}^{\omega+1}(\mathcal{G}_K)$ were equal to $\bar{\Delta}_{\mathbb{Q}}^{\omega+1}(\mathcal{G}_K)$, then clearly the right augmentation powers $\bar{\Delta}_{\mathbb{Q}}^{\alpha}(\mathcal{G}_K)$ would also stabilize at ω and equal $\Delta_{\mathbb{Q}}(\langle b \rangle)\mathbb{Q}[\mathcal{G}_K]$ for all $\alpha \geq \omega$, and so we would have, in particular, $\bar{\Delta}_{\mathbb{Q}}^{2\omega}(\mathcal{G}_K) = \Delta_{\mathbb{Q}}(\langle b \rangle)\mathbb{Q}[\mathcal{G}_K]$. However, we claim that

$$\bar{\Delta}_{\mathbb{Q}}^{2\omega}(\mathcal{G}_K) = \Delta_{\mathbb{Q}}^2(\langle b \rangle)\mathbb{Q}[\mathcal{G}_K] \ (\neq \Delta_{\mathbb{Q}}(\langle b \rangle)\mathbb{Q}[\mathcal{G}_K]).$$

For, let $z \in \bar{\Delta}_{\mathbb{Q}}^{2\omega}(\mathcal{G}_K)$. Then, for every natural number $n \geq 1$, we can write $z = (1 - b)z_n$ with $z_n \in \Delta_{\mathbb{Q}}^n(\mathcal{G}_K)$. Since the element b is of infinite order, we must have $z_n = z_m$ for all m, n. Therefore $z \in (\Delta_{\mathbb{Q}}^{\omega}(\mathcal{G}_K))^2 = \Delta_{\mathbb{Q}}^2(\langle b \rangle)\mathbb{Q}[\mathcal{G}_K]$ and so it follows that $\bar{\Delta}_{\mathbb{Q}}^{2\omega}(\mathcal{G}_K) = \Delta_{\mathbb{Q}}^2(\langle b \rangle)\mathbb{Q}[\mathcal{G}_K]$. Hence

$$\Delta_{\mathbb{Q}}^{\omega+1}(\mathcal{G}_K) \neq \bar{\Delta}_{\mathbb{Q}}^{\omega+1}(\mathcal{G}_K).$$

The group $\mathcal{G}_K$ provides an example to show that Proposition 4.13 does not hold, in general, for the transfinite augmentation powers $\Delta_{\mathbb{Q}}^{\alpha}(G)$.

To see this, let

$$1 \rightarrow N \rightarrow F \rightarrow \mathcal{G}_K \rightarrow 1$$

be a free presentation of $\mathcal{G}_K$ with $F = \langle a, b \mid \emptyset \rangle$ and N the normal closure of $aba^{-1}b$ in F. We have examined the augmentation powers of $\mathcal{G}_K$ in the preceding example. Proceeding as above, one can show that

$$\bar{\Delta}_{\mathbb{Q}}^{n\omega}(\mathcal{G}_K) = \Delta_{\mathbb{Q}}^n(\langle b \rangle)\mathbb{Q}[\mathcal{G}_K] \neq 0$$

for all natural numbers $n \geq 1$; whereas,

$$\bar{\Delta}_{\mathbb{Q}}^{\omega^2}(\mathcal{G}_K) = \Delta_{\mathbb{Q}}^{\omega}(\langle b \rangle)\mathbb{Q}[\mathcal{G}_K] = 0.$$

Consequently it follows, by Proposition 4.13, that $\gamma_{\alpha, \mathbb{Q}}(F/[N, N]) \neq 1$ for $\alpha < \omega^2$ and $\gamma_{\omega^2, \mathbb{Q}}(F/[N, N]) = 1$. On the other hand,

$$\Delta_{\mathbb{Q}}^{\omega^2}(\mathcal{G}_K) = \Delta_{\mathbb{Q}}^{\omega}(\mathcal{G}_K) \neq 0.$$

Proposition 4.15 *Let F be a finitely generated free group and N its normal subgroup such that F/N is nilpotent. Then $F/\gamma_2(N)$ is transfinitely nilpotent if and only if $F/\gamma_2(N)$ is residually nilpotent.*

Proof. The assertion follows from Proposition 4.13 and the property:

$$\mathfrak{g}^\omega = \bar{\Delta}^{\omega+1}(G),$$

which holds for any finitely generated nilpotent group G (see [Pas79, Theorem 5.3, p. 102]). $\square$

4.5 Schur Multiplicator

In this section we discuss various subgroups of Schur multiplicator which provide a homological approach to the identification of subgroups determined by two-sided ideals in group rings.

Generalized polynomial 2-cocycles

Let $\mathfrak{a} \subseteq \mathfrak{g}$ be a two-sided ideal of the integral group ring $\mathbb{Z}[G]$ and let M be a trivial G-module. Consider the following classes of maps on $G \times G$. A normalized 2-cocycle $f : G \times G \to M$ is said to be a *left (resp: right) $\mathfrak{a}$-2-cocycle* if the linear extentions to $\mathbb{Z}[G]$ of the maps $l_y : G \to M$, $y \in G$, (resp: $r_x : G \to M$, $x \in G$), defined by $l_y(x) = f(x,y)$, $x \in G$ (resp: $r_x(y) = f(x,y)$, $y \in G$) vanish on $\mathfrak{a}$. We denote by $P_\mathfrak{a}(G, M)_l$ (resp: $P_\mathfrak{a}(G, M)_r$) the subgroup of $H^2(G, M)$, the second cohomology group of G with coefficients in M, consisting of the cohomology classes represented by the left (resp: right) $\mathfrak{a}$-2-cocycles. Furthermore, we denote by $P_{\mathfrak{a},\mathfrak{a}}(G, M)$ the subgroup consisting of those cohomology classes which possess representative 2-cocycles which are both left as well as right $\mathfrak{a}$-2-cocyles.

As $\mathfrak{a}$ varies over the augmentation powers $\mathfrak{g}^\alpha$, α any ordinal number, we obtain an increasing filtration of $H^2(G, M)$:

$$0 = P_0(G, M) \subseteq P_1(G, M) \subseteq \ldots P_\alpha(G, M) \subseteq \cdots \subseteq H^2(G, M), \quad (4.14)$$

where $P_{\alpha-1}(G, M) = P_{\mathfrak{g}^\alpha}(G, M)_l$. This filtration has been studied in [Pas74] for finite ordinals.

We begin by considering the case when M is a divisible abelian group. In this case the two subgroups $P_\mathfrak{a}(G, M)_l$ and $P_\mathfrak{a}(G, M)_r$ of $H^2(G, M)$ coincide. The proof is essentially the same as given in [Pas74] where it is proved that, for all integers $n \geq 1$, $P_{n-1}(G, \mathbb{T})$ is equal to the image of the homomorphism

$$\mathrm{Ext}^1_G(\mathfrak{g}/\mathfrak{g}^n,\, \mathbb{T}) \to \mathrm{Ext}^1_G(\mathfrak{g}, M)\ (\simeq H^2(G,\, \mathbb{T}))$$

induced by the natural projection $\mathfrak{g} \to \mathfrak{g}/\mathfrak{g}^n$, where $\mathbb{T}$ denotes the additive group of rationals modulo 1. More precisely, we have:

Theorem 4.16 (Mikhailov-Passi [Mik04]). *Let G be a group, $\mathfrak{a} \subseteq \mathfrak{g}$ an arbitrary two-sided ideal of $\mathbb{Z}[G]$ and M a divisible abelian group regarded as a trivial G-module. Let $\delta : \mathrm{Ext}^1_G(\mathfrak{g}/\mathfrak{a},\, M) \to \mathrm{Ext}^1_G(\mathfrak{g},\, M)$ be the map induced by the natural projection $\mathfrak{g} \to \mathfrak{g}/\mathfrak{a}$. Then*

$$P_\mathfrak{a}(G,\, M)_l = P_\mathfrak{a}(G,\, M)_r = \mathrm{im}(\delta).$$

Proof. The short exact sequences

$$0 \to \mathfrak{a} \to \mathfrak{g} \to \mathfrak{g}/\mathfrak{a} \to 0,$$

$$0 \to \mathfrak{g} \to \mathbb{Z}[G] \to \mathbb{Z} \to 0$$

yield the following commutative diagram with exact rows and columns:

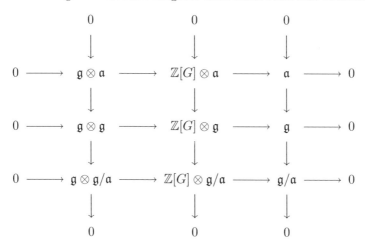

where all tensor products are over $\mathbb{Z}$ and the G-action is diagonal. On applying the functor $\mathrm{Hom}_G(-,\, M)$ to this diagram, we obtain the following commutative diagram, again with exact rows and columns:

where $S = \text{im}\{\text{Hom}_G(\mathfrak{g} \otimes \mathfrak{g}/\mathfrak{a}, M) \to \text{Ext}^1_G(\mathfrak{g}/\mathfrak{a}, M)\}$. From the spectral sequences argument, as in [Pas74], it follows that $\text{Ext}^1_G(\mathbb{Z}[G] \otimes \mathfrak{g}/\mathfrak{a}, M) = 0$. Therefore, the map γ_3 is an epimorphism. The map γ_2 is an epimorphism due to the step-by-step resolution, as observed in [Pas74], while the map γ_1 is an epimorphism by construction. It then follows easily that

$$P_{\mathfrak{a}}(G, M)_r = \ker(\alpha) = \text{im}(\delta).$$

In view of the isomorphism $\mathfrak{g} \otimes \mathfrak{g}/\mathfrak{a} \simeq \mathfrak{g}/\mathfrak{a} \otimes \mathfrak{g}$, it similarly follows that $P_{\mathfrak{a}}(G, M)_l$ also equals $\text{im}(\delta)$, and the result is proved. $\square$

In view of the preceding Theorem we drop the suffix and write simply $P_{\mathfrak{a}}(G, M)$ instead of $P_{\mathfrak{a}}(G, M)_l$ or $P_{\mathfrak{a}}(G, M)_r$.

Let $\mathfrak{a} \subseteq \mathfrak{g}$ be a two-sided ideal of the group ring $\mathbb{Z}[G]$. Denote by $D_{\mathfrak{a}}(G)$ the normal subgroup of G determined by $\mathfrak{a}$:

$$D_{\mathfrak{a}}(G) = G \cap (1 + \mathfrak{a}).$$

If $\mathfrak{a} = \mathfrak{g}^\alpha$, then we write $D_\alpha(G)$ for $D_{\mathfrak{g}^\alpha}$. For a trivial G-module M, let $\psi_{\mathfrak{a}}(G, M)$ denote the image of the inflation homomorphism $H^2(G/D_{\mathfrak{a}}(G), M) \to H^2(G, M)$ induced by $G \to G/D_{\mathfrak{a}}(G)$:

$$\psi_{\mathfrak{a}}(G, M) = \text{im}(\inf : H^2(G/D_{\mathfrak{a}}(G), M) \to H^2(G, M)). \tag{4.15}$$

It is easy to see that $\psi_{\mathfrak{a}}(G, M) = P_{\Delta(D_{\mathfrak{a}}(G))\mathbb{Z}[G]}(G, M) \supseteq P_{\mathfrak{a}}(G, M)$.

Theorem 4.17 Let $\mathfrak{a}$ be an ideal of $\mathbb{Z}[G]$ contained in $\mathfrak{g}$ and $\bar{\mathfrak{a}}$ its image under the natural map $\mathfrak{g} \to \Delta(G/D_{\mathfrak{a}}(G))$. Then

(a) $P_{\bar{\mathfrak{a}}}(G/D_{\mathfrak{a}}(G), \mathbb{T}) = H^2(G/D_{\mathfrak{a}}(G), \mathbb{T})$ implies that

$$D_{\mathfrak{a}\mathfrak{g}}(G).D_{\mathfrak{g}\mathfrak{a}}(G) \subseteq [D_{\mathfrak{a}}(G), G];$$

(b) $P_{\bar{\mathfrak{a}}, \bar{\mathfrak{a}}}(G/D_{\mathfrak{a}}(G), \mathbb{T}) = H^2(G/D_{\mathfrak{a}}(G), \mathbb{T})$ implies that

$$D_{\mathfrak{a}\mathfrak{g}+\mathfrak{g}\mathfrak{a}}(G) = [D_{\mathfrak{a}}(G), G].$$

Proof. It is easy to see that $[D_{\mathfrak{a}}(G), G] \subseteq D_{\mathfrak{g}\mathfrak{a}+\mathfrak{a}\mathfrak{g}}(G)$. Let us assume that $[D_{\mathfrak{a}}(G), G] = 1$; it clearly suffices to consider this case. Suppose $a \in D_{\mathfrak{a}\mathfrak{g}}(G)$ in case (a) (resp: $a \in D_{\mathfrak{a}\mathfrak{g}+\mathfrak{g}\mathfrak{a}}(G)$ in case (b)), $a \neq 1$. Then $a \in D_{\mathfrak{a}}(G)$ and we can find a homomorphism

$$\alpha : D_{\mathfrak{a}}(G) \to \mathbb{T}, \quad \alpha(a) \neq 0.$$

Consider the following commutative diagram induced by α, where the rows are central extensions of groups:

$$1 \longrightarrow D_{\mathfrak{a}}(G) \longrightarrow G \longrightarrow G/D_{\mathfrak{a}}(G) \longrightarrow 1$$

$$\alpha \downarrow \qquad\qquad \downarrow \qquad\qquad \|$$

$$1 \longrightarrow \mathbb{T} \longrightarrow E \longrightarrow G/D_{\mathfrak{a}}(G) \longrightarrow 1,$$

Choose a set of representatives $w(g) \in G$ for the elements $g \in G/D_{\mathfrak{a}}(G)$ and let

$$W : G/D_{\mathfrak{a}}(G) \times G/D_{\mathfrak{a}}(G) \to D_{\mathfrak{a}}(G)$$

be the corresponding 2-cocycle:

$$W(g_1,\, g_2) = w(g_1 g_2)^{-1} w(g_1) w(g_2) \quad g_1,\, g_2 \in G/D_{\mathfrak{a}}(G).$$

By hypothesis the 2-cocycle $\alpha(W(g_1, g_2)) : G/D_{\mathfrak{a}}(G) \times G/D_{\mathfrak{a}}(G) \to \mathbb{T}$ is cohomologous to a normalized 2-cocycle $f : G/D_{\mathfrak{a}}(G) \times G/D_{\mathfrak{a}}(G) \to \mathbb{T}$ whose linear extension to $\mathbb{Z}[G/D_{\mathfrak{a}}(G)] \times \mathbb{Z}[G/D_{\mathfrak{a}}(G)]$ vanishes on $\bar{\mathfrak{a}} \times \mathbb{Z}[G/D_{\mathfrak{a}}(G)]$ in case (a) and on $(\bar{\mathfrak{a}} \times \mathbb{Z}[G/D_{\mathfrak{a}}(G)]) \cup (\mathbb{Z}[G/D_{\mathfrak{a}}(G)] \times \bar{\mathfrak{a}})$ in case (b). We can therefore extend the homomorphism α to a map $\varphi : G \to \mathbb{T}$ whose linear extension to $\mathbb{Z}[G]$ vanishes on $\mathfrak{a}g$ in case (a) and on $\mathfrak{a}g + g\mathfrak{a}$ in case (b). Indeed, if $\chi : G/D_{\mathfrak{a}}(G) \to \mathbb{T}$ is a correcting normalized coboundary, i.e., $\chi(1) = 0$ and

$$\alpha(W(g_1, g_2)) = f(g_1, g_2) + \chi(g_1 - 1)(g_2 - 1) \ (g_1,\, g_2 \in G/D_{\mathfrak{a}}(G)),$$

then set

$$\varphi(zw(g)) = \alpha(z) - \chi(g) \ (g \in G/D_{\mathfrak{a}}(G),\ z \in D_{\mathfrak{a}}(G)).$$

However, for such a map φ, we have $\varphi(a) = 0$, and also $\varphi(a) = \alpha(a) \neq 0$, a contradiction. Hence $D_{\mathfrak{a}g}(G) = 1$ in case (a) and $D_{\mathfrak{a}g + g\mathfrak{a}}(G) = 1$ in case (b).

Since $P_{\bar{\mathfrak{a}}}(G/D_{\mathfrak{a}}(G), \mathbb{T})_l = P_{\bar{\mathfrak{a}}}(G/D_{\mathfrak{a}}(G), \mathbb{T})_r$ (Theorem 4.16), in case (a) it similarly follows that $D_{g\mathfrak{a}}(G) = 1$. $\square$

For an arbitrary ordinal α taking $\mathfrak{a} = \mathfrak{g}^\alpha$ in Theorem 4.17(b), we have the following:

Corollary 4.18 *If* $P_{\Delta^\alpha(G/D_\alpha(G)), \Delta^\alpha(G/D_\alpha(G))}(G/D_\alpha(G), \mathbb{T}) = H^2(G/D_\alpha (G), \mathbb{T})$, *then*

$$D_{\alpha+1}(G) = [D_\alpha(G),\, G].$$

Analogously working with $\mathbb{T}_p$, the p-torsion subgroup of $\mathbb{T}$, instead of $\mathbb{T}$ in the proof of the Theorem 4.17, we have the following:

Corollary 4.19 *If* G *is a p-group and* $P_{\Delta^\alpha(G/D_\alpha(G)), \Delta^\alpha(G/D_\alpha(G))}(G/D_\alpha(G), \mathbb{T}_p) = H^2(G/D_\alpha(G), \mathbb{T}_p)$, *then*

$$D_{\alpha+1}(G) = [D_\alpha(G),\, G].$$

Let us call an ideal $\mathfrak{a}$ of $\mathbb{Z}[G]$ *cohomologically concordant (CC-ideal)* with respect to an abelian group M (regarded as a trivial G-module) if

$$P_{\mathfrak{a}}(G,\, M) = \psi_{\mathfrak{a}}(G,\, M).$$

The preceding results show usefulness of this notion in the study of normal subgroups determined by ideals of $\mathbb{Z}[G]$. If N is a normal subgroup of a group G, then the two-sided ideal $\mathfrak{n}\mathbb{Z}[G]$ is a CC-ideal with respect to every abelian group M. Observe that $\mathfrak{g}^2$ is a CC-ideal with respect to $\mathbb{T}$ for any group G ([Pas79], p. 66). However, there exist groups G for which $\mathfrak{g}^3$ is not a CC-ideal with respect to $\mathbb{T}$.

Example 4.20

Let Π be a group with the following properties: Π is nilpotent of class 3 and $D_4(\Pi) \neq 1$, i.e., Π is a nilpotent group of class three without dimension property (for the existence of such groups see Chapter 2). Let $G = \Pi/\gamma_3(\Pi)$. Then $D_3(G) = \gamma_3(G) = 1$, and therefore $\psi_{\mathfrak{g}^3}(G,\, \mathbb{T}) = H^2(G,\, \mathbb{T})$, whereas $P_{\mathfrak{g}^3}(G,\, \mathbb{T}) \neq H^2(G,\, \mathbb{T})$ as can be seen from Theorem 4.17 applied to Π with $\mathfrak{a} = \Delta^3(\Pi)$.

The following easy result relates the CC-property for an ideal $\mathfrak{a}$ in $\mathbb{Z}[G]$ with respect to a abelian group M to the CC-property for its image in $\mathbb{Z}[G/D_{\mathfrak{a}}(G)]$.

Proposition 4.21 *Let $\mathfrak{a}$ be a two-sided ideal in $\mathfrak{g}$ and $\bar{\mathfrak{a}}$ its image in $\Delta(G/D_{\mathfrak{a}}(G))$, then the natural map*

$$\psi_{\bar{\mathfrak{a}}}(G/D_{\mathfrak{a}}(G),\, M)/P_{\bar{\mathfrak{a}}}(G/D_{\mathfrak{a}}(G),\, M) \to \psi_{\mathfrak{a}}(G,\, M)/P_{\mathfrak{a}}(G,\, M)$$

is an epimorphism for any abelian group M regarded as a trivial G-module.

Proof. It is immediate from definitions that

$$H^2(G/D_{\mathfrak{a}}(G),\, M) = \psi_{\bar{\mathfrak{a}}}(G/D_{\mathfrak{a}}(G),\, M).$$

The result follows from the following commutative diagram:

$$
\begin{array}{ccc}
H^2(G/D_{\mathfrak{a}}(G),\, M) & \longrightarrow & \psi_{\mathfrak{a}}(G,\, M) \\
\downarrow & & \downarrow \\
\psi_{\bar{\mathfrak{a}}}(G/D_{\mathfrak{a}}(G),\, M)/P_{\bar{\mathfrak{a}}}(G/D_{\mathfrak{a}}(G),\, M) & \longrightarrow & \psi_{\mathfrak{a}}(G,\, M)/P_{\mathfrak{a}}(G,\, M). \;\square
\end{array}
$$

As an immediate consequence, we have the following.

Corollary 4.22 *If $\bar{\mathfrak{a}}$ is a CC-ideal in $\mathbb{Z}[G/D_{\mathfrak{a}}(G)]$ with respect to the abelian group M, then $\mathfrak{a}$ is a CC-ideal in $\mathbb{Z}[G]$ with respect to M.*

Let G be a group, $\mathfrak{a}$ a two-sided ideal in $\mathbb{Z}[G]$ and M an abelian group. The following result gives a characterization of the elements of $P_{\mathfrak{a},\mathfrak{a}}(G, M)$ when M is divisible.

Proposition 4.23 *The cohomology class classifying the central extension*

$$1 \to M \to \Pi \xrightarrow{p} G \to 1$$

lies in $P_{\mathfrak{a},\mathfrak{a}}(G, M)$ *if and only if*

$$M \cap (1 + \tilde{\mathfrak{a}}\Delta(\Pi) + \Delta(\Pi)\tilde{\mathfrak{a}} + \mathfrak{m}\Delta(\Pi)) = 1,$$

where $\tilde{\mathfrak{a}}$ *is the pre-image of* $\mathfrak{a}$ *in* $\Delta(\Pi)$ *under the homomorphism* $\mathbb{Z}[\Pi] \to \mathbb{Z}[G]$ *induced by* p, *provided* M *is divisible.*

Proof. Let $w(g) \in \Pi$ be a set of representatives in Π for the elements $g \in G$ so that $p(w(g)) = g$ and let $W(g_1, g_2) : G \times G \to M$ be the corresponding 2-cocycle:

$$w(g_1)w(g_2) = w(g_1 g_2)W(g_1, g_2), \quad g_1,\ g_2 \in G.$$

Suppose $M \cap (1 + \tilde{\mathfrak{a}}\Delta(\Pi) + \Delta(\Pi)\tilde{\mathfrak{a}} + \mathfrak{m}\Delta(\Pi)) = 1$. Then, since M is divisible, there exists a map $\varphi : \Pi \to M$ whose linear extension to $\mathbb{Z}[\Pi]$ vanishes on $\tilde{\mathfrak{a}}\Delta(\Pi) + \Delta(\Pi)\tilde{\mathfrak{a}} + \mathfrak{m}\Delta(\Pi)$ and $\varphi|_M$ is the identity map. Let $\chi : G \to M$ be the map defined by $\chi(g) = \varphi(w(g))$, $g \in G$, and extend it to $\mathbb{Z}[G]$ by linearity. Let $f : G \times G \to M$ be the 2-cocycle given by

$$f(g_1, g_2) = W(g_1, g_2) + \chi((g_1 - 1)(g_2 - 1)), \quad g_1,\ g_2 \in G.$$

A simple calculation shows that $\varphi((x_1 - 1)(x_2 - 1)) = f(g_1, g_2)$, for $x_i \in \Pi$, $p(x_i) = g_i$, $i = 1, 2$. It is then clear that the 2-cocycle f vanishes on $(\mathbb{Z}[\Pi] \times \tilde{\mathfrak{a}}) \cup (\tilde{\mathfrak{a}} \times \mathbb{Z}[\Pi])$, and therefore the cohomology class defined by the 2-cocycle W lies in $P_{\mathfrak{a},\mathfrak{a}}(G, M)$.

Conversely, suppose that the 2-cocycle $W : G \times G \to M$ represents a cohomology class in $P_{\mathfrak{a},\mathfrak{a}}(G, M)$. Then there exists a map $\chi : G \to M$ such that the 2-cocycle $f(g_1, g_2) = W(g_1, g_2) + \chi((g_1 - 1)(g_2 - 1))$ $(g_1,\ g_2 \in G)$ vanishes on $(\mathbb{Z}[G] \times \mathfrak{a}) \cup (\mathfrak{a} \times \mathbb{Z}[G])$. Define $\varphi : \Pi \to M$ by setting $\varphi(w(g)m) = m + \chi(g)$ $(g \in G,\ m \in M)$. It is easy to see that the linear extension of φ to $\mathbb{Z}[\Pi]$ vanishes on $\tilde{\mathfrak{a}}\Delta(\Pi) + \Delta(\Pi)\tilde{\mathfrak{a}} + \mathfrak{m}\Delta(\Pi)$ and $\varphi|_M$ is the identity map. Therefore, it follows that $M \cap (1 + \tilde{\mathfrak{a}}\Delta(\Pi) + \Delta(\Pi)\tilde{\mathfrak{a}} + \mathfrak{m}\Delta(\Pi)) = 1$. $\square$

For any two-sided ideal $\mathfrak{a} \subseteq \mathfrak{g}$, and arbitrary abelian group M, the inflation homomorphism inf : $H^2(G/D_{\mathfrak{a}^{(1)}}(G), M) \to H^2(G, M)$ clearly maps $P_{\bar{\mathfrak{a}},\bar{\mathfrak{a}}}(G/D_{\mathfrak{a}^{(1)}}(G), M)$ into $P_{\mathfrak{a},\mathfrak{a}}(G, M)$, where $\bar{\mathfrak{a}}$ is the image of $\mathfrak{a}$ under the homomorphism induced by the natural projection $p : G \to G/D_{\mathfrak{a}^{(1)}}(G)$. In fact, we have:

Proposition 4.24 *The map* $p_{\mathfrak{a}} : P_{\bar{\mathfrak{a}},\bar{\mathfrak{a}}}(G/D_{\mathfrak{a}^{(1)}}(G), M) \to P_{\mathfrak{a},\mathfrak{a}}(G, M)$ *induced by the inflation is a monomorphism.*

Proof. Suppose we have the following commutative diagram of central extensions:

$$1 \longrightarrow M \longrightarrow N \longrightarrow G \longrightarrow 1$$

$$\left\| \qquad \downarrow \qquad p\downarrow \right.$$

$$1 \longrightarrow M \longrightarrow \Pi \longrightarrow \bar{G} \longrightarrow 1,$$

where $\bar{G} = G/D_{\mathfrak{a}^{(1)}}(G)$ and p is the natural projection, the class of the lower central extension lies in $P_{\bar{\mathfrak{a}}, \bar{\mathfrak{a}}}(\bar{G}, M)$ and the upper central extension splits. We need to show that, under these conditions, the lower central extension must also split.

Let $f : \bar{G} \times \bar{G} \to M$ be a normalized 2-cocycle representing the lower central extension with the property that its linear extension to $\mathbb{Z}[\bar{G}] \times \mathbb{Z}[\bar{G}]$ vanishes on $(\bar{\mathfrak{a}} \times \mathbb{Z}[\bar{G}]) \cup (\mathbb{Z}[\bar{G}] \times \bar{\mathfrak{a}})$. Since the upper central extension splits, there exists a map $\chi : G \to M$ such that $f(p(x), p(y)) = \chi((x-1)(y-1))$, $x, y \in G$.

Define $\eta : \bar{G} \to M$ by setting $\eta(\bar{g}) = \chi(g)$, where $p(g) = \bar{g}$. Observe that η is well-defined; for, if $g' = gd$, $d \in D_{\mathfrak{a}^{(1)}}(G)$, then $\chi(g' - g) = \chi(d - 1 + (g-1)(d-1))$. Since $d - 1 \in \mathfrak{a}^{(1)}$, $f(p(x), p(y)) = \chi((x-1)(y-1))$ and f vanishes on $(\mathbb{Z}[\bar{G}] \times \bar{\mathfrak{a}}) \cup (\bar{\mathfrak{a}} \times \mathbb{Z}[\bar{G}])$, it follows that $\chi(g' - g) = 0$. It is now clear that the 2-cocycle f is equal to the coboundary of η, and hence the lower central extension splits. $\square$

We next consider the inflation map $H^2(G/D_{\mathfrak{a}}(G), M) \to H^2(G, M)$.

Theorem 4.25 *For any two-sided ideal $\mathfrak{a} \subseteq \mathfrak{g}$ and divisible abelian group M, there is an exact sequence:*

$$0 \to \mathrm{Hom}((D_{\mathfrak{a}}(G) \cap G')/D_{\mathfrak{a}^{(1)}}(G), M) \xrightarrow{i} P_{\bar{\mathfrak{a}}, \bar{\mathfrak{a}}}(G/D_{\mathfrak{a}}(G), M) \xrightarrow{j_{\mathfrak{a}}} P_{\mathfrak{a}, \mathfrak{a}}(G, M),$$

where $\bar{\mathfrak{a}}$ is the image of $\mathfrak{a}$ in $\Delta(G/D_{\mathfrak{a}}(G))$ under the homomorphism induced by the natural projection $j : G \to G/D_{\mathfrak{a}}(G)$.

Proof. It is clear that the inflation homomorphism

$$\mathrm{inf} : H^2(G/D_{\mathfrak{a}}(G), M) \to H^2(G, M)$$

maps $P_{\bar{\mathfrak{a}}, \bar{\mathfrak{a}}}(G/D_{\mathfrak{a}}(G), M)$ into $P_{\mathfrak{a}, \mathfrak{a}}(G, M)$; let $j_{\mathfrak{a}}$ be its restriction to $P_{\bar{\mathfrak{a}}, \bar{\mathfrak{a}}}$ $(G/D_{\mathfrak{a}}(G), M)$.

Let

$$\beta : \mathrm{Hom}(D_{\mathfrak{a}}(G)/D_{\mathfrak{a}^{(1)}}(G), M) \to H^2(G/D_{\mathfrak{a}}(G), M)$$

be the restriction of the transgression map

$$\mathrm{trans} : \mathrm{Hom}(D_{\mathfrak{a}}(G)/[D_{\mathfrak{a}}(G), G], M) \to H^2(G/D_{\mathfrak{a}}(G), M).$$

We claim that β takes values in $P_{\bar{\mathfrak{a}},\bar{\mathfrak{a}}}(G/D_{\mathfrak{a}}(G), M)$. Indeed suppose that we have the following commutative diagram:

$$
\begin{array}{ccccccccc}
1 & \longrightarrow & D_{\mathfrak{a}}(G)/D_{\mathfrak{a}^{(1)}}(G) & \longrightarrow & \bar{G} & \longrightarrow & G/D_{\mathfrak{a}}(G) & \longrightarrow & 1 \\
& & {\scriptstyle \alpha}\downarrow & & \downarrow & & \| & & \\
1 & \longrightarrow & M & \longrightarrow & \Pi & \longrightarrow & G/D_{\mathfrak{a}}(G) & \longrightarrow & 1,
\end{array}
$$

where $\bar{G} = G/D_{\mathfrak{a}^{(1)}}(G)$, in which the lower row is the central extension induced by α. Observe that we have a monomorphism

$$
D_{\mathfrak{a}}(G)/D_{\mathfrak{a}^{(1)}}(G) \to \mathbb{Z}[\bar{G}]/\tilde{\mathfrak{a}}^{(1)},
$$

$x D_{\mathfrak{a}^{(1)}}(G) \mapsto x - 1 + \mathfrak{a}^{(1)}$ ($x \in D_{\mathfrak{a}}(G)$), where $\tilde{\mathfrak{a}}$ is the image of $\mathfrak{a}$ under the natural projection $\mathbb{Z}[G] \to \mathbb{Z}[G/D_{\mathfrak{a}^{(1)}}(G)]$. Therefore α extends to a map $\varphi : \bar{G} \to M$ whose linear extension to $\mathbb{Z}[\bar{G}]$ vanishes on $\tilde{\mathfrak{a}}^{(1)}$. A standard argument then shows that the lower central extension belongs to $P_{\bar{\mathfrak{a}},\bar{\mathfrak{a}}}(G/D_{\mathfrak{a}}(G), M)$. Thus we have a homomorphism

$$
\beta : \mathrm{Hom}(D_{\mathfrak{a}}(G)/D_{\mathfrak{a}^{(1)}}(G), M) \to P_{\bar{\mathfrak{a}},\bar{\mathfrak{a}}}(G/D_{\mathfrak{a}}(G), M).
$$

Since $\inf \circ \mathrm{trans} = 0$, we have $j_{\mathfrak{a}} \circ \beta = 0$. We assert that $\ker(j_{\mathfrak{a}}) = \mathrm{im}(\beta)$.

Let $1 \to M \to \Pi \to G/D_{\mathfrak{a}}(G) \to 1$ be an extension whose class, ξ say, lies in the $\ker(j_{\mathfrak{a}})$. Since $\ker(\inf) = \mathrm{im}(\mathrm{trans})$, we have a commutative diagram:

$$
\begin{array}{ccccccccc}
1 & \longrightarrow & D_{\mathfrak{a}}(G)/[D_{\mathfrak{a}}(G), G] & \longrightarrow & G/[D_{\mathfrak{a}}(G), G] & \longrightarrow & G/D_{\mathfrak{a}}(G) & \longrightarrow & 1 \\
& & {\scriptstyle \gamma}\downarrow & & \downarrow & & \| & & \\
1 & \longrightarrow & M & \longrightarrow & \Pi & \longrightarrow & G/D_{\mathfrak{a}}(G) & \longrightarrow & 1
\end{array}
$$

Since ξ belongs to $P_{\bar{\mathfrak{a}},\bar{\mathfrak{a}}}(G/D_{\mathfrak{a}}(G), M)$ the homomorphism γ must vanish on $D_{\mathfrak{a}^{(1)}}(G)/[D_{\mathfrak{a}}(G), G]$, as can be checked with help of Proposition 4.24. Thus we have the exact sequence

$$
\mathrm{Hom}(D_{\mathfrak{a}}(G)/D_{\mathfrak{a}^{(1)}}(G), M) \xrightarrow{\beta} P_{\bar{\mathfrak{a}},\bar{\mathfrak{a}}}(G/D_{\mathfrak{a}}(G), M) \xrightarrow{j_{\mathfrak{a}}} P_{\mathfrak{a},\mathfrak{a}}(G, M).
$$

Observe that the kernel of β is precisely the subgroup $\mathrm{Hom}(D_{\mathfrak{a}}(G)/(D_{\mathfrak{a}}(G) \cap G'), M)$, where G' is the derived subgroup of G. Hence we have the exact sequence

$$
0 \to \mathrm{Hom}((D_{\mathfrak{a}}(G) \cap G')/D_{\mathfrak{a}^{(1)}}(G), M) \xrightarrow{i} P_{\bar{\mathfrak{a}},\bar{\mathfrak{a}}}(G/D_{\mathfrak{a}}(G), M) \xrightarrow{j_{\mathfrak{a}}} P_{\mathfrak{a},\mathfrak{a}}(G, M),
$$

where i is the homomorphism induced by β. $\square$

Notation. For a given ordinal τ, let $\mathfrak{P}_{\tau}$ denote the class of group homomorphisms $f : G \to H$ for which the induced homomorphism $H_1(G) \to H_1(H)$

on the first integral homology groups is an isomorphism and $P_{\mathfrak{h}^\tau}(H, \mathbb{T}) \to$ $P_{\mathfrak{g}^\tau}(G, \mathbb{T})$ is a monomorphism.

The foregoing considerations, when applied to augmentation powers, yield the following:

Theorem 4.26 *If n is a natural number and $f : G \to H$ a $\mathfrak{P}_n$-homomorphism, then the induced homomorphism $G/D_{n+1}(G) \to H/D_{n+1}(H)$ is an isomorphism.*

Proof. We proceed by induction on n. Since $D_2(\Pi) = \gamma_2(\Pi)$ for every group Π, the assertion holds for $n = 1$. Assume that $n > 1$, the Theorem holds for $n - 1$, and in addition

$$j_n : P_{\mathfrak{h}^n}(H, \mathbb{T}) \to P_{\mathfrak{g}^n}(G, \mathbb{T})$$

is a monomorphism; in that case note that

$$j_{n-1} : P_{\mathfrak{h}^{n-1}}(H, \mathbb{T}) \to P_{\mathfrak{g}^{n-1}}(G, \mathbb{T})$$

too is a monomorphism and therefore, by induction hypothesis, f induces an isomorphism $G/D_n(G) \simeq H/D_n(H)$. Theorem 4.25 (applied to $\mathfrak{a} = \mathfrak{g}^n$ and $\mathfrak{h}^n$) yields the following commutative diagram:

$$0 \to \mathrm{Hom}(D_n(G)/D_{n+1}(G), \mathbb{T}) \longrightarrow P_{\Delta^n(G/D_n(G))}(G/D_n(G), \mathbb{T}) \xrightarrow{\mathrm{inf}_1} P_{\mathfrak{g}^n}(G, \mathbb{T})$$

$$\qquad\qquad f_1^* \uparrow \qquad\qquad\qquad\qquad f_2^* \uparrow \qquad\qquad\qquad\qquad f_3^* \uparrow$$

$$0 \to \mathrm{Hom}(D_n(H)/D_{n+1}(H), \mathbb{T}) \longrightarrow P_{\Delta^n(H/D_n(H))}(H/D_n(H), \mathbb{T}) \xrightarrow{\mathrm{inf}_2} P_{\mathfrak{h}^n}(H, \mathbb{T}),$$

where the vertical maps are induced by f, and f_2^* is an isomorphism. Since f_3^* is a monomorphism, therefore $\mathrm{im}(\mathrm{inf}_2) \to \mathrm{im}(\mathrm{inf}_1)$ is also a monomorphism and f_1^* is an epimorphism. It follows that f_1^* is an isomorphism, and consequently f induces an isomorphism

$$D_n(G)/D_{n+1}(G) \simeq D_n(H)/D_{n+1}(H).$$

Hence f induces an isomorphism $G/D_{n+1}(G) \simeq H/D_{n+1}(H)$ and the proof is complete. $\square$

Proposition 4.27 *For every natural number n, the class $loc(\mathfrak{P}_n)$ is the quasi-variety consisting of the groups G with $D_{n+1}(G) = 1$.*

[See Section 1.7 for the definition of $loc(\mathfrak{P}_n)$.]

Proof. Let G be a group with $D_{n+1}(G) = 1$ and let $f : X \to Y$ be a $\mathfrak{P}_n$-homomorphism. Consider an arbitrary homomorphism $h : X \to G$. By Theorem 4.26 f induces an isomorphism $\bar{f} : X/D_{n+1}(X) \to Y/D_{n+1}(Y)$. Since $D_{n+1}(G) = 1$, h induces a homomorphism $\bar{h} : X/D_{n+1}(X) \to G$. It is easy to see that the correspondence $h \mapsto \bar{h} \circ \bar{f} \circ p$, where p is the natural projection $Y \to Y/D_{n+1}(Y)$, is a bijection, and it follows that G is $\mathfrak{P}_n$-local.

Conversely, let G be a $\mathfrak{P}_n$-local group. By Proposition 4.24, the natural projection $p_n : G \to G/D_{n+1}(G)$ is a $\mathfrak{P}_n$-homomorphism. Therefore the identity homomorphism $\mathrm{id}_G : G \to G$ must factor through the projection p_n, and consequently $D_{n+1}(G) = 1$. $\square$

Corollary 4.28 *For every group G and natural number $n \geq 1$, the homomorphism $G \to G/D_{n+1}(G)$ is, up to isomorphism, the unique $\mathfrak{P}_n$-localization of G.*

Proof. Propositions 4.24 and 4.27 imply that the natural projection $G \to G/D_{n+1}(G)$ is a $\mathfrak{P}_n$-localization of G. The uniqueness follows from Proposition 4.27 and Theorem 4.26. $\square$

We now apply the method developed above for identifying subgroups determined by two-sided ideals in group rings to the study of transfinite dimension subgroups. It is not known whether for a nilpotent group G, $D_\omega(G) = 1$. It has been shown by Hartley [Har82c] that $D_\omega(G) = D_\omega(T)$, where T is the torsion subgroup of G. We prove that $D_{\omega+1}(T) = 1$ for any torsion nilpotent group T; this comes out as a consequence of the following:

Theorem 4.29 *If G is a nilpotent p-group with $D_\omega(G) = 1$, then*

$$P_{\mathfrak{g}^\omega, \mathfrak{g}^\omega}(G, \mathbb{T}_p) = H^2(G, \mathbb{T}_p).$$

Proof. By Theorem 4.8 (see also ([GR], §3), ([Pa], p. 99)),

$$\mathfrak{g}^\omega = \Delta(G(p))\mathfrak{g},$$

where $G(p)$ is the subgroup consisting of elements of infinite p-height in G. Since $G(p)$ is contained in the centre of G, we thus need to show that if

$$1 \to \mathbb{T}_p \xrightarrow{i} E \xrightarrow{j} G \to 1$$

is a central extension, then there exists a choice of representative of G in M such that the corresponding 2-cocycle $f : G \times G \to \mathbb{T}_p$, when extended to $\mathbb{Z}[G] \times \mathbb{Z}[G]$ by linearity, vanishes on $\Delta(G(p))\mathfrak{g} \times \mathbb{Z}[G]$.

Let $N = j^{-1}(G(p))$ and note that N is central in E. Therefore the extension

$$1 \to \mathbb{T}_p \xrightarrow{i} N \xrightarrow{j} G(p) \to 1$$

of abelian groups splits, since $\mathbb{T}_p$ is divisible. Thus there exists a choice of representatives $u(z) \in E$ for the elements $z \in G(p)$ such that $u(z) \mapsto z$ is a homomorphism. Write $H = G/G(p)$ and pick a set of representatives $\{w(h)\}_{h \in H}$ in G for the elements of H. Then every element $g \in G$ is uniquely expressible as $g = w(h)z$ $(h \in H$, $z \in G(p)$. Now choose representatives $\varphi(g)$ $(g \in G)$ for the elements of G by picking $\varphi(w(h))$ $(h \in H)$ arbitrarily and taking, for $g = w(h)z$, $\varphi(g) = \varphi(w(h))u(z)$. It is then straightforward to check that the resulting 2-cocycle $f : G \times G \to \mathbb{T}_p$ satisfies $f(x, g) = 0$ for $x \in G(p), g \in G$, and hence has the desired property. $\square$

Corollary 4.30 *Let T be a torsion nilpotent group, then $D_{\omega+1}(T) = 1$.*

Proof. Observe first that it suffices to prove the result for nilpotent p-groups. Let T be a nilpotent p-group. By Corollary 4.19, Theorem 4.29 and Corollary 4.5, $D_{\omega+1}(T) = [D_\omega(T), T] = 1$. $\square$

Notation. For subgroups H, K of the group G and natural number n define $[H, {}_n K]$ inductively as follows:

$$[H, {}_1 K] = [H, K], \quad [H, {}_{(n+1)} K] = [[H, {}_n K], K].$$

We need the following:

Lemma 4.31 *Let N be a normal subgroup in G, then*

$$G \cap (1 + \mathfrak{g}^n \mathfrak{n} + \mathfrak{n}\mathfrak{g}) = [N, N] \cdot [N, {}_n G],$$

for all integers $n \geq 1$.

Proof. It is easy to see that

$$[N, N] \cdot [N, {}_n G] \subseteq G \cap (1 + \mathfrak{g}^n \mathfrak{n} + \mathfrak{n}\mathfrak{g}).$$

To see the reverse inclusion, we may clearly assume without loss of generality that

$$[N, N] \cdot [N, {}_n G] = 1.$$

Let $H = G/N$. Suppose $x \in G \cap (1 + \mathfrak{g}^n \mathfrak{n} + \mathfrak{n}\mathfrak{g})$ and $x \neq 1$. Since N is abelian, there exists a homomorphism $\alpha : N \to \mathbb{T}$ such that $\alpha(x) \neq 0$. Choose a set of representatives $w(h) \in G$ for the elements $h \in H$. Then every element $g \in G$ can be uniquely written as $g = zw(h)$ $(h \in H$, $z \in N)$; define $\varphi : G \to \mathbb{T}$ by setting $\varphi(g) = \alpha(z)$. It is then easy to check that the extension of φ by linearity to $\mathbb{Z}[G]$ vanishes on $\mathfrak{g}^n \mathfrak{n} + \mathfrak{n}\mathfrak{g}$, and so we have $\alpha(x) = \varphi(x) = \varphi(x - 1) = 0$, a contradiction. Hence

$$G \cap (1 + \mathfrak{g}^n \mathfrak{n} + \mathfrak{n}\mathfrak{g}) = 1,$$

and the assertion is proved. $\square$

Theorem 4.32 *If G is a nilpotent group of class c, T its torsion subgroup and $\mathfrak{a} = \mathfrak{t}^\omega \mathbb{Z}[G]$, then $D_{\mathfrak{a}(c)}(G) = 1$.*

Proof. We first consider the case when $T = G_p$, the p-torsion subgroup of G.

Note that $\mathfrak{a} \subseteq \Delta(G_p(p))\mathbb{Z}[G]$, where $G_p(p)$ denotes the subgroup consisting of elements of infinite p-height in G_p (see [Pas79], p.84). Therefore $\mathfrak{a}^{(c)} \subseteq (\Delta(G_p(p))\mathbb{Z}[G])^{(c)} \subseteq \mathfrak{g}^c \Delta(G_p(p)) + \Delta(G_p(p))\mathfrak{g}$. Consequently, by Lemma 4.31,

$$D_{\mathfrak{a}^{(c)}}(G) \subseteq [G_p(p), G_p(p)][G_p(p), {}_cG] = 1,$$

since G is nilpotent of class c and $G_p(p)$ lies in the centre of $G_p(p)$.

The general result follows from the case considered above by applying it to the quotients $G/G_{p'}$, where $G_{p'}$ denotes the subgroup of G consisting of elements of order prime to p. $\square$

The following result is an immediate consequence of Theorem 4.32 and Theorem 4.7.

Corollary 4.33 *Let G be a nilpotent group of class c and T its torsion subgroup. If there is no prime p such that*

(i) *G/T contains a non-trivial element of infinite p-height and*

(ii) *T contains an element of order p,*

then $D_{\omega+c}(G) = 1$.

We next review some results on the filtration $\{P_n H^2(G, \mathbb{T})\}_{0 \le n < \infty}$.

Proposition 4.34 (Passi-Vermani [Pas83]). *Let G be a group, D a divisible abelian group regarded as a trivial G-module, $\xi \in H^2(G, D)$ and*

$$1 \to D \to E \to G \to 1$$

a central extension classified by ξ. Then $\xi \in P_n H^2(G, D)$ if and only if

$$D \cap (1 + \mathfrak{e}^{n+2} + \mathfrak{e}\mathfrak{d}) = 1.$$

Proposition 4.35 (Passi-Vermani [Pas83]). *Let G be a finitely generated nilpotent group. Then there exists an integer $n \ge 1$ such that*

$$P_n(G, \mathbb{T}) = H^2(G, \mathbb{T}).$$

Proof. Let $1 \to R \to F \to G \to 1$ be a free presentation of G with F free of finite rank. Write $\overline{F} = F/[F, R]$ and $\overline{R} = R/[F, R]$. Now observe that $\Delta(\overline{F})$ is a polycentral ideal of $\mathbb{Z}[\overline{F}]$, and therefore it satisfies the weak Artin-Rees property (see [Pas77b, Chapter XI, Theorem 2.8]; for a discussion of applications of Artin-Rees property to the questions of localization, see Section 4.9). Thus, there exists an integer n such that

$$\Delta^{n+2}(\overline{F}) \cap \Delta(\overline{R})\mathbb{Z}[\overline{F}] \subseteq \Delta(\overline{R})\Delta(\overline{F}).$$

For this n, we then have

$$\overline{R} \cap (1 + \Delta^{n+2}(\overline{F}) + \Delta(\overline{R})\Delta(\overline{F})) = (1).$$

Hence, by [Pas74, Cor. 3.2], $P_n(G, \mathbb{T}) = H^2(G, \mathbb{T})$. $\square$

The same conclusion as in Theorem 4.35 holds in case G is a nilpotent group which is either torsion-free or divisible. In fact, we have, more precisely, the following:

Theorem 4.36 (Passi-Sucheta [Pas87a]). (*i*) *If G is a torsion-free nilpotent group of class c, then $P_{3c+1}H^2(G, \mathbb{T}) = H^2(G, \mathbb{T})$.*
(*ii*) *If G is a divisible nilpotent group of class c, then $P_cH^2(G, \mathbb{T}) = H^2(G, \mathbb{T})$.*

Proposition 4.37 *If G is a nilpotent group such that the quotient $G/\zeta(G)$ by its centre $\zeta(G)$ is either torsion-free or finitely generated, then G has finite dimension series.*

Proof. Let $G/\zeta(G) = H$. Then, by hypothesis, H is either torsion-free or finitely generated nilpotent group; in either case, H has finite dimension series and, by Theorems 4.35 and 4.36, there exists an integer n such that

$$P_n H^2(H, \mathbb{T}) = H^2(H, \mathbb{T}). \tag{4.16}$$

Since $D_{s(H)}(H) = 1$ (see p. 148), it follows that for $m \geq s(H)$, $D_m(G) \subseteq \zeta(G)$. In view of (4.16) every homomorphism $\alpha : \zeta(G) \to \mathbb{T}$ can be extended to a polynomial map $\varphi : G \to \mathbb{T}$ of degree $\leq n+1$. Therefore $\zeta(G) \cap D_{n+2}(G) = 1$. Hence $s(G) \leq \max\{s(H), n+2\}$. $\square$

Corollary 4.38 (Plotkin [Plo73]). *If in a periodic nilpotent group G the centre is of finite index, then the group has finite dimension series.*

Theorem 4.39 (Passi-Vermani [Pas83]). *The integral dimension series of every nilpotent group terminates with identity if and only if, for every nilpotent group G, $P_n H^2(G, \mathbb{T}) = H^2(G, \mathbb{T})$ for some n.*

Theorem 4.40 (Passi-Vermani [Pas94]; Mikhailov-Passi [Mik04]). *For any nilpotent group G of class $c \geq 1$, $P_c H^2(G, \mathbb{Q}) = H^2(G, \mathbb{Q})$.*

Theorem 4.41 (Passi-Sucheta-Tahara [Pas87b]). *If G is a finite 2-group of class 2, then $P_2(G, \mathbb{T}) = H^2(G, \mathbb{T})$, provided the rank of G_{ab} is at most 3.*

Remark. Let Π be a 2-group of class 3 such that $D_4(\Pi) \neq 1$, and $G = \Pi/\gamma_3(\Pi)$. Then $P_2 H^2(G, \mathbb{T}) \neq H^2(G, \mathbb{T})$. There exist such groups with G_{ab} of rank 4 (see Section 2.1); for example, we can take Π to be the Rip's counter-example to the dimension conjecture [Rip72].

It has been shown in [Pas83] that Sjögren's theorem implies the existence of constants d_1, d_2, ... such that

$$d_n H^2(G,\, \mathbb{T}) \subseteq P_n H^2(G,\, \mathbb{T}) \tag{4.17}$$

for every nilpotent group G of class $\leq n$. The converse also is true, i.e., if there exist constants d_n, $n \geq 1$, satisfying (4.17), then there exist constants c_n, $n \geq 1$, satisfying

$$D_n(G)^{c_n} \subseteq \gamma_n(G). \tag{4.18}$$

Sjögren's theorem 2.17 has a direct impact on the Schur multiplicator.

Theorem 4.42 (Passi-Vermani [Pas94]). *Let F be a free group, R a normal subgroup of F and D a divisible abelian group. Then*

$$d(n,\, k) H^2(F/\gamma_{n+1}(F)R(k),\, D) \subseteq P_n H^2(F/\gamma_{n+1}(F)R(k),\, D)$$

for all integers $n \geq k \geq 1$, where

$$d(n,\, k) = \prod_{i=1}^{n-k} b(k+1)^{\binom{n-k}{i}}.$$

In particular, for all $n \geq 1$,

(i) $H^2(F/\gamma_{n+1}(F)R(n),\, D) = P_n H^2(F/\gamma_{n+1}(F)R(n),\, D)$,
(ii) $H^2(F/\gamma_{n+1}(F)R,\, D) = P_n H^2(F/\gamma_{n+1}(F)R,\, D)$,

provided D is torsion-free.

Proof. Let integers $n \geq k \geq 1$ be given. Define series of normal subgroups

$$H_1 \supseteq H_2 \supseteq \ldots \quad \text{and} \quad K_1 \supseteq K_2 \supseteq \ldots$$

of F by setting

$$H_m = R(m+k), \quad K_m = \gamma_{m+k}(F), \quad m \geq 1.$$

For $1 \leq m \leq l$, let

$$D_{m,l} = F \cap (1 + \mathfrak{f}^{l+k} + \mathfrak{r}(m+k)).$$

By Lemma 2.20, $D_{m,m+1} = H_m K_{m+1}$ for all $m \geq 1$. It is easy to see that

$$H_m K_l \subseteq D_{m,l} \quad \text{and} \quad D_{m,l+1} \subseteq D_{m,l}.$$

By Lemma 2.19,

$$(K_{l+m} \cap D_{l,l+m+1})^{b(k+l)} \subseteq D_{l+1,l+m+1} H_l.$$

Therefore, by Lemma 2.21,

$$D_{1,\,n-k+2}^{a(1,\,n-k+2)} \subseteq H_1 K_{n-k+2},$$

where $a(1,\, n-k+2) = \prod_{i=1}^{n-k} b(k+i)^{\binom{n-k}{i}} = d(n,\, k)$, i.e.,

$$F \cap (1 + \mathfrak{f}^{n+2} + \mathfrak{r}(k+1))^{d(n,\,k)} \subseteq \gamma_{n+2}(F)R(k+1).$$

In particular,

$$(S \cap (1 + \mathfrak{f}^{n+2} + \mathfrak{r}(k+1))^{d(n,\,k)} \subseteq [S,\, F], \tag{4.19}$$

where $S = \gamma_{n+1}(F)R(k)$.

Let $\xi \in H^2(F/S,\, D)$ and $1 \to D \to M \to F/S \to 1$ be a central extension corresponding to ξ. There exists then a commutative diagram

$$
\begin{array}{ccccccccc}
1 & \longrightarrow & S/[S,\, F] & \overset{i}{\longrightarrow} & F/[S,\, F] & \longrightarrow & F/S & \longrightarrow & 1 \\
 & & \alpha \downarrow & & \beta \downarrow & & \| & & \\
1 & \longrightarrow & D & \longrightarrow & M & \longrightarrow & F/S & \longrightarrow & 1
\end{array}
$$

in which the upper row is the central extension of F/S with i the inclusion map. Let $1 \to D \to M \to F/S \to 1$ be a central extension corresponding to $d(n,\, k)\xi$. Then we have a commutative diagram

$$
\begin{array}{ccccccccc}
1 & \longrightarrow & S/[S,\, F] & \overset{i}{\longrightarrow} & F/[S,\, F] & \longrightarrow & F/S & \longrightarrow & 1 \\
 & & d(n,\,k)\alpha \downarrow & & \gamma \downarrow & & \| & & \\
1 & \longrightarrow & D & \longrightarrow & M & \longrightarrow & F/S & \longrightarrow & 1.
\end{array}
$$

Since D is divisible abelian and (4.19) holds, the homomorphism $d(n,\, k)\alpha$ can be extended to a map $F/[S,\, F] \to D$ the linear extension of which to the integral group ring $\mathbb{Z}[F/[S,\, F]$ vanishes on

$$\Delta^{n+2}(F/[S,\, F]) + \Delta(F/[S,\, F])\Delta(S/[S,\, F]).$$

Hence, by Proposition 4.34, $d(n,\, k)\xi \in P_n H^2(F/S,\, D)$ and the proof is complete.

4.6 Relative Dimension Subgroups

If E is a group and $N \triangleleft E$, then

$$D_n(E,\, N) := E \cap \left(1 + \mathfrak{n}\mathfrak{e} + \mathfrak{e}^n\right)$$

is a normal subgroup of E; this normal subgroup is called the nth *dimension subgroup of E relative to N*. Relative dimension subgroups provide a generalization of the usual dimension subgroups, since $D_n(E, N) = D_n(E)$ in case $N = \{1\}$ or $\gamma_{n-1}(E)$. We give here a brief account of a homological approach to the investigate these subgroups; for more details see ([Har96a], [Har98], [Har08], [Kuz96]).

Let K be a commutative unitary ring and A an augmented K-algebra with augmentation ideal $\bar{A}$.

Definition 4.43 Let $n \geq 0$. The polynomial bar construction of degree n over A, $(P_n B(A), \bar{\delta})$, is defined by $P_n B_0(A) = 0$ and

$$P_n B_i(A) = (A/\bar{A}^{n+1}) \otimes_K (\bar{A}/\bar{A}^{n+1})^{\otimes i-1} \otimes_K (\bar{A}/\bar{A}^{n+2})$$

for $i \geq 1$, and the differential $\bar{\delta}_i : P_n B_i(A) \to P_n B_{i-1}(A)$ is given by

$$\bar{\delta}_i(\overline{a_0} \otimes \cdots \otimes \overline{a_i}) = \sum_{j=0}^{i-1} (-1)^j \overline{a_0} \otimes \cdots \otimes \overline{a_j a_{j+1}} \otimes \cdots \otimes \overline{a_i}$$

for $i \geq 2$. For left (resp. right) $(A/\bar{A}^{n+1})$-modules M (resp. N), define *polynomial (co)homology of degree n of A* by

$$P_n H^i(A, M) = H^i(\mathrm{Hom}_{A/\bar{A}^{n+1}}(P_n B(A), M)),$$

$$P_n H_i(A, M) = H_i(N \otimes_{A/\bar{A}^{n+1}} P_n B(A)).$$

There exist natural maps (see [Har08])

$$\begin{aligned} \rho_n^* &: P_n H^i(A, M) \to H^i(A, M) \\ \rho_{n*} &: H_i(A, M) \to P_n H_i(A, M) \end{aligned} \tag{4.20}$$

which, for $i = 2$, are injective resp. surjective.

The K-modules $P_n H^i(A, M)$ and $P_n H_i(A, M)$, for fixed $i \geq 2$ and varying $n \geq 0$, are related by chains of natural maps

$$0 = P_0 H^i(A, M) \to \ldots \to P_n H^i(A, M) \xrightarrow{\sigma_n^*} P_{n+1} H^i(A, M) \to \ldots \xrightarrow{\rho^*} H^i(A, M)$$

$$H_i(A, N) \xrightarrow{\rho_*} \ldots \to P_{n+1} H_i(A, N) \xrightarrow{\sigma_{n*}} P_n H_i(A, N) \to \ldots \to P_0 H_i(A, N) = 0 \tag{4.21}$$

commuting with the maps ρ^* and ρ_* where $\sigma_n : P_{n+1} B(A) \twoheadrightarrow P_n B(A)$ is the tensor product of the canonical projections. For $i = 2$ the maps σ_n^* and σ_{n*} are injective resp. surjective, so identifying $P_n H^2(A, M)$ with its isomorphic image $\rho_n^* P_n H^2(A, M)$ in $H^2(A, M)$ provides a *natural ascending filtration* of $H^2(A, M)$,

$$0 = P_0 H^i(A, M) \subset \ldots \subset P_n H^i(A, M) \subset P_{n+1} H^i(A, M) \subset \ldots \subset H^i(A, M)$$

Dually, the maps in (4.21) being surjective for $i = 2$ they can be interpreted as a *natural cofiltration* of $H_2(A, N)$ which in turn gives rise to a *natural descending filtration*

$$H_2(A, N) = \ker \rho_{0*} \supset \ldots \supset \ker \rho_{n*} \supset \ker \rho_{n+1*} \supset \ldots \supset 0 \qquad (4.22)$$

Given a group G, define polynomial (co)homology of degree n by applying the above constructions to the group ring $\mathbb{Z}[G]$: for left (resp. right) $(n+1)$-step nilpotent G-modules M (resp. N), let

$$P_n H^i(G, M) = P_n H^i(\mathbb{Z}[G], M),$$

$$P_n H_i(G, N) = P_n H_i(\mathbb{Z}[G], N).$$

It turns out that $\rho_n^* P_n H^i(G, M)$ is the subgroup of $H^i(G, M)$ consisting of elements representable by multipolynomial cocycles of degree $\leq n$ in the first $i - 1$ variables and of degree $\leq n + 1$ in the last variable. For $i = 2$, note that

$$\mathrm{coker}\, \mathbb{Z} \otimes \bar{\delta}_3 \,:\, \mathbb{Z} \otimes_G P_n B_3(\mathbb{Z}(G)) \to \mathbb{Z} \otimes_G P_n B_2(\mathbb{Z}(G))$$

$$\cong P_n(G) \otimes_G P_{n+1}(G)$$

$$\cong P_n(G) \otimes_G P_n(G). \qquad (4.23)$$

Thus, if M is a *trivial* G-module, $\rho_n^* P_n H^2(G, M)$ is the subgroup of $H^2(G, M)$ consisting of elements representable by bipolynomial cocycles of degree $\leq n$ in both variables, so $P_n H^2(G, M)$ is isomorphic via ρ_n^* with the polynomial cohomology groups defined in (4.14).

Let us denote

$$\rho_{n*}^G := \rho_{n*} \,:\, H_2(G) \to P_n H_2(G)$$

abbreviating $H_2(G) = H_2(G, \mathbb{Z})$ and $P_n H_2(G) = P_n H_2(G, \mathbb{Z})$.

One of the main results in [Har08] which is helpful in analysing the relative dimension subgroups is the following

Theorem 4.44 *Let* $e \,:\, C \overset{i}{\hookrightarrow} E \overset{q}{\twoheadrightarrow} G$ *be a central group extension and suppose that* $\gamma_n(E) = 1$. *Then*

$$D_n(E, C) \cap C = \kappa\Big(\ker \rho_{n-2*}^G \,:\, H_2(G) \twoheadrightarrow P_{n-2} H_2(G) \Big)$$

where $\kappa \,:\, H_2(G) \to C$ *is adjoint to the cohomology class of* e *under the Kronecker pairing* $H^2(G, C) \times H_2(G) \to C$.

As a consequence of the above theorem, one has the following

Corollary 4.45 *Let E be an $(n-1)$-step nilpotent group and C a central subgroup of E. Then $D_n(E, C) \cap C\gamma_{n-1}(E)$ is a homomorphic image of $\ker \rho^G_{n-2*}$ for the $(n-2)$-step nilpotent group $G = E/C\gamma_{n-1}(E)$.*

It may be noted that the following well-known result can be easily deduced from Theorem 4.44.

Corollary 4.46 *For any group Γ and $n \leq 3$, $D_n(\Gamma) = \gamma_n(\Gamma)$.*

For an abelian group A, let $L(A)$, $T(A)$, $S(A)$ denote the free Lie algebra, the tensor algebra and the symmetric algebra over A, respectively. (These functors will be studied in the next Chapter.) The natural maps of graded abelian groups $L(A) \overset{l}{\rightarrowtail} T(A) \overset{s}{\twoheadrightarrow} S(A)$ are the injection into the universal enveloping algebra and the canonical projection, respectively. Thus l_n sends an n-fold Lie bracket in $L_n(A)$ to the corresponding tensor commutator in $T_n(A) = A^{\otimes n}$. In particular, $L_2(A) = A \wedge A$, the exterior square of A, and $l_2(a \wedge b) = a \otimes b - b \otimes a$ for $a, b \in A$.

For the rest of this section, let G be a 2-*step nilpotent group*. The surjective homomorphism $c_2 : G_{ab} \wedge G_{ab} \twoheadrightarrow G'$ is defined by $c_2(\bar{a} \wedge \bar{b}) = [a, b]$ for $a, b \in G$. For $x \in G_{ab}$ and $k \in \mathbb{Z}$ such that $kx = 0$, choose elements $\tilde{x} \in G$ and $f_k x \in G_{ab} \wedge G_{ab}$ such that $\tilde{x}G' = x$ and $c_2(f_k x) = \tilde{x}^k$.

The main ingredients for calculating $\ker \rho^G_{2*}$ are the structure theorems describing $H_2(G)$ and $P_2(G) \otimes_G P_2(G)$ for 2-step nilpotent G, due to M. Hartl ([Har96b], [Har95]):

Theorem 4.47 *If G is a 2-step nilpotent group, then there are natural exact sequences*

$$\mathrm{Tor}(G_{ab}, G_{ab}) \overset{\delta}{\to} \frac{L_3(G_{ab})}{[G_{ab}, \ker c_2] + V} \overset{\nu i}{\to} H_2(G) \overset{\sigma}{\to} G_{ab} \wedge G_{ab} \overset{c_2}{\to} G' \to 1 \quad (4.24)$$

$$\mathrm{Tor}(G_{ab}, G_{ab}) \overset{\delta'}{\to} \frac{(G_{ab})^{\otimes i - 1}}{l_2 \ker c_2 \otimes G_{ab} + G_{ab} \otimes l_2 \ker c_2} \overset{i'}{\to} P_2(G) \otimes_G P_2(G)$$

$$\overset{\sigma'}{\to} G_{ab} \otimes G_{ab} \to 0 \quad (4.25)$$

where

$$\delta\langle x_1, k, x_2 \rangle = q\left([x_1, f_k x_2] + [x_2, f_k x_1] + \binom{k}{2}[x_1 + x_2, [x_1, x_2]]\right)$$

$$\delta'\langle x_1, k, x_2 \rangle = q'\left(x_1 \otimes (l_2 f_k x_2) - (l_2 f_k x_1) \otimes x_2 + \binom{k}{2}(x_1 \otimes x_1 \otimes x_2 - x_1 \otimes x_2 \otimes x_2)\right)$$

with q, q' being the canonical projections, and $i[\bar{a}, [\bar{b}, \bar{c}]] = \overline{[a, [b, c]]}$, σ :

$$H_2(G) \overset{\cong}{\underset{\nu^{-1}}{\to}} \frac{R \cap F'}{[F, R]} \overset{\nu''}{\hookrightarrow} \frac{[F, F]}{[F, R]} \overset{\nu''}{\to} G_{ab} \wedge G_{ab} \text{ with } \nu''(\overline{[a, b]}) = \bar{a} \wedge \bar{b}, \ i'(\bar{a} \otimes \bar{b} \otimes \bar{c}) =$$

$\overline{(a-1)(b-1)} \otimes \overline{(c-1)}$, $\sigma'(\overline{(a-1)} \otimes \overline{(b-1)}) = \bar{a} \otimes \bar{b}$, for a, b, $c \in G$. Finally, V denotes the subgroup of $L_3(G_{ab})$ generated by the elements $[x, f_{o(x)}x]$ where x ranges over the elements of finite even order $o(x)$ of G_{ab}. $\square$

Note that for any torsion element x of G_{ab}, $\delta\langle x, o(x), x \rangle = 2[x, f_{o(x)}x]$, so if $o(x)$ is odd, $[x, f_{o(x)}x] \in \mathrm{im}\, \delta$. Thus V can be replaced by the subgroup V' generated by the elements $[x, f_{o(x)}x]$ for any torsion elements $x \in G_{ab}$. Now if $kx = 0$ for $k \in \mathbb{Z}$ then $[x, f_k x] \in V'$; this shows that the map

$$\delta_1 : G_{ab} \overset{\wedge}{*} G_{ab} \longrightarrow \frac{L_3(G_{ab})}{[G_{ab}, \ker c_2] + V + \mathrm{im}\, \delta}$$

defined by $\delta_1(\langle x_1, k, x_2 \rangle) = \overline{[x_2, f_k x_1]}$ is well defined where $G_{ab} \overset{\wedge}{*} G_{ab}$ denotes the exterior torsion square of G_{ab}. Moreover, define homomorphisms

$$G_{ab} \otimes G' \overset{\delta_2}{\longleftarrow} G_{ab} \overset{\wedge}{*} G_{ab} \overset{\delta_3}{\longrightarrow} SP^3(G_{ab})$$

by $\delta_2(\langle x_1, k, x_2 \rangle) = x_1 \otimes \tilde{x}_2^k - x_2 \otimes \tilde{x}_1^k$ and $\delta_3(\langle x_1, k, x_2 \rangle) = \binom{k}{2} s(x_1 \otimes x_1 \otimes x_2 - x_1 \otimes x_2 \otimes x_2)$.

Theorem 4.48 (Hartl-Mikhailov-Passi [Har08]). *If G is a 2-step nilpotent, group, then*

$$\ker \rho_{2*}^G = \overline{\nu i} \delta_1(\ker \delta_2 \cap \ker \delta_3).$$

An immediate consequence of the above Theorem is the following

Corollary 4.49 *For G 2-step nilpotent one has $2 \ker \rho_{2*}^G = 0$ and $\ker \rho_{2*}^G \subset 2 \mathrm{im}\, \overline{\nu i} \subset 2H_2(G)$.*

In view of Corollaries 4.45 and 4.46, the above approach reproves the result stating that for every group E and a central subgroup N of E, $D_4(E, C)/\gamma_4(E)$ is of exponent 1 or 2.

4.7 A Characterization of Para-free Groups

The para-free groups (see Chapter 1, p. 63) can be characterized in terms of the filtration $\{P_n H^2(G, \mathbb{T})\}_{0 \le n < \infty}$. Recall that if G is an abelian group, then $P_1 H^2(G, \mathbb{T}) = H^2(G, \mathbb{T})$, and $P_n H^2(F/\gamma_{n+1}(F), \mathbb{T}) = H^2(F/\gamma_{n+1}(F), \mathbb{T})$ for a free group F.

Theorem 4.50 (Passi-Stammabch [Pas74]). *A residually nilpotent group G is para-free if and only if*

(i) $G/\gamma_2(G)$ is free abelian and
(ii) $P_n(G, \mathbb{T}) = 0$ for all $n \ge 0$.

Proof. Let G be a group satisfing (i) and (ii). Then we have a free group F and a homomorphism $\varphi : F \to G$ which induces the isomorphism $F/\gamma_2(F) \simeq G/\gamma_2(G)$. The five-term sequences arising from the exact sequences

$$1 \to \gamma_2(G) \to G \to G/\gamma_2(G) \to 1,$$

$$1 \to \gamma_2(F) \to F \to F/\gamma_2(F) \to 1$$

yield the following commutative diagram

$$
\begin{array}{ccccc}
0 \to \mathrm{Hom}(\gamma_2(G)/\gamma_3(G), \mathbb{T}) & \xrightarrow{trans} & H^2(G/\gamma_2(G), \mathbb{T}) & \xrightarrow{inf} & H^2(G, \mathbb{T}) \\
\downarrow u & & \downarrow v & & \downarrow w \\
0 \to \mathrm{Hom}(\gamma_2(F)/\gamma_3(F), \mathbb{T}) & \xrightarrow{trans} & H^2(F/\gamma_2(F), \mathbb{T}) & \xrightarrow{inf} & H^2(F, \mathbb{T}).
\end{array}
$$

Since $P_1 H^2(G/\gamma_2(G), \mathbb{T}) = H^2(G/\gamma_2(G), \mathbb{T})$,

$$\mathrm{im}(\inf : H^2(G/\gamma_2(G), \mathbb{T}) \to H^2(G, \mathbb{T})) = P_1 H^2(G, \mathbb{T}) = 0.$$

Hence the left hand vertical homomorphism is an isomorphism and consequently φ induces an isomorphism

$$\gamma_2(F)/\gamma_3(F) \simeq \gamma_2(G)/\gamma_3(G).$$

It therefore follows that the induced map $F/\gamma_3(F) \to G/\gamma_3(G)$ is an isomorphism. Induction and the fact that $P_n H^2(F/\gamma_{n+1}(F), \mathbb{T}) = H^2(F/\gamma_{n+1}(F), \mathbb{T})$ for all $n \geq 1$ gives the isomorphisms $F/\gamma_n(F) \simeq G/\gamma_n(G)$ for all $n \geq 1$. Hence G is para-free.

Conversely, let G be a para-free group. Then $P_n H^2(G/\gamma_{n+1}(G), \mathbb{T}) = H^2(G/\gamma_{n+1}, \mathbb{T})$ for all $n \geq 1$ and $\inf : H^2(G/\gamma_{n+1}(G), \mathbb{T}) \to H^2(G, \mathbb{T})$ is zero for all $n \geq 1$. Hence $P_n(G, \mathbb{T}) = 0$ for all $n \geq 0$. $\square$

4.8 τ-para-free Groups

Motivated by the results in [Pas74], [Bou77], [Dwy75], and [Sta65], we are interested in finding, for a given group G and an ordinal number τ, (co)homological conditions in terms of $P_{\mathfrak{g}^\tau}$-filtration of the Schur multiplicator $H^2(G, \mathbb{T})$ under which (i) G is τ-para-free; (ii) a group homomorphism $f : G \to H$ induces an isomorphism

$$L_\tau(f) : L(G)/\gamma_\tau(L(G)) \to L(H)/\gamma_\tau(L(H)),$$

where L is the HZ-localization functor on the category of groups (see Chapter 1, p. 77).

We need the following:

Lemma 4.51 *Let α be an ordinal number, G, H HZ-local groups and $f :$*
$G \to H$ *a $\mathfrak{P}_\alpha$-homomorphism. If*

$$P_{\Delta^\tau(G/\gamma_\tau(G))}(G/\gamma_\tau(G), \mathbb{T}) = H^2(G/\gamma_\tau(G), \mathbb{T}) \text{ for all } \tau \le \alpha,$$

then the induced map $G/\gamma_{\alpha+1}(G) \to H/\gamma_{\alpha+1}(H)$ is an isomorphism.

Proof. We proceed by transfinite induction on α; for $\alpha = 1$ the Lemma holds
trivially.

Suppose that the Lemma holds for an ordinal α and we have the hypothesis
for $\alpha + 1$. Then, since $P_{\Delta^{\alpha+1}(G/\gamma_{\alpha+1}(G))}(G/\gamma_{\alpha+1}(G), \mathbb{T}) = H^2(G/\gamma_{\alpha+1}(G), \mathbb{T})$
and f induces an isomorphism $f_{\alpha+1} : G/\gamma_{\alpha+1}(G) \simeq H/\gamma_{\alpha+1}(H)$ by induction
hypothesis, we have

$$P_{\Delta^{\alpha+1}(H/\gamma_{\alpha+1}(H))}(H/\gamma_{\alpha+1}(H), \mathbb{T}) = H^2(H/\gamma_{\alpha+1}(H), \mathbb{T}),$$

and hence the following commutative diagram with exact rows:

$$
\begin{array}{ccccc}
\mathrm{Hom}(\gamma_{\alpha+1}(H)/\gamma_{\alpha+2}(H), \mathbb{T}) & \xrightarrow{\text{ trans }} & H^2(H/\gamma_{\alpha+1}(H), \mathbb{T}) & \xrightarrow{\text{ inf }} & P_{\mathfrak{h}^{\alpha+1}}(H, \mathbb{T}) \\
\downarrow{\scriptstyle u} & & \downarrow{\scriptstyle v} & & \downarrow{\scriptstyle w} \\
\mathrm{Hom}(\gamma_{\alpha+1}(G)/\gamma_{\alpha+2}(G), \mathbb{T}) & \xrightarrow{\text{ trans }} & H^2(G/\gamma_{\alpha+1}(G), \mathbb{T}) & \xrightarrow{\text{ inf }} & P_{\mathfrak{g}^{\alpha+1}}(G, \mathbb{T})
\end{array}
$$

where u, v and w are the homomorphisms induced by f. Clearly then w is
an epimorphism and, since w is a monomorphism by hypothesis, it follows
that it is an isomorphism. Consequently, u is an isomorphism, and so f in-
duces induces an isomorphism $\gamma_{\alpha+1}(G)/\gamma_{\alpha+2}(G) \simeq \gamma_{\alpha+1}(H)/\gamma_{\alpha+2}(H)$. Hence
f induces an isomorphism

$$G/\gamma_{\alpha+2}(G) \simeq H/\gamma_{\alpha+2}(H),$$

and the Lemma holds for $\alpha + 1$.
In view of limit property of HZ-local groups, the transfinite induction is
complete and the Lemma is proved. $\square$

Theorem 4.52 *Let α be an ordinal number and $f : G \to H$ a $\mathfrak{P}_\alpha$-
homomorphism. If*

$$P_{\Delta^\tau(L(G)/\gamma_\tau(L(G)))}(L(G)/\gamma_\tau(L(G)), \mathbb{T}) = H^2(L(G)/\gamma_\tau(L(G)), \mathbb{T})$$

for all $\tau \le \alpha$, then f induces an isomorphism

$$L(G)/\gamma_{\alpha+1}(L(G)) \simeq L(H)/\gamma_{\alpha+1}(L(H)).$$

Proof. Observe that for every group Γ, by definition of the Bousfield localization, the homomorphism $H_2(\Gamma) \to H_2(L(\Gamma))$ induced by $L : \Gamma \to L(\Gamma)$ is an epimorphism, and therefore the induced homomorphism

$$H^2(L(\Gamma), \mathbb{T}) \to H^2(\Gamma, \mathbb{T})$$

is a monomorphism. Consequently, the induced homomorphism

$$P_{\Delta^\alpha(L(\Gamma))}(L(\Gamma), \mathbb{T}) \to P_{\Delta^\alpha(\Gamma)}(\Gamma, \mathbb{T})$$

is also a monomorphism. Thus, under the given hypothesis, f induces the following commutative diagram:

$$
\begin{array}{ccc}
P_{\Delta^\alpha(L(H))}(L(H), \mathbb{T}) & \longrightarrow & P_{\Delta^\alpha(L(G))}(L(G), \mathbb{T}) \\
\downarrow & & \downarrow \\
P_{\mathfrak{h}^\alpha}(H, \mathbb{T}) & \longrightarrow & P_{\mathfrak{g}^\alpha}(G, \mathbb{T})
\end{array}
$$

in which the vertical maps are monomorphisms, and the lower horizontal map is a monomorphism, since f is a $\mathfrak{P}_\alpha$-homomorphism. Hence the upper horizontal map is a monomorphism and the assertion follows by Lemma 4.51. $\qquad\square$

It follows from ([Bou77], Lemma 3.7) that *if G is a group such that $H_1(G)$ is free abelian and $H^2(G, \mathbb{T}) = 0$, then G is τ-para-free for every ordinal number τ.* This result naturally raises the following:

Problem 4.53 *Is it true that if G is τ-para-free for all countable ordinals τ, then $H^2(G, \mathbb{T}) = 0$?*

The above problem may be compared with the *Para-free Conjecture* (Problem 1.101).

Next we give a characterization of $(\omega + 1)$-para-free groups.

Theorem 4.54 *A group G is $(\omega + 1)$-para-free if and only if $H_1(G)$ is free abelian and $P_{\Delta^\omega(L(G))}(L(G), \mathbb{T}) = 0$.*

Proof. Let G be an $(\omega + 1)$-para-free group. Then there exists a homomorphism $f : F \to G$, where F is a free group, which induces an isomorphism $L(F)/\gamma_{\omega+1}(L(F)) \simeq L(G)/\gamma_{\omega+1}(L(G))$. It follows that $L(F)/\gamma_\omega(L(F)) \simeq L(G)/\gamma_\omega(L(G))$, and $H_1(L(F)) \simeq H_1(L(G))$ showing that $H_1(G)$ is free abelian. Consider the following commutative diagram with exact rows:

$$
\begin{array}{ccccc}
\mathrm{Hom}(\gamma_\omega(L(G))/\gamma_{\omega+1}(L(G)), \mathbb{T}) & \xrightarrow{\ \mathrm{trans}\ } & H^2(L(G)/\gamma_\omega L(G), \mathbb{T}) & \xrightarrow{\ \mathrm{inf}\ } & H^2(L(G), \mathbb{T}) \\
\Big\| & & \Big\| & & \downarrow \\
\mathrm{Hom}(\gamma_\omega(L(F))/\gamma_{\omega+1}(L(F)), \mathbb{T}) & \xrightarrow{\ \mathrm{trans}\ } & H^2(L(F)/\gamma_\omega L(F), \mathbb{T}) & \xrightarrow{\ \mathrm{inf}\ } & H^2(L(F), \mathbb{T}).
\end{array}
$$

Observe that $P_{\Delta^\omega(L(G))}(L(G), \mathbb{T})$ is contained in the image of the inflation homomorphism and $H^2(L(F), \mathbb{T}) = 0$. Therefore we have

$$P_{\Delta^\omega(L(G))}(L(G), \mathbb{T}) = 0.$$

Conversely, suppose that G is a group such that $H_1(G)$ is free abelian and $P_{\Delta^\omega(L(G))}(L(G), \mathbb{T}) = 0$. We can then clearly construct a homomorphism $f : F \to G$, with F a free group, which induces an isomorphism $H_1(F) \simeq H_1(G)$. Note that $L(F)/\gamma_\omega(L(F))$ is residually torsion-free nilpotent, and therefore $\Delta^\omega(L(F)/\gamma_\omega(L(F))) = 0$. Consequently,

$$H^2(L(F)/\gamma_\omega(L(F)), \mathbb{T}) = P_{\Delta^\omega(L(F)/\gamma_\omega(L(F)))}(L(F)/\gamma_\omega(L(F)), \mathbb{T}).$$

Since $L(F)/\gamma_n(L(F)) \simeq F/\gamma_n(F)$ for every natural number n, we have, by ([Pas68a], Theorem 6.10),

$$H^2(L(F)/\gamma_n(L(F)), \mathbb{T}) = P_{\Delta^n(L(F)/\gamma_n(L(F)))}(L(F)/\gamma_n(L(F)), \mathbb{T}).$$

Hence it follows from Lemma 4.51 that G is $(\omega + 1)$-para-free. $\square$

As an immediate consequence of the above result, we have

Corollary 4.55 *If G is a group such that $H_1(G)$ is free abelian and $P_{\mathfrak{g}^\omega}(G, \mathbb{T}) = 0$, then G is $(\omega + 1)$-para-free.*

Remark. Note that the argument of the first part of the proof of the preceeding theorem, in fact, works for arbitrary ordinal numbers τ, i.e., the following holds:

If G is a τ-para-free group, then $P_{\Delta^\tau(L(G))}(L(G), \mathbb{T}) = 0$.

There is an obvious difficulty in extending the second part of the proof to ordinals $\tau \geq \omega + 2$. Our next result addresses the case $\tau = \omega + 2$. We need the following:

Lemma 4.56 *Let G be a group and N a central subgroup in G. If the restriction map $\mathrm{res} : H^2(G, \mathbb{T}) \to H^2(N, \mathbb{T})$ is the zero map, then $P_{\mathfrak{n}\mathfrak{g}}(G, \mathbb{T}) = H^2(G, \mathbb{T})$.*

Proof. Let

$$1 \to \mathbb{T} \to M \to G \to 1$$

be a central extension. Since res is the zero map, we can choose a set of representatives $u(n) \in M$ for elements $n \in N$ such that $u(n_1 n_2) = u(n_1)u(n_2)$ for all $n_1, n_2 \in N$. Let $H = G/N$ and choose a set of representatives $v(h) \in G$ for elements $h \in H$. Then every element of G can be uniquely written as $g = v(h)n$ ($h \in H, n \in N$). Now choose representatives $w(v(h)) \in M$, $h \in H$, arbitrarily and set $w(g) = u(n)w(v(h))$. Then $\{w(g)\}$ is a choice of representatives of the elements g of G in M for which the corresponding 2-cocycle $W(g_1, g_2) : G \times G \to \mathbb{T}$, when extended to $\mathbb{Z}[G] \times \mathbb{Z}[G]$ by linearity, is easily seen to vanish on $\mathfrak{n}\mathfrak{g} \times \mathbb{Z}[G]$. $\square$

Proposition 4.57 *Let G be a group with $\Delta^\omega(L(G)/\gamma_\omega(L(G))) = 0$ and $H^2(L(G)/\gamma_{\omega+1}(L(G)), \mathbb{T}) \to H^2(\gamma_\omega(L(G))/\gamma_{\omega+1}(L(G)), \mathbb{T})$ the zero map. If $f : G \to H$ is an $\mathfrak{P}_{\omega+1}$-homomorphism, then f induces an isomorphism*

$$L(G)/\gamma_{\omega+2}(L(G)) \simeq L(H)/\gamma_{\omega+2}(L(H)).$$

Proof. Since $\Delta^\omega(L(G)/\gamma_\omega(L(G))) = 0$, we have

$$\Delta^\omega(L(G)/\gamma_{\omega+1}(L(G))) =$$
$$\Delta(\gamma_\omega(L(G))/\gamma_{\omega+1}(L(G)))\mathbb{Z}[L(G)/\gamma_{\omega+1}(L(G))].$$

Therefore, $\gamma_\omega(L(G))/\gamma_{\omega+1}(L(G))$ being central in $L(G)/\gamma_{\omega+1}(L(G))$, we have

$$\Delta^{\omega+1}(L(G)/\gamma_{\omega+1}(L(G))) =$$
$$\Delta(\gamma_\omega(L(G))/\gamma_{\omega+1}(L(G)))\Delta(L(G)/\gamma_{\omega+1}(L(G))).$$

It then follows from Lemma 2, applied to $L(G)/\gamma_{\omega+1}(L(G))$ with the central subgroup $\gamma_\omega(L(G))/\gamma_{\omega+1}(L(G))$ that

$$P_{\Delta^{\omega+1}(L(G)/\gamma_{\omega+1}(L(G)))}(L(G)/\gamma_{\omega+1}(L(G)), \mathbb{T}) = H^2(L(G)/\gamma_{\omega+1}(L(G)), \mathbb{T}),$$

and the Proposition follows from Theorem 3. $\square$

The above Proposition suggests the following:

Problem 4.58 *Is it true that, for every free group F, the restriction map $H^2(L(F)/\gamma_{\omega+1}(L(F)), \mathbb{T}) \to H^2(\gamma_\omega(L(F))/\gamma_{\omega+1}(L(F)), \mathbb{T})$ is the zero map?*

If the answer is yes, then it is clear from the discussion above that groups G with $P_{\mathfrak{g}^{\omega+1}}(G, \mathbb{T}) = 0$ and $H_1(G)$ free abelian are $(\omega + 2)$-para-free; more precisely, a group G with free abelianization is $(\omega + 2)$-para-free if and only if $P_{\Delta^{\omega+1}(L(G))}(L(G), \mathbb{T}) = 0$.

4.9 Homological Localization of $\mathbb{Z}[G]$-modules

Let $f : M \to N$ be a map between $\mathbb{Z}[G]$-modules. The map f is called an HZ-*map* if it induces

   (i) an isomorphism $f_0 : H_0(G, M) \to H_0(G, N)$ and

   (ii) an epimorphism $f_1 : H_1(G, M) \to H_1(G, N)$.

The following commutative diagram

$$
\begin{array}{ccccccc}
H_1(G,\, M) & \longrightarrow & H_1(G,\, M/\mathfrak{g}M) & \longrightarrow & \mathfrak{g}M/\mathfrak{g}^2 M & \longrightarrow & 0 \\
\downarrow{\scriptstyle f_1} & & \downarrow{\scriptstyle f_0^\cdot} & & \downarrow & & \\
H_1(G,\, N) & \longrightarrow & H_1(G,\, N/\mathfrak{g}N) & \longrightarrow & \mathfrak{g}N/\mathfrak{g}^2 N & \longrightarrow & 0
\end{array}
$$

implies that an HZ-map $f : M \to N$ induces an isomorphism

$$
M/\mathfrak{g}^2 M \simeq N/\mathfrak{g}^2 N.
$$

A straight-forward induction argument then yields the following

Proposition 4.59 *An HZ-map* $f : M \to N$ *of* $\mathbb{Z}[G]$-*modules implies isomorphisms*

$$
M/\mathfrak{g}^n M \simeq N/\mathfrak{g}^n N, \quad n \geq 1,
$$

of $\mathbb{Z}[G]$-*modules.*

We denote by $\mathcal{HZ}$ the class of HZ-maps $f : M \to N$ of $\mathbb{Z}[G]$-modules. The construction of $\mathcal{HZ}$-localization in the category of $\mathbb{Z}[G]$-modules is due to Bousfield (see [Bou77]); it is defined to be an HZ-map $E : M \to E(M)$ of $\mathbb{Z}[G]$-modules such that $E(M)$ is $\mathcal{HZ}$-local. In analogy with the class of HZ-local groups (see Chapter 1, Section 1.7), the class $loc(\mathcal{HZ})$ can be described (see [Bou77], Theorem 8.9) as the smallest class of $\mathbb{Z}[G]$-modules such that

(i) the class contains the zero $\mathbb{Z}[G]$-module;
(ii) the class is closed under inverse limits;
(iii) if Y is in the class and $0 \to W \to X \to Y \to 0$ is an extension of $\mathbb{Z}[G]$-modules with W trivial, then X is in the class.

The homotopical meaning of the class of HZ-local groups and HZ-local modules can be explained in terms of homological localizations in the category $\mathcal{H}o$ of pointed spaces. For any $X \in \mathcal{H}o$, and a generalized homology theory h_*, Bousfield constructed the homological localization $X \to X_{h_*}$, which is terminal h_*-homological equivalence going out of X. The class of $H(-, \mathbb{Z})$-local spaces can be characterized algebraically as follows [Bou77]:

X is an $H(-, \mathbb{Z})$-local space if and only if $\pi_1(X)$ is a HZ-local group and $\pi_n(X)$ is an HZ-local $\mathbb{Z}[\pi_1(X)]$-module.

This characterization was the main motivation for the study of HZ-localization theory for groups and modules over group rings.

For any $\mathbb{Z}[G]$-module M, the HZ-localization $E(M)$ can be constructed as the inverse limit of a transfinite tower of maps. The *HZ-tower* for a $\mathbb{Z}[G]$-module M is a transfinite tower of maps η_α of $\mathbb{Z}[G]$-modules:

$$M \xleftarrow{\ id\ } M \xleftarrow{\ id\ } \cdots \xleftarrow{\ id\ } M \xleftarrow{\ id\ } M \longleftarrow \cdots$$

$$\eta_1 \downarrow \qquad \eta_2 \downarrow \qquad\qquad \eta_\alpha \downarrow \qquad \eta_{\alpha+1} \downarrow$$

$$T_1 M \xleftarrow{\ t_1\ } T_2 M \xleftarrow{\ t_2\ } \cdots \longleftarrow T_\alpha M \xleftarrow{\ t_\alpha\ } T_{\alpha+1} M \longleftarrow \cdots,$$

such that:

(i) $T_1 M = 0$, and for each $\alpha \geq 1$, the map t_α is surjective with kernel a trivial $\mathbb{Z}[G]$-module;

(ii) for each limit ordinal number τ, the map $T_\tau M \to \varprojlim_{\alpha<\tau} T_\alpha M$ carries $T_\tau M$ isomorphically into the HZ-closure in $\varprojlim_{\alpha<\tau} T_\alpha M$ of the image of the map $M \to \varprojlim_{\alpha<\tau} T_\alpha M$;

(iii) for each $\alpha \geq 1$, $H_0(G, \operatorname{coker}(\eta_\alpha)) = 0$, and the map $H_1(G, \eta_{\alpha+1}) \to H_1(G, \eta_\alpha)$ is zero.

Then the HZ-localization of M can be defined as the inverse limit of an HZ-tower:

$$E(M) = \varprojlim_\alpha T_\alpha M.$$

Artin-Rees property. A group G is said to have the *Artin-Rees property* if $\mathbb{Z}[G]$ is Noetherian and for every finitely generated (left) G-module M and every submodule N, the $\mathfrak{g}$-adic topology on N coincides with the restriction to N of the $\mathfrak{g}$-adic topology on M (see [Bro75], [Ros79], [Smi82]).

Theorem 4.60 (Nouazé-Gabriel [Nou67]). *Any finitely generated nilpotent group has the Artin-Rees property.*

The Artin-Rees property has been investigated from homological point of view by K. S. Brown and E. Dror [Bro75]. We mention some of their results in this direction.

Notation. For a G-module M, let $Z_\infty^G(M)$ denote the completion $\varprojlim M/\mathfrak{g}^n M$ of M with respect to the $\mathfrak{g}$-adic topology.

Proposition 4.61 (Brown-Dror, [Bro75]). *Let G be a finitely generated nilpotent group. Then:*

(i) The $\mathfrak{g}$-adic completion functor is exact on the category of finitely generated $\mathbb{Z}[G]$-modules.

(ii) For any finitely generated $\mathbb{Z}[G]$-module M, there is the natural isomorphism $Z_\infty^G(\mathbb{Z}[G]) \otimes_{\mathbb{Z}[G]} M \simeq Z_\infty^G(M)$.

Theorem 4.62 (Brown-Dror, [Bro75]) *Let G be a finitely generated nilpotent group and M a finitely generated $\mathbb{Z}[G]$-module. Then the $\mathfrak{g}$-adic completion map $M \to Z_\infty^G(M)$ induces isomorphisms $H_*(G, M) \simeq H_*(G, Z_\infty^G(M))$.*

One of the applications of the Artin-Rees property is the description of HZ-localization for finitely generated modules over pre-nilpotent groups. Recall that a group G is called *pre-nilpotent* if $\gamma_n(G) = \gamma_{n+1}(G)$ for some $n \geq 1$.

Theorem 4.63 (Brown-Dror [Bro75]). *Let G be a pre-nilpotent group and M a finitely generated $\mathbb{Z}[G]$-module. Then the map $E(M) \to Z_\infty^G(M)$ is an isomorphism.*

Proof. The module $Z_\infty^G(M)$ is HZ-local, being an inverse limit of nilpotent $\mathbb{Z}[G]$-modules. The uniqueness of HZ-localization implies that $E(M)$ is naturally isomorphic to $Z_\infty^G(M)$ if and only if the $\mathfrak{g}$-adic completion map $M \to Z_\infty^G(M)$ is an HZ-map.

Let n be such that $\gamma_n(G) = \gamma_{n+1}(G)$. Consider the $\mathbb{Z}[G/\gamma_n(G)]$-module $N = H_0(\gamma_n(G), M)$. Then, clearly, $Z_\infty^G(M) \simeq Z_\infty^{G/\gamma_n(G)}(N)$. The projections $G \to G/\gamma_n(G)$ and $M \to N$ induce the following commutative diagram:

$$
\begin{array}{ccc}
H_i(G, M) & \xrightarrow{\ \alpha_i\ } & H_i(G, Z_\infty^G(M)) \\[2mm]
\beta_i \Big\downarrow & & \beta_i^* \Big\downarrow \\[2mm]
H_i(G/\gamma_n(G), N) & \xrightarrow{\ \alpha_i'\ } & H_i(G/\gamma_n(G), Z_\infty^{G/\gamma_n(G)}(N)).
\end{array}
$$

The map α_0' is an isomorphism by Theorem 4.62. Clearly, β_0 and β_0^* are also isomorphisms; hence α_0 is an isomorphism. The homological spectral sequence for the group extension $1 \to \gamma_n(G) \to G \to G/\gamma_n(G) \to 1$ implies that the map β_1 is an epimorphism. Applying the same type of spectral sequence for homology $H_*(-, Z_\infty^G(M)) = H_*(-, Z_\infty^{G/\gamma_n(G)}(N))$, we get the following exact sequence:

$$
H_0(G/\gamma_n(G), H_1(\gamma_n(G), Z_\infty^{G/\gamma_n(G)}(N))) \to
$$
$$
H_1(G, Z_\infty^G(M)) \xrightarrow{\beta_1^*} H_1(G/\gamma_n(G), Z_\infty^{G/\gamma_n(G)}(N)) \to 0.
$$

By Proposition 4.61,

$$
H_1(\gamma_n(G)) \otimes_{\mathbb{Z}[G/\gamma_n(G)]} Z_\infty^{G/\gamma_n(G)}(\mathbb{Z}[G/\gamma_n(G)]) = Z_\infty^{G/\gamma_n(G)}(H_1(\gamma_n(G)))) = 0. \tag{4.26}
$$

Since the $\mathfrak{g}$-adic completion functor is right exact for $\mathbb{Z}[G/\gamma_n(G)]$-modules, (4.26) implies

$$
H_0(G/\gamma_n(G), H_1(\gamma_n(G), Z_\infty^{G/\gamma_n(G)}(N))) \simeq H_1(\gamma_n(G)) \otimes_{\mathbb{Z}[G/\gamma_n(G)]} Z_\infty^{G/\gamma_n(G)}(N)
$$
$$
= 0.
$$

Hence β_1^* is an isomorphism and α_1 is an epimorphism. $\square$

In the case of finitely presented groups, the kernel of HZ-localization can be described explicitly.

Theorem 4.64 (Dwyer [Dwy78]). *Let G be a finitely presented group and M a $\mathbb{Z}[G]$-module. Then there is a natural short exact sequence:*

$$0 \to \varprojlim_n^1 \mathrm{Tor}_1^{\mathbb{Z}[G]}(\mathbb{Z}[G]/\mathfrak{g}^n, \mathfrak{g}^n M) \to E(M) \to Z_\infty^G(M) \to 0.$$

Proposition 4.65 (Dwyer [Dwy78]). *Let G be a finitely presented group. Then the map $E(M) \to \mathcal{L}_0 Z_\infty^G(M)$ is an isomorphism for all $\mathbb{Z}[G]$-modules M if and only if*

$$\varprojlim_n^1 H_2(G, \mathbb{Z}[G]/\mathfrak{g}^n) = 0. \tag{4.27}$$

Proof. The natural map $E(M) \to \mathcal{L}_0 Z_\infty^G(M)$ is an isomorphism if and only if the natural map $M \to Z_\infty^G(M)$ is an HZ-map. Since the *zero*th derived functor of the $\mathfrak{g}$-adic completion and HZ-localization functors are right exact, it is enough to consider free $\mathbb{Z}[G]$-modules.

Let F be a free $\mathbb{Z}[G]$-module. Since $\mathfrak{g}$ is a finitely generated $\mathbb{Z}[G]$-module, we have the following short exact sequences:

$$0 \to \varprojlim_n^1 H_1(G, F/\mathfrak{g}^n F) \to H_0(G, Z_\infty^G(F)) \to \varprojlim_n H_0(G, F/\mathfrak{g}^n F) \to 0, \tag{4.28}$$

$$0 \to \varprojlim_n^1 H_2(G, F/\mathfrak{g}^n F) \to H_1(G, Z_\infty^G(F)) \to \varprojlim_n H_1(G, F/\mathfrak{g}^n F) \to 0. \tag{4.29}$$

Clearly, $H_1(G, F/\mathfrak{g}^n F) = \mathfrak{g}^n F/\mathfrak{g}^{n+1} F$, therefore,

$$\varprojlim_n H_1(G, F/\mathfrak{g}^n F) = 0, \quad \varprojlim_n^1 H_1(G, F/\mathfrak{g}^n F) = 0.$$

In view of (4.28) the $\mathfrak{g}$-adic completion map induces the natural isomorphisms

$$H_0(G, F) = \varprojlim_n H_0(G, F/\mathfrak{g}^n F) = H_0(G, Z_\infty^G(F)). \tag{4.30}$$

Taking $F = \mathbb{Z}[G]$, (4.29) implies the isomorphism

$$\varprojlim_n^1 H_2(G, \mathbb{Z}[G]/\mathfrak{g}^n) \simeq H_1(G, Z_\infty^G(\mathbb{Z}[G])) \tag{4.31}$$

and therefore, the condition (4.27) is necessary for the $\mathfrak{g}$-adic completion to be the HZ-localization.

Now suppose that the condition (4.27) holds. Proposition A.20 implies that the spectrum $\{H_2(G, \mathbb{Z}[G]/\mathfrak{g}^n)\}_{n \geq 1}$ is Mittag-Leffler (see Appendix, p. 330). Clearly, the direct sum of Mittag-Leffler spectra is Mittag-Leffler; hence

$$\varprojlim_n^1 H_2(G, F/\mathfrak{g}^n F) = 0.$$

Therefore, $H_1(G, Z_\infty^G(F)) = 0$ by (4.31). Thus the isomorphism (4.30) implies that the $\mathfrak{g}$-adic completion of F is an HZ-map and the assertion follows. $\square$

The following result generalizes Theorem 4.63 and presents necessary and sufficient conditions for the equivalence of the HZ-localization and the $\mathfrak{g}$-adic completion functors.

Theorem 4.66 (Dwyer [Dwy78]). *Let G be a finitely presented group. Then the map $E(M) \to Z_\infty^G(M)$ is an isomorphism for all $\mathbb{Z}[G]$-modules M if and only if*

(i) $\varprojlim_n^1 \phi_n^{(1)}(G) = 0$;

(ii) $\varprojlim_n^1 H_3(G/\gamma_n(G)) = 0$.

The condition (i) from Theorem 4.66 can be used for construction of examples of HZ-localizations, which are not equivalent to the $\mathfrak{g}$-adic completion. For any group G with $\varprojlim_n^1 \phi_n^{(1)}(G) \neq 0$, the HZ-localization of the group ring $\mathbb{Z}[G]$ is not equivalent to the $\mathfrak{g}$-adic completion $\varprojlim_n \mathbb{Z}[G]/\mathfrak{g}^n$.

Example 4.67 (Dwyer [Dwy78]).

Let G be a group

$$G = \langle x_1, x_2, x_3 \mid [x_1, x_2] = 1, \ x_3 x_1 x_3^{-1} x_1 = 1, \ x_3 x_2 x_3^{-1} x_2 = 1 \rangle.$$

Then $\varprojlim_n \phi_n^{(1)}(G) \neq 0$.

The theory of residual nilpotence of central extensions, described in Chapter 1, can also be used for construction of groups with $\varprojlim_n \phi_n^{(1)}(G) \neq 0$.

Example 4.68

Let G be a free abelian extension of the cyclic group of order 3:

$$G = \langle a, b \mid [a, b^3] = 1, \ [a, b, a] = 1 \rangle.$$

As noted in Example 1.85, $G \in \mathcal{J} \setminus \tilde{\mathcal{J}}$, hence

$$\varprojlim_n \phi_n^{(1)}(G) \neq 0,$$

by Proposition 1.86.

Problem 4.69 *Construct a group G with a non-trivial $\varprojlim_n^1 H_3(G/\gamma_n(G))$.*

Proposition 4.70 (Burns-Ellis [Bur97]). *Let G be a group given by the free presentation $G = F/R$. Then, for all $n \geq 1$, there is a natural isomorphism*

$$H_2(G, \mathbb{Z}[G]/\mathfrak{g}^n) \simeq \frac{R \cap \gamma_{n+1}(F)}{[R, R] \cap \gamma_{n+1}(F).[R, {}_nF]}.$$

Proof. Dimension shifting argument shows that there is a natural isomorphism of the kernel of Magnus embedding map tensored with $\mathbb{Z}[G]/\mathfrak{g}^n$:

$$H_2(G, \mathbb{Z}[G]/\mathfrak{g}^n) \simeq \ker\{\mathfrak{r}/\mathfrak{fr} \otimes_{\mathbb{Z}[G]} \mathbb{Z}[G]/\mathfrak{g}^n \to \mathfrak{f} \otimes_{\mathbb{Z}[F]} \mathbb{Z}[G]/\mathfrak{g}^n\}.$$

We have the following exact sequences of abelian groups:

$$\mathfrak{r}/\mathfrak{f}\mathfrak{r} \otimes_{\mathbb{Z}[G]} \mathfrak{f}^n/\mathfrak{f}^n \cap \mathfrak{r} \to \mathfrak{r}/\mathfrak{f}\mathfrak{r} \to \mathfrak{r}/\mathfrak{f}\mathfrak{r} \otimes_{\mathbb{Z}[G]} \mathbb{Z}[G]/\mathfrak{g}^n \to 0,$$

$$\mathfrak{f} \otimes_{\mathbb{Z}[F]} (\mathfrak{f}^n + \mathfrak{r}) \to \mathfrak{f} \to \mathfrak{f} \otimes_{\mathbb{Z}[F]} \mathbb{Z}[G]/\mathfrak{g}^n \to 0,$$

which imply the isomorphisms

$$\mathfrak{r}/\mathfrak{f}\mathfrak{r} \otimes_{\mathbb{Z}[G]} \mathbb{Z}[G]/\mathfrak{g}^n \simeq \mathfrak{r}/\mathfrak{f}\mathfrak{r} + \mathfrak{r}\mathfrak{f}^n,$$

$$\mathfrak{f} \otimes_{\mathbb{Z}[F]} \mathbb{Z}[G]/\mathfrak{g}^n \simeq \mathfrak{f}/\mathfrak{f}\mathfrak{r} + \mathfrak{f}^{n+1}.$$

Consequently,

$$H_2(G,\, \mathbb{Z}[G]/\mathfrak{g}^n) \simeq \ker\{\mathfrak{r}/\mathfrak{f}\mathfrak{r} + \mathfrak{r}\mathfrak{f}^n \to \mathfrak{f}/\mathfrak{f}\mathfrak{r} + \mathfrak{f}^{n+1}\} \simeq$$

$$\frac{\mathfrak{r} \cap (\mathfrak{f}\mathfrak{r} + \mathfrak{f}^n)}{\mathfrak{r}\mathfrak{f}^n + \mathfrak{f}\mathfrak{r} \cap \mathfrak{f}^{n+1}} \simeq \frac{R \cap \gamma_{n+1}(F)}{[R,\,R] \cap \gamma_{n+1}(F).[R,\,_nF]}$$

(see Lemma 4.31). $\square$

Proposition 4.70 implies that there is a natural short exact sequence:

$$0 \to \frac{[R,\,F] \cap \gamma_{n+1}(F)}{[R,\,R] \cap \gamma_{n+1}(F).[R,\,_nF]} \to H_2(G,\, \mathbb{Z}[G]/\mathfrak{g}^n) \to \phi_n^{(1)}(G) \to 0,$$

which induces the short exact sequence of corresponding spectra (see 1.4). Hence, the condition (4.27) implies the condition (i) of Theorem 4.66 for any group G.

Clearly, there is a natural exact sequence of abelian groups:

$$0 \to \frac{[R,\,R] \cap \gamma_{n+1}(F)}{[R,\,R] \cap [R,\,_nF]} \to M^{(n)}(G) \to \frac{R \cap \gamma_{n+1}(F)}{[R,\,R] \cap \gamma_{n+1}(F).[R,\,_nF]} \to 0$$

$$\tag{4.32}$$

In [Dwy75], the standard composition of functors spectral sequence implies that for every finitely-presented group G there exists the following exact sequence:

$$0 \to \varprojlim_n \Psi_n(G) \to \varprojlim_n H_0(G,\, \mathrm{Tor}_2^{\mathbb{Z}[G]}(\mathbb{Z}[G]/\mathfrak{g}^n,\, \mathbb{Z})) \to \phi_\omega(G) \to$$

$$\varprojlim_n^1 \Psi_n(G) \to \varprojlim_n^1 H_0(G,\, \mathrm{Tor}_2^{\mathbb{Z}[G]}(\mathbb{Z}[G]/\mathfrak{g}^n,\, \mathbb{Z})) \to \varprojlim_n^1 \phi_n^{(1)}(G) \to 0, \quad (4.33)$$

where $\Psi_n(G) = \mathrm{coker}\{H_3(G) \to H_3(G/\gamma_n(G))\}$, $n \geq 2$. The sequence (4.33) is the main argument for the proof of the Theorem 4.66. The existence of the canonical anti-automorphism of the ring $\mathbb{Z}[G]$ implies that

$$\varprojlim_n H_0(G,\, \mathrm{Tor}_2^{\mathbb{Z}[G]}(\mathbb{Z}[G]/\mathfrak{g}^n,\, \mathbb{Z})) = \varprojlim_n H_0(G,\, \mathrm{Tor}_2^{\mathbb{Z}[G]}(\mathbb{Z},\, \mathbb{Z}[G]/\mathfrak{g}^n)) =$$

$$\varprojlim_n H_0(G,\, H_2(G,\, \mathbb{Z}[G]/\mathfrak{g}^n)),$$

$$\varprojlim\nolimits_n^1 H_0(G, \operatorname{Tor}_2^{\mathbb{Z}[G]}(\mathbb{Z}[G]/\mathfrak{g}^n, \mathbb{Z})) = \varprojlim\nolimits_n^1 H_0(G, \operatorname{Tor}_2^{\mathbb{Z}[G]}(\mathbb{Z}, \mathbb{Z}[G]/\mathfrak{g}^n)) =$$
$$\varprojlim\nolimits_n^1 H_0(G, H_2(G, \mathbb{Z}[G]/\mathfrak{g}^n)).$$

Since $\varprojlim\nolimits_n^1 \Psi_n = \varprojlim\nolimits_n^1 H_3(G/\gamma_n(G))$, we have the following exact sequence:

$$0 \to \varprojlim\nolimits_n \Psi_n(G) \to \varprojlim\nolimits_n H_0(G, H_2(G, \mathbb{Z}[G]/\mathfrak{g}^n)) \to \phi_\omega(G) \to$$
$$\varprojlim\nolimits_n^1 H_3(G/\gamma_n(G)) \to \varprojlim\nolimits_n^1 H_0(G, H_2(G, \mathbb{Z}[G]/\mathfrak{g}^n)) \to \varprojlim\nolimits_n^1 \phi_n^{(1)}(G) \to 0.$$
$$(4.34)$$

Proposition 4.70 implies the following exact sequence of abelian groups

$$0 \to \frac{[R, F] \cap \gamma_{n+1}(F)}{([R, R] \cap \gamma_{n+1}(F))[R \cap \gamma_{n+1}(F), F][R, \, _nF]} \to$$
$$H_0(G, H_2(G, \mathbb{Z}[G]/\mathfrak{g}^n)) \to \phi_n^{(1)}(G) \to 0. \quad (4.35)$$

Exact sequences (4.34) and 4.35 imply the following isomorphism

$$\varprojlim\nolimits_n^1 H_3(G/\gamma_n(G)) \simeq \varprojlim\nolimits_n^1 \frac{[R, F] \cap \gamma_{n+1}(F)}{([R, R] \cap \gamma_{n+1}(F))[R \cap \gamma_{n+1}(F), F][R, \, _nF]},$$

which takes place for every finitely-presented group F/R. Clearly, there is an epimorphism

$$M^{(n)}(F/[F, R]) \to \frac{[R, F] \cap \gamma_{n+1}(F)}{([R, R] \cap \gamma_{n+1}(F))[R \cap \gamma_{n+1}(F), F][R, \, _nF]},$$

which implies the epimorphism

$$\varprojlim\nolimits_n^1 M^{(n)}(F/[F, R]) \to \varprojlim\nolimits_n^1 H_3(G/\gamma_n(G)).$$

Proposition 4.71 *Let G be a finitely presented group with finite $H_2(G)$. Then the natural map $E(M) \to \mathcal{L}_0 Z_\infty^G(M)$ is an isomorphism for any $\mathbb{Z}[G]$-module M.*

Proof. By Theorem 1.74, all Baer invariants $M^{(n)}(G)$ are finite for $n \geq 1$. Hence, $\varprojlim\nolimits_n^1 M^{(n)}(G) = 0$ and therefore, $\varprojlim\nolimits_n^1 H_2(G, \mathbb{Z}[G]/\mathfrak{g}^n) = 0$ due to Proposition 4.70 and sequence (4.32). The statement then follows from Theorem 4.65. $\square$

Let G be a group and X a $\mathbb{Z}[G]$-module. To summarize the discussion on HZ-localization, we see that this construction can be viewed as the transfinite extension of $\mathfrak{g}$-adic completion of X; it can be constructed in analogy with HZ-localization of a group, described in Chapter 1. The HZ-localization of $\mathbb{Z}[G]$-modules is interesting and important from various points of view:

- The Artin-Rees property can be applied for the description of HZ-localization for modules over finitely generated nilpotent groups.
- The theory of HZ-localization introduces the *Dwyer's condition* ((1.53), case $k = 1$): $\varprojlim_n^1 \phi_n^{(1)}(G) = 0$, which plays important role in the theory of residual nilpotence.
- HZ-localization tower (or HZ-tower) defines natural transfinite extension of the functor of $\mathfrak{g}$-adic completion.
- There are homotopical applications.

Chapter 5
Homotopical Aspects

The purpose of this Chapter is to develop a connection between lower cenral and dimension series of groups, simplicial homotopy theory, and derived functors of non-additive functors in the sense of Dold-Puppe. A basic role in this study is played by two spectral sequences investigated by E. Curtis. Our analysis leads to interesting homotopical applications.

For various notions about simplicial objects and the tools required from simplicial homotopy theory, the reader is referred to the Appendix.

5.1 The Associated Graded Ring of a Group Ring

Let R commutative ring with identity, and M an R-module.

Tensor Algebra. Let $M^{\otimes i}$ denote the i-fold tensor product $\underbrace{M \otimes_R \cdots \otimes_R M}_{i \ terms}$ of M over R. Then the direct sum

$$\mathcal{T}_R(M) := R \oplus \bigoplus_{i \geq 1} M^{\otimes i},$$

with the multiplication induced by setting

$$m_1 \otimes \cdots \otimes m_i . n_1 \otimes \cdots \otimes n_j = m_1 \otimes \cdots \otimes m_i \otimes n_1 \otimes \cdots \otimes n_j,$$

is called the *tensor algebra* of M over R. The assignment $M \mapsto \mathcal{T}_R(M)$ is a functor from the category Mod_R of R-modules to the category Alg_R of R-algebras. This functor has the following universal property:

Given any associateve R-algebra $\mathcal{A}$ and an R-homomorphism $\varphi : M \to \mathcal{A}$, there exists a unique R-algebra homomorphism $\eta : \mathcal{T}_R(M) \to \mathcal{A}$ such that $\eta \circ i = \varphi$, where i is the inclusion map $i : M \to \mathcal{T}_R(M)$.

Universal Enveloping Algebra. Let L be an R-Lie algebra, and $\mathcal{T}_R(L)$ its tensor algebra (when L is viewed as an R-module). Then the *universal*

R. Mikhailov, I.B.S. Passi, *Lower Central and Dimension Series of Groups,*
Lecture Notes in Mathematics 1952,
© Springer-Verlag Berlin Heidelberg 2009

enveloping algebra of L over R, denoted $\mathcal{U}_R(L)$, is the R-associative algebra defined by

$$\mathcal{U}_R(L) = \mathcal{T}_R(L)/I,$$

where I is the two-sided ideal of $\mathcal{T}_R(L)$, generated by the elements

$$a \otimes b - b \otimes a - [a, b], \ a, b \in L,$$

where $[a, b]$ is the Lie-product in L. Recall that any R-associative algebra $\mathcal{A}$ can be viewed as an R-Lie algebra under the operation

$$[a, b] = ab - ba, \ a, b \in \mathcal{A};$$

we denote this R-Lie algebra by $\mathcal{A}^{Lie}$. The universal enveloping algebra construction is a functor

$$\mathcal{U}_R : L \mapsto \mathcal{U}_R(L),$$

with the following universal property:

> The map $i : L \to \mathcal{U}_R(L)^{Lie}$, $x \mapsto x + I$, $x \in L$, is a homomorphism of R-Lie algebras, and for any R-associative algebra $\mathcal{A}$ and a homomorphism of R-Lie algebras $\psi : L \to \mathcal{A}^{Lie}$, there exists a unique homomorphism $\mu : \mathcal{U}_R(L) \to \mathcal{A}$ of associative R-algebras such that $\mu \circ i = \psi$.

Free Lie Ring Generated by a Module. Let $\mathcal{L}_R(M)$ be the sub-Lie ring of $\mathcal{T}_R(M)^{Lie}$ generated by M. The Lie ring $\mathcal{L}_R(M)$ is called the *free R-Lie ring generated by the R-module M*; it has the following universal property:

> Given any R-Lie algebra L and an R-homomorphism $\alpha : M \to L$, there exists a unique homomorphism $\theta : \mathcal{L}_R(M) \to L$ of R-Lie algberas such that $\theta \circ i = \alpha$, where $i : M \to \mathcal{L}_R(M)$ is the inclusion map, provided the ring R has the property that the map $i : L \to \mathcal{U}_R(L)^{Lie}$ is always a monomorphism.

The assignment $M \mapsto \mathcal{L}_R(M)$ is clearly a functor from the catgory Mod_R to the category Lie_R of R-Lie algebras.

It may be mentioned that Poincaré-Birkhoff-Witt theorem (see [Car56], p. 271) implies:

Theorem 5.1 *The homomorphism $i : L \to \mathcal{U}_R(L)$ is injective, provided L is R-free.*

Furthermore, the homomorphism $i : L \to \mathcal{U}_R(L)$ is known to be injective if R is a principal ideal domain [Laz54]; this map is, however, not always injective [Coh63] (see Grievel [Gri04] for a detailed account of Poincare-Birkhoff-Witt theorem).

Recall that the *coequalizer* $\mathrm{coeq}(f_1, f_2)$ of two R-homomorphisms $f_1, f_2 : M_1 \to M_2$ is the quotient M_2/N, where N is the submodule of M_2 generated by the elements $f_1(x) - f_2(x)$, $x \in M_1$. Let

$$q : M_2 \to \mathrm{coeq}(f_1, f_2)$$

be the natural projection. We will later need the following property of the functors discussed above.

Proposition 5.2 *The functors $\mathcal{L}_R$, $\mathcal{T}_R$ and $\mathcal{U}_R$ preserve coequalizers.*

Proof. Let $f_1 : M_1 \to M_2$, $f_2 : M_1 \to M_2$ be two maps between R-modules, and $q : M_2 \to \mathrm{coeq}(f_1,\ f_2)$ their coequalizer map. Then, for $i = 1,\ 2$, we have the following commutative diagram:

$$
\begin{array}{ccccc}
M_1 & \xrightarrow{\ f_i\ } & M_2 & \xrightarrow{\ q\ } & \mathrm{coeq}(f_1,\ f_2) \\
{\scriptstyle i_R}\downarrow & & {\scriptstyle i_R}\downarrow & & {\scriptstyle i_R}\downarrow \\
\mathcal{L}_R(M_1) & \xrightarrow{\ \mathcal{L}_R(f_i)\ } & \mathcal{L}_R(M_2) & \xrightarrow{\ \mathcal{L}_R(q)\ } & \mathcal{L}_R(\mathrm{coeq}(f_1,\ f_2)).
\end{array}
$$

Since $q \circ f_1 = q \circ f_2$, we have $\mathcal{L}_R(q) \circ \mathcal{L}_R(f_1) = \mathcal{L}_R(q) \circ \mathcal{L}_R(f_2)$. Therefore, there exists a map

$$e : \mathrm{coeq}(\mathcal{L}_R(f_1),\ \mathcal{L}_R(f_2)) \to \mathcal{L}_R(\mathrm{coeq}(f_1,\ f_2))$$

such that the following diagram is commutative

$$
\mathcal{L}_R(M_1) \overset{\mathcal{L}_R(f_1)}{\underset{\mathcal{L}_R(f_2)}{\rightrightarrows}} \mathcal{L}_R(M_2) \xrightarrow{\quad\quad \mathcal{L}_R(q) \quad\quad} \mathcal{L}_R(\mathrm{coeq}(f_1,\ f_2))
$$

$$
\searrow \qquad \nearrow{\scriptstyle e}
$$

$$
\mathrm{coeq}(\mathcal{L}_R(f_1),\ \mathcal{L}_R(f_2))
$$

$$(5.1)$$

The fact that $\mathcal{L}_R(q)$ is an epimorphism implies that e is also an epimorphism. On the other hand, $\ker(\mathcal{L}_R(q))$ lies inside the ideal generated by elements $\mathcal{L}_R(f_1)(x) - \mathcal{L}_R(f_2)(x)$, $x \in \mathcal{L}_R(M_1)$. Hence, e is also a monomorphism and therefore an isomorphism:

$$\mathrm{coeq}(\mathcal{L}_R(f_1),\ \mathcal{L}_R(f_2)) \simeq \mathcal{L}_R(\mathrm{coeq}(f_1,\ f_2)).$$

Analogous statements about the functors $\mathcal{T}^R$ and $\mathcal{U}_R$ can be proved in a similar manner. $\square$

The Associated Graded Lie Ring of a Dimension Series

Let G be a group and $\{D_{n,\,R}(G)\}_{n\geq 0}$ its series of dimension subgroups over the commutative ring R which is a principal ideal domain. We can naturally associate with this series the following graded R-Lie algebra:

$$L_R^{dim}(G) := \bigoplus_{i \geq 1} D_{i,R}(G)/D_{i+1,R}(G) \otimes_{\mathbb{Z}} R$$

with the Lie multiplication induced by

$$aD_{k+1,R}(G) \circ bD_{l+1,R}(G) = [a,b]D_{k+l+1,R}(G), \ a \in D_{k,R}(G), \ b \in D_{l,R}(G).$$

Associated with the lower central series $\{\gamma_n(G)\}_{n \geq 1}$, we have the Lie ring

$$L_R(G) = \oplus_{n \geq 1} \gamma_n(G)/\gamma_{n+1}(G) \otimes_{\mathbb{Z}} R,$$

where, as above, the Lie bracket is induced via commutators in G. The R-homomorphism

$$i : G_{ab} \otimes R \to L_R(G)$$

implies the existence of a homomorphism of R-Lie algebras:

$$\theta : \mathcal{L}_R(G_{ab} \otimes R) \to L_R(G) \tag{5.2}$$

which is clearly surjective. Since $\gamma_n(G) \subseteq D_{n,R}(G)$ for all $n \geq 1$, we also have a homomorphism of R-Lie algebras

$$\varphi : L_R(G) \to L_R^{dim}(G).$$

Theorem 5.3 (Magnus [Mag37], Witt [Wit37]). *For a free group G,*

$$\mathcal{L}_{\mathbb{Z}}(G_{ab}) \simeq L_{\mathbb{Z}}(G) = L_{\mathbb{Z}}^{dim}(G).$$

A canonical object associated with a group G and an arbitrary commutative ring R with identity is the associative graded R-algebra

$$\mathcal{A}_R(G) = R \oplus \bigoplus_{i \geq 1} \Delta_R^i(G)/\Delta_R^{i+1}(G),$$

where $\Delta_R(G)$ is the augmentation ideal of the group ring $R[G]$.

The R-homomorphism

$$i : G_{ab} \otimes_{\mathbb{Z}} R \to \Delta_R(G)/\Delta_R^2(G),$$
$$g\gamma_2(G) \otimes r \mapsto r(g-1) + \Delta_R^2(G), \ g \in G, \ r \in R,$$

implies the existence of an R-algebra homomorphism:

$$\eta : \mathcal{T}_R(G_{ab} \otimes R) \to \mathcal{A}_R(G), \tag{5.3}$$

which is clearly surjective.

Theorem 5.4 (see [Pas79], p. 116). *The homomorphism (5.3)*

$$\eta : \mathcal{T}_R(G_{ab} \otimes R) \to \mathcal{A}_R(G)$$

is an isomorphism in case G is a free group.

For every group G, the map

$$aD_{n+1, R}(G) \mapsto a - 1 + \Delta_R^{n+1}(G), \ a \in D_{n, R}(G),$$

induces an R-Lie homomorphism

$$L_R^{dim}(G) \to \mathcal{A}_R(G)^{Lie},$$

which implies the existence of a natural surjective homomorphism of associative R-algebras:

$$\mu : \mathcal{U}_R(L_R^{dim}(G)) \to \mathcal{A}_R(G).$$

Theorem 5.5 (Quillen [Qui68]). *The homomorphism $\mu : \mathcal{U}_R(L_R^{dim}(G)) \to \mathcal{A}_R(G)$ is an isomorphism, provided R is a field of characteristic zero.*

An appropriate analogous version of the above theorem, due to Quillen (loc. cit.) holds for fields of positive characteristic (see Passi [Pas79]).

5.2 Spectral Sequences

Throughout this section, let R be a subring of $\mathbb{Q}$, containing $\mathbb{Z}$.

Curtis Spectral Sequences

Let X be a simplicial group. The lower central series filtration in X gives rise to the long exact sequence

$$\cdots \to \pi_{i+1}(X/\gamma_n(X)) \to \pi_i(\gamma_n(X)/\gamma_{n+1}(X)) \to$$
$$\pi_i(X/\gamma_{n+1}(X)) \to \pi_i(X/\gamma_n(X)) \to \cdots$$

of simplicial homotopy groups. This exact sequence defines a graded exact couple which gives rise to the natural spectral sequence $E(X)$ with the initial terms

$$E_{n, m}^1(X) = \pi_m(\gamma_n(X)/\gamma_{n+1}(X)).$$

and the differentials

$$d^i : E^i_{n,m}(X) \to E^i_{n+i, m-1}(X), \ i \geq 1.$$

This spectral sequence naturally comes into play in homotopy theory. One of the major results about this sequence is the following:

Theorem 5.6 (Cutis [Cur71]). *Let K be a connected and simply connected simplicial set, $G = GK$ its Kan's construction. Then the spectral sequence $E(G)$ converges to $E^\infty(G)$ and $\oplus_r E^\infty_{r,q}$ is the graded group associated with the filtration on $\pi_q(GK) = \pi_{q+1}(|K|)$. The groups $E^1(K)$ are homology invariants of K.*

As in the case of the filtration arising from the lower central series, the filtration of $\mathbb{Z}[X]$ provided by the augmentation powers leads to the long exact sequence

$$\cdots \to \pi_{i+1}(\mathbb{Z}[X]/\Delta^n(X)) \to \pi_i(\Delta^n(X)/\Delta^{n+1}(X)) \to$$
$$\pi_i(\mathbb{Z}[X]/\Delta^{n+1}(X)) \to \pi_i(\mathbb{Z}[X]/\Delta^n(X)) \to \cdots.$$

This exact sequence, in turn, defines a graded exact couple which gives rise to the spectral sequences $\overline{E}(X)$ with the initial terms

$$\overline{E}^1_{n,m}(X) = \pi_m(\Delta^n(X)/\Delta^{n+1}(X))$$

and the differentials

$$\bar{d}^i : \overline{E}^i_{n,m}(X) \to \overline{E}^i_{n+i, m-1}(X).$$

The natural map $k : X \to \Delta(X)$, $x \mapsto 1 - x$, $x \in X_n$, $n \geq 0$, induces a map from the spectral $E(X)$ to the spectral sequence $\overline{E}(X)$:

$$\kappa_* : E(X) \to \overline{E}(X).$$

A convergence result analogous to Theorem 5.6, which is more convenient for applications, holds over the prime field $\mathbb{F}_p$ of p elements. Consider the p-analog of the above spectral sequences:

$$E^1_{n,m}(X, \mathbb{F}_p) = \pi_m(\gamma_{n,p}(X)/\gamma_{n+1,p}(X)),$$
$$\overline{E}^1_{n,m}(X, \mathbb{F}_p) = \pi_m(\Delta^n_p(X)/\Delta^{n+1}_p(X)),$$

where $\Delta_p(G)$ is the augmentation ideal in group algebra $\mathbb{F}_p[G]$. The following theorem is due to Rector and Curtis ([Rec66], [Cur71]):

Theorem 5.7 *Let K be a connected and simply connected simplicial set, $G = GK$. Then $E^i(G, \mathbb{F}_p)$ converges to $E^\infty(G, \mathbb{F}_p)$, which is the graded filtration on $\pi_*(K)$ modulo the subgroup of elements of finite order prime to p.*

$\overline{E}^i(G, \mathbb{F}_p)$ converges to $\overline{E}^\infty(G, \mathbb{F}_p)$, which is the graded group associated with a filtration on $H_*(GK, \mathbb{F}_p)$.

The natural map $\kappa_p : G \to \mathbb{F}_p[G]$ induces maps of spectral sequences $\kappa_p^i : E^i(G, \mathbb{F}_p) \to \overline{E}^i(G, \mathbb{F}_p)$, such that

$$\kappa_p^\infty : E^\infty(G, \mathbb{F}_p) \to \overline{E}^\infty(G, \mathbb{F}_p)$$

is induced by the Hurewicz homomorphism $\pi_*(GK) \to H_*(GK)$.

Proposition 5.2 implies that for the coequalizer

$$L_1 \overset{f_1,\, f_2}{\rightrightarrows} L_2 \longrightarrow \operatorname{coeq}(f_1,\, f_2)$$

of two maps $f_1,\, f_2 : L_1 \to L_2$ between Lie algebras, there exists the following diagram of universal enveloping algebras:

$$\mathcal{U}_R(L_1) \overset{U_R(f_1),\, U_R(f_2)}{\rightrightarrows} \mathcal{U}_R(L_2) \longrightarrow \mathcal{U}_R(\operatorname{coeq}(f_1,\, f_2)). \tag{5.4}$$

Theorem 5.8 (Gruenenfelder, [Gru80]). *Let X be a free simplicial group. Then*

(i) $\quad E_{n,\,m}^1(X,\, R) = \pi_m(\mathcal{L}_{R,\,n}(X_{ab} \otimes R))$,

$\qquad \overline{E}_{n,\,m}^1(X,\, R) = \pi_m(\mathcal{T}_{R,\,n}(X_{ab} \otimes R))$,

(ii) $\quad \kappa_{n,\,0}^1 : E_{n,\,0}^1(X,\, R) \to \overline{E}_{n,\,0}^1(X,\, R)$ *is injective.*

(iii) $\quad E_{n,\,0}^\infty(X,\, R) = E_{n,\,0}^n(X,\, R) = \gamma_n(\pi_0(X))/\gamma_{n+1}(\pi_0(X)) \otimes R$,

$\qquad \overline{E}_{n,\,0}^\infty(X,\, R) = \overline{E}_{n,\,0}^n(X,\, R) = \Delta_R^n(\pi_0(X))/\Delta_R^{n+1}(\pi_0(X))$.

[Recall the assumption on R ar the beginning of §5.2

Proof. (Sketch) Since X is free, we have

$$\gamma_n(X)/\gamma_{n+1}(X) \otimes R = \mathcal{L}_R^n(X_{ab} \otimes R)$$
$$\Delta^n(X)/\Delta^{n+1}(X) \otimes R = \mathcal{T}_R^n(X_{ab} \otimes R).$$

and (i) follows immediately. Proposition 5.2 implies that

$$\pi_0(\mathcal{L}_R^n(X_{ab} \otimes R)) = \mathcal{L}_R^n(\pi_0(X_{ab}) \otimes R),$$
$$\pi_0(\mathcal{T}_R^n(X_{ab} \otimes R)) = \mathcal{T}_R^n(\pi_0(X_{ab}) \otimes R).$$

Since universal enveloping functor $\mathcal{U}_R$ preserve coequalizers (5.4), we have the map

$$\mathcal{L}_R(\pi_0(X_{ab}) \otimes R) \to \mathcal{T}_R(\pi_0(X_{ab}) \otimes R) = \mathcal{U}_R(\mathcal{L}_R(\pi_0(X_{ab}) \otimes R)),$$

which is injective by Poincaré–Birhoff–Witt Theorem and where the gradation gives the maps $\kappa^1_{n,0}$, $n \geq 1$. Hence, $\kappa^1_{n,0}$ is injective for all $n \geq 1$ and (ii) follows.

The statement (iii) follows from standard spectral sequence arguments and the fact that

$$\gamma_n(\pi_0(X))/\gamma_{n+1}(\pi_0(X))) \otimes R = \ker\{\pi_0(X/\gamma_{n+1}(X) \otimes R) \to \pi_0(X/\gamma_n(X) \otimes R)\},$$

$$\Delta^k(\pi_0(X))/\Delta^{k+1}(\pi_0(X)) \otimes R = $$
$$\ker\{\pi_0(\Delta(X)/\Delta^{n+1}(X) \otimes R) \to \pi_0(\Delta(X)/\Delta^n(X) \otimes R)\}. \quad \square$$

For a free simplicial group X, the first differentials d^1 and $\bar{d}^1$ can be described as follows. For $n \geq 2$, consider the free abelian simplicial group $\gamma_n(X)/\gamma_{n+1}(X)$ with face maps

$$\partial^n_i : \gamma_n(X_j)/\gamma_{n+1}(X_j) \to \gamma_n(X_{j-1})/\gamma_{n+1}(X_{j-1}), \ 0 \leq i \leq j,$$

induced by the face maps $\partial_i : X_j \to X_{j-1}$, $0 \leq i \leq j$. Then $E^1_{n,m}(X, \mathbb{Z})$ is the mth homology of the chain complex

$$\ldots \gamma_n(X_{m+1})/\gamma_{n+1}(X_{m+1}) \xrightarrow{\bar{\partial}^n_{m+1}} \gamma_n(X_m)/\gamma_{n+1}(X_m)$$
$$\xrightarrow{\bar{\partial}^n_m} \gamma_n(X_{m-1})/\gamma_{n+1}(X_{m-1}) \to \ldots,$$

where $\bar{\partial}^n_m = \sum^m_{i=0}(-1)^i \partial^n_i$. Let $a \in \ker(\bar{\partial}^n_m)$. Write a as a coset $a = b\gamma_{n+1}(X_m)$, $b \in \gamma_n(X_m)$. Define

$$c := \left(\sum^m_{i=0}(-1)^i \partial^n_i\right)(b) \in \gamma_{n+1}(X_{m-1}).$$

Looking at the simplicial group $\gamma_n(X)/\gamma_{n+2}(X)$, we conclude that

$$\left(\sum^{m-1}_{i=0}(-1)^i \partial^{n+1}_i\right)(c) \in \gamma_{n+2}(X_{m-1}).$$

Therefore the map

$$a.\mathrm{im}(\bar{\partial}^n_{m+1}) \mapsto \left(\sum^m_{i=0}(-1)^i \partial^n_i\right)(b)\gamma_{n+2}(X_{m-1}), \ a \in \ker(\bar{\partial}^n_m)$$

defines the map

$$\pi_m(\gamma_n(X)/\gamma_{n+1}(X)) \to \pi_{m-1}(\gamma_{n+1}(X)/\gamma_{n+2}(X)),$$

which is exactly the first differential d^1. The differential $\bar{d}^1$ can be defined in the same way.

Let G be a group, and $F \to G$ a free simplicial resolution:

$$\pi_0(F) = G, \ \pi_i(F) = 0, \ i > 0.$$

Denote by $E(G, R)$ the spectral sequence $E(F, R)$. Since F_* is dimension-wise free, all the preceding results from this section hold. In particular, the following theorem follows from Theorem 5.8 (ii).

Theorem 5.9 *The map* $\kappa_{n,0}^1 : E_{n,0}^1(G, \mathbb{Z}) \to \overline{E}_{n,0}^1(G, \mathbb{Z})$ *is a monomorphism.*

Let $1 \to N \to F \to G \to 1$ be a free presentation of the group G. Define $\mathfrak{n}(1) = \mathbb{Z}[F](N - 1)$, and

$$\mathfrak{n}(k) = \sum_{i+j=k} \mathfrak{f}^i \mathfrak{n}(1) \mathfrak{f}^j, \ k \geq 1, \ i, \ j \geq 0.$$

It is easy to show that we have

Proposition 5.10

$$E_{n,0}^1(G, \mathbb{Z}) = \frac{\gamma_n(F)}{[N, \ _{n-1}F]\gamma_{n+1}(F)},$$

$$\overline{E}_{n,0}^1(G, \mathbb{Z}) = \frac{\mathfrak{f}^n}{\mathfrak{n}(n-1) + \mathfrak{f}^{n+1}}.$$

Proof. To see this, it is enough to look at the long exact sequences

$$\pi_1(F_*/\gamma_n(F_*))(G) \to \pi_0(\gamma_n(F_*)/\gamma_{n+1}(F_*))(G) \to$$
$$\pi_0(F_*/\gamma_{n+1}(F_*))(G) \to \pi_0(F_*/\gamma_n(F_*))(G) \to 0, \quad (5.5)$$

$$\pi_1(\mathfrak{f}_*/\mathfrak{f}_*^{n+1})(G) \to \pi_1(\mathfrak{f}_*/\mathfrak{f}_*^n)(G) \to \pi_0(\mathfrak{f}_*^n/\mathfrak{f}_*^{n+1})(G) \to$$
$$\pi_0(\mathfrak{f}_*/\mathfrak{f}_*^{n+1})(G) \to \pi_0(\mathfrak{f}_*/\mathfrak{f}_*^n)(G) \to 0 \quad (5.6)$$

and observe that the first derived functor of the lower central quotient functors is the corresponding Baer invariant. □

Corollary 5.11 (Sjögren [Sjo79]). *For any free group* F, *its normal subgroup* N *and* $n \geq 1$, *we have*

$$F \cap (1 + \mathfrak{n}(n-1) + \mathfrak{f}^n) = [N, \ _{n-1}F]\gamma_n(F).$$

Remark. Since $\mathfrak{f}^n/\mathfrak{f}^{n+1} = \mathcal{J}_n(F_{ab})$ for any free group F, we can apply the Künneth formulae to get the precise expressions of the derived functors of the augmentation quotients. First of all, we have

$$\pi_0(\mathfrak{f}_*^n/\mathfrak{f}_*^{n+1})(G) = \frac{\mathfrak{f}^n}{\mathfrak{n}(n-1) + \mathfrak{f}^{n+1}} = (G_{ab})^{\otimes n}, \; n \geq 1.$$

In general, the initial terms of $\overline{E}(G)$ can be described by standard simplicial arguments using the Künneth formula and the Eilenberg-Zilber equivalence. In particular, there exists the following exact sequence

$$0 \to (H_1(G) \otimes H_2(G))^{\oplus 2} \to \overline{E}_{2,1}^1(G) \to \mathrm{Tor}(H_1(G), H_1(G)) \to 0 \qquad (5.7)$$

and, in general,

$$0 \to T_n(G) \to \overline{E}_{n,1}^1(G) \to \mathrm{Tor}_1(\underbrace{H_1(G), \ldots, H_1(G)}_{n \text{ terms}}) \to 0. \qquad (5.8)$$

where $T_1(G) = H_2(G)$ and $T_{k+1}(G) = H_1(G) \otimes T_k(G) \oplus T_k(G) \otimes H_1(G)$, $k \geq 1$ and for abelian groups $B_1, \ldots, B_n$, the group $\mathrm{Tor}_i(B_1, \ldots, B_n)$ denotes the ith homology group of the complex $P_1 \otimes \cdots \otimes P_n$, where P_j is a $\mathbb{Z}$-flat resolution of B_j for $j = 1, \ldots, n$. We clearly have

$$\mathrm{Tor}_0(B_1, \ldots, B_n) = B_0 \otimes \cdots \otimes B_n, \; \mathrm{Tor}_i(B_1, \ldots, B_n) = 0, \; i \geq n.$$

For a description of the initial terms of the spectral sequence $E(G)$, one needs more complicated theory of derived functors of polynomial functors. In the quadratic case such a theory was developed by Baues and Pirashvili [Bau00]; their theory implies that the terms $E_{2,m}^1$ ($m \geq 0$) can be described explicitly. Let X be a simplicial group which is free abelian in each degree. Then there exists [[Bau00], (4.1)] a natural short exact sequence of graded abelian groups

$$0 \to \mathrm{Sq}^\otimes(\pi_*(X)) \to \pi_*(\wedge^2 X) \to \mathrm{Sq}^*(\pi_*(X))[-1] \to 0 \qquad (5.9)$$

where $\pi_*(X)$ and $\pi_*(\wedge^2 X)$ are the graded homotopy groups of X and $\wedge^2 X$ respectively (see Section A.14 for definitions). The sequence (A.31) gives the following functorial description of the term $E_{2,1}^1$: There exists a natural short exact sequence:

$$0 \to H_1(G) \otimes H_2(G) \to E_{2,1}^1(G) \to \Omega(H_1(G)) \to 0. \qquad (5.10)$$

[For the definition of the functor Ω, see Section A.14.]

Proposition 5.12 *There exist the following natural isomorphisms:*

$$E_{n,0}^2(G, \mathbb{Z}) \simeq \frac{\gamma_n(F)}{([N, {}_{n-2}F] \cap \gamma_n(F))\gamma_{n+1}(F)},$$

$$\overline{E}_{n,0}^2(G, \mathbb{Z}) \simeq \frac{\mathfrak{f}^n}{\mathfrak{n}(n-2) \cap \mathfrak{f}^n + \mathfrak{f}^{n+1}}.$$

Proof. Let X be a free simplicial resolution of G with X_0 a free group F on generators $\{x_i\}_{i \in I}$, X_1 a free group $F * F(y_j \mid j \in J)$ and $\partial_0 : x_i \mapsto x_i$, $y_j \mapsto 1$, $\partial_1 : x_i \mapsto x_i$, $y_j \mapsto r_j$, where $r_j \in F$ is a set of generators of N, as a normal subgroup of F, and $F/N = G$. We use the above description of differentials d^1 and $\bar{d}^1$.

Let $b \in \gamma_n(F_1)$ be such that $\partial_0(b)\partial_1(b)^{-1} \in \gamma_{n+1}(F_0)$. Write b as a product of the form

$$b = b_1 b_2 b_3, \quad b_1 \in F, \quad b_2 \in \ker(\partial_0), \quad b_3 \in \gamma_{n+1}(F_1).$$

Then $\partial_0(b_1)\partial_1(b_1)^{-1} = 1$. Clearly,

$$\partial_0 \partial_1^{-1}(\gamma_{n+1}(F_1)) \subseteq [N, {}_nF]\gamma_{n+2}(F).$$

Therefore,

$$\partial_0(b)\partial_1(b)^{-1} \equiv \partial_1(b_2) \quad \mathrm{mod}\ [N, {}_nF]\gamma_{n+2}(F),$$

$$\partial_1(b_2) \in [N, {}_{n-1}F] \cap \gamma_{n+1}(F).$$

It is also clear that any element $x \in [N, {}_{n-1}F] \cap \gamma_{n+1}(F)$ can be obtained in this a way, i.e. there exists an element $b(x) \in F_1$ with $\partial_0(b(x))\partial_1(b(x))^{-1} \equiv x$ mod $[N, {}_nF]\gamma_{n+2}(G)$. The above description of the differential d^1 shows that

$$\mathrm{im}(d_{n,1}^1) = \frac{([N, {}_{n-1}F] \cap \gamma_{n+1}(F))\gamma_{n+2}(F)}{([N, {}_nF] \cap \gamma_{n+1}(F))\gamma_{n+2}(F)}$$

and the asserted description of $E_{n,0}^2(G, \mathbb{Z})$ follows.

The description of $\overline{E}_{n,0}^2(G, \mathbb{Z})$ can be obtained by a similar argument applied to the augmentation quotients. $\square$

Remark. A similar analysis provides the structure of the terms $E_{n,0}^k(G, \mathbb{Z})$ and $\overline{E}_{n,0}^k(G, \mathbb{Z})$, $k \geq 2$:

$$E_{n,0}^k(G, \mathbb{Z}) = \frac{\gamma_n(F)}{([N, {}_{n-k}F] \cap \gamma_n(F))\gamma_{n+1}(F)}, \tag{5.11}$$

$$\overline{E}_{n,0}^k(G, \mathbb{Z}) = \frac{\mathfrak{f}^n}{\mathfrak{n}(n-k) \cap \mathfrak{f}^n + \mathfrak{f}^{n+1}}. \tag{5.12}$$

In particular, the stable terms are the lower central and the augmentation quotients.

We now discuss some of the applications of the preceding simplicial techniques.

Theorem 5.13 (Grünenfelder [Gru80]). *Let G be a group with $H_1(G, R)$ torsion-free and $H_2(G, R)$ torsion. Then we have the following isomorphisms of associative and Lie R-algebras respectively:*

$$\mathcal{T}_R(G_{ab} \otimes R) \simeq \oplus_{n \geq 0} \mathfrak{g}^n / \mathfrak{g}^{n+1} \otimes R, \tag{5.13}$$

$$\mathcal{L}_R(G_{ab} \otimes R) \simeq \oplus_{n \geq 0} \gamma_n(G) / \gamma_{n+1}(G) \otimes R. \tag{5.14}$$

Proof. Let X be a free simplicial resolution of G. Then

$$\overline{E}^1_{n,m}(G, R) = \pi_m(\mathcal{T}^n_R(R \otimes X_{ab})).$$

Künneth formula then gives unnatural decompositions

$$\overline{E}^1_{n,m}(G, R) = \bigoplus_{i+j=m} \overline{E}^1_{n-1,i}(G, R) \otimes_R \overline{E}^1_{1,i}(G, R) \oplus$$

$$\bigoplus_{i+j=m-1} \operatorname{Tor}^R(\overline{E}^1_{n-1,i}(G, R), \overline{E}^1_{i,j}(G, R)). \tag{5.15}$$

In particular,

$$\overline{E}^1_{n,1}(G, R) = \overline{E}^1_{n-1,1}(G, R) \otimes_R E^1_{1,0}(G, R) \oplus$$

$$\overline{E}^1_{n-1,0} \otimes_R \overline{E}^1_{1,1}(G, R) \oplus \operatorname{Tor}^R(\overline{E}^1_{n-1,0}(G, R), \overline{E}^1_{1,0}(G, R)) =$$

$$\overline{E}^1_{n-1,1}(G, R) \otimes G_{ab} \oplus (G_{ab})^{\otimes n-1} \otimes H_2(G, R) \oplus \operatorname{Tor}^R((G_{ab})^{\otimes n-1} \otimes R, G_{ab} \otimes R)$$

Since G_{ab} is torsion-free, the Tor-part vanishes. The induction on n then shows that all terms $\overline{E}^1_{n,1}(G, R)$ and hence all terms $\overline{E}^r_{n,1}(G, R)$, $r \geq 1$ are R-torsion. By induction on r, we conclude that $\overline{E}^r_{n,0}(G, R)$ are R-torsion-free and the maps

$$\overline{d}^r : \overline{E}^r_{n,1}(G, R) \to \overline{E}^r_{n+r,0}(G, R)$$

are zero maps. It follows that

$$\mathfrak{g}^n / \mathfrak{g}^{n+1} \otimes R = \overline{E}^\infty_{n,0}(G, R) = \overline{E}^1_{n,0}(G, R) = (G_{ab})^{\otimes n} \otimes R = \mathcal{T}^n_R(G_{ab} \otimes R)$$

and the statement (5.13) follows.

Now the commutativity of the diagram

$$
\begin{array}{ccc}
E^r_{n,1}(G, R) & \xrightarrow{\ d^r\ } & E^r_{n+r,0}(G, R) \\
\downarrow{\scriptstyle \kappa^r_{n,1}} & & \downarrow{\scriptstyle \kappa^r_{n+r,0}} \\
\overline{E}^r_{n,1}(G, R) & \xrightarrow{\ \overline{d}^r\ } & \overline{E}^r_{n+r,0}(G, R)
\end{array}
$$

and injectivity of the map $\kappa_{n,0}^1 : E_{n,0}^1(G,\,R) \to \overline{E}_{n,0}^1(G,\,R)$ (Theorem 5.8 (ii)) implies that the maps

$$d^r : E_{n,1}^r(G,\,R) \to E_{n+r,0}^r(G,\,R)$$

are zero maps. Therefore,

$$\gamma_n(G)/\gamma_{n+1}(G) \otimes R = E_{n,0}^\infty(G,\,R) = E_{n,0}^1(G,\,R) = \mathcal{L}_R^n(G_{ab} \otimes R)$$

and the statement (5.14) follows. $\square$

Theorem 5.14 (see also Ellis [Ell02]).
(*i*) *Let G be a group with $H_1(G)$ torsion-free and $H_2(G) = 0$, then the Baer invariant $M^{(k)}(G) = 0$, for all $k \geq 1$.*
(*ii*) *Let G be a group with $H_2(G)$ a torsion group, then $M^{(k)}(G)$ is a torsion group for all $k \geq 1$.*

Proof. Let X be a free simplicial resolution of G. Choose a set X_0 of elements x_i, $i \in I$, such that their images $\{g_i\gamma_2(G)\}_{i \in I}$ form a basis of G_{ab}. Let F be the constant free simplicial group with $F_n = F(y_i \mid i \in I)$. Then $\pi_0(F) = F_0$, $\pi_i(F) = 0$, $i \geq 1$. Consider the homomorphism

$$f : F_0 \to G, \ y_i \mapsto g_i.$$

Then f can be extended to the simplicial map

$$\gamma : F \to X,$$

such that $\pi_0(\gamma) = f$ by Theorem A.11. Consider the abelianization of γ:

$$\gamma_{ab} : F_{ab} \to X_{ab}.$$

It follows that γ_{ab} induces isomorphisms

$$\pi_0(\gamma_{ab}) : \pi_0(F_{ab}) \to \pi_0(X_{ab}), \ \pi_1(\gamma_{ab}) : \pi_1(F_{ab}) \to \pi_1(X_{ab}),$$

since $\pi_1(X_{ab}) = H_2(G) = 0$. Therefore, γ_{ab} induces epimorphism

$$\pi_1(\mathcal{L}^n(\gamma_{ab})) : \pi_1(\mathcal{L}^n(F_{ab})) \to \pi_1(\mathcal{L}^n(X_{ab}))$$

by Theorem A.12 (the case $R = S = \mathbb{Z}$). Since F_{ab} is constant, $\pi_1(\mathcal{L}_n(F_{ab})) = 0$, hence

$$\pi_1(\mathcal{L}^n(X_{ab})) = 0, \ n \geq 1.$$

Now the statement (i) follows by induction on n from the exact sequence:

$$\pi_1(\mathcal{L}^n(X_{ab})) \longrightarrow \pi_1(X/\gamma_{n+1}(X)) \longrightarrow \pi_1(X/\gamma_n(X))$$

$$\|\qquad\qquad\qquad\|\qquad\qquad\qquad\|$$

$$\pi_1(\mathcal{L}^n(X_{ab})) \longrightarrow \qquad M^{(n)}(G) \qquad \longrightarrow \quad M^{(n-1)}(G).$$

The statement (ii) can be proved similalry by tensoring with $\mathbb{Q}$. $\square$

For one more application of Theorem A.12, consider higher universal quadratic functors in the sense of Ellis ([Ell89], [Ell91]). It is shown in [Ell89] that, for any group G, there is the following exact sequence of abelian groups:

$$\cdots \to H_{n+1}(G) \to \overline{\Gamma}_n(G) \to J_n(G) \to H_n(G) \to$$
$$\cdots \to H_3(G) \to \overline{\Gamma}_2(G_{ab}) \to J_2(G) \to H_2(G) \to 0, \quad (5.16)$$

where $\overline{\Gamma}_2(G) = \Gamma_2(G_{ab})$ is Whitehead's universal quadratic functor (see A.14), $J_2(G) = \ker\{G \otimes G \to G\}$, where $G \otimes G$ is the nonabelian tensor square [Bro87], $\overline{\Gamma}_n(G)$ is the $(n-2)$th derived functor of $\Gamma_2(G_{ab})$, $J_n(G)$ is the $(n-2)$th derived functor of the non-abelian tensor square.

Due to Theorem A.12 it is easy to show that the sequence (5.16) is a homological invariant of the group G. More precisely, we have the following

Theorem 5.15 (Ellis [Ell91]). *Let $f : G \to H$ be a group homomorphism, which induces an isomorphisms $H_i(G) \to H_i(H)$, $i = 1, \ldots, k$. Then f induces isomorphism of sequences:*

$$H_s(G) \longrightarrow \overline{\Gamma}_{s-1}(G) \longrightarrow J_{s-1}(G) \longrightarrow H_{s-1}(G)$$

$$\|\qquad\qquad\qquad\|\qquad\qquad\qquad\|\qquad\qquad\qquad\|$$

$$H_s(H) \longrightarrow \overline{\Gamma}_{s-1}(H) \longrightarrow J_{s-1}(H) \longrightarrow H_{s-1}(H)$$

for all $s = 3, \ldots, k$.

Proof. Let

$$F^1 \to G, \ \pi_0(F^1) = G, \ \pi_i(F^1) = 0, \ i > 0,$$
$$F^2 \to H, \ \pi_0(F^2) = H, \ \pi_i(F^2) = 0, \ i > 0$$

be simplicial resolutions of groups G and H, and $\theta : F^1 \to F^2$ a homomorphism of simplicial groups, such that $\pi_0(\theta) = f$, which exists by Theorem A.11. Since the homologies are derived functors of the functor of abelianization, θ induces isomorphisms

$$F_i : \pi_i(F^1_{ab}) \to \pi_i(F^2_{ab}), \ i = 0, \ldots, k-1.$$

By Theorem A.12 (the case $R = S = \mathbb{Z}$), after applying the functor Γ for the abelianization of resolutions F^1 and F^2, we get an isomorphism

$$\overline{\Gamma}_{k-1} : \pi_{k-1}(\Gamma(F_{ab}^1)) \to \pi_{k-1}(\Gamma(F_{ab}^2)).$$

Functors $\overline{\Gamma}_s(G)$ are, by definition, $\pi_{s-2}(\Gamma(F_{ab}^1))$, thus, it follows that f induces isomorphisms $\overline{\Gamma}_s(G) \to \overline{\Gamma}_s(H)$, $s \le k - 1$, and therefore, due to the exactness of (5.16), we have the isomorphism of whole sequence (5.16) up to terms with numbers s. $\square$

5.3 Applications to Dimension Subgroups

We next consider application of spectral sequences discussed in this section to the theory of dimension subgroups. Clearly, these spectral sequences must contain a lot of information about dimension subgroups, since for any group G, one has

$$\frac{\gamma_n(G) \cap D_{n+1}(G)}{\gamma_{n+1}(G)} = \ker\{E_{n,0}^\infty(G, \mathbb{Z}) \to \overline{E}_{n,0}^\infty(G, \mathbb{Z})\}$$

by Theorem 5.8 (iii).

For every $n \ge 3$ and $1 \le k \le n - 1$, one has the following commutative diagram

$$
\begin{array}{ccccccc}
E_{n-k,1}^k(G, \mathbb{Z}) & \xrightarrow{d_{n-k,1}^k} & E_{n,0}^k(G, \mathbb{Z}) & \longrightarrow & E_{n,0}^{k+1}(G, \mathbb{Z}) & \longrightarrow & 0 \\
\kappa_{n-k,1}^k \downarrow & & \kappa_{n,0}^k \downarrow & & \kappa_{n,0}^{k+1} \downarrow & & \\
\overline{E}_{n-k,1}^k(G, \mathbb{Z}) & \xrightarrow{\overline{d}_{n-k,1}^k} & \overline{E}_{n,0}^k(G, \mathbb{Z}) & \longrightarrow & \overline{E}_{n,0}^{k+1}(G, \mathbb{Z}) & \longrightarrow & 0.
\end{array}
$$

which has the form

$$
\begin{array}{ccccccc}
E_{n-1,1}^1(G, \mathbb{Z}) & \xrightarrow{d_{n-1,1}^1} & E_{n,0}^1(G, \mathbb{Z}) & \longrightarrow & E_{n,0}^2(G, \mathbb{Z}) & \longrightarrow & 0 \\
\kappa_{n-1,1}^1 \downarrow & & \kappa_{n,0}^1 \downarrow & & \kappa_{n,0}^2 \downarrow & & \\
\overline{E}_{n-1,1}^1(G, \mathbb{Z}) & \xrightarrow{\overline{d}_{n-1,1}^1} & \overline{E}_{n,0}^1(G, \mathbb{Z}) & \longrightarrow & E_{n,0}^2(G, \mathbb{Z}) & \longrightarrow & 0
\end{array}
\tag{5.17}
$$

for $k = 1$, and

$$
\begin{array}{ccccccc}
E_{1,1}^{n-1}(G, \mathbb{Z}) & \xrightarrow{d_{1,1}^{n-1}} & E_{n,0}^{n-1}(G, \mathbb{Z}) & \longrightarrow & \gamma_n(G)/\gamma_{n+1}(G) & \longrightarrow & 0 \\
\kappa_{1,1}^{n-1} \downarrow & & \kappa_{n,0}^{n-1} \downarrow & & \kappa_{n,0}^n \downarrow & & \\
\overline{E}_{1,1}^{n-1}(G, \mathbb{Z}) & \xrightarrow{\overline{d}_{1,1}^{n-1}} & \overline{E}_{n,0}^{n-1}(G, \mathbb{Z}) & \longrightarrow & \mathfrak{g}^n/\mathfrak{g}^{n+1} & \longrightarrow & 0
\end{array}
\tag{5.18}
$$

for $k = n - 1$. Diagrams (5.17) and (5.18) together with the snake lemma imply the following exact sequences of abelian groups:

$$0 \to \ker(\kappa^2_{n,0}) \to \mathrm{coker}\{\mathrm{im}(d^1_{n-1,1}) \to \mathrm{im}(\bar{d}^1_{n-1,1})\} \to$$

$$\mathrm{coker}(\kappa^1_{n,0}) \to \mathrm{coker}(\kappa^2_{n,0}) \to 0 \quad (5.19)$$

and

$$0 \to \ker\{\mathrm{im}(d^{n-1}_{1,1}) \to \mathrm{im}(\bar{d}^{n-1}_{1,1})\} \to \ker(\kappa^{n-1}_{n,0}) \to$$

$$(\gamma_n(G) \cap D_{n+1}(G))/\gamma_{n+1}(G) \to \mathrm{coker}\{\mathrm{im}(d^{n-1}_{1,1}) \to \mathrm{im}(\bar{d}^{n-1}_{1,1})\} \to$$

$$\mathrm{coker}(\kappa^{n-1}_{n,0}) \to \mathrm{coker}(\kappa^n_{n,0}) \to 0 \quad (5.20)$$

As a result we obtain the following diagram which we will use later:

$$\ker(\kappa^2_{n,0})\,\ker(\kappa^3_{n,0}) \longrightarrow \cdots \longrightarrow \ker(\kappa^{n-1}_{n,0}) \qquad (5.21)$$

$$\mathrm{coker}(\mathrm{im}(d^1_{n-1,1}) \to \mathrm{im}(\bar{d}^1_{n-1,1})) \qquad\qquad \frac{\gamma_n(G)\cap D_{n+1}(G)}{\gamma_{n+1}(G)}$$

For $n = 2$, the diagram (5.18) has the following form:

$$
\begin{array}{ccccccc}
H_2(G) & \xrightarrow{d^1} & E^1_{2,0} & \longrightarrow & \gamma_2(G)/\gamma_3(G) & \longrightarrow & 0 \\
\| & & \kappa^1_{2,0}\downarrow & & \kappa^2_{2,0}\downarrow & & \\
H_2(G) & \xrightarrow{\bar{d}^1} & \bar{E}^1_{2,0} & \longrightarrow & \mathfrak{g}^2/\mathfrak{g}^3 & \longrightarrow & 0.
\end{array}
$$

The map $\kappa^1_{2,0}$ is injective by Proposition 5.9, hence the map $\kappa^2_{2,0}$ is injective. Therefore,

$$\gamma_2(G) \cap D_3(G) = \gamma_3(G).$$

Since $D_2(G) = \gamma_2(G)$, we have the well-known equality

$$D_3(G) = \gamma_3(G) \qquad (5.22)$$

for every group G.

Lemma 5.16 *Let $n \geq 4$, $w - 1 \in \mathfrak{n}(2) + \mathfrak{f}^n$, then there exists $r \in [N, F]$ such that $w^2 r - 1 \in \mathfrak{n}(3) + \mathfrak{f}^n$.*

Proof. Let $\theta : \mathbb{Z}[F] \to \mathbb{Z}[F]$ be the linear extension of the bijective map $g \to g^{-1}$ mapping F onto F. Then

$$\theta(xy - x - y + 1) = y^{-1}x^{-1} - x^{-1} - y^{-1} + 1$$

implies

$$\theta((x-1)(y-1)) = (y^{-1} - 1)(x^{-1} - 1)$$

One can easily verify that for all $t \geq 2$,

$$\theta((f_1 - 1)(f_2 - 1) \ldots (f_t - 1)) = (f_t^{-1} - 1) \ldots (f_2^{-1} - 1)(f_1^{-1} - 1), \ f_i \in F.$$

Therefore, the ideals $\mathfrak{n}(3)$ and $\mathfrak{f}^n$ are invariant under θ. We may thus assume that w satisfies the following congruence :

$$w - 1 \equiv v \quad \mathrm{mod} \ \mathfrak{n}(3) + \mathfrak{f}^n, \tag{5.23}$$

where

$$v = \sum_{i<j} a_{ij}(r_i - 1)(z_j - 1) + b_{ij}(z_i - 1)(r_j - 1), \ r_t \in N, \ z_t \in F, \ a_{ij}, \ b_{ij} \in \mathbb{Z}$$

Since

$$\theta(v) = \sum_{ij} a_{ij}(z_j^{-1} - 1)(r_i^{-1} - 1) + b_{ij}(r_j^{-1} - 1)(z_i^{-1} - 1)$$

$$\equiv \sum_{i<j} a_{ij}(z_j - 1)(r_i - 1) + b_{ij}(r_j - 1)(z_i - 1) \quad \mathrm{mod} \ \mathfrak{n}(3) + \mathfrak{f}^n,$$

we have the congruence

$$v - \theta(v) \equiv \prod_{i<j} [r_i, z_j]^{a_{ij}} [z_i, r_j]^{b_{ij}} - 1 \quad \mathrm{mod} \ \mathfrak{n}(3) + \mathfrak{f}^n. \tag{5.24}$$

Now applying θ to both sides of (12) and subtracting (12) yields

$$(w - 1) - (w^{-1} - 1) \equiv v - \theta(v) \quad \mathrm{mod} \ \mathfrak{n}(3) + \mathfrak{f}^n,$$

which, in turn, yields upon using (13) the desired congruence

$$w^2 - 1 \equiv r - 1 \quad \mathrm{mod} \ \mathfrak{n}(3) + \mathfrak{f}^n,$$

where

$$r = \prod_{i<j} [r_i, z_j]^{a_{ij}} [z_j, r_j]^{b_{ij}} \in [N, F].$$

The assertion then follows. $\square$

We next give an alternate proof of another well known result.

Theorem 5.17 *For every group G, $D_4(G)/\gamma_4(G)$ has exponent 1 or 2.*

Proof. We have the following commutative diagram (the diagram 5.18 for $n = 3$):

$$
\begin{array}{ccccccc}
E^2_{1,1}(G,\mathbb{Z}) & \xrightarrow{d^2_{1,1}} & E^2_{3,0}(G,\mathbb{Z}) & \longrightarrow & \gamma_3(G)/\gamma_4(G) & \longrightarrow & 0 \\
\| & & \kappa^2_{3,0}\downarrow & & \kappa^3_{3,0}\downarrow & & \\
\overline{E}^2_{1,1}(G,\mathbb{Z}) & \xrightarrow{\overline{d}^2_{1,1}} & \overline{E}^2_{3,0}(G,\mathbb{Z}) & \longrightarrow & \mathfrak{g}^3/\mathfrak{g}^4 & \longrightarrow & 0.
\end{array}
\tag{5.25}
$$

with exact rows. The fact that the map $\kappa^1_{1,1} : E^2_{1,1}(G,\mathbb{Z}) \to \overline{E}^2_{1,1}(G,\mathbb{Z})$ is an isomorphism follows from the fact that the map $\kappa^1_{2,0}$ is a monomorphism. Let $x \in \ker(\kappa^3_{3,0})$ and $y \in E^2_{3,0}$ be any its pre-image, i.e. $y.\mathrm{im}(d^2_{1,1}) = x$. Clearly,

$$
y \in \ker(\kappa^2_{3,0}).
\tag{5.26}
$$

It is enough to show that $2y = 0$. By Proposition 5.12, y can be presented as a coset

$$
y = z([N, F] \cap \gamma_3(F))\gamma_4(F), \ z \in N \cap \gamma_3(F).
$$

Proposition (5.12) and condition (5.26) imply that

$$
z - 1 \in \mathfrak{n}(2) \cap \mathfrak{f}^3 + \mathfrak{f}^4.
$$

Lemma 5.16 (the case $n = 4$) implies that there exists $r \in [F, R]$, such that $z^2 r - 1 \in \mathfrak{n}(3) + \mathfrak{f}^4$. Therefore,

$$
z^2 r \in [N, F, F]\gamma_4(F),
$$

by Corollary 5.11. Hence $z^2 \in [N, F]\gamma_4(F)$ and therefore $2y = 0$. In view of (5.22) the result follows. $\square$

The general Sjögren's Theorem follows from the description of terms $E^k_{n,0}$, $\overline{E}^k_{n,0}$ (see 5.11 and 5.12) and the following

Lemma 5.18 *Let $w \in \gamma_n(F)$, $n \geq 2$, be such that $w - 1 \in \mathfrak{f}^{n+1} + \mathfrak{n}(k)$ for some k, $1 \leq k \leq n$. Then $w^{b(k)} - 1 \equiv f_k - 1 \mod \mathfrak{f}^{n+1} + \mathfrak{n}(k+1)$ for some $f_i \in N(k)$.*

The diagram (5.25) can be analyzed more precisely. We need an identification theorem, which we present next.

Theorem 5.19 (Hartl-Mikhailov-Passi [Har08]). *If F is a free group and R a normal subgroup of F, then*

$$F \cap (1 + \mathfrak{f}(R \cap F' - 1) + (R \cap F' - 1)\mathfrak{f} + \mathbf{r}(2) + \mathfrak{f}^4) = [R \cap F', F]\gamma_4(F),$$

where F' is the derived subgroup of F.

We need the following

Lemma 5.20 *Let F be a free group with a basis $\{x_1, \dots, x_m\}$, u an element of*
$(F' - 1)\mathfrak{f}$, *such that*
$$w - 1 \equiv u + c.v \quad \mathrm{mod}\ \mathfrak{f}^4,$$
for some $c > 0$ and $w \in \gamma_3(F)$, $v \in \mathfrak{f}^3$. Then
$$u \equiv c.v_1 \quad \mathrm{mod}\ \mathfrak{f}^4,$$
where $v_1 \in (F' - 1)\mathfrak{f}$.

Proof. We can view $\mathfrak{f}^3/\mathfrak{f}^4$ as a free abelian group isomorphic to $F_{ab}^{\otimes 3}$ with a basis $\{x_i \otimes x_j \otimes x_k \mid i, j, k = 1, \dots, m\}$. Modulo $\mathfrak{f}^4$, the group $(F' - 1)\mathfrak{f}$ is generated by elements

$$([x_i, x_j] - 1)(x_k - 1), \ i, j, k = 1, \dots, m.$$

Let
$$u \equiv \sum_{i, j, k} d_{i, j, k}([x_i, x_j] - 1)(x_k - 1) \quad \mathrm{mod}\ \mathfrak{f}^4.$$

For a given triple (i, j, k), the sum of all coefficients in u, which contribute to $x_i \otimes x_j \otimes x_k$ (and S_3-permutations) must be divided by c. We have in $\mathfrak{f}^3/\mathfrak{f}^4$:

$$([x_i, x_j] - 1)(x_k - 1) \mapsto x_i \otimes x_j \otimes x_k - x_j \otimes x_i \otimes x_k,$$

$$([x_k, x_i] - 1)(x_j - 1) \mapsto x_k \otimes x_i \otimes x_j - x_i \otimes x_k \otimes x_j,$$

$$([x_j, x_k] - 1)(x_i - 1) \mapsto x_j \otimes x_k \otimes x_i - x_k \otimes x_j \otimes x_i$$

and no more terms can contribute to $x_i \otimes x_j \otimes x_k$ (and permutations). Clearly only the product of commutators of the form $[x_i, x_j, x_k]$ and $[x_k, x_i, x_j]$ with different powers can contribute from the element w. We have:

$$1 - [x_i, x_j, x_k]^{f_{i,j,k}} \mapsto$$
$$f_{i,j,k}(x_i \otimes x_j \otimes x_k - x_j \otimes x_i \otimes x_k + x_k \otimes x_j \otimes x_i - x_k \otimes x_i \otimes x_j)$$

$$1 - [x_k, x_i, x_j]^{f_{k,i,j}} \mapsto$$
$$f_{k,i,j}(x_k \otimes x_i \otimes x_j - x_i \otimes x_k \otimes x_j + x_j \otimes x_i \otimes x_k - x_j \otimes x_k \otimes x_i)$$

Now we have

$$d_{i,j,k}(x_i \otimes x_j \otimes x_k - x_j \otimes x_i \otimes x_k) + d_{k,i,j}(x_k \otimes x_i \otimes x_j - x_i \otimes x_k \otimes x_j) +$$

$$d_{j,k,i}(x_j \otimes x_k \otimes x_i - x_k \otimes x_j \otimes x_i) +$$

$$f_{i,j,k}(x_i \otimes x_j \otimes x_k - x_j \otimes x_i \otimes x_k + x_k \otimes x_j \otimes x_i - x_k \otimes x_i \otimes x_j) +$$

$$f_{k,i,j}(x_k \otimes x_i \otimes x_j - x_i \otimes x_k \otimes x_j + x_j \otimes x_i \otimes x_k - x_j \otimes x_k \otimes x_i)$$

$$= x_i \otimes x_j \otimes x_k(d_{i,j,k} + f_{i,j,k}) + x_j \otimes x_i \otimes x_k(-d_{i,j,k} - f_{i,j,k} + f_{k,i,j}) +$$

$$x_k \otimes x_i \otimes x_j(d_{k,i,j} - f_{i,j,k} + f_{k,i,j}) + x_i \otimes x_k \otimes x_j(-d_{k,i,j} - f_{k,i,j}) +$$

$$x_j \otimes x_k \otimes x_i(d_{j,k,i} - f_{k,i,j}) + x_k \otimes x_j \otimes x_i(-d_{j,k,i} + f_{i,j,k}) =$$

$$c.v(x_i, x_j, x_k), \quad (5.27)$$

for some element $v(x_i, x_j, x_k)$ from $\mathfrak{f}^3/\mathfrak{f}^4$. Suppose first that all i, j, k are different. Then (5.27) implies that

$$d_{i,j,k} + f_{i,j,k}, \; -d_{i,j,k} - f_{i,j,k} + f_{k,i,j}, \; d_{k,i,j} - f_{i,j,k} + f_{k,i,j},$$

$$-d_{k,i,j} - f_{k,i,j}, \; d_{j,k,i} - f_{k,i,j}, \; -d_{j,k,i} + f_{i,j,k}$$

must be divided by c. Therefore, all $d_{i,j,k}, d_{k,i,j}, d_{j,k,i}, f_{i,j,k}, f_{k,i,j}$ must be divisible by c.

Suppose $i = j$. Then we reduce all the expression to the element

$$d_{i,k,i}([x_i, x_k] - 1)(x_i - 1)$$

and a bracket $1 - [x_i, x_k, x_i]^{f_{i,k,i}}$. We then have

$$- f_{i,k,i}x_i \otimes x_i \otimes x_k + (2f_{i,k,i} + d_{i,k,i})x_i \otimes x_k \otimes x_i +$$

$$(-f_{i,k,i} - d_{i,k,i})x_k \otimes x_i \otimes x_i = c.v(x_i, x_k)$$

and we again conclude that $d_{i,k,i}$ and $f_{i,k,i}$ must be divisible by c. $\square$

Next, let $w \in \gamma_3(F)$ and

$$w - 1 \equiv u \mod \mathfrak{f}^4,$$

where

$$u \in (R \cap F' - 1)\mathfrak{f} + \mathfrak{f}(R \cap F' - 1) + \mathfrak{r}(2).$$

We claim that

$$w = w_1 w_2 \quad \text{mod } [R \cap F', F][R, F, F]\gamma_4(F),$$

such that

$$w_1 - 1 \in (R \cap F' - 1)\mathfrak{f} + \mathfrak{f}(R \cap F' - 1) + \mathfrak{f}^4,$$
$$w_2 - 1 \in \mathbf{r}(2) + \mathfrak{f}^4,$$

that is, we can divide the problem of identification of the subgroup

$$F \cap (1 + (R \cap F' - 1)\mathfrak{f} + \mathfrak{f}(R \cap F' - 1) + \mathbf{r}(2) + \mathfrak{f}^4)$$

into two parts:
(i) identification of

$$F \cap (1 + (R \cap F' - 1)\mathfrak{f} + \mathfrak{f}(R \cap F' - 1) + \mathfrak{f}^4)$$

and
(ii) identification of

$$F \cap (1 + \mathbf{r}(2) + \mathfrak{f}^4)$$

Let $u = u_1 + u_2$, where

$$u_1 \in (R \cap F' - 1)\mathfrak{f} + \mathfrak{f}(R \cap F' - 1), \ u_2 \in \mathbf{r}(2).$$

Since we can work modulo $[R \cap F', F]$, we can assume that

$$u_1 \in (R \cap F' - 1)\mathfrak{f}.$$

Now, using the argument of Gupta from Lemma 1.5(B) [Gup87c], we can easily conclude that

$$u_2 \equiv e_m.v_2 \quad \text{mod } \mathfrak{f}^4,$$

where v_2 has an entrance of elements x_m. Hence, by Lemma 5.20, we conclude that $u_1 = e_m.v_3$, where v_3 contains some nontrivial entries of the element x_m. Therefore, we have

$$u = e_m.v \quad \text{mod } \mathfrak{f}^4,$$

where v is an element from $(R \cap F' - 1)\mathfrak{f} + \mathfrak{f}(R \cap F' - 1) + \mathbf{r}(2)$. Since u is a Lie element, v is again a Lie element and we conclude that

$$w \equiv w'^{e_m} \quad \text{mod } [R \cap F', F][R, F, F]\gamma_4(F).$$

Now we can delete all the brackets from w' with entries of x_m and make an induction. The induction argument shows the following: *let $w \in \gamma_3(F)$, such that*

$$w - 1 \in (R \cap F' - 1)\mathfrak{f} + \mathfrak{f}(R \cap F' - 1) + \mathbf{r}(2) + \mathfrak{f}^4$$

then

$$w = w_1 w_2 \quad \mod [R \cap F', F][R, F, F]\gamma_4(F),$$

such that

$$w_1 - 1 \in (R \cap F' - 1)\mathfrak{f} + \mathfrak{f}(R \cap F' - 1) + \mathfrak{f}^4,$$

$$w_2 - 1 \in \mathbf{r}(2) + \mathfrak{f}^4$$

and Theorem 5.19 follows.

Now we return to the fourth dimension subgroup.

Since $D_4(G) \subseteq \gamma_3(G)$, the dimension quotient $D_4(G)/\gamma_4(G)$ is exactly the kernel of the map

$$\kappa_{3,0}^3 : E_{3,0}^3(G) \to \overline{E}_{3,0}^3(G). \tag{5.28}$$

For $n = 3$ the sequence (5.20) reduces to the following

$$0 \to \ker(\mathrm{im}(d_{1,1}^2) \to \mathrm{im}(\overline{d}_{1,1}^2)) \to \ker(\kappa_{3,0}^2) \to D_4(G)/\gamma_4(G) \to 1$$

for every group G. The sequence (5.19) has the following form:

$$0 \to \ker(\kappa_{3,0}^2) \to \mathrm{coker}(\eta) \to \mathrm{coker}(\kappa_{3,0}^1) \to \mathrm{coker}(\kappa_{3,0}^2) \to 0,$$

where $\eta : \mathrm{im}(d_{1,1}^1) \to \mathrm{im}(\overline{d}_{1,1}^1)$. From the exact sequences (5.7) and (5.10), we have a commutative diagram

$$
\begin{array}{ccccc}
H_1(G) \otimes H_2(G) & \rightarrowtail & E_{2,1}^1(G) & \twoheadrightarrow & \Omega(H_1(G)) \\
\downarrow & & \kappa_{2,1}^1 \downarrow & & T \downarrow \\
[6pt](H_1(G) \otimes H_2(G))^{\oplus 2} & \rightarrowtail & \overline{E}_{2,1}^1(G) & \twoheadrightarrow & \mathrm{Tor}(H_1(G), H_1(G))
\end{array}
$$

$$\tag{5.29}$$

where T is as in (A.26).

Every element $x \in H_2(G) = H_2(F/R)$ can be presented in the form $x \equiv \prod_{i=1}^k [f_1^{(i)}, f_2^{(i)}] \mod \gamma_3(F)$, with $\prod_{i=1}^k [f_1^{(i)}, f_2^{(i)}] \in R$. Then the map $d_{1,1}^1$ restricted to the component $H_2(G) \otimes H_1(G)$ is given by

$$x \otimes \bar{g} \mapsto \prod_{i=1}^k [f_1^{(i)}, f_2^{(i)}, g].[R, F, F]\gamma_4(F), \quad x \in H_2(G), \ \bar{g} \in G_{ab},$$

where $\bar{f}$ is the image of $f \in F$ in G_{ab}. On the other hand, it is easy to see that the map $\bar{d}^1_{1,1}$ restricted to $(H_2(G) \otimes H_1(G))^{\oplus 2}$ is induced by

$$(x \otimes \bar{g}_1, x \otimes \bar{g}_2) \mapsto \sum_{i=1}^k \bar{f}_1^{(i)} \otimes \bar{f}_2^{(i)} \otimes \bar{g}_1 + \sum_{i=1}^k \bar{g}_2 \otimes \bar{f}_1^{(i)} \otimes \bar{f}_2^{(i)}, \ x \in H_2(G), \ \bar{g}_1, \bar{g}_2 \in G_{ab}.$$

Hence we have the following diagrams:

$$
\begin{array}{ccccc}
H_1(G) \otimes H_2(G) & \rightarrowtail & E^1_{1,1} & \longrightarrow & \Omega(H_1(G)) \\
\downarrow & & \ \downarrow d^1_{1,1} & & \downarrow \\
\dfrac{[F, R \cap \gamma_2(F)]\gamma_4(F)}{[R, F, F]\gamma_4(F)} & \rightarrowtail & \dfrac{\gamma_3(F)}{[R, F, F]\gamma_4(F)} & \longrightarrow & \dfrac{\gamma_3(F)}{[F, R \cap \gamma_2(F)]\gamma_4(F)} \\
& & \downarrow & & \downarrow \\
& & E^2_{3,0} & =\!=\!=\!= & E^2_{3,0}
\end{array}
$$

$$
\begin{array}{ccccc}
(H_1(G) \otimes H_2(G))^{\oplus 2} & \rightarrowtail & \overline{E}^1_{1,1} & \longrightarrow & \mathrm{Tor}(H_1(G), H_1(G)) \\
\downarrow & & \ \downarrow \overline{d}^1_{1,1} & & \downarrow \\
\dfrac{(R \cap \gamma_2(F)-1)(1)+\mathfrak{r}(2)+\mathfrak{f}^4}{\mathfrak{r}(2)+\mathfrak{f}^4} & \rightarrowtail & \dfrac{\mathfrak{f}^3}{\mathfrak{r}(2)+\mathfrak{f}^4} & \longrightarrow & \dfrac{\mathfrak{f}^3}{(R \cap \gamma_2(F)-1)(1)+\mathfrak{r}(2)+\mathfrak{f}^4} \\
& & \downarrow & & \downarrow \\
& & \overline{E}^2_{3,0} & =\!=\!=\!= & \overline{E}^2_{3,0}
\end{array}
$$

and

$$
\begin{array}{ccccccc}
\Omega(G_{ab}) & \xrightarrow{\ ^1 d^1_{1,1}\ } & \dfrac{\gamma_3(F)}{[R \cap \gamma_2(F), F]\gamma_4(F)} & \longrightarrow & E^2_{3,0}(F) & \longrightarrow & 0 \\
\downarrow T & & \downarrow {^1 \kappa^1_{3,0}} & & \downarrow \kappa^2_{3,0} & & \\
\mathrm{Tor}(G_{ab}, G_{ab}) & \xrightarrow{\ ^1 \overline{d}^1_{1,1}\ } & \dfrac{\mathfrak{f}^3}{(R \cap F'-1)(1)+\mathfrak{r}(2)+\mathfrak{f}^4} & \longrightarrow & \overline{E}^2_{3,0}(F) & \longrightarrow & 0
\end{array}
$$

(5.30)

where $^1 d^1_{1,1}$ and $^1 \overline{d}^1_{1,1}$ are induced by $d^1_{1,1}$ and $\overline{d}^1_{1,1}$ respectively.

Now Theorem 5.19 implies the following commutative diagram

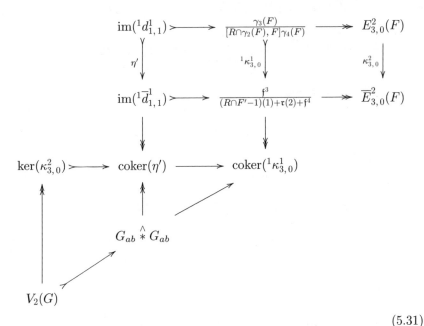

$$(5.31)$$

where η' is the map induced by T and

$$V_2(G) = \ker\{G_{ab} \overset{\wedge}{*} G_{ab} \to \operatorname{coker}({}^1\kappa^1_{3,0})\}.$$

Hence for $n = 3$ the diagram (5.21) has a simple form and we have the following:

Theorem 5.21 (Hartl-Mikhailov-Passi [Har08]). *There exists the following natural system of monomorphisms and epimorphsism*

$$
\begin{array}{ccc}
\ker(\kappa^2_{3,0}) & \longrightarrow & D_4(G)/\gamma_4(G) \\
\uparrow & \nearrow & \\
V_2(G) & \rightarrowtail & G_{ab} \overset{\wedge}{*} G_{ab}
\end{array}
$$

As an immediate consequence of the above result, we have various conditions for the fourth dimension quotient of a group G to be trivial.

Corollary 5.22 *If either* $\ker(\kappa^2_{3,0})$, *or* $V_2(G)$, *or* $G_{ab} \overset{\wedge}{*} G_{ab}$ *is trivial, then* $D_4(G) = \gamma_4(G)$.

We now proceed to generalize the above construction to higher dimensions. For any endo-functor $F : \mathsf{Ab} \to \mathsf{Ab}$, let $\mathfrak{L}_i F = L_i(F, 0)$ be the ith derived functor of F at level 0 in the sense of Dold and Puppe [Dol61] (see

Appendix A.15). Clearly, the map $\kappa_{n-1,1}^1$ can be presented in a natural diagram

$$
\begin{array}{ccccccccc}
0 & \longrightarrow & K_{n-1}(G) & \longrightarrow & E_{n-1,1}^1 & \longrightarrow & \mathfrak{L}_1\mathcal{L}^n(G_{ab}) & \longrightarrow & 0 \\
& & \downarrow & & {\scriptstyle\kappa_{n-1,1}^1}\downarrow & & \downarrow & & \\
0 & \longrightarrow & T_{n-1}(G) & \longrightarrow & \overline{E}_{n-1,1}^1 & \longrightarrow & \mathfrak{L}_1\otimes^n(G_{ab}) & \longrightarrow & 0
\end{array}
$$

for some functor $K_{n-1} : \mathsf{Gr} \to \mathsf{Gr}$. Denote

$$
{}^1E_{n,0}^1 = \mathrm{coker}(K_{n-1} \overset{d_{n-1,1}^1}{\to} E_{n,0}^1), \quad {}^1\overline{E}_{n,0}^1 = \mathrm{coker}(T_{n-1} \overset{\overline{d}_{n-1,1}^1}{\to} \overline{E}_{n,0}^1)
$$

In this notation we obtain the following natural diagram:

$$
\begin{array}{ccccccc}
\mathfrak{L}_1\mathcal{L}^{n-1}(G_{ab}) & \xrightarrow{{}^1d_{n-1,1}^1} & {}^1E_{n,0}^1 & \longrightarrow & E_{n,0}^2 & \longrightarrow & 0 \\
\downarrow & & {\scriptstyle{}^1\kappa_{n,0}^1}\downarrow & & {\scriptstyle\kappa_{n,0}^2}\downarrow & & \\
\mathfrak{L}_1\otimes^{n-1}(G_{ab}) & \xrightarrow{{}^1\overline{d}_{n-1,1}^1} & {}^1\overline{E}_{n,0}^1 & \longrightarrow & \overline{E}_{n,0}^2 & \longrightarrow & 0
\end{array}
$$

where the maps ${}^1d_{n-1,1}^1$ and ${}^1\overline{d}_{n-1,1}^1$ are induced by the maps $d_{n-1,1}^1$ and $\overline{d}_{n-1,1}^1$ respectively.

Define the functor $S_n : \mathsf{Ab} \to \mathsf{Ab}$ by setting

$$
S_n(A) = \mathrm{coker}(\mathcal{L}^n(A) \to \otimes^n(A)), \ A \in \mathsf{Ab}.
$$

Clearly, $\mathfrak{L}_1 S_n(A) = \mathrm{coker}\{\mathfrak{L}_1\mathcal{L}^n(A) \to \mathfrak{L}_1\otimes^n(A)\}$. In this notation, snake lemma implies the following analog of the diagram (5.21):

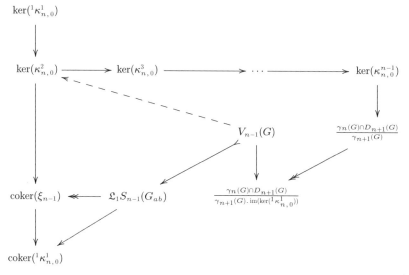

$$(5.32)$$

where $\xi_{n-1} : \mathrm{im}(^1d^1_{n-1,1}) \to \mathrm{im}(^1\overline{d}^1_{n-1,1})$ and $V_{n-1}(G)$ is the kernel of the composition of the natural maps $L_1 S_{n-1}(G_{ab}) \to \mathrm{coker}(\xi) \to \mathrm{coker}(^1\kappa^1_{n,0})$. Hence, for every group G and $n \geq 3$, there is a natural subgroup of $\mathfrak{L}_1 S_{n-1}(G_{ab})$, namely $V_{n-1}(G)$, which maps canonically to the quotient $\frac{\gamma_n(G) \cap D_{n+1}(G)}{\gamma_{n+1}(G). \, \mathrm{im}(\ker(^1\kappa_{n,0}))}$. Denote this map by

$$v_n : V_{n-1}(G) \to \frac{\gamma_n(G) \cap D_{n+1}(G)}{\gamma_{n+1}(G). \, \mathrm{im}(\ker(^1\kappa^1_{n,0}))}$$

For $n = 3$, $\mathfrak{L}_1 S_2(G_{ab}) = G_{ab} \overset{\wedge}{*} G_{ab}$ and the map v_n is an epimorphism (see Theorem 5.21).

Remark 5.23

It may be noted that, heuristically speaking, the contribution of the functor $\mathfrak{L}_1 S_n(G_{ab})$ to the structure of the $(n+1)$-st dimension quotient becomes smaller and smaller as n increases.

Remark 5.24

The map $\overline{d}^1_{n-1,1} : T_{n-1}(G) \to \overline{E}^1_{n,0}$ can be described explicitly (see, for example, the above description of $\overline{d}^1_{1,1}$) and the exact sequence (5.8) implies that

$$\mathrm{im}(\ker(^1\kappa^1_{n,0})) = R\gamma_n(F) \cap (1 + (R \cap F' - 1)(n-2) + \mathfrak{f}^{n+1})/R\gamma_{n+1}(F)$$

In particular, $\mathrm{im}(\ker(^1\kappa^1_{n,0})) = 0$ provided $H_2(G) = 0$ by Corollary 5.11.

5.4 Homotopical Applications

For $n \geq 1$, consider the simplicial n-sphere:

$$S^n_0 = \{*\}, \quad \ldots, \quad S^n_{n-1} = \{*\}, \quad S^n_n = \{*, \sigma\}, \quad S^n_{n+1} = \{*, s_0\sigma, \ldots, s_n\sigma\}, \quad \ldots$$

and the corresponding Milnor's $F[S^n]$-construction (see A.10). Clearly, $F[S^n]_k$ is a free group with $\binom{n+k}{k}$ generators, for $k \geq n$, and $F[S^n]_k = 1$ for $k < n$. Then there is a weak homotopy equivalence

$$|F[S^n]| \simeq \Omega S^{n+1} \text{ (loop space)}$$

and, hence, the isomorphisms of homotopy groups:

$$\pi_i(F[S^n]) \simeq \pi_{i+1}(S^{n+1}), \; i \geq 1.$$

Moreover, it has been shown by Wu [Wu,01] that for every simplicial group X, for which X_k is a free group of rank $\binom{n}{k}$ for $k \geq n$ and $X_k = 1$ for $k < n$, there is a weak homotopy equivalence $F[S^n] \to X$.

Theorem 5.6 implies that the Curtis spectral sequence with initial terms

$$E^1_{p,q} = \pi_q(\gamma_p(F[S^n])/\gamma_{p+1}(F[S^n])) \tag{5.33}$$

converges to a filtration of the homotopy group of the $(n+1)$-sphere. On the other hand, it may be observed that the abelianization $F[S^n]_{ab}$ is exactly the simplicial group $N^{-1}(\mathbb{Z}[n])$ (see A.6), here by $\mathbb{Z}[n]$ we mean the chain complex $\cdots \to 0 \to \mathbb{Z} \to 0 \to \cdots$, where $\mathbb{Z}$ occurs in degree n. In particular, the free abelian simplicial group $F[S^n]_{ab}$ has the following homotopy groups:

$$\pi_i(F[S^n]_{ab}) = \begin{cases} \mathbb{Z}, & i = n, \\ 0, & i \neq n \end{cases} \tag{5.34}$$

Hence, the initial terms of the Curtis spectral sequence can also be written as

$$E^1_{p,q} = \mathfrak{L}_q \mathcal{L}^p(\mathbb{Z}, n),$$

where $\mathfrak{L}_q \mathcal{L}^p(-, n)$ denotes the qth derived functor of the pth Lie functor at level n, in the sense of Dold and Puppe (see A.15). It follows that the structure of homotopy groups of spheres is based on the theory of Lie functors. In this section we study the homotopical aspect of the theory of Lie functors.

The derived functors of Lie functors have highly non-trivial structure. Recall the result of Schlesinger [Sch66]: if p is an odd prime then

$$\mathfrak{L}_{n+k}\mathcal{L}^p(\mathbb{Z}, n) = \begin{cases} \mathbb{Z}_p, & k = 2i(p-1) - 1, \ i = 1, 2, \ldots, [n/2] \\ 0, & \text{otherwise} \end{cases} \tag{5.35}$$

However, for a non-prime m, the computation of derived functors $\mathfrak{L}_i \mathcal{L}^m(\mathbb{Z}, n)$ is much more complicated.

The description of homotopy groups (5.34) is a motivation to introduce the notation $K(\mathbb{Z}, n) = F[S^n]_{ab}$, which we will use in this section.

Graded Functors

For $A \in \mathsf{Ab}$, define the commutative graded ring $\wedge(A)$ as a quotient of the tensor ring $\mathcal{T}(A)$ by the ideal generated by the elements $x \otimes x$, $x \in A$. It defines the series of *exterior power functors*

$$\wedge^k : \mathsf{Ab} \to \mathsf{Ab}, \ k = 0, 1, \ldots$$

as kth degree component of $\wedge$.

For $A \in \mathsf{Ab}$, the commutative graded ring $\mathrm{SP}(A)$ is the quotient of the tensor ring $\mathfrak{T}(A)$ by the ideal generated by elements $x \otimes y - y \otimes x$, $x, y \subset A$. The kth degree component of SP ($k = 0, 1, \dots$) defines a functor

$$\mathrm{SP}^k : \mathsf{Ab} \to \mathsf{Ab}.$$

By definition, we have the natural epimorphisms

$$l_k : \otimes^k(A) \to \wedge^k(A),$$
$$s_k : \otimes^k(A) \to \mathrm{SP}^k(A)$$

which we denote by

$$l_k : a_1 \otimes \cdots \otimes a_k \mapsto a_1 \wedge \cdots \wedge a_k, \; a_1, \dots, a_k \in A$$
$$s_k : a_1 \otimes \cdots \otimes a_k \mapsto a_1 \dots a_k, \; a_1, \dots, a_k \in A.$$

For $A \in \mathsf{Ab}$, define the commutative graded ring $\Gamma(A)$, called *the divided power ring* over A, generated by the elements $\gamma_t(x)$ for each $x \in A$ and each non-negative integer t, of degree t, subject to the relations

$$\gamma_0(x) = 1, \tag{5.36}$$

$$\gamma_t(rx) = r^t \gamma_t(x), \tag{5.37}$$

$$\gamma_t(x + y) = \sum_{i+j=t} \gamma_i(x)\gamma_j(y), \tag{5.38}$$

$$\gamma_s(x)\gamma_t(x) = \binom{s+t}{s} \gamma_{s+t}(x). \tag{5.39}$$

The kth degree component of Γ defines a functor

$$\Gamma_k : \mathsf{Ab} \to \mathsf{Ab}$$

for $k = 1, 2, \dots$. For $k = 2$, the functor Γ_2 coincides with the universal quadratic functor due to Whitehead (see (A.14)).

For an abelian group A with basis $\{x_1, \dots, x_s\}$, the group $\Gamma_k(A)$ is also a free abelian group generated by elements

$$\{\gamma_{i_1}(x_1) \dots \gamma_{i_s}(x_s) \mid \sum_{j=1}^{s} i_j = k, \; i_j \geq 0\}$$

There is a natural map

$$g_k : \Gamma_k(A) \to \otimes^k(A)$$

defined by setting

$$g_k : \gamma_{r_1}(x_1) \dots \gamma_{r_s}(x_s) \mapsto \sum_{(i_1, \dots, i_k)} x_{i_1} \otimes \cdots \otimes x_{i_k},$$

where $(i_1, \ldots, i_k)$ in the definition of g_k ranges over the set of integral n-tuples in which j appears r_j times for all $1 \leq j \leq s$.

Let $f : P \to Q$ be a homomorphism of abelian groups. For $n \geq 1$ and any $k = 0, \ldots, n - 1$, consider the Koszul map

$$d_{k+1} : \wedge^{k+1}(P) \otimes \mathrm{SP}^{n-k-1}(Q) \to \wedge^k(P) \otimes \mathrm{SP}^{n-k}(Q)$$

defined by setting

$$d_{k+1} : p_1 \wedge \cdots \wedge p_{k+1} \otimes q_{k+2} \ldots q_n \mapsto$$

$$\sum_{i=1}^{k+1}(-1)^{k+1-i}p_1 \wedge \cdots \wedge \hat{p}_i \wedge \cdots \wedge p_{k+1} \otimes f(p_i)q_{k+2} \ldots q_n,$$

$$p_1, \ldots, p_{k+1} \in P, \ q_{k+2}, \ldots, q_n \in Q.$$

The complex

$$\mathrm{Kos}^n(f) : \quad 0 \to \wedge^n(P) \xrightarrow{d_n} \wedge^{n-1}(P) \otimes Q \xrightarrow{d_{n-1}} \ldots$$

$$\to P \otimes \mathrm{SP}^{n-1}(Q) \xrightarrow{d_1} \mathrm{SP}^n(Q) \to 0 \quad (5.40)$$

is called the Koszul complex. Analogously one can define a map

$$\bar{d}_{k+1} : \Gamma_{k+1}(P) \otimes \wedge^{n-k-1}(Q) \to \Gamma_k(P) \otimes \wedge^{n-k}(Q), \ k = 0, \ldots, n-1$$

and obtain the following complex of abelian groups:

$$\overline{\mathrm{Kos}}^n(f) : \quad 0 \to \Gamma_n(P) \xrightarrow{\bar{d}_n} \Gamma_{n-1}(P) \otimes Q \xrightarrow{\bar{d}_{n-1}} \ldots$$

$$\to P \otimes \wedge^{n-1}(Q) \xrightarrow{\bar{d}_1} \wedge^n(Q) \to 0 \quad (5.41)$$

The complexes $\mathrm{Kos}^n(f)$ and $\overline{\mathrm{Kos}}^n(f)$ are the total degree n-component complexes of the Koszul complexes $\wedge(P) \otimes \mathrm{SP}(Q)$ and $\Gamma(P) \otimes \wedge(Q)$ attached to the map $f : P \to Q$. It turns out that for a free abelian group P and the identity homomorphism $f : P \to P$, the complexes $\mathrm{Kos}^n(f)$ and $\overline{\mathrm{Kos}}^n(f)$ are acyclic. The above Koszul complexes are considered in, for example, [Qui68], [Ill71, Chapter 1, Section 4.3]. The above constructions imply the canonical isomorphisms:

$$\mathcal{L}_i \, \mathrm{SP}^k(A, n) \simeq \mathcal{L}_{i-k} \wedge^k (A, n-1), \quad (5.42)$$

$$\mathcal{L}_i \wedge^k (A, n) \simeq \mathcal{L}_{i-k}\Gamma_k(A, n-1), \ n \geq 1, \ k \geq 1, \ i \geq k, \quad (5.43)$$

called the Bousfield-Quillen décalage formulae (see [Bre99]).

The 2-sphere

Consider the simplest homotopical application of the above spectral sequence applied to the case of the 2-sphere S^2. It is one of the deep problems of algebraic topology to compute homotopy groups $\pi_n(S^2)$. In low dimensions one has (see [Tod62]):

n	2	3	4	5	6	7	8	9	10	11	12
$\pi_n(S^2)$	$\mathbb{Z}$	$\mathbb{Z}$	$\mathbb{Z}_2$	$\mathbb{Z}_2$	$\mathbb{Z}_4 \oplus \mathbb{Z}_3$	$\mathbb{Z}_2$	$\mathbb{Z}_2$	$\mathbb{Z}_3$	$\mathbb{Z}_{15}$	$\mathbb{Z}_2$	$\mathbb{Z}_2 \oplus \mathbb{Z}_2$

n	13	14	15	16	17	18	19
$\pi_n(S^2)$	$\mathbb{Z}_4 \oplus \mathbb{Z}_6$	$\mathbb{Z}_2^{\oplus 2} \oplus \mathbb{Z}_{84}$	$\mathbb{Z}_2^{\oplus 2}$	$\mathbb{Z}_6$	$\mathbb{Z}_{30}$	$\mathbb{Z}_{30}$	$\mathbb{Z}_2^{\oplus 2} \oplus \mathbb{Z}_3$

The simplicial group $K(\mathbb{Z}, 1)$ has $K(\mathbb{Z}, 1)_n$ a free abelian group of rank n with generators $x_0^{(n)}, \ldots, x_{n-1}^{(n)}$, written as

$$x_i^{(n)} = s_{n-1} \ldots s_{i+1} \hat{s}_i s_{i-1} \ldots s_0 \sigma \tag{5.44}$$

with degeneracy maps $s_i : K(\mathbb{Z}, 1)_{n-1} \to K(\mathbb{Z}, 1)_n$, $i = 0, \ldots, n-1$ defined via standard simplicial laws. Then the Theorem 5.6 implies that that the lower central series spectral sequence with initial terms

$$E_{p,q}^1 = \mathfrak{L}_q \mathcal{L}^p(\mathbb{Z}, 1) \tag{5.45}$$

converges to the graded group associated with a filtration of $\pi_*(S^2)$.

Consider first the simplicial group $\otimes^r K(\mathbb{Z}, 1)$ constructed by applying the rth tensor power functor to each dimension of $K(\mathbb{Z}, 1)$. The Eilenberg-Zilber equivalence together with Kunneth formula (A.15) for tensor products of chain complexes imply that

$$\mathfrak{L}_r \otimes^r (\mathbb{Z}, 1) \simeq \mathbb{Z}, \quad \mathfrak{L}_i \otimes^r (\mathbb{Z}, 1) = 0, \quad i \neq r.$$

That is, the rth tensor power $\otimes^r K(\mathbb{Z}, 1)$ gives a model of $K(\mathbb{Z}, r)$. The definition of the 'shuffle' map (see (307)) shows that the generator of the cyclic group $\mathfrak{L}_r \otimes^r (\mathbb{Z}, 1)$ is the coset of the element

$$\theta_r := \sum_{(i_1, \ldots, i_r) \in \Sigma_r} (-1)^{sign(i_1, \ldots, i_r)} x_{i_1}^{(r)} \otimes \cdots \otimes x_{i_r}^{(r)} \in \otimes^r K(\mathbb{Z}, 1)_r.$$

The following well known fact now follows:

Proposition 5.25 *The homotopy groups $\pi_n(S^2)$ of the 2-sphere are finite for $n \geq 4$.*

Proof. The natural map $s_r : \otimes^r K(\mathbb{Z}, 1)_r \to \mathcal{L}^r K(\mathbb{Z}, 1)_r$ sends θ_r to the element

$$s_r(\theta_r) = \sum_{(i_1,\dots,i_r)\in\Sigma_r} (-1)^{sign(i_1,\dots,i_r)}[x_{i_1}^{(r)},\dots,x_{i_r}^{(r)}],$$

which is trivial for $r > 2$ due to Jacobi identity. Since the r-multliplication in $\mathcal{L}^r K(\mathbb{Z}, 1)$ can be defined as the composition map

$$\mathcal{L}^r K(\mathbb{Z}, 1) \to \otimes^r K(\mathbb{Z}, 1) \to \mathcal{L}^r K(\mathbb{Z}, 1),$$

coming to the level of homotopy groups, we conclude that every element of $\mathcal{L}_* \mathcal{L}^r(\mathbb{Z}, 1)$ is annihilated by multiplication by r, for $r > 2$.

Hence, the initial terms of the spectral sequence (5.45) are finite for $p > 2$. Clearly, $E^1_{1,q} = \pi_q(K(\mathbb{Z}, 1)) = 0$, $q > 1$. The terms $E^1_{2,q} = \mathcal{L}_q \mathcal{L}^2(\mathbb{Z}, 1)$ are also finite for $q > 2$ (moreover, they are all zero); one can see this by arguing as above and using the fact that $\otimes^2 K(\mathbb{Z}, 1)$ is a model of $K(\mathbb{Z}, 2)$. The needed statement then follows from the convergence of (5.45) to the graded group associated with $\pi_*(S^2)$ by Theorem 5.6. □

Remark 5.26

Using group-theoretical properties of the simplicial groups $\mathcal{L}^r K(\mathbb{Z}, n)$ for the Eilenberg-MacLane spaces $K(\mathbb{Z}, n)$, Schlesinger gave algebraic proof of the well-known Serre's Theorem (see [Sch66]): for $q > 0$,

$$\pi_{n+q}(S^n) = \begin{cases} \mathbb{Z} \oplus \text{finite group}, & q = 2n - 1, \ n \text{ is even,} \\ \text{finite group}, & \text{otherwise.} \end{cases}$$

The main ingredient of the proof of Theorem 5.6 is the following group-theoretical result due to Curtis [Cur63]:

Theorem 5.27 *Let X be a free simplicial group, which is m-connected ($m \geq 0$). Then $\gamma_r(X)/\gamma_{r+1}(X)$ is $\{m + \log_2 r\}$-connected, where $\{a\}$ is the least integer $\geq a$.*

The proof is based on the following decomposition of the Lie functors.

Curtis Decomposition

Let $\{*\}$ be the set with the single element $*$. Define $T(1) = \{*\}$. Suppose we defined $T(k)$, $1 \leq k \leq n - 1$. Then define

$$T(n) = \bigcup_{1 \leq k \leq n-1} \{T(k) \times T(n - k)\},$$

where $\times$ is a formal operation. Set $T = \cup_{n \geq 1} T(n)$. Then T is the free (non-associative) monoid generated by $*$. The first terms $T(n)$ can be written as

$$T(1) = \{*\},$$
$$T(2) = \{**\},$$
$$T(3) = \{(**)*, *(**)\},$$
$$T(4) = \{(**)(**), (((**)*)*), (*((**)*)), (*(*(**))), ((*(**))*)\},$$

$\dots$

The set T is called *the set of types*. The type $t \in T(n)$ is of *weight* n. We will write $\mathrm{wt}(t) = n$ in this case.

Define a pre-order in the set T. Let $t, s \in T$. We set $t < s$ if $\mathrm{wt}(t) < \mathrm{wt}(s)$. Suppose now that the pre-order is defined for elements of weight less than n and $\mathrm{wt}(t) = \mathrm{wt}(s) = n$. The types t and s can be written uniquely as products $t = t_1 t_2$, $s = s_1 s_2$. We set $t \leq s$ if either $\max\{s_1, s_2\} > \max\{t_1, t_2\}$, or $\min\{t_1, t_2\} \leq \min\{s_1, s_2\}$ and $\max\{t_1, t_2\} \leq \max\{s_1, s_2\}$, and $\max\{s_1, s_2\} \leq \max\{t_1, t_2\}$. In the case $t \leq s$, but not $s \leq t$, we set $t < s$. For example, we have

$$(**)* \leq *(**), \quad *(**) \leq (**)*,$$
$$(**)(**) < (((**)*)*), \quad \dots$$

Define the set of *basic types* $B = \cup_{n \geq 1} B(n)$, a subset in T, inductively. Let $B(1) = T(1) = \{*\}$. Suppose we defined $B(k)$, $k < n$. Then $B(n)$ consists of the types $t_1 t_2$, where t_1, t_2 are basic types and $t_2 \leq t_1$, and if $t_1 = uv$ then $v \leq t_2$.

It can be checked that the pre-order in T induces an order in the set of basic types B. In low dimensions the terms of B are:

$$B(1) = \{*\},$$
$$B(2) = \{**\},$$
$$B(3) = \{***\},$$
$$B(4) = \{(**)(**) < ****\},$$
$$B(5) = \{(***)(**) < *****\},$$
$$B(6) = \{(**)(**)(**) < (***)(***) < (****)(**) < ******\},$$
$$B(7) = \{(**)(**)(***) < (****)(***) < (***)(**)(**) <$$
$$(*****)(**) < *******\},$$
$$B(8) = \{((**)(**))((**)(**)) < (****)(****) < (***)(**)(***)$$
$$< (*****)(***) < (**)(**)(**)(**) < (****)(**)(**)$$
$$< (******)(**) < ********\}$$

$\dots$

For any R-module M, the set of basic types gives a filtration of the Lie functor $\mathcal{L}(M)$. For a basic type $t \in B(n)$, define the R-submodule $F^t(M)$ in $\mathcal{L}(M)$ as the submodule generated by Lie brackets with the configuration s, where $s \in T(n)$, $s \le t$. For example,

$$F^{****}(M) = \mathcal{L}^4(M), \quad F^{(**)(**)}(M) = \mathcal{L}^2\mathcal{L}^2(M).$$

For every $t \in B(n)$ denote by $G^t L^n(M)$ the quotient $F^t(M)/F^s(M)$, where s is the immediate predecessor of t in $B(n)$.

Notation. If t is the greatest element in $B(n)$, i.e., the element $* \cdots *$ (n terms), denote $G^t L^n(M)$ by $J^n(M)$.

It is easy to see that $J^n(M)$ is the *metabelianization* of $\mathcal{L}^n$, i.e.,

$$J^n(M) = \mathcal{L}^n(M)/(\mathcal{L}^n(M) \cap \mathcal{L}^2\mathcal{L}^2(M)).$$

Clearly, the nth Lie functor can be presented as a direct sum

$$\mathcal{L}^n(M) = \bigoplus_{t \in B(n)} G^t L^n(M).$$

This is a functorial decomposition, since for every $t \in B(n)$, G^t can be viewed as a functor in the category of R-modules.

For a given R-module, let $\mathrm{SP}^n(M)$ denote the symmetric n-fold tensor power of M.

Proposition 5.28 (Schlesinger [Sch66]). *Let M be a free $\mathbb{Z}$-module and $n \ge 2$. There exists the following natural exact sequence:*

$$0 \to J^n(M) \xrightarrow{\ i\ } M \otimes \mathrm{SP}^{n-1}(M) \xrightarrow{\ r_n\ } \mathrm{SP}^n(M) \to 0,$$

where r_n is the natural multiplication map, and the inclusion i is given by setting

$$i : [m_1, \ldots, m_n] \mapsto m_1 \otimes (m_2, \ldots, m_n) - m_2 \otimes (m_1, m_3, \ldots, m_n), \quad m_j \in M.$$

Every basic type $t \in B(n)$ can be written uniquely as a sequence of products

$$t = t_1 \ldots t_k, \tag{5.46}$$

where all t_j are certain basic types. Suppose there exists $0 < a < k$, such that

$$t_k = t_{k-1} = \cdots = t_{k-a+1} > t_{k-a},$$

and $t_0 t_1 \ldots t_{k-a} > t_{k-a+1}$. Then we will write $t \in B^1(n; a)$. If there exists $0 < a < k$, such that $t_k = \cdots = t_{k-a+1} = t_0 \ldots t_1 \ldots t_{k-a}$, we will write $t \in B^2(n; a)$. We will denote by t^{gr} the greatest type in $B(n)$.

It is easy to see that

$$B(n) = t^{gr} \cup \bigcup_{a>0} (B^1(n; a) \cup B^2(n; a)).$$

The following Theorem shows that one can get complete description of the values of the Lie functor from the description of the functor J^n together with symmetric tensor products.

Theorem 5.29 (Curtis [Cur63]). *Let M be a free $\mathbb{Z}$-module.*

1. *Let $t \in B^1(n; a)$ and let (5.46) be its presentation as a product. Then*

$$G^t L^n(M) \simeq G^{t_0 \dots t_{k-a}} L^{\mathrm{wt}(t_0 \dots t_{k-a})}(M) \otimes \mathrm{SP}^a (G^{t_k} L^{\mathrm{wt}(t_k)}(M)).$$

2. *Let $t \in B^2(n; a)$ and let (5.46) be its presentation as a product. Then*

$$G^t L^n(M) \simeq J^{a+1}(G^{t_k} L^{\mathrm{wt}(t_k)}(M)).$$

For example, in low degrees, we have the following decompositions (we assume M is a free $\mathbb{Z}$-module):

$$\mathcal{L}^2(M) = J^2(M),$$
$$\mathcal{L}^3(M) = J^3(M),$$
$$\mathcal{L}^4(M) = J^2 J^2(M) \oplus J^4(M),$$
$$\mathcal{L}^5(M) = (J^3(M) \otimes J^2(M)) \oplus J^5(M),$$
$$\mathcal{L}^6(M) = J^3 J^2(M) \oplus J^2 J^3(M) \oplus (J^4(M) \otimes J^2(M)) \oplus J^6(M)$$
$$\mathcal{L}^7(M) = (J^3(M) \otimes \mathrm{SP}^2 J^2(M)) \oplus (J^5(M) \otimes J^2(M)) \oplus (J^2 J^2(M) \otimes J^3(M))$$
$$\qquad \oplus (J^4(M) \otimes J^3(M)) \oplus J^7(M)$$
$$\mathcal{L}^8(M) = J^2 J^2 J^2(M) \oplus J^2 J^4(M) \oplus (J^3(M) \otimes J^2(M) \otimes J^3(M))$$
$$\qquad \oplus J^5(M) \otimes J^3(M) \oplus J^4 J^2(M) \oplus (J^4(M) \otimes \mathrm{SP}^2 J^2(M))$$
$$\qquad \oplus (J^6(M) \otimes J^2(M)) \oplus J^8(M)$$

$$\dots$$

We will refer to such decompositions as the *Curtis decompositions* of the Lie functors.

Remark 5.30

In the above decomposition the direct sum of functors does not mean that it is preserved by homotopy. For example, the exact sequence

$$\mathcal{L}_6 \mathcal{L}^2 \mathcal{L}^2(\mathbb{Z}, 2) \to \mathcal{L}_6 \mathcal{L}^4(\mathbb{Z}, 2) \to \mathcal{L}_6 J^4(\mathbb{Z}, 2)$$

is not split:

$$\mathfrak{L}_6 \mathcal{L}^2 \mathcal{L}^2(\mathbb{Z},\,2) = \mathbb{Z}_4, \ \ \mathfrak{L}_6 \mathcal{L}^4(\mathbb{Z},\,2) = 0, \ \ \mathfrak{L}_6 J^4(\mathbb{Z},\,2) = 0$$

and the boundary map

$$\mathfrak{L}_7 J^4(\mathbb{Z},\,2) \rightarrow \mathfrak{L}_6 \mathcal{L}^2 \mathcal{L}^2(\mathbb{Z},\,2)$$

is an isomorphism.

The Sandling-Tahara Decomposition

An alternative way to reduce the homotopy theory of the Lie functors to the
homotopy theory of the symmetric powers and metabelian Lie functors is as
follows.

Let G be a group with torsion-free lower central quotients, for example a
free group, then there is the following decomposition [San79]:

$$\mathfrak{g}^n / \mathfrak{g}^{n+1} = \sum_{(a_1,\,\dots,\,a_n)} \bigotimes_{i=1}^n \mathrm{SP}^{a_i}(\gamma_i(G)/\gamma_{i+1}(G)),$$

where the sum runs over all non-negative integers $a_1,\,\dots,\,a_n$ such that
$\sum_{i=1}^n i a_i = n$ (here $\mathrm{SP}^0(M) = \mathbb{Z}$ for an abelian group M). For a free $\mathbb{Z}$-
module M, therefore, one has the following decomposition:

$$\otimes^n(M) = \sum_{(a_1,\,\dots,\,a_n)} \bigotimes_{i=1}^n \mathrm{SP}^{a_i} \mathcal{L}^i(M).$$

For example, in the case $n = 6$, we have the following decomposition:

$$\otimes^6(M) = \mathcal{L}^6(M) \oplus (M \otimes \mathcal{L}^5(M)) \oplus (\mathcal{L}^2(M) \otimes \mathcal{L}^4(M)) \oplus \mathrm{SP}^3 \mathcal{L}^2(M)$$
$$\oplus (M \otimes \mathcal{L}^2(M) \otimes \mathcal{L}^3(M)) \oplus (\mathrm{SP}^2(M) \otimes \mathrm{SP}^2 \mathcal{L}^2(M))$$
$$\oplus (\mathrm{SP}^3(M) \otimes \mathcal{L}^3(M)) \oplus \mathrm{SP}^6(M).$$

We will, however prefer to use the Curtis decomposition as it is more convenient for our purpose.

The Simplicial Groups $J^n K(\mathbb{Z},\,1)$

Let G be a simplicial abelian group. Define CG, the *cone over* G, as the
simplicial abelian group with

$$CG_n = G_n \oplus G_{n-1} \oplus \cdots \oplus G_0,$$
$$\partial_i : (a_n, \ldots, a_0) \mapsto$$
$$(\partial_i a_n, \partial_{i-1} a_{n-1}, \ldots, \partial_1 a_{n-i+1}, \partial_0 a_{n-i} + a_{n-i-1}, a_{n-i-2}, \ldots, a_0), \ i < n,$$
$$\partial_n : (a_n, \ldots, a_0) \mapsto (\partial_n a_n, \ldots, \partial_1 a_1),$$
$$s_i : (a_n, \ldots, a_0) \mapsto (s_i a_n, \ldots, s_0 a_{n-i}, 0, a_{n-i-1}, \ldots, a_0).$$

For every simplicial abelian group G, there is a natural injection $i : G \to CG$. The cokernel of the injection i is called the *suspension* of G. The suspension defines a functor, so -called *suspension functor*

$$\Sigma : \mathcal{S}\mathsf{Ab} \to \mathcal{S}\mathsf{Ab},$$

which has the property

$$H_{n+1}(N\Sigma G) = \pi_{n+1}(\Sigma G) = \pi_n(G) = H_n(NG),$$

for a given simplicial abelian group G.

For a given endofunctor T on the category Ab of abelian groups, consider its cross-effects

$$T_n : \mathsf{Ab}^n \to \mathsf{Ab},$$

which can be determined by the property

$$T(M_1 \oplus \cdots \oplus M_n) = \bigoplus_{(\sigma_1, \ldots, \sigma_s)} T_s(M_1, \ldots, M_s),$$

where the direct sum is taken over all non-empty ordered subsets $(\sigma_1, \ldots, \sigma_s) \subseteq \{1, \ldots, n\}$.

Given free modules $K_1, \ldots, K_n$, consider the module $K = K_1 \oplus \cdots \oplus K_n$. Suppose we have an ordered basis of each of the modules K_i. Define the natural order on the resulting basis of K by setting all the basis elements of K_{i+1} to be greater than the basis elements of K_i. Clearly, the $\mathbb{Z}$-basis of $J^r(K_1, \ldots, K_n)$ consists of the elements

$$[x_1, \ldots, x_r], \tag{5.47}$$

such that $x_1 > x_2 \leq x_3 \leq \cdots \leq x_r$, with the property that for each $i = 1, \ldots, n$, we have at least one element x_j from K_j in (5.47).

Proposition 5.31 (Curtis [Cur63]) *For any free modules $K_1, \ldots, K_n$, there is a direct sum decomposition*

$$J_n^r(K_1, \ldots, K_n) \simeq \bigoplus_R J^{r_1}(K_1) \otimes \mathrm{SP}^{r_2}(K_2) \otimes \cdots \otimes \mathrm{SP}^{r_n}(K_n) +$$

$$\bigoplus_{R'} K_m \otimes \mathrm{SP}^{r_1}(K_1) \otimes \cdots \otimes \mathrm{SP}^{r_n}(K_n),$$

where the first direct sum is taken over

$$R := \{\{r_1, \ldots, r_n\} \mid r_i \in \mathbb{Z}, \, r_1 \geq 2, \, r_i \geq 1, \, \sum r_i = r\},$$

the second sum is taken over

$$R' = \{\{m; r_1, \ldots, r_n\} \mid 2 \leq m \leq n, \, r_m \geq 0, \, r_i \geq 1, \, for \, i \neq m, \, \sum r_i = r - 1\}.$$

In particular, in the simplest case $n = 2$, one has

$$J^2(K_1, K_2) = K_1 \otimes K_2.$$

The main tool in the theory of suspended simplicial abelian groups is the following spectral sequence.

Theorem 5.32 (Dold - Puppe [Dol61]). *For a given functor T and a simplicial abelian group G, there is a spectral sequence $\{E^i_{p,q}\}$, which converges to the graded group associated with a filtration of $\pi_*(T\Sigma G)$. Its first term is given by*

$$E^1_{p,q} = \pi_q(T_p(G, \ldots, G)),$$

with the natural differential

$$d^1 : \pi_q(T_p(G, \ldots, G)) \to \pi_q(T_{p-1}(G, \ldots, G)).$$

In the case T is a quadratic functor, i.e. $T(K_1, K_2, K_3) = 0$ for any abelian groups K_1, K_2, K_3, then the above spectral sequence degenerates to the following long exact sequence:

$$\cdots \to \pi_n(T(G, G)) \to \pi_n(T(G)) \to \pi_{n+1}(T\Sigma G) \to$$

$$\pi_{n-1}(T(G, G)) \to \pi_{n-1}(T(G)) \to \cdots \quad (5.48)$$

The free simplicial abelian group X modelling $K(\mathbb{Z}, 1)$ is the suspension over the group $K(\mathbb{Z}, 0)$, the constant simplicial group with the identity simplicial homomorphisms. The decompositions of the Lie functors given by Theorem 5.29 reduces the general question about the structure of the spectral sequence (5.45) to the homotopical properties of the functors J^n and SP^n. For $n = 2$, $J^2 = \mathcal{L}^2$ is a quadratic functor, hence $J^2(M, M, M) = J^2(M, M, \ldots, M) = 0$. Therefore, for a given simplicial module M, the spectral sequence in Theorem 5.32 degenerates to the exact sequence (5.48).

The following result follows from the sequence (A.31). However, we will provide a proof using the sequence (5.48).

Proposition 5.33 (Schlezinger, Curtis [[Sch66], [Cur63]]). *For any $n \geq 1$, $q \geq 0$, one has*

$$\mathcal{L}_{n+q}\mathcal{L}^2(\mathbb{Z},\, n) = \mathcal{L}_{n+q}J^2(\mathbb{Z},\, n) = \begin{cases} \mathbb{Z}, & \text{if } n \text{ is odd and } q = n \\ \mathbb{Z}_2, & \text{if } q = 1, 3, \ldots, 2[n/2] - 1, \\ 0 & \text{otherwise} \end{cases}$$

Proof. The proof is by induction on $n \geq 1$. Observe that $J^2(A,\, B) = A \otimes B$ and therefore, $J^2 K(\mathbb{Z},\, n) \simeq K(\mathbb{Z},\, 2n)$. The sequence (5.48) has the following form in our case:

$$\cdots \to \pi_n(K(\mathbb{Z},\, 2n)) \to \mathcal{L}_n J^2(\mathbb{Z},\, n) \to \mathcal{L}_{n+1} J^2(\mathbb{Z},\, n+1) \to$$
$$\pi_{n-1}(K(\mathbb{Z},\, 2n)) \to \mathcal{L}_{n-1} J^2(\mathbb{Z},\, n) \to \ldots \quad (5.49)$$

For $n = 1$, clearly, the simplicial groups $J^2 K(\mathbb{Z},\, 1)$ and $K(\mathbb{Z},\, 2)$ have the same homotopy groups. Suppose by induction that the statement is proved for $n \geq 1$. Consider the following cases.

Case I: n=2k. In this case $\mathcal{L}_{n+q}J^2(\mathbb{Z},\, n) = 0$ for $q \geq n$ and

$$\mathcal{L}_{n+q}J^2(\mathbb{Z},\, n) = \mathcal{L}_{n+q+1}J^2(\mathbb{Z},\, n+1) = \mathbb{Z}_2, \quad q = 1, 3, \ldots, 2k - 1.$$

Also we have one more nontrivial term, coming from $\pi_{2n}(K(\mathbb{Z},\, 2n))$:

$$\mathcal{L}_{2n+2}J^2(\mathbb{Z},\, n+1) = \pi_{2n}(K(\mathbb{Z},\, 2n)) = \mathbb{Z}.$$

All other homotopy groups are zero.

Case II: n=2k+1. In this case we have $\mathcal{L}_{n+q}J^2(\mathbb{Z},\, n) = 0$, $q \geq n+1$ and again

$$\mathcal{L}_{n+q}J^2(\mathbb{Z},\, n) = \mathcal{L}_{n+q+1}J^2(\mathbb{Z},\, n+1) = \mathbb{Z}_2, \quad q = 1, 3, \ldots, 2k - 1.$$

The difference with the Case I is that there is an exact sequence

$$0 \to \mathcal{L}_{2n+2}J^2(\mathbb{Z},\, n+1) \to \pi_{2n}(K(\mathbb{Z},\, 2n)) \xrightarrow{\alpha}$$
$$\mathcal{L}_{2n}J^2(\mathbb{Z},\, n) \to \mathcal{L}_{2n+1}J^2(\mathbb{Z},\, n+1) \to 0,$$

where the map α is induced by the natural projection

$$\otimes^2 K(\mathbb{Z},\, n) \to J^2 K(\mathbb{Z},\, n).$$

Observe that for $K_i = \mathbb{Z}$, $i = 1, \ldots, n$, one has

$$J_n^r(K_1, \ldots, K_n) = \sum_{R'} K_m \otimes \mathrm{SP}^{r_1} K_1 \otimes \cdots \otimes \mathrm{SP}^{r_n} K_n, \quad (5.50)$$

since $J^l(K_i) = 0$, $l \geq 2$, R' is the same set as in the Proposition 5.31 and the result follows. $\square$

Remark 5.34 *It may be noted that the case $n = 1$ of Proposition 5.33 implies that $\pi_i(\mathcal{L}^2(X)) = 0$, $i = 0, 1$ for every free simplicial abelian group X with $\pi_0(X) = 0$.*

Proposition 5.35 (Curtis [Cur63]). *For $r \geq 3$, $\mathcal{L}_i J^r(\mathbb{Z}, 1) = 0$, $i \geq 0$.*

Proof. Set $K = K(\mathbb{Z}, 0)$. Using the decomposition (5.50), we get

$$\pi_i(J_n^r(K, \ldots, K)) = 0, \; i \geq 1$$

and

$$\pi_0(J_n^r(K, \ldots, K)) = \mathbb{Z}, \; n \geq 2.$$

Therefore, the groups $\pi_i(J^r K(\mathbb{Z}, 1))$ are equal to the homology of the complex

$$\cdots \xrightarrow{d_4} \pi_0(J_3^r(K, K, K)) \xrightarrow{d_3} \pi_0(J_2^r(K, K)) \to 0,$$

which consists of free abelian groups. Observe that $J_{r+1}^r(K, \ldots, K) = 0$. The homomorphism

$$d_n : \pi_0(J_n^r(K, \ldots, K)) \to \pi_0(J_{n-1}^r(K, \ldots, K))$$

is induced by the map

$$\sum_{i=1}^{n-1}(-1)^i \alpha_i : J_n^r(K, \ldots, K) \to J_{n-1}^r(K, \ldots, K),$$

where α_i is induced by the map

$$K^{\oplus n} \to K^{\oplus n-1}, \; (a_1, \ldots, a_n) \mapsto (a_1, \ldots, a_{i-1}, a_i + a_{i+1}, a_{i+2}, \ldots, a_n).$$

The group $J_n^r(K, \ldots, K)$ is a free abelian group, with a basis consisting of the elements

$$k \otimes k^{r_1} \otimes \cdots \otimes k^{r_n} \in K_m \otimes \mathrm{SP}^{r_1} K_1 \otimes \cdots \otimes \mathrm{SP}^{r_n} K_n, \; \{m; r_1, \ldots, r_n\} \in R',$$

where all K_i are infinite cyclic and we write k^{r_i} for the element $k \otimes \cdots \otimes k$ (r_i times) in $\mathrm{SP}^{r_i}(K)$. Since all K_i are infinite cyclic, we can say that $J_n^r(K, \ldots, K)$ is a free abelian group with basis R'. We can identify the group $\pi_0(J_n^r(K, \ldots, K))$ with $J_n^r(K_1, \ldots, K_n)$, since the simplicial group we consider is constant. The maps d_n are defined for the basis elements as follows:

$$d_n : (m; r_1, \ldots, r_n) \mapsto \sum_{i=1}^{m-1} (-1)^i (m-1; r_1, \ldots, r_i + r_{i+1}, \ldots, r_n) +$$

$$\sum_{i=m}^{n-1} (-1)^i (m; r_1, \ldots, r_i + r_{i+1}, \ldots, r_n), \quad m \geq 3,$$

$$(2; r_1, \ldots, r_n) \mapsto \sum_{i=2}^{n-1} (-1)^i (2; r_1, \ldots, r_i + r_{i+1}, \ldots, r_n).$$

For example, for $n = 3$, we have the following map:

$$d_3 : (2; r_1, r_2, r_3) \mapsto -(2; r_1, r_2 + r_3),$$
$$(3; r_1, r_2, r_3) \mapsto (2; r_1 + r_2, r_3) - (2; r_1, r_2 + r_3).$$

Now consider the map

$$\Phi_n : J_n^r(K_1, \ldots, K_n) \to J_{n+1}^r(K_1, \ldots, K_{n+1}), \quad n \geq 2$$

defined as follows.

Case I. Let $n \geq 2$, $r_n = 0$, then

$$\Phi_n : \{n; r_1, \ldots, r_{n-1}, 0\} \mapsto \begin{cases} \{n; r_1, \ldots, r_{n-1}-1, 0, 1\} - \\ \qquad \{n+1; r_1, \ldots, r_{n-1}-1, 1, 0\}, \quad r_{n-1} \geq 2, \\ 0, \quad r_{n-1} = 1 \end{cases} \quad .$$

Case II. Let $n \geq 2$, $r_n \neq 0$:

$$\Phi\{m; r_1, \ldots, r_n\} \mapsto \begin{cases} \{m; r_1, \ldots, r_{n-1}, r_n - 1, 1\}, \quad r_n \geq 2 - \delta(m, n), \\ 0, \quad m \neq n, \; r_n = 1, \end{cases} \quad .$$

where $\delta(m, n)$ is the Kronecher symbol: $\delta(m, n) = 1$, $m = n$, $\delta(m, n) = 0$, $m \neq n$.

We have the composite map

$$\Phi_{n-1} \circ d_n : J_n^r(K_1, \ldots, K_n) \to J_n^r(K_1, \ldots, K_n),$$

defined as follows:

$$\Phi_{n-1} \circ d_n : \{m; r_1, \ldots, r_n\} \mapsto \sum_{i=1}^{m-1} (-1)^i (m-1; r_1, \ldots, r_i + r_{i+1}, \ldots, r_n - 1, 1) +$$

$$\sum_{i=m}^{n-1} (-1)^i (m; r_1, \ldots, r_i + r_{i+1}, \ldots, r_n - 1, 1),$$

for $r_n \geq 0$, $r_n \geq 2 - \delta(m, n)$, and

$$\Phi_{n-1} \circ d_n : \{n; r_1, \ldots, r_{n-1}, 0\} \mapsto$$

$$\sum_{i=1}^{n-1}(-1)^i(n-1; r_1, \ldots, r_i + r_{i+1}, \ldots, r_{n-1} - 1, 0, 1)+$$

$$\sum_{i=1}^{n-1}(-1)^{i+1}\{n; r_1, \ldots, r_{n-1} - 1, 1, 0\},$$

if $r_{n-1} \neq 1$, with

$$\Phi_{n-1} \circ d_n(\{n; r_1, \ldots, r_{n-2}, 1, 0\}) = (-1)^{n-1}\{m; r_1, \ldots, r_{n-2}, 1, 0\}.$$

On the other hand, we have the composite map

$$d_{n+1} \circ \Phi_n : J_n^r(K_1, \ldots, K_n) \to J_n^r(K_1, \ldots, K_n),$$

defined as follows:

$$d_{n+1} \circ \Phi_n : \{m; r_1, \ldots, r_n\} \mapsto$$

$$\sum_{i=1}^{m-1}(-1)^i(m-1; r_1, \ldots, r_i + r_{i+1}, \ldots, r_n - 1, 1)+$$

$$\sum_{i=m}^{n}(-1)^i(m; r_1, \ldots, r_i + r_{i+1}, \ldots, r_n - 1, 1),$$

for $r_n \geq 0$, $r_n \geq 2 - \delta(m, n)$, and

$$d_{n+1} \circ \Phi_n : \{n; r_1, \ldots, r_{n-1}, 0\} \mapsto$$
$$(-1)^n\{n; r_1, \ldots, r_n\}+$$

$$\sum_{i=1}^{n-1}\{n-1; r_1, \ldots, r_i + r_{i+1}, \ldots, r_{n-1}, 0, 1\}+$$

$$\sum_{i=1}^{n}(-1)^{i+1}\{n; r_1, \ldots, r_i + r_{i+1}, \ldots, r_{n-1} - 1, 1, 0\}$$

for $r_{n-1} \neq 1$, with $d_{n+1} \circ \Phi_n(\{n; r_1, \ldots, r_{n-2}, 1, 0\}) = 0$. Now it is easy to observe that the map

$$(-1)^{n-1}\Phi_{n-1} \circ d_{n-1} + (-1)^n d_{n+1} \circ \Phi_n : J_n^r(K_1, \ldots, K_n) \to J_n^r(K_1, \ldots, K_n)$$

is the identity map. Therefore, the maps Φ_n present the contracting homotopy of the complex $J = (J_n^r(K_1, \ldots, K_n), d_n)$, hence the homology of J are trivial and $\mathfrak{L}_i J^r(\mathbb{Z}, 1) = 0$, $r \geq 3$. $\square$

Corollary 5.36 (Dold - Puppe [Dol61]). *Let $n \geq 2$. Then the simplicial group* $SP^n K(\mathbb{Z}, 1)$ *is contractible, i.e.* $\mathfrak{L}_i SP^n(\mathbb{Z}, 1) = 0$, $i \geq 0$.

Proof. The result easily follows from Propositions 5.28 and 5.35. □

Theorem 5.29 and Proposition 5.35 imply the following:

Corollary 5.37 *Let* $k \geq 1$, *then the simplicial group* $\mathcal{L}^{2k+1}K(\mathbb{Z}, 1)$ *is contractible, i.e.* $\mathcal{L}_i\mathcal{L}^{2k+1}(\mathbb{Z}, 1) = 0$, $i \geq 0$.

Remark 5.38

The topological reason for the validity of Corollary 5.37 is the existence of the Hopf fibration $S^3 \to S^2$, which induces the isomorphism of homotopy groups in dimensions ≥ 3.

The following result is proved in [Dol58b]: for $r \geq 2$,

$$\mathcal{L}_i \, SP^r(\mathbb{Z}, 2) = \begin{cases} \mathbb{Z}, \ i = 2n, \\ 0, \ i \neq 2n \end{cases}$$

This description follows also from Bousfield-Quillen décalage formulae (5.42) and (5.43). As shown in [Dol61], for n even, the homomorphism

$$\pi_{rn}(SP^{r-1}\,K(\mathbb{Z}, n) \otimes K(\mathbb{Z}, n)) \to \mathcal{L}_{rn}\,SP^r(\mathbb{Z}, n)$$

is the r-multiplication $\mathbb{Z} \xrightarrow{r} \mathbb{Z}$. This information is enough to compute the homotopy groups of the simplicial abelian groups $J^r K(\mathbb{Z}, 2)$ for $r \geq 2$. However, we give another way of computation, using the spectral sequence from Theorem 5.32.

Proposition 5.39 *Let* $r \geq 2$. *Then*

$$\mathcal{L}_i J^r(\mathbb{Z}, 2) = \begin{cases} \mathbb{Z}_r, \ i = 2r - 1 \\ 0, \ i \neq 2r - 1 \end{cases}$$

Proof. Consider the functor $J^r(K_1, \ldots, K_n)$ applied to the simplicial group $K(\mathbb{Z}, 1)$. Since all symmetric powers $SP^k K(\mathbb{Z}, 1)$ ($k \geq 2$) are contractible, we have

$$\pi_*(J^r(K_1, \ldots, K_n)) = 0, \ n \neq r, r - 1,$$

where $K_i = K(\mathbb{Z}, 1)$. For the nontrivial components we have

$$\pi_r J^r(K_1, \ldots, K_{r-1}) = \pi_r(J^2(K_1) \otimes K_2 \otimes \ldots \otimes K_{r-1}+$$
$$K_2 \otimes K_1 \otimes K_2 \otimes \cdots \otimes K_{r-1} + \cdots + K_{r-1} \otimes K_1 \otimes \cdots \otimes K_{r-1}) \simeq \mathbb{Z}^{\oplus r-1};$$

$$\pi_r J^r(K_1, \ldots, K_r) =$$
$$\pi_r(K_2 \otimes K_1 \otimes K_3 \otimes \cdots \otimes K_r + \cdots + K_r \otimes K_1 \otimes K_2 \otimes \cdots \otimes K_{r-1}) \simeq \mathbb{Z}^{\oplus r-1},$$

$$\pi_j J^r(K_1, \ldots, K_r) = \pi_j J^r(K_1, \ldots, K_{r-1}) = 0, \ j \neq r.$$

The spectral sequence degenerates to the following form:

$$0 \to \mathcal{L}_{2r} J^r(\mathbb{Z}, 2) \to \pi_r(J^r(K_1, \dots, K_r)) \xrightarrow{\rho}$$
$$\pi_r(J^r(K_1, \dots, K_{r-1})) \to \mathcal{L}_{2r-1} J^r(\mathbb{Z}, 2) \to 0. \quad (5.51)$$

Denote

$$\xi(i_1, \dots, i_r) = \sum_{\sigma=(\sigma_1, \dots, \sigma_r) \in \Sigma_r} (-1)^{sign(\sigma)} [x_{\sigma_1}^{(i_1)}, \dots, x_{\sigma_r}^{(i_r)}].$$

For every $r \geq 3$, the elements

$$\xi(2, 1, 3, \dots, r), \; \xi(3, 1, 2, \dots, r), \; \dots, \; \xi(r, 1, 2, \dots, r-1)$$

represent the basis elements of the group $\pi_r J^r(K_1, \dots, K_r) \simeq \mathbb{Z}^{\oplus r-1}$. The homotopy group $\pi_r J^r(K_1, \dots, K_{r-1})$ has the basis represented by the following elements:

$$c_i = \xi(i, 1, 2, \dots, r-1), \; i = 2, \dots, r-1$$

together with the following element

$$c_1 = \sum_{\sigma=(\sigma_1, \dots, \sigma_r) \in A_r} [x_{\sigma_1}^{(1)}, x_{\sigma_2}^{(1)}, x_{\sigma_3}^{(2)}, \dots, x_{\sigma_r}^{(r-1)}],$$

where A_r is the alternating subgroup of Σ_r. We have the following values for the map ρ defined on the basis of $\pi_r J^r(K_1, \dots, K_r)$:

$$\rho : \xi(2, 1, 3, \dots, r) \mapsto 2c_1 - c_2;$$
$$- \xi(3, 1, 2, \dots, r) \mapsto c_1 + c_2 - c_3;$$
$$\xi(4, 1, 2, \dots, r) \mapsto c_1 + c_3 - c_4;$$
$$\cdots$$
$$\xi(2k, 1, 2, \dots, r) \mapsto c_1 + c_{2k-1} - c_{2k};$$
$$- \xi(2k+1, 1, 2, \dots, r) \mapsto c_1 + c_{2k} - c_{2k+1};$$
$$\cdots$$
$$(-1)^r \xi(r, 1, 2, \dots, r-1) \mapsto c_1 + c_{r-1}.$$

Clearly, the map ρ is a monomorphism, hence $\mathcal{L}_{2r} J^r(\mathbb{Z}, 2) = 0$ by (5.51). Taking the quotient of $\pi_r J^r(K_1, \dots, K_{r-1})$ by the image of the homomorphism ρ, we conclude that

$$\mathcal{L}_{2r-1} J^r(\mathbb{Z}, 2) = \operatorname{coker}(\rho) = \mathbb{Z}_r$$

with a generator represented by the image of the element c_1. □

Corollary 5.40 *Let p be a prime. Then $\mathcal{L}_{2p-1} \mathcal{L}^{2p}(\mathbb{Z}, 1) = \mathbb{Z}_p$.*

Proof. Using the Curtis decomposition and Proposition 5.35, we conclude that only the terms of the form $J^p J^2 K(\mathbb{Z}, 1)$ give a homotopical contribution in the dimension $2p$. Since $J^2 K(\mathbb{Z}, 1)$ has the same homotopy groups as $K(\mathbb{Z}, 2)$, we have

$$\mathfrak{L}_{2p-1} J^p J^2(\mathbb{Z}, 1) = \mathfrak{L}_{2p-1} \mathcal{L}^{2p}(\mathbb{Z}, 1) = \mathbb{Z}_p$$

by Proposition 5.39. $\square$

5.5 Computations and Connectivity Results

Homotopy Groups of $\mathcal{L}^8 K(\mathbb{Z}, 1)$

Proposition 5.33 implies that

$$\mathfrak{L}_i J^2 J^2 J^2(\mathbb{Z}, 1) = \begin{cases} \mathbb{Z}_2, & \text{if } i = 4,\, 5,\, 7, \\ \mathbb{Z}_4, & \text{if } i = 6, \\ 0 & \text{otherwise} \end{cases} \tag{5.52}$$

Proposition 5.35 together with the Curtis decomposition of the functor $\mathcal{L}^8$ implies that only the functor $J^2 J^2 J^2 = \wedge^2 \wedge^2 \wedge^2$ gives the contribution to the groups $\mathfrak{L}_i \mathcal{L}^8(\mathbb{Z}, 1)$ for $i = 4,\, 5$.

   The description (5.52) can be obtained using the suspension spectral sequence from Theorem 5.32. The cross-effects of the functor $\wedge^2 \wedge^2$ can be easily computed as follows:

$$\wedge^2 \wedge^2(A|B) = \wedge^2(A) \otimes A \otimes B \oplus \wedge^2(B) \otimes A \otimes B \oplus$$
$$\wedge^2(A) \otimes \wedge^2(B) \oplus \wedge^2(A \otimes B)$$
$$\wedge^2 \wedge^2(A|B|C) = \wedge^2(A) \otimes B \otimes C \oplus \wedge^2(B) \otimes C \otimes A \oplus \wedge^2(C) \otimes A \otimes B \oplus$$
$$A \otimes B \otimes B \otimes C \oplus A \otimes B \otimes A \otimes C \oplus A \otimes C \otimes B \otimes C$$
$$\wedge^2 \wedge^2(A|B|C|D) = A \otimes B \otimes C \otimes D \oplus A \otimes C \otimes B \otimes D \oplus A \otimes D \otimes B \otimes C.$$

Set $K_1 = K_2 = K_3 = K_4 = K(\mathbb{Z}, 1)$. The initial terms of the suspension spectral sequence for $\wedge^2 \wedge^2 K(\mathbb{Z}, 2)$ are of the form:

$$E_{1,3}^1 = \pi_3(\wedge^2 \wedge^2(K, 1)) = \mathbb{Z}_2$$
$$E_{2,3}^1 = \pi_3(\wedge^2 \wedge^2(K_1|K_2)) = \mathbb{Z}_2$$
$$E_{2,4}^1 = \pi_4(\wedge^2 \wedge^2(K_1|K_2)) = \mathbb{Z}^{\oplus 3}$$
$$E_{3,4}^1 = \pi_4(\wedge^2 \wedge^2(K_1|K_2|K_3)) = \mathbb{Z}^{\oplus 6}$$
$$E_{4,4}^1 = \pi_4(\wedge^2 \wedge^2(K_1|K_2|K_3|K_4)) = \mathbb{Z}^{\oplus 3}$$

with all other terms equal to zero. The differential $E^1_{1,3} \leftarrow E^1_{2,3}$ turns out to be trivial. According to the above decompositions of cross-effects of the functor $\wedge^2 \wedge^2$, denote the generators of $E^1_{4,4}$ by α^1, α^2, α^3, generators of $E^2_{3,4}$ by $\beta^1, \ldots, \beta^6$, generators of $E^1_{2,4}$ by γ^1, γ^2, γ^3. The differentials in the complex

$$E^1_{2,4} \xleftarrow{d^{3,4}_1} E^1_{3,4} \xleftarrow{d^{4,4}_1} E^1_{4,4} \tag{5.53}$$

are defined on the generators as follows:

$$d^{4,4}_1 : \alpha^1 \mapsto 2\beta^1 - 2\beta^2 + 2\beta^3$$

$$\alpha^2 \mapsto \beta^5 - \beta^4 + \beta^6$$

$$\alpha^3 \mapsto \beta^5 - 2\beta^2 + \beta^6$$

$$d^{3,4}_1 : \beta^1 \mapsto \gamma^1 - 2\gamma^3$$

$$\beta^2 \mapsto \gamma^1 - \gamma^2$$

$$\beta^3 \mapsto 2\gamma_3 - \gamma_2$$

$$\beta^4 \mapsto 2\gamma_1 - 2\gamma_2$$

$$\beta^5 \mapsto 2\gamma_1$$

$$\beta^6 \mapsto 2\gamma_2.$$

Simple computations show that the homology groups of the complex (5.53) are $\mathbb{Z}_4$ and $\mathbb{Z}_2$ and the description (5.52) follows from a natural weak equivalence $\wedge^2 K(\mathbb{Z}, 1) \simeq K(\mathbb{Z}, 2)$.

Recall the following simplicial construction from (Remark 3.5 [Bau00]). Let X be a free abelian simplicial group. There is a natural maps

$$\gamma : \pi_i(X) \to \pi_{2i}(\wedge^2 X),$$

given by

$$[x] \mapsto \sum_{(a_1, \ldots, a_i; b_1, \ldots, b_i), \ a_1 = 0} (-1)^{sign(a_1, \ldots, b_i)} [s_{b_i} \ldots s_{b_1} x, s_{a_i} \ldots s_{a_1} x], \quad x \in X_i,$$

where the sum is taken over all (i, i)-shuffles $(a_1 < \cdots < a_i, b_1 < \cdots < b_i)$ which are permutations of $\{0, \ldots, 2i-1\}$ with $a_1 = 0$. The homomorphism γ is a part of the left homomorphism in (A.31), considered in the appropriate dimension. In particular, the homotopy group $\pi_6(J^2 J^2 J^2 K(\mathbb{Z}, 1)) = \Gamma(\mathbb{Z}_2) = \mathbb{Z}_4$ is generated by the image of the map γ, applied to the generator of $\pi_3(J^2 J^2 K(\mathbb{Z}, 1))$. Using the generators (5.44) for $K(\mathbb{Z}, 1)$, we conclude that the generator of $\mathcal{L}_6 J^2 J^2 J^2(\mathbb{Z}, 1)$ is the homotopy class of the element

$$\chi - [[[x_0^{(6)}, x_2^{(6)}], [x_1^{(6)}, x_2^{(6)}]], [[x_3^{(6)}, x_5^{(6)}], [x_4^{(6)}, x_5^{(6)}]]] -$$
$$[[[x_0^{(6)}, x_3^{(6)}], [x_1^{(6)}, x_3^{(6)}]], [[x_2^{(6)}, x_5^{(6)}], [x_4^{(6)}, x_5^{(6)}]]] +$$
$$[[[x_0^{(6)}, x_4^{(6)}], [x_1^{(6)}, x_4^{(6)}]], [[x_2^{(6)}, x_5^{(6)}], [x_3^{(6)}, x_5^{(6)}]]] -$$
$$[[[x_0^{(6)}, x_5^{(6)}], [x_1^{(6)}, x_5^{(6)}]], [[x_2^{(6)}, x_4^{(6)}], [x_3^{(6)}, x_4^{(6)}]]] +$$
$$[[[x_0^{(6)}, x_3^{(6)}], [x_2^{(6)}, x_3^{(6)}]], [[x_1^{(6)}, x_5^{(6)}], [x_4^{(6)}, x_5^{(6)}]]] -$$
$$[[[x_0^{(6)}, x_4^{(6)}], [x_2^{(6)}, x_4^{(6)}]], [[x_1^{(6)}, x_5^{(6)}], [x_3^{(6)}, x_5^{(6)}]]] +$$
$$[[[x_0^{(6)}, x_5^{(6)}], [x_2^{(6)}, x_5^{(6)}]], [[x_1^{(6)}, x_4^{(6)}], [x_3^{(6)}, x_4^{(6)}]]] +$$
$$[[[x_0^{(6)}, x_4^{(6)}], [x_3^{(6)}, x_4^{(6)}]], [[x_1^{(6)}, x_5^{(6)}], [x_2^{(6)}, x_5^{(6)}]]] +$$
$$[[[x_0^{(6)}, x_5^{(6)}], [x_4^{(6)}, x_5^{(6)}]], [[x_1^{(6)}, x_3^{(6)}], [x_2^{(6)}, x_3^{(6)}]]] -$$
$$[[[x_0^{(6)}, x_5^{(6)}], [x_3^{(6)}, x_5^{(6)}]], [[x_1^{(6)}, x_4^{(6)}], [x_2^{(6)}, x_4^{(6)}]]].$$

The Curtis decomposition of the fourth Lie functor is the following:

$$0 \to J^2 J^2(M) \to \mathcal{L}^4(M) \to J^4(M) \to 0$$

(we assume that the abelian group M is free). Therefore, for every free simplicial group X, we have a boundary homomorphism

$$\vartheta_i : \pi_{i+1} J^4(X) \to \pi_i J^2 J^2(X). \tag{5.54}$$

Consider the element

$$\mu =$$

$$\sum_{(i_1, \ldots, i_6) \in A_5, \ i_1 < i_2, \ i_3 < i_4, \ i_3 < i_5, \ i_5 < i_6} [[x_0^{(7)}, x_{i_1}^{(7)}], [x_0^{(7)}, x_{i_2}^{(7)}], [x_{i_3}^{(7)}, x_{i_4}^{(7)}], [x_{i_5}^{(7)}, x_{i_6}^{(7)}]]$$

$$\in J^4 J^2 K(\mathbb{Z}, 1)_7 \tag{5.55}$$

The element μ is presented as the sum of 30 brackets. It is straightforward to check check that $\mu \in J^4 J^2 K(\mathbb{Z}, 1)_7$. The element $\vartheta_6 \mu$ also can be viewed as a sum of 30 brackets, however, 10 brackets from $\vartheta_6 \mu$ can be pair-wisely deleted. Other 20 brackets with the help of Jacobi identity can be transformed to a sum of 10 brackets, which defines exactly the element χ. Since $\mathfrak{L}_7 J^4 J^2(\mathbb{Z}, 1) = \mathbb{Z}_4$ by Proposition 5.39, we can conclude that the element μ defines the generator of $L_7 J^4 J^2(\mathbb{Z}, 1)$ and the map ϑ_6 is an isomorphism.

Now the Curtis decomposition of the functor $\mathcal{L}^8$ implies the following description of derived functors:

$$\mathfrak{L}_i \mathcal{L}^8(\mathbb{Z}, 1) = \begin{cases} \mathbb{Z}_2, & \text{if } i = 4, 5, 7, \\ 0 & \text{otherwise.} \end{cases}$$

The Curtis decomposition of the functor $\mathcal{L}^{2^n}$ implies that only the functors $J^2 \ldots J^2$ (n times) give a contribution to the groups $\mathcal{L}_{n+1}\mathcal{L}^{2^n}(\mathbb{Z}, 1)$ and therefore the simple induction on n gives the following

Proposition 5.41 (Curtis [Cur63]). *For $n \geq 2$, $\mathcal{L}_{n+1}\mathcal{L}^{2^n}(\mathbb{Z}, 1) = \mathbb{Z}_2$.*

Homotopy Groups of $\mathcal{L}^{10}K(\mathbb{Z}, 1)$

The only nontrivial component of the Curtis decomposition of the 10th Lie functor for the simplicial group $K(\mathbb{Z}, 1)$ is $J^5 J^2 K(\mathbb{Z}, 1) = K(\mathbb{Z}_5, 9)$. That is,

$$\mathcal{L}_i \mathcal{L}^{10}(\mathbb{Z}, 1) = \begin{cases} \mathbb{Z}_5, & i = 9 \\ 0, & i \neq 9 \end{cases}$$

by Proposition 5.39.

Homotopy Groups of $\mathcal{L}^{12}K(\mathbb{Z}, 1)$

The nontrivial components of the Curtis decomposition of the 12th Lie functor for the simplicial group $K(\mathbb{Z}, 1)$ are the following:

$$J^2 J^3 J^2, \quad J^3 J^2 J^2, \quad J^4 J^2 \otimes J^2 J^2, \quad J^6 J^2$$

Proposition 5.39 implies that

$$\mathcal{L}_i J^6 J^2(\mathbb{Z}, 1) = \begin{cases} \mathbb{Z}_6, & i = 11 \\ 0, & i \neq 11 \end{cases}$$

The homotopy groups of $J^2 J^3 J^2 K(\mathbb{Z}, 1) = J^2 K(\mathbb{Z}_3, 5)$ can be described due to the universal coefficient sequence for quadratic functors (A.31). We get that $J^2 J^3 J^2 K(\mathbb{Z}, 1)$ is $K(\mathbb{Z}_3, 10)$. The simplicial abelian group $J^4 J^2 K(\mathbb{Z}, 1) \otimes J^2 J^2 K(\mathbb{Z}, 1) = K(\mathbb{Z}_4, 7) \otimes K(\mathbb{Z}_2, 3)$ has the following homotopy groups:

$$\pi_i(J^4 J^2 K(\mathbb{Z}, 1) \otimes J^2 J^2 K(\mathbb{Z}, 1)) = \begin{cases} \mathbb{Z}_2, & i = 10, 11 \\ 0, & i \neq 10, 11 \end{cases}$$

Since for a free abelian group A, the composition map

$$\mathcal{L}^3(A) \to \otimes^3(A) \to \mathcal{L}^3(A)$$

is the multiplication with 3, and $\otimes^3 J^2 J^2 K(\mathbb{Z}, 1) \simeq \otimes^3 K(\mathbb{Z}_2, 3)$ has trivial homotopy groups in dimensions ≤ 8, we conclude that

$$\mathfrak{L}_i J^3 J^2 J^2(\mathbb{Z}, 1) = 0, \ i \leq 8.$$

Homotopy Groups of $\mathcal{L}^{14} K(\mathbb{Z}, 1)$

Using Proposition 5.39, we obtain the contractibility of the functors

$$J^3 J^2 \otimes \mathrm{SP}^2 J^2 J^2, \ J^5 J^2 \otimes J^2 J^2, \ J^2 J^2 J^2 \otimes J^3 J^2, \ J^4 J^2 \otimes J^3 J^2$$

applied to the simplicial group $K(\mathbb{Z}, 1)$. Thus, we have

$$\mathfrak{L}_i \mathcal{L}^{14}(\mathbb{Z}, 1) = \mathfrak{L}_i J^7 J^2(\mathbb{Z}, 1) = \begin{cases} \mathbb{Z}_7, \ i = 13 \\ 0, \ i \neq 13 \end{cases}$$

Homotopy Groups of $\mathcal{L}^{16} K(\mathbb{Z}, 1)$

Nontrivial components of the Curtis decompositions of the 16th Lie functor for $K(\mathbb{Z}, 1)$ are:

$$J^2 J^2 J^2 J^2, \ J^2 J^4 J^2, \ J^4 J^2 J^2, \ J^4 J^2 \otimes \mathrm{SP}^2 J^2 J^2, \ J^6 J^2 \otimes J^2 J^2, \ J^8 J^2.$$

The exact sequence A.31 implies

$$\mathfrak{L}_i J^2 J^2 J^2 J^2(\mathbb{Z}, 1) = \begin{cases} \mathbb{Z}_2, & i = 15 \\ \mathbb{Z}_4 \oplus \mathbb{Z}_2, & i = 14 \\ \mathbb{Z}_4 \oplus \mathbb{Z}_2 \oplus \mathbb{Z}_2 \oplus \mathbb{Z}_2, & i = 13 \\ \mathbb{Z}_2 \oplus \mathbb{Z}_2 \oplus \mathbb{Z}_2 \oplus \mathbb{Z}_2 \oplus \mathbb{Z}_2, & i = 12 \\ \mathbb{Z}_2 \oplus \mathbb{Z}_2 \oplus \mathbb{Z}_2 \oplus \mathbb{Z}_2 \oplus \mathbb{Z}_2 \oplus \mathbb{Z}_2, & i = 11 \\ \mathbb{Z}_4 \oplus \mathbb{Z}_2 \oplus \mathbb{Z}_2 \oplus \mathbb{Z}_2 \oplus \mathbb{Z}_2, & i = 10 \\ \mathbb{Z}_2 \oplus \mathbb{Z}_2 \oplus \mathbb{Z}_2 \oplus \mathbb{Z}_2 \oplus \mathbb{Z}_2, & i = 9 \\ \mathbb{Z}_2 \oplus \mathbb{Z}_2 \oplus \mathbb{Z}_2 \oplus \mathbb{Z}_2, & i = 8 \\ \mathbb{Z}_2 \oplus \mathbb{Z}_2 \oplus \mathbb{Z}_2, & i = 7 \\ \mathbb{Z}_2 \oplus \mathbb{Z}_2, & i = 6 \\ \mathbb{Z}_2, & i = 5 \\ 0, & \text{otherwise} \end{cases}$$

together with other homotopy groups in dimensions ≥ 9. Also the sequence (A.31) implies the complete description of the homotopy groups of

$J^2 J^4 J^2 K(\mathbb{Z}, 1)$:

$$\mathcal{L}_i J^2 J^4 J^2(\mathbb{Z}, 1) = \begin{cases} \mathbb{Z}_2, & i = 8, 9, 10, 11, 12, 13, 15, \\ \mathbb{Z}_8, & i = 14, \\ 0 & \text{otherwise.} \end{cases}$$

Proposition 5.39 and sequence (A.31) imply that

$$\pi_i(J^4 J^2 K(\mathbb{Z}, 1) \otimes \mathrm{SP}^2 J^2 J^2 K(\mathbb{Z}, 1)) = \begin{cases} \mathbb{Z}_2, & i = 12, 15 \\ \mathbb{Z}_2 \oplus \mathbb{Z}_2, & i = 13, 14 \\ 0, & \text{otherwise} \end{cases}$$

$$\pi_i(J^6 J^2 K(\mathbb{Z}, 1) \otimes J^2 J^2(K(\mathbb{Z}, 1))) = \begin{cases} \mathbb{Z}_2, & i = 14, 15 \\ 0, & \text{otherwise} \end{cases}$$

Theorem 5.44 implies that there is an isomorphism

$$\mathcal{L}_{10} \mathrm{SP}^4(\mathbb{Z}_2, 3) \simeq \mathcal{L}_9 J^4 J^2 J^2(\mathbb{Z}, 1) \simeq \mathbb{Z}_2,$$

moreover,

$$\mathcal{L}_i J^4 J^2 J^2(\mathbb{Z}, 1) = 0, \ i \leq 8.$$

The arguments, used in the proof that ϑ_6 is an isomorphism now imply that the map

$$\mathcal{L}_8 J^4 J^2 J^2(\mathbb{Z}, 1) \to \mathcal{L}_7 J^2 J^2 J^2 J^2(\mathbb{Z}, 1)$$

is nontrivial and its image is cyclic. We obtain the following description of homotopy groups

$$\mathcal{L}_i \mathcal{L}^{16}(\mathbb{Z}, 1) = \begin{cases} \mathbb{Z}_2, & i = 5 \\ \mathbb{Z}_2 \oplus \mathbb{Z}_2, & i = 6, 7 \end{cases}$$

with certain other non-trivial homotopy groups in dimensions ≥ 8.

Homotopy Groups of $\mathcal{L}^{18} K(\mathbb{Z}, 1)$

It is easy to see that the only contribution to the homotopy groups of $\mathcal{L}^{18} K(\mathbb{Z}, 1)$ at dimensions ≤ 8 comes from the functor $J^3 J^3 J^2$ in the Curtis decomposition. To detect certain elements of the homotopy groups of $\mathcal{L}^{18} K(\mathbb{Z}, 1)$, we need an information about the structure of the simplicial abelian group $J^p J^p J^2 K(\mathbb{Z}, 1)$ (for $p = 3$).

Simplicial Group $J^p J^p J^2 K(\mathbb{Z}, 1)$

Let $K_i = K(A, m)$ for some $m \geq 2$. Then

$$\pi_i(J^{r_1}(K_1) \otimes \mathrm{SP}^{r_2}(K_2) \otimes \cdots \otimes \mathrm{SP}^{r_n}(K_n)) = 0,$$

$$i < 2r_1 + m - 3 + \sum_{i=2}^{n}(2r_i + m - 2) = 2r + n(m - 2) - 1$$

$$\pi_i(K_i \otimes \mathrm{SP}^{r_1}(K_1) \otimes \cdots \otimes \mathrm{SP}^{r_n}(K_n)) = 0,$$

$$i < m + \sum_{i=1}^{n}(2r_i - 2 + m) = 2(r - 1) + (m - 2)n + m.$$

Thus, for $m > 3$, we have

$$\pi_i(J^r(K_1, \ldots, K_n)) = 0, \; i < 2r + n(m - 2) - 1.$$

Therefore, for $i \leq 2r + 2(m - 2) - 1$, we have

$$\mathfrak{L}_{i+1} J^r(A, m + 1) = \mathfrak{L}_i J^r(A, m).$$

Therefore,

$$\mathfrak{L}_i J^r(A, m) = \mathfrak{L}_{i-1} J^r(A, m - 1), \; i \leq 2r + 2(m - 3),$$

provided $m > 3$. For the particular case p an odd prime and $K(A, m) = K(\mathbb{Z}_p, 2p - 1)$, we have

$$\mathfrak{L}_{4p-4} J^p(\mathbb{Z}_p, 2p - 1) = \mathfrak{L}_{2p} J^p(\mathbb{Z}_p, 3),$$
$$\mathfrak{L}_{4p-3} J^p(\mathbb{Z}_p, 2p - 1) = \mathfrak{L}_{2p+1} J^p(\mathbb{Z}_p, 3),$$
$$\mathfrak{L}_{4p-2} J^p(\mathbb{Z}_p, 2p - 1) = \mathfrak{L}_{2p+2} J^p(\mathbb{Z}_p, 3),$$
$$\mathfrak{L}_{4p-1} J^p(\mathbb{Z}_p, 2p - 1) = \mathfrak{L}_{2p+3} J^p(\mathbb{Z}_p, 3), \; p > 3$$

The following result is an immediate consequence of the exact sequence for the functor SP^2 analogous to (A.31) given by [Bau00]:

Proposition 5.42 *Let X be a free abelian simplicial group with $\pi_i(X) = 0$, $i < k$ and $\pi_j(X) \otimes \mathbb{Z}_2 = \mathrm{Tor}(\pi_j(X), \mathbb{Z}_2) = 0$ for all j. Then $\pi_i(\mathrm{SP}^2(X)) = 0$, $i < 2k + 1$.*

Let X be a free abelian simplicial group with $\pi_i(X) = 0$, $i < k$. Then, clearly, $\pi_i(\otimes^n(X)) = 0$, $i < nk$. Since the composition

$$\mathcal{L}^t(M) \to \otimes^t(M) \to \mathcal{L}^t(M)$$

is just the multiplication with t, we have that

$$t\pi_i \mathcal{L}^t(X) = 0, \ t < nk.$$

Using the induction, we conclude:

Proposition 5.43 *Let X be a free abelian simplicial group with $\pi_i(X) = 0$, $i < k$ and $\pi_j(X)$ are finite p-torsion free for all j. Then $\pi_i(\mathrm{SP}^p(X)) = 0$, $i < pk$.*

Let p be an odd prime. We have $\pi_i(\mathrm{SP}^{p-1} K(\mathbb{Z}_p, 3) \otimes K(\mathbb{Z}_p, 3)) = 0$, $i < 3p$. Hence

$$\mathcal{L}_{4p-4} J^p(\mathbb{Z}_p, 2p-1) = \mathcal{L}_{2p} J^p(\mathbb{Z}_p, 3) = \mathcal{L}_{2p+1} \mathrm{SP}^p(\mathbb{Z}_p, 3),$$
$$\mathcal{L}_{4p-3} J^p(\mathbb{Z}_p, 2p-1) = \mathcal{L}_{2p+1} J^p(\mathbb{Z}_p, 3) = \mathcal{L}_{2p+2} \mathrm{SP}^p(\mathbb{Z}_p, 3),$$
$$\mathcal{L}_{4p-2} J^p(\mathbb{Z}_p, 2p-1) = \mathcal{L}_{2p+2} J^p(\mathbb{Z}_p, 3) = \mathcal{L}_{2p+3} \mathrm{SP}^p(\mathbb{Z}_p, 3),$$
$$\mathcal{L}_{4p-1} J^p(\mathbb{Z}_p, 2p-1) = \mathcal{L}_{2p+3} J^p(\mathbb{Z}_p, 3) = \mathcal{L}_{2p+4} \mathrm{SP}^p(\mathbb{Z}_p, 3), \ p > 3$$

The following statements are due to Dold and Puppe:

Theorem 5.44 (Dold - Puppe [Dol61, Satz 12.10]). *Let X be a free simplicial abelian group with $\pi_i(X) = 0$, $i < k$ ($k > 1$). Then*
(1) *the suspension homomorphism*

$$\pi_i(\mathrm{SP}^n(X)) \to \pi_{i+1}(\mathrm{SP}^n(\Sigma X))$$

is an isomorphism for $i < 2k + 2n - 4$ and epimorphism for $i = 2k + 2n - 4$,
(2) *if p a prime, $n = p^r$, $r > 0$*

$$\pi_{k+2n-2}(\mathrm{SP}^n(X)) \simeq \pi_k(X) \otimes \mathbb{Z}_p.$$

(3) *if p a prime, $n = p^r$, $r > 0$,*

$$p\pi_i(\mathrm{SP}^n(X)) = 0, \ i < 2k + 2n - 4.$$

Taking $n = p$, $X = K(\mathbb{Z}_p, 3)$, Theorem 5.44 2) implies that

$$\mathcal{L}_{2p+1} \mathrm{SP}^p(\mathbb{Z}_p, 3) = \mathbb{Z}_p.$$

Hence,
$$\mathcal{L}_{4p-4} J^p(\mathbb{Z}_p, 2p-1) = \mathbb{Z}_p. \tag{5.56}$$

In particular, $\mathcal{L}_8 J^3 J^3 J^2(\mathbb{Z}, 1) = \mathbb{Z}_3$.

We can complete the results of the above analysis in the following table of the lower terms of the Curtis spectral sequence:

n	1	2	4	6	8	10	12	14	16	18
$\mathfrak{L}_8\mathcal{L}^n(\mathbb{Z},1)$	0	0	0	0	0	0	0	0	?	$\mathbb{Z}_3$
$\mathfrak{L}_7\mathcal{L}^n(\mathbb{Z},1)$	0	0	0	0	$\mathbb{Z}_2$	0	0	0	$\mathbb{Z}_2\oplus\mathbb{Z}_2$	0
$\mathfrak{L}_6\mathcal{L}^n(\mathbb{Z},1)$	0	0	0	0	0	0	0	0	$\mathbb{Z}_2\oplus\mathbb{Z}_2$	0
$\mathfrak{L}_5\mathcal{L}^n(\mathbb{Z},1)$	0	0	0	$\mathbb{Z}_3$	$\mathbb{Z}_2$	0	0	0	$\mathbb{Z}_2$	0
$\mathfrak{L}_4\mathcal{L}^n(\mathbb{Z},1)$	0	0	0	0	$\mathbb{Z}_2$	0	0	0	0	0
$\mathfrak{L}_3\mathcal{L}^n(\mathbb{Z},1)$	0	0	$\mathbb{Z}_2$	0	0	0	0	0	0	0
$\mathfrak{L}_2\mathcal{L}^n(\mathbb{Z},1)$	0	$\mathbb{Z}$	0	0	0	0	0	0	0	0
$\mathfrak{L}_1\mathcal{L}^n(\mathbb{Z},1)$	$\mathbb{Z}$	0	0	0	0	0	0	0	0	0

As a corollary we get the low dimensional homotopy groups of the 2-sphere:

$$\pi_2(S^2) = \mathbb{Z},$$
$$\pi_3(S^2) = \mathbb{Z},$$
$$\pi_4(S^2) = \mathbb{Z}_2,$$
$$\pi_5(S^2) = \mathbb{Z}_2.$$

Also we have that the group $\pi_6(S^2)$ has 12 elements. Another well-known result, which easily follows by the described methods is that the groups $\pi_7(S^2)$ and $\pi_8(S^2)$ are 2-groups. The simplest non-trivial differential in the Curtis spectral sequence is given by

$$d_8 : \mathfrak{L}_7\mathcal{L}^8(\mathbb{Z},1) \to \mathfrak{L}_6\mathcal{L}^{16}(\mathbb{Z},1).$$

p-torsion in $\pi_*(S^2)$

The following well-known result, due to Serre, gives some information about p-torsion in homotopy groups of S^2:

Theorem 5.45 (Serre [Ser51]). *The p-torsion of groups $\pi_i(S^2)$ is $\mathbb{Z}_p$ for $i = 2p$, $4p - 3$ and trivial for $i < 2p$ and $2p < i < 4p - 3$.*

Here we deduce this result as a consequence of a simple analysis of the Curtis spectral sequence.

Lemma 5.46 *Let p be an odd prime. Then $\mathfrak{L}_i\mathcal{L}^j(\mathbb{Z},1)$ does not contain p-torsion for $i < 2p - 1$.*

Proof. Suppose $\mathfrak{L}_i\mathcal{L}^j(\mathbb{Z},1)$ contains a subgroup $\mathbb{Z}_p$ for some $i < 2p-1$. Then the Curtis decomposition implies that there is a sequence $(i_1, \dots, i_m)$, such that

$$\mathfrak{L}_i J^{i_m} J^{i_{m-1}} \dots J^{i_1}(\mathbb{Z},1) \supseteq \mathbb{Z}_p. \tag{5.57}$$

Proposition 5.35 implies that $i_1 = 2$. Then $J^{i_2} J^{i_1} K(\mathbb{Z},1)$ is $K(\mathbb{Z}_{i_2}, 2i_2 - 1)$. Suppose $(i_2, p) = 1$. In this case one has the triviality of the p-torsion part $\mathrm{Tor}(\mathfrak{L}_i J^{i_m} J^{i_{m-1}} \ldots J^{i_1}(\mathbb{Z}, 1), \mathbb{Z}_p) = 0$ and we have a contradiction with (5.57). Hence $i_2 = p \cdot c$, $c \geq 1$. In this case $J^{i_2} J^{i_1} K(\mathbb{Z}, 1)$ is $(2p \cdot c - 2)$-connected. Therefore, the simplicial group $J^{i_m} J^{i_{m-1}} \ldots J^{i_1} K(\mathbb{Z}, 1)$ is also $(2p \cdot c - 2)$-connected, which contradicts (5.57). Hence the assumption is not possible and the statement is proved. $\square$

Analogous analysis of the triple composition functors $J^{i_3} J^{i_2} J^{i_1} K(\mathbb{Z}, 1)$ implies the following:

Lemma 5.47 *Let p be an odd prime. Then $\mathfrak{L}_i \mathcal{L}^j(\mathbb{Z}, 1)$ does not contain p-torsion for $2p - 1 < i < 4p - 4$.*

One can easily check that $\mathrm{Tor}(\mathfrak{L}_{4p-4} \mathcal{L}^n(\mathbb{Z}, 1), \mathbb{Z}_p) = 0$ for $n \neq 2p^2$ and $\mathrm{Tor}(\mathfrak{L}_{4p-3} \mathcal{L}^n(\mathbb{Z}, 1), \mathbb{Z}_p) = 0$ for $n < 2p^2$. Now Theorem 5.45 follows from Corollary 5.40, isomorphism (5.56), Lemmas 5.46, 5.47 and the spectral sequence argument.

Derived Series Filtration

One can easily see the analogy between Curtis decomposition of Lie functors and derived series methods used for the proof of Theorem 3.2 (compare definition of $\gamma_{(n)}(F^{(k)})$ with definition of the metabelian Lie functors $J^n : \mathsf{Ab} \to \mathsf{Ab}$). This analogy gives a natural motivation to define a derived series filtration for homotopy groups of simplicial groups.

For a given simplicial group G, consider the following filtration of the graded abelian groups:

$$\pi_*(G) \supseteq \pi_*^{(1)}(G) \supseteq \pi_*^{(2)}(G) \supseteq \ldots,$$

where

$$\pi_*^{(i)}(G) = \ker\{\pi_*(G) \to \pi_*(G/\delta_i(G))\}, \quad i = 1, 2, \ldots$$

and the dual filtration:

$$\bar{\pi}_*^{(1)}(G) \subseteq \bar{\pi}_*^{(2)}(G) \subseteq \cdots \subseteq \pi_*(G),$$

defined as $\bar{\pi}_*^{(i)}(G) = \pi_*(G) \setminus \pi_*^{(i)}(G)$, $i = 1, 2, \ldots$.

For the free simplicial group $F[S^1]$, the above analysis shows that

$$\bar{\pi}_1^{(1)}(F[S^1]) = \pi_1(F[S^1]), \quad \bar{\pi}_i^{(1)}(F[S^1]) = 0, \ i > 1,$$

$$\bar{\pi}_2^{(2)}(F[S^1]) = \pi_2(F[S^1]), \quad \bar{\pi}_i^{(2)}(F[S^1]) = 0, \ i > 2,$$

$$\bar{\pi}_{2p-1}^{(3)}(F[S^1]) \ni \text{Serre's } p\text{-element, which defines the } p\text{-torsion of } \pi_{2p}(S^2),$$

$\bar{\pi}_{4p-4}^{(4)}(F[S^1]) \ni$ Serre's p-element, which defines the p-torsion of $\pi_{4p-3}(S^2)$,

$\bar{\pi}_5^{(5)}(F[S^1]) \ni$ 2-torsion element of $\pi_6(S^2)$,

$\bar{\pi}_6^{(5)}(F[S^1]) \ni$ generator of $\pi_7(S^2)$.

Roughly speaking, the ith term of this derived series filtration consists of the homotopy elements which come from derived functors of components of the Lie functor which can be decomposed as a composition of $(i-1)$ metabelian Lie functors, i.e., the J-functors.

Connectivity Results

Consider the Koszul complexes (5.40) and (5.41). Define the functors

$$V_{i,n}, \ \overline{V}_{i,n} : \mathsf{Ab} \to \mathsf{Ab}, \ i = 1, \dots, n$$

by setting

$$V_{i,n}(P) = \ker(d_i) = \ker\{\wedge^i(P) \otimes \mathrm{SP}^{n-i}(P) \to \wedge^{i-1}(P) \otimes \mathrm{SP}^{n-i+1}(P)\}$$
$$\overline{V}_{i,n}(P) = \ker(\bar{d}_i) = \ker\{\Gamma_i(P) \otimes \wedge^{n-i}(P) \to \Gamma_{i-1}(P) \otimes \wedge^{n-i+1}(P)\}$$

The functors $V_{i,n}$, $\overline{V}_{i,n}$ are particular cases of the so-called Schur functors. By definition, $V_{1,n}(P) = J^n(P)$.

For a free group P and the identity homomorphism $f : P \to P$, the Koszul complexes $\mathrm{Kos}^n(f)$, $\overline{\mathrm{Kos}}^n(f)$ are acyclic. Hence, for every free abelian group P, we have the following sequence of short exact sequences:

$$0 \to V_{1,n}(P) \to \mathrm{SP}^{n-1}(P) \otimes P \to \mathrm{SP}^n(P) \to 0,$$
$$0 \to V_{2,n}(P) \to \mathrm{SP}^{n-2}(P) \otimes \wedge^2(P) \to V_{1,n}(P) \to 0,$$
$$\cdots$$
$$0 \to V_{n,n}(P) \to \wedge^n(P) \to V_{n-1,n}(P) \to 0,$$
$$0 \to \overline{V}_{1,n}(P) \to \wedge^{n-1}(P) \otimes P \to \wedge^n(P) \to 0,$$
$$0 \to \overline{V}_{2,n}(P) \to \wedge^{n-2}(P) \otimes \Gamma_2(P) \to \overline{V}_{1,n}(P) \to 0,$$
$$\cdots,$$
$$0 \to \overline{V}_{n,n}(P) \to \Gamma_n(P) \to \overline{V}_{n-1,n}(P) \to 0.$$

Hence, for every free simplicial abelian group X, we obtain the following exact sequences of homotopy groups:

$$\cdots \to \pi_{k+1}(\mathrm{SP}^n(X)) \to \pi_k(V_{1,n}(X))$$
$$\to \pi_k(\mathrm{SP}^{n-1}(X) \otimes X) \to \pi_k(\mathrm{SP}^n(X)) \to \dots,$$

$$\cdots \to \pi_{k+1}(V_{1,n}(X)) \to \pi_k(V_{2,n}(X))$$
$$\to \pi_k(\mathrm{SP}^{n-2}(X) \otimes \wedge^2(X)) \to \pi_k(V_{1,n}(X)) \to \cdots,$$

$$\cdots$$

$$\cdots \to \pi_{k+1}(V_{n-1,n}(X)) \to \pi_k(V_{n,n}(X))$$
$$\to \pi_k(\wedge^n(X)) \to \pi_k(V_{n-1,n}(X)) \to \cdots,$$

$$\cdots \to \pi_{k+1}(\wedge^n(X)) \to \pi_k(\overline{V}_{1,n}(X))$$
$$\to \pi_k(\wedge^{n-1}(X) \otimes X) \to \pi_k(\wedge^n(X)) \to \cdots,$$

$$\cdots \to \pi_{k+1}(\overline{V}_{1,n}(X)) \to \pi_k(\overline{V}_{2,n}(X))$$
$$\to \pi_k(\wedge^{n-2}(X) \otimes \Gamma_2(X)) \to \pi_k(\overline{V}_{1,n}(X)) \to \cdots,$$

$$\cdots$$

$$\cdots \to \pi_{k+1}\overline{V}_{n-1,n}(X) \to \pi_k(\overline{V}_{n,n}(X))$$
$$\to \pi_k(\Gamma_n(X)) \to \pi_k(\overline{V}_{n-1,n}(X)) \to \cdots.$$

Hence, the inductive argument and (A.15) imply the following:

Theorem 5.48 (Dold - Puppe [Dol61, Satz 12.1]). *Let X be a free abelian simplicial group and $k \geq 1$, $n \geq 2$ be integers. If $\pi_i(X) = 0$, $i < k$, then*

$$\pi_i(\mathrm{SP}^n(X)) = 0, \begin{cases} \text{for } i < n, \text{ provided } k = 1, \\ \text{for } i < k + 2n - 2, \text{ provided } k > 1. \end{cases}$$

The above result yields the following

Corollary 5.49 *Let X be a free abelian simplicial group and $r \geq 2$, $k \geq 1$ be integers. If $\pi_i(X) = 0$, $i < m$, then*

$$\pi_i(J^r(X)) = 0, \begin{cases} \text{for } i < r - 1, \text{ provided } m = 1, \\ \text{for } i < m + 2r - 3, \text{ provided } m > 1. \end{cases}$$

Proof. Proposition 5.28 implies that there exists an exact sequence of homotopy groups

$$\cdots \to \pi_{i+1}(\mathrm{SP}^r(X)) \to \pi_i(J^r(X)) \to \pi_i(X \otimes \mathrm{SP}^{r-1}(X)) \to \cdots$$

In view of Theorem 5.48 and Künneth Theorem (see (A.15)), if $k > 1$, we have

$$\pi_i(J^r(X)) = 0, \ i < \min\{m + 2r - 3, \ 2m + 2r - 4\} = m + 2r - 3;$$

furthermore, if $m = 1$ then $\pi_i(J^r(X)) = 0, \ i < r - 1$. $\square$

Proof of Theorem 5.27 In view of Kan's version of Hurewicz Theorem for free simplicial groups (see A.10), it suffices to prove the following:

Let X be a free abelian simplicial group with $\pi_i(X) = 0$, $i \leq m$, then $\pi_i(\mathcal{L}^r(X)) = 0$, $i \leq \{m + \log_2 r\}$, where for an integer a, $\{a\}$ is the least integer greater or equal than a.

In view of the Curtis decomposition of the functor $\mathcal{L}^r(X)$ (see Theorem 5.29), it is enough to prove the above statement for the components $G^t L^r(X)$, $t \in B(r)$. For a given $t \in B(r)$ there is a unique decomposition $t = t_1 \ldots t_{h(t)}$, where all t_i-s are basic types. Observe that $h(t) \geq \log_2 r$. Therefore, it suffices to prove the following

Let X be a free abelian simplicial group with $\pi_i(X) = 0$, $i \leq m$, then $\pi_i(G^t L^r(X)) = 0$, $i \leq m + h(t)$.

We proceed by induction on $r \geq 2$. For $r = 2$ the statement follows from Remark 5.34 and Corollary 5.49. Suppose $r > 2$ and the statement holds for all $2 \leq r' < r$.
Case I: Let $t \in B^1(r; a)$. Then

$$G^t L^r(X) \simeq G^{t_0 \ldots t_{k-a}} L^{\mathrm{wt}(t_0 \ldots t_{k-a})}(X) \otimes \mathrm{SP}^a(G^{t_k} L^{\mathrm{wt}(t_k)}(X))$$

for some integers k, a and $\mathrm{wt}(t_k) \geq 2$. By induction hypothesis,

$$\pi_i(G^{t_0 \ldots t_{k-a}} L^{\mathrm{wt}(t_0 \ldots t_{k-a})}(X)) = 0, \ i \leq m + k - a.$$

The case $a = 1$ clearly follows from the Künneth formula (A.15). Suppose $a > 1$. Applying Corollary 5.49 and induction hypothesis, we conclude that

$$\pi_i(\mathrm{SP}^a(G^{t_k} L^{\mathrm{wt}(t_k)}(X))) = 0, \ i \leq \{m + \log_2 \mathrm{wt}(t_k)\} + 2a - 2.$$

The Künneth formula (A.15) then shows that

$$\pi_i(G^t L^r(X)) = 0, \ i \leq m + k + a - 1 + \{m + \log_2 \mathrm{wt}(t_k)\} (\geq m + h(t))$$

and the needed statement follows.
Case II: Let $t \in B^2(r; a)$. Then

$$G^t L^r(X) \simeq J^{a+1}(G^{t_k} L^{\mathrm{wt}(t_k)}(X))$$

for some integers k, a. By induction hypothesis,

$$\pi_i(G^{t_k} L^{\text{wt}(t_k)}(X)) = 0, \ i \leq m + h(t) - a.$$

Since $h(t) - a \geq 1$, the needed statement follows from Corollary 5.49.
Case III: Let $t = t^{gr}$. The statement follows from Corollary 5.49 and Remark 5.34. $\square$

p-Local Version

Proposition 5.50 *Let X be a free abelian simplicial group with $\pi_i(X) = 0$, $i < m$, p a prime, $r < p$, then $\text{Tor}(\pi_i(\text{SP}^r(X)), \mathbb{Z}_p) = 0$, $i < rm$.*

Proof. For $r = 2$ the statement follows from (A.31) and the exact sequence

$$0 \to \wedge^2(X) \to \otimes^2(X) \to \text{SP}^2(X) \to 0.$$

Suppose the statement is true for all $s < p$. If $s + 1 < p$, then consider the following composition map

$$\text{SP}^{s+1}(X) \to \text{SP}^s(X) \otimes X \to \text{SP}^{s+1}(X),$$

which is induced by multiplication with $\binom{s+1}{1} = s + 1$ (see [Dol61], Korollar 10.9) Since $\text{Tor}(\pi_i(\text{SP}^s(X)), \mathbb{Z}_p) = 0$, $i < sm$, we get

$$\text{Tor}(\pi_i(\text{SP}^s(X) \otimes X), \mathbb{Z}_p) = 0, \ i < (s+1)m.$$

Therefore, $(s + 1)\pi_i(\text{SP}^{s+1}(X)) = 0$, $i < (s+1)m$ and the needed statement follows. $\square$

Theorem 5.51 *Let p be a prime, X a free simplicial group, such that $\pi_i(X) = 0$ for $i \leq m$ $(m \geq p)$. Then $\pi_i(\gamma_r(X)/\gamma_{r+1}(X))$ is p-torsion free for $i \leq \{m + (2p - 3)\log_p r\}$.*

Proof. The case $p = 2$ follows from Theorem 5.27. We assume $p > 2$. Again, in view of the Curtis decomposition of the functor $\mathcal{L}^r(X)$ (see Theorem 5.29), it is enough to prove the above statement for the components $G^t L^r(X)$, $t \in B(r)$. For a given $t \in B(r)$ consider its decomposition $t = t_1 \ldots t_k$, where all t_i-s are basic types.
Case I. First let $t = t^{gr}$ and $G^t L^r(X) = J^r(X)$. For $r \geq p$ the statement follows from Corollary 5.49, since

$$2r - 3 \geq (2p - 3)\log_p r \ (r \geq p).$$

Suppose that $r < p$. We have the following exact sequence

$$\cdots \to \pi_{i+1}(\text{SP}^r(X)) \to \pi_i(J^r(X)) \to \pi_i(\text{SP}^{r-1}(X) \otimes X) \to \cdots$$

and we conclude that $\mathrm{Tor}(\pi_i(J^r(X)), \mathbb{Z}_p) = 0$, $i \leq rm-1$ due to Proposition 5.50. We have

$$m + (2p-3)\log_p r \leq rm - 1 \ (m \geq p)$$

and the needed statement follows.

Case II. Let $t \in B^1(r; a)$. Then

$$G^t L^r(X) \simeq G^{t_0 \ldots t_{k-a}} L^{\mathrm{wt}(t_0 \ldots t_{k-a})}(X) \otimes \mathrm{SP}^a(G^{t_k} L^{\mathrm{wt}(t_k)}(X))$$

for some integers k, a and $\mathrm{wt}(t_k) \geq 2$. By induction hypothesis,

$$\mathrm{Tor}(\pi_i(G^{t_0 \ldots t_{k-a}} L^{\mathrm{wt}(t_0 \ldots t_{k-a})}(X)), \mathbb{Z}_p) = 0,$$
$$i \leq \{m + (2p-3)\log_p \mathrm{wt}(t_0 \ldots t_{k-a})\},$$

$$\mathrm{Tor}(\pi_i(G^{t_k} L^{\mathrm{wt}(t_k)}), \mathbb{Z}_p) = 0, \ i \leq \{m + (2p-3)\log_p \mathrm{wt}(t_k)\}.$$

The case $a = 1$ follows from the Künneth formula. Suppose $a > 1$. Applying Theorem 5.48 induction hypothesis and Künneth formula, we conclude that

$$\mathrm{Tor}(\pi_i(G^t L^r(X)), \mathbb{Z}_p) = 0,$$
$$i \leq \{m + (2p-3)\log_p \mathrm{wt}(t_0 \ldots t_{k-a})\} + 2a - 2 + \{m + (2p-3)\log_p \mathrm{wt}(t_k)\}$$
$$(\geq m + (2p-3)\log_p r)$$

and the needed statement follows.

Case III. Now suppose that $t \in B^2(r; a)$ and

$$G^t L^r(X) \simeq J^{a+1}(G^{t_k} L^{\mathrm{wt}(t_k)}(X))$$

for some integers k, a. By induction hypothesis,

$$\mathrm{Tor}(\pi_i(G^{t_k} L^{\mathrm{wt}(t_k)}(X)), \mathbb{Z}_p) = 0, \ i \leq m + (2p-3)\log_p \mathrm{wt}(t_k).$$

The needed statement follows from Case I. $\square$

Remark 5.52

The statement of Theorem 5.51 can not be generalized for all $m \geq 0$ with the same valuation. For example, one has

$$\pi_4 J^3 K(\mathbb{Z}_3, 1) \supseteq \mathbb{Z}_3.$$

Remark 5.53

Take $A = K(\mathbb{Z}, 2)$. Then for a prime $p > 2$, one has

$$\pi_i(\underbrace{J^p \ldots J^p}_{n \ terms} A) = 0, \ i < (2p-3)n$$

by Corollary 5.49. The exact sequence

$$\pi_{i+1}(\mathrm{SP}^p \underbrace{J^p \ldots J^p}_{n-1 \ terms} A) \to \pi_i(\underbrace{J^p \ldots J^p}_{n-1 \ terms} A) \to$$

$$\pi_i(\mathrm{SP}^{p-1} \underbrace{J^p \ldots J^p}_{n-1 \ terms} A \otimes \underbrace{J^p \ldots J^p}_{n-1 \ terms} A)$$

and Proposition 5.50 imply the isomorphism

$$\pi_{2p-2}(\mathrm{SP}^p \underbrace{J^p \ldots J^p}_{n-1 \ terms} A) \simeq \pi_{2p-3}(\underbrace{J^p \ldots J^p}_{n \ terms} A).$$

The induction argument together with Theorem 5.44 2) imply that

$$\pi_{(2p-3)n}(\underbrace{J^p \ldots J^p}_{n \ terms} A) = \mathbb{Z}_p.$$

Analyzing the connectivity of components in the Curtis decomposition of $\mathcal{L}^{p^n}$, one can see that only the component $\underbrace{J^p \ldots J^p}_{n \ terms}$ gives a contribution at dimension $(2p-3)n$ and there is an isomorphism

$$\pi_{(2p-3)n}(\mathcal{L}^{p^n} A) = \mathbb{Z}_p, \ n \geq 1.$$

Suspensions

Here we recall from [Ell08] a simple group-theoretical interpretation of nilpotence of stable homotopy elements.

Consider $n \geq 1$ and Milnor's construction $F[S^n]$. We have the Curtis spectral sequence (5.33), which converges to homotopy groups of the $(n+1)$-sphere S^{n+1}. Theorem 5.27 and the sequence (A.31) imply the following:

$$\pi_{n+1}(\mathcal{L}^i K(\mathbb{Z}, n)) = \begin{cases} \mathbb{Z}, & i = 2, \ n = 1 \\ \mathbb{Z}_2, & i = 2, \ n \geq 1 \\ 0, & i > 2 \end{cases}$$

Therefore,

$$\pi_1^S = \pi_{n+2}(S^{n+1}) = \mathbb{Z}_2, \ n > 1.$$

Recall the argumets from [Wu,01, Example 2.21], where the fact that $\pi_1^S = \mathbb{Z}_2$ was proved using commutator calculus in $F[S^n]$. Consider the lower terms of the simplicial group $F[S^n]$:

$$F[S^n]_n - F(\sigma),$$
$$F[S^n]_{n+1} = F(s_0\sigma, \ldots, s_n\sigma),$$
$$F[S^n]_{n+2} = F(s_j s_i \sigma \mid i < j),$$

$$\cdots$$

In the simplicial group $F[S^n]$ the following relations take place for $i + 1 < j \leq n$:

$$\partial_k([s_{j-1}s_i\sigma, s_{j+1}s_i\sigma]) = \begin{cases} [s_i\sigma, s_j\sigma], & k = j \\ 1, & k \neq j \end{cases} \tag{5.58}$$

$$\partial_k([s_{i+2}s_{i+1}\sigma, s_{i+3}s_i\sigma]) = \begin{cases} [s_{i+1}\sigma, s_{i+2}\sigma], & k = i+1 \\ [s_{i+1}\sigma, s_i\sigma], & k = i+3 \\ 1 \text{ otherwise} \end{cases} \tag{5.59}$$

$$\partial_k([s_{i+2}s_i\sigma, s_{i+3}s_{i+1}\sigma]) = \begin{cases} [s_{i+1}\sigma, s_{i+2}\sigma], & k = i+1 \\ [s_i\sigma, s_{i+2}\sigma], & k = i+2 \\ [s_i\sigma, s_{i+1}\sigma], & k = i+3 \\ 1 \text{ otherwise} \end{cases} \tag{5.60}$$

Recall now the Homotopy Addition Lemma (see [Cur71], Theorem 2.4):

Theorem 5.54 *Let G be a simplicial group, $y_i \in G_n$, $0 \leq i \leq n+1$, $d_j y_i = 1$, $0 \leq j \leq n$. Then the following relation holds in $\pi_n(G)$*

$$[y_0] - [y_1] + [y_2] - \cdots + (-1)^{n+1}[y_{n+1}] = 0$$

(the operation is written additively) if and only if there exists an element $y \in G_{n+1}$ such that $d_i y = y_i$, $0 \leq i \leq n+1$.

Theorem 5.54 together with (5.58) imply that

$$[s_i\sigma, s_j\sigma] \in B(F[S^n]), \ i+1 < j. \tag{5.61}$$

Theorem 5.54 together with (5.59) imply that

$$[s_{i+1}\sigma, s_{i+2}\sigma][s_{i+1}\sigma, s_i\sigma] \in B(F[S^n]). \tag{5.62}$$

Analogously, (5.60) imply

$$[s_{i+1}\sigma, s_{i+2}\sigma][s_i\sigma, s_{i+2}\sigma]^{-1}[s_i\sigma, s_{i+1}\sigma] \in B(F[S^n]).$$

Hence,

$$[s_{i+1}\sigma, s_{i+2}\sigma][s_i\sigma, s_{i+1}\sigma] \in B(F[S^n]) \tag{5.63}$$

in view of (5.61). Now (5.62) and (5.63) imply that

$$[s_i\sigma, s_{i+1}\sigma]^2 \in B(F[S^n]).$$

Since the natural map $F[S^n] \to F[S^n]/\gamma_3(F[S^n])$ of simplicial groups induces isomorphism

$$\pi_{n+1}(F[S^n]) \to \pi_{n+1}(F[S^n]/\gamma_3(F[S^n])),$$

we can choose the generators of $F[S^n]_{n+1}$ which define the non-trivial class in $\pi_{n+1}(F[S^n])$ as $[s_i\sigma, s_{i+1}\sigma]$ for every $0 \le i \le n-1$.

It is easy to see that we can take also the element $[s_0\sigma, s_1\sigma]$ in $F[S^1]_2$ as a generator of $\pi_2(F[S^1]) = \pi_3(S^2) = \mathbb{Z}$. Let us denote

$$F[S^1]_k = F_k = F(x_0, \ldots, x_{k-1})$$

with $x_i = s_{k-1} \ldots \hat{s}_i \ldots s_0\sigma$. Following [Wu,01], define another set of generators of F_k by setting:

$$y_j = x_j x_{j+1}^{-1}, \; j = 0, \ldots, k-2, \; y_{k-1} = x_{k-1}.$$

It directly follows from the definition of homotopy groups of $F[S^1]$, that $\pi_k(F[S^1])$ can be identified with a certain quotient of the intersection subgroup $\langle y_0 \rangle^{F_k} \cap \ldots \cap \langle y_{k-2} \rangle^{F_k} \cap \langle y_{-1} \rangle^{F_k}$, where $y_{-1} = (y_0 \ldots y_{k-1})^{-1}$. As it was shown in [Wu,01], this quotient is exactly the one, given in (A.21), i.e. the boundaries in $F[S^1]_k$ can be identified with normal subgroup $[[y_{-1}, \ldots, y_{k-1}]]$.

Now the generators of $F[S^n]_{n+1}$ which define nontrivial homotopy class in the first stable homotopy group π_1^S define the elements from $F[S^1]_k$, which correspond the homotopy classes of composition maps

$$S^{k+1} \to S^k \to \cdots \to S^3 \to S^2,$$

where every map is viewed as a suspension over the Hopf fibration. Consider these elements.

1. First, let $F_2 = F(y_0, y_1)$, then the element

$$[y_0, y_1] \notin [[y_{-1}, y_0, y_1]]$$

corresponds to the homotopy class of the Hopf fibration $S^3 \to S^2$.

2. Let $F_3 = F(y_0, y_1, y_2)$, then the element

$$[[y_0, y_1], [y_0, y_1y_2]] \notin [[y_{-1}, y_0, y_1, y_2]]$$

corresponds to the homotopy class of the composition map $S^4 \to S^3 \to S^2$.

3. Let $F_4 = F(y_0, y_1, y_2, y_3)$, then the element

$$[[[y_0, y_1], [y_0, y_1y_2]], [[y_0, y_1], [y_0, y_1y_2y_3]]] \notin [[y_{-1}, y_0, y_1, y_2, y_3]]$$

corresponds to the homotopy class of the composition map $S^5 \to S^4 \to S^3 \to S^2$.

4. Let $F_5 = F(y_0, y_1, y_2, y_3, y_4)$, then the element

$$[[[[y_0, y_1], [y_0, y_1y_2]], [[y_0, y_1], [y_0, y_1y_2y_3]]],$$
$$[[[y_0, y_1], [y_0, y_1y_2]], [[y_0, y_1], [y_0, y_1y_2y_3y_4]]]]$$
$$\notin [[y_{-1}, y_0, y_1, y_2, y_3, y_4]]$$

corresponds to the homotopy class of the composition map $S^6 \to S^5 \to S^4 \to S^3 \to S^2$.

5. Let $F_6 = F(y_0, y_1, y_2, y_3, y_4, y_5)$, then the element

$$[[[[[y_0, y_1], [y_0, y_1y_2]], [[y_0, y_1], [y_0, y_1y_2y_3]]],$$
$$[[[y_0, y_1], [y_0, y_1y_2]], [[y_0, y_1], [y_0, y_1y_2y_3y_4]]]],$$
$$[[[[y_0, y_1], [y_0, y_1y_2]], [[y_0, y_1], [y_0, y_1y_2y_3]]],$$
$$[[[y_0, y_1], [y_0, y_1y_2]], [[y_0, y_1], [y_0, y_1y_2y_3y_4y_5]]]]]$$
$$\in [[y_{-1}, y_0, y_1, y_2, y_3, y_4, y_5]]$$

corresponds to the (trivial !) homotopy class of the composition map

$$S^7 \to S^6 \to S^5 \to S^4 \to S^3 \to S^2.$$

The triviality of this map can be proved using standard methods in homotopy theory [Tod62]. This is the simplest case of the Nilpotence Theorem, due to Nishida [Nis73], which states that every element in the ring of stable homotopy groups of spheres is nilpotent.

Chapter 6
Miscellanea

In this Chapter we present assorted examples involving the group ring construction.

6.1 Power-closed Groups

Modular dimension subgroup play a fundamental role in understanding the power structure of p-groups (see e.g., [Sco91], [Wil03]).

Let G be a p-group. Then G is said to be *k-power closed*, $k \geq 1$, if every product $x_1^{p^k} \ldots x_n^{p^k}$ $x_i \in G$, $n \geq 1$, can be written as y^p for some $y \in G$.

Theorem 6.1 (Wilson [Wil03]). *If G is a p-group of nilpotency class less than p^k, then G is k-power closed.*

This result is proved by carrying out an extensive study of the modular dimension subgroups $D_{n, \mathbb{F}_p}(G)$. Note that, for every $x \in G$ and $k \geq 1$, $x^{p^k} \in D_{p^k, \mathbb{F}_p}(G)$; therefore, Theorem 6.1 follows from the following:

Theorem 6.2 (Wilson [Wil03]). *Let G be a finite p-group such that $\gamma_{p^k}(G) \subseteq D_{p^{k+1}, \mathbb{F}_p}(G)$ for some k. Then $D_{p^{k+\ell-1}, \mathbb{F}_p}(G) \subseteq \{x^{p^\ell} \mid x \in G\}$ for positive integers ℓ.*

An immediate consequence of Theorem 6.2 is the following:

Corollary 6.3 (Wilson [Wil03]). *Let G be a finite p-group of nilpotency class c. Let k be the minimal integer such that $c < p^{k+1}$. Then $D_{p^{k+\ell}, \mathbb{F}_p}(G) \subseteq \{x^{p^\ell} \mid x \in G\}$ for positive integers l.*

6.2 Braid Groups

The lower central series of pure braid groups (see §1.2, p. 15) plays an important role in the theory of braid invariants. A *singular pure braid* is a

R. Mikhailov, I.B.S. Passi, *Lower Central and Dimension Series of Groups*,
Lecture Notes in Mathematics 1952,
© Springer-Verlag Berlin Heidelberg 2009

pure braid with a finite number of transversal intersections. Any invariant of braids which takes values in some ring R can be viewed as a collection of maps $P_n \to R$, $n \geq 2$. Let $v : P_n \to R$ be an invariant of pure braids. Then we can extend v to be defined on singular braids, by the following rule (so-called *Vassiliev skein relation*):

$$v(\times\!\!\!\!\bullet\,) = v(\times) - v(\times),$$

where the above diagrams represent braids which differ by one intersection inside a ball and completely identical outside the ball. Clearly, this rule makes it possible to extend the invariant v to be defined on singular pure braids. An invariant v of pure braids is said to be an *invariant of type k* if its extension vanishes on all singular braids with more than k double points. We say that two braids B_1 and B_2 are *k-equivalent* if $v(B_1) = v(B_2)$ for any invariant v of type less than k.

One can formally view a singular pure braid with n strands as an element of the integral group ring $\mathbb{Z}[P_n]$ by setting

$$\times\!\!\!\!\bullet = \times - \times \in \mathbb{Z}[P_n]. \qquad (6.1)$$

Then the extension of the invariant v defines a $\mathbb{Z}$-linear map

$$\bar{v} : \mathbb{Z}[P_n] \to R.$$

Clearly, any singular braid with exactly one double point defines an element of the augmentation ideal $\Delta(P_n)$ of $\mathbb{Z}[P_n]$, since it is a "difference" of two pure braids. It is easy to see that any singular braid with exactly two double points can be drawn as a composition of two singular braids with exactly one double point each. In general, any singular braid with k double points can be written as a composition of k singular braids with one double point each. With composition of braids corresponding to the multiplication in the group ring $\mathbb{Z}[P_n]$, any singular braid with n strands and more than k double points represents an element from the kth power of the augmentation ideal of $\mathbb{Z}[P_n]$. On the other hand, any pure braid can be deformed to the trivial one by the sequence of crossed moves:

$$\times \to \times , \quad \times \to \times$$

This implies that the augmentation ideal of $\mathbb{Z}[P_n]$ is the $\mathbb{Z}$-linear closure of elements of the form (6.1), i.e., of singular braids with n strands. Similarly, we conclude that the kth power of the augmentation ideal of $\mathbb{Z}[P_n]$ is the $\mathbb{Z}$-linear closure of singular braids with n strands and not less than k double points.

Let p_1 and p_2 be pure braids with n strands. For $k \geq 1$, the above argument shows that there is an invariant v of type k which differs on p_1 and p_2 if and only if $p_1 - p_2$ determines a nontrivial element in the quotient $\mathbb{Z}[P_n]/\Delta^k(P_n)$;

this is equivalent to saying that $1 - p_1 p_2^{-1} \notin \Delta^k(P_n)$, i.e., $p_1 p_2^{-1} \notin D_k(P_n)$, the kth dimension subgroup of P_n. It is easy to see, in view of (Chapter 1, 1.6), that the lower central series and the dimension series are identical for the pure braid groups. Hence, we have the following

Proposition 6.4 *Two pure braids* p_1, p_2 *with* n *strands are* k-*equivalent if and only if* $p_1 p_2^{-1} \in \gamma_k(P_n)$.

A similar equivalence occurs in the case of classical knots. Every knot is the closure of some braid. However, different braids can determine isotopical knots. In analogy with singular braids, one can define the singular knots and type k invariants as knot isotopy invariants which vanish for singular knots with more than k double points. As for braids, we say that two knots K_1 and K_2 are k-*equivalent* if $v(K_1) = v(K_2)$ for any invariant v of type less than k.

Theorem 6.5 (Stanford [Sta98]). *Let* K_1 *and* K_2 *be knots. Then* K_1 *and* K_2 *are* k-*equivalent if and only if there exists a braid* $b \in B_n$ *and a pure braid* $p \in P_n$ *for some* n, *such that* K_1 *is the closure of* b, *but* K_2 *is the closure of* bp.

Remark. It may be noted that the residual nilpotence of the pure braid groups implies that non-equal (non-isotopical) braids always differ by some invariant of finite type. However, the same result for knots does not follow immediately and the conjecture about completeness of invariants of finite type for knots is still open.

6.3 3-dimensional Surgery

The applications of the dimension subgroup theory to the 3-dimensional surgery was discovered by G. Massuyeau [Mas07]. Here we recall the construction from [Mas07].

Let S be a surface. The mapping class group $\mathcal{M}(S)$ of S is the group of all isotopy classes of orientation-preserving homeomorphisms of S to itself. There is a natural action of $\mathcal{M}(S)$ on the first homology group of S, hence there is a natural homomorphism

$$\Psi : \mathcal{M}(S) \to \mathrm{Aut}(H_1(S)).$$

The kernel of Ψ is called *Torelli group* of S and denoted by $\mathcal{I}(S)$. The homeomorphisms of S to itself acting trivially on homology are called *Torelli automorphisms*.

Let M be a compact oriented 3-dimensional manifold and $H \subset M$ a handlebody. Consider a Torelli automorphism $h : \partial(M) \to \partial(M)$. Then one can construct a new 3-dimensional manifold M_h in the following way:

$$M_h = (M \setminus \text{int}(H)) \cup_h H.$$

The transformation

$$M \rightsquigarrow M_h$$

is called a *Torelli surgery*. One can naturally generalize this definition for the case

$$M \rightsquigarrow M_I$$

where I is a set of pairwise disjoint handlebodies in M with selected Torelli automorphisms.

Following M. Goussarov and K. Habiro, given $k \geq 1$, call two compact oriented 3-manifolds M and N, Y_k-*equivalent* if there exists a Torelli automorphism h, which belongs to the k-th term of lower central series of the Torelli group $\partial(H)$, such that

$$M \rightsquigarrow M_h = N.$$

Let A be an abelian group and f a topological invariant of compact oriented 3-manifolds with values in A. We call f an invariant of degree at most d if, for any manifold M and every set Γ of pairwise disjoint handlebodies H_i, $i \in \Gamma$ with selected Torelli automorphisms $h_i : \partial(H_i) \to \partial(H_i)$, $i \in \Gamma$, the following identity holds:

$$\sum_{\Gamma' \subset \Gamma} (-1)^{|\Gamma'|} \cdot f(M_{\Gamma'}) = 0 \in A.$$

Two Y_{k+1}-equivalent manifolds are not distinguished by invariants of degree at most k [Mas07]. The converse statement is proved for integral homology 3-spheres by M. Goussarov and K. Habiro [Hab00], [Gou99]; however, in general, the converse statement is not true [Mas07]. The special interest of the equivalence of the above equivalence relations is in the case of homology cylinders. Given an oriented surface Σ, the homology cylinder over Σ is a cobordism M between Σ and $-\Sigma$, which can be obtained from $\Sigma \times [1, -1]$ by a Torelli surgery. Homology cylinders form a natural monoid $\text{Cyl}(\Sigma)$, where the product is the composition of cobordisms. It is shown in [Hab00] and [Gou99] that the quotient of the monoid $\text{Cyl}(\Sigma)$ by the Y_{k+1}-equivalence relation is a group.

In [Hab00] and [Gou99] the following filtration of the monoid $\text{Cyl}(\Sigma)$ is introduced:

$$\text{Cyl}(\Sigma) = \text{Cyl}_1(\Sigma) \supseteq \text{Cyl}_2(\Sigma) \supseteq \text{Cyl}_3(\Sigma) \supseteq \dots,$$

where

$$\text{Cyl}_k(\Sigma) = \{M \in \text{Cyl}(\Sigma) \mid M \text{ is } Y_k - \text{equivalent to } \Sigma \times [1, -1]\}.$$

For $1 \leq k \leq l$, the quotients $\text{Cyl}_k(\Sigma)/Y_l$ are finitely generated subgroups of $\text{Cyl}(\Sigma)/Y_l$ and for $1 \leq k_1 + k_2 \leq l$, one has

$$[\mathrm{Cyl}_{k_1}(\Sigma)/Y_l, \mathrm{Cyl}_{k_2}(\Sigma)/Y_l] \subseteq \mathrm{Cyl}_{k_1+k_2}(\Sigma)/Y_l$$

(see [Hab00], [Gou99]). The following result provides a connection between the above equivalence relations and the dimension subgroup theory.

Theorem 6.6 (Massuyeau [Mas07]). *Let* $1 \leq d \leq k$. *The following statements are equivalent:*

(1) *The homology cylinders over a surface* Σ *are* Y_{d+1}*-equivalent if and only if the* $\mathbb{Z}$*-valued invariants of degree* $\leq d$ *do not separate them.*
(2) $D_{d+1}(\mathrm{Cyl}(\Sigma)/Y_{k+1}) = \mathrm{Cyl}_{d+1}(\Sigma)/Y_{k+1}$.

Note that the problem of description of dimension subgroups $D_{d+1}(\mathrm{Cyl}(\Sigma)/Y_{k+1})$ seems to be highly non-trivial. It is shown in [Mas03] that the group $\mathrm{Cyl}(\Sigma)/Y_2$ contains elements of order 2.

6.4 Vanishing Sums of Roots of Unity

We mention next an interesting application, due to T. Y. Lam and K. H. Leung [Lam00], to a problem in number theory.

Given a natural number m, the problem asks for the computation of the set $W(m)$ of all the possible integers n for which there exist mth roots of unity $\alpha_1, \ldots, \alpha_n$ in the field $\mathbb{C}$ of complex numbers such that $\alpha_1 + \cdots + \alpha_n = 0$.

Let $G = \langle z \rangle$ be a cyclic group of order m. Let $m = p_1^{a_1} \ldots p_r^{a_r}$ be the prime factorization of m with $p_1 < \ldots < p_r$ and $\zeta = \zeta_m$ a primitive mth root of unity. Let $\mathbb{N}[G]$ be the subgroup of $\mathbb{Z}[G]$ consisting of elements $\alpha \in \mathbb{Z}[G]$ with coefficients in $\mathbb{N}$, the set of non-negative integers. Consider the ring homomorphisms

$$\varphi : \mathbb{Z}[G] \to \mathbb{Z}[\zeta], \quad \epsilon : \mathbb{Z}[G] \to \mathbb{Z} \tag{6.2}$$

defined by $z \mapsto \zeta$ and $z \mapsto 1$ respectively.

Since, for every prime p and a primitive pth root ζ of unity,

$$1 + \zeta + \cdots + \zeta^{p-1} = 0,$$

it is easy to see that

$$\mathbb{N}p_1 + \cdots + \mathbb{N}p_r \subseteq W(m). \tag{6.3}$$

That equality holds in (6.3) follows from the following

Theorem 6.7 (Lam - Leung [Lam00]). *For every* $\alpha \in \mathbb{N}[G] \cap \ker \varphi$, $\epsilon(\alpha) \in \sum_{i=1}^r \mathbb{N}p_i$.

Call a nonzero element $x \in \mathbb{N}[G] \cap \ker \varphi$ to be *minimal* if it cannot be decomposed as a sum of two nonzero elements in $\mathbb{N}[G] \cap \ker \varphi$. For any group

H, let

$$\sigma(H) := \sum_{h \in H} h \in \mathbb{N}[H].$$

Let P_i be the unique subgroup of G of order p_i. The elements $g.\sigma(P_i)$ with $g \in G$ and $i-1, \ldots, r$, are clearly all minimal elements in $\mathbb{N}[G] \cap \ker(\varphi)$; call these elements *symmetric minimal elements*. The crux of the argument in the proof of Theorem 6.7 is the following

Theorem 6.8 (Lam - Leung [Lam00]). *For any minimal $x \in \mathbb{N}[G] \cap \ker(\varphi)$, either*

    (A) x is symmetric, or
    (B) $r \geq 3$ and $\epsilon(x) \geq \epsilon_0(x) \geq p_1(p_2 - 1) + p_3 - p_2 > p_3$, where $\epsilon_0(x)$ denotes the cardinality of the support of x.

To deduce Theorem 6.7 from Theorem 6.8, note that it clearly suffices to consider minimal elements in $\mathbb{N}[G] \cap \ker(\varphi)$. By Theorem 6.8, such an element is either symmetric or $r \geq 3$ and $\epsilon(x) \geq p_1(p_2 - 1) + p_3 - p_2$. Thus either $\epsilon(x) = p_i$ for some i, or

$$\epsilon(x) > p_1(p_2 - 1) > (p_1 - 1)(p_2 - 1),$$

and consequently $\epsilon(x) \in \mathbb{N}p_1 + \mathbb{N}p_2$. $\square$

We thus have

Theorem 6.9 (Lam - Leung [Lam00]). *For any natural number m,*

$$W(m) = \mathbb{N}p_1 + \cdots + \mathbb{N}p_r,$$

where $p_1, \ldots, p_r$ are all the distinct prime divisors of m.

The above result in turn has an application to representation theory of finite groups.

Theorem 6.10 (Lam - Leung [Lam00]). *Let χ be the character of a representation of a finite group G over a field F of characteristic 0. Let $g \in G$ be an element of order $m = p_1^{a_1} \ldots p_r^{a_r}$ (where $p_1 < p_2 < \ldots p_r$) such that $\chi(g) \in \mathbb{Z}$, and let $t := \chi(1) + |\chi(g)|$. If $\chi(g) \leq 0$, then $t \in \sum \mathbb{N}p_i$. If $\chi(g) > 0$ and t is odd, then $t \geq \ell$, where ℓ ($= p_1$ or p_2) is the smallest odd prime dividing m.*

6.5 Fundamental Groups of Projective Curves

Let k be an algebraically closed field of characteristic $p > 0$. For a projective curve D over k, let $\pi_A(D)$ denote the set of isomorphism classes of finite groups occurring as Galois groups of un-ramified Galois covers of D. A group

G occurs as a quotient of the fundamental group $\pi_1(D)$ if and only if it lies in $\pi_A(D)$. For an integer $g \geq 0$, let $\pi_A(g)$ denote the set of finite groups G for which there exists a curve D of genus g such that G lies in $\pi_A(D)$. Let $d(G)$ denote the minimal number of generators of the group G, and let $t(G)$ denote the number of generators of the augmentation ideal $\mathfrak{g}_k$, of the group algebra $k[G]$, as a $k[G]$ module.

Theorem 6.11 (Stevenson [Ste98]). *Let $g \geq 2$ be a positive integer and let G be a finite group with normal Sylow p-subgroup P, such that $d(G/P) \leq g$. Then G lies in $\pi_A(g)$ if and only if $t(G) \leq g$.*

Appendix A
Simplicial Methods

A.1 Chain Complexes

For any commutative ring R, a *chain complex* K of R-modules is a family $\{K_n,\ d_n\}$ of R-modules K_n and R-homomorphisms $d_n : K_n \to K_{n-1}$, defined for all integers n such that $d_n d_{n+1} = 0$. An *n-cycle* of K is an element of the submodule $C_n(K) = \ker d_n$, and an *n-boundary* is an element of $d_{n+1}(K_{n+1})$. The *homology* of K, denoted $H(K)$, is the family of modules

$$H_n(K) = \ker d_n / \operatorname{im} d_{n+1}.$$

If K and K' are complexes, a *chain transformation* $f : K \to K'$ is a family of module homomorphisms $f_n : K_n \to K'_n$, such that for all n

$$d'_n f_n = f_{n-1} d_n.$$

Denote by $\mathcal{C}h(R)$ the category of chain complexes of R-modules, i.e., the category whose class $Ob(\mathcal{C}h(R))$ of objects consists of the chain complexes K, and the set $\operatorname{Hom}_{\mathcal{C}h(R)}(K,\ K')$ of morphisms between two objects $K,\ K'$ is the set of all chain transformations $f : K \to K'$. Every chain transformation $f \in \operatorname{Hom}_{\mathcal{C}h(R)}(K,\ K')$ induces a family of homomorphisms

$$H_n(f) : H_n(K) \to H_n(K')$$

defined by

$$H_n(f)(c + dK_{n+1}) = f(c) + dK'_{n+1}, \quad c \in \ker d_n.$$

A *chain homotopy* s between two chain transformations $f,\ g \in \operatorname{Hom}_{\mathcal{C}h(R)}$ $(K,\ K')$, denoted $s : f \simeq g$, is a family of module homomorphisms

$$s_n : K_n \to K'_{n+1}$$

such that

$$d'_{n+1}s_n + s_{n-1}d_n = f_n - g_n.$$

Theorem A.1 *If $s : f \simeq g : K \to K'$, then*

$$H_n(f) = H_n(g) : H_n(K) \to H_n(K'), \quad -\infty < n < \infty.$$

A chain transformation $f \in \mathrm{Hom}_{Ch(R)}(K, K')$ is said to be a *chain equivalence* if there exists a chain transformation $h \in \mathrm{Hom}_{Ch(R)}(K', K)$ and homotopies $hf \simeq 1_K$, $fh \simeq 1_{K'}$.

Corollary A.2 *If $f \in \mathrm{Hom}_{Ch(R)}(K, K')$ is a chain equivalence, then the induced map $H_n(f) : H_n(K) \to H_n(K')$ is an isomorphism for each n.*

A.2 Simplicial Objects

Let $\mathcal{C}$ be a category. A simplicial object X_* in $\mathcal{C}$ is a family $\{X_i\}_{i \geq 0}$, $X_i \in Ob(\mathcal{C})$ together with two families of morphisms

$$d_i \in \mathrm{Hom}_{\mathcal{C}}(X_q, X_{q-1}), \quad s_i \in \mathrm{Hom}_{\mathcal{C}}(X_q, X_{q+1}), \quad 0 \leq i \leq q,$$

called the *face* and the *degeneracy* maps respectively, which satisfy the following identities:

$$
\begin{aligned}
d_i d_j &= d_{j-1} d_i, \ i < j, \\
s_i s_j &= s_{j+1} s_i, \ i \leq j, \\
d_i s_j &= s_{j-1} d_i, \ i < j, \\
d_j s_j &= d_{j+1} s_j = id, \\
d_i s_j &= s_j d_{i-1}, \ i > j + 1.
\end{aligned}
\tag{A.1}
$$

A simplicial morphism $f : X_* \to Y_*$ is a family $f_i \in \mathrm{Hom}_{\mathcal{C}}(X_i, Y_i)$, $i \geq 0$, of morphisms compatible with the face and the degeneracy maps. The category of simplicial objects in $\mathcal{C}$ will be denoted by $\mathcal{SC}$.

The *simplicial category* (also called *ordinal number category*) Δ consists of the objects

$$Ob(\Delta) = \{[n] := \{0, 1, \ldots, n\}\}$$

and order preserving maps $\{f : [n] \to [m]\}$ as elements of $\mathrm{Hom}_{\Delta}([n], [m])$. In particular, there are the following morphisms, called the face and degeneracy maps, in this category:

$$\delta_i : [n-1] \to [n], \ 0 \le i \le n,$$
$$\sigma_i : [n+1] \to [n], \ 0 \le i \le n,$$
$$\delta_i : \{0,1,\dots,n-1\} \to \{0,1,\dots,i-1,i+1,\dots,n\},$$
$$\sigma_i : \{0,1,\dots,n+1\} \to \{0,1,\dots,i,i,\dots,n\}.$$

It is easy to check that the maps δ_i, σ_i satisfy the following *cosimplicial relations*:

$$\begin{aligned}
\delta_j \delta_i &= \delta_i \delta_{j-1}, \ i < j, \\
\sigma_j \sigma_i &= \sigma_i \sigma_{j+1}, \ i \le j, \\
\sigma_j \delta_i &= \delta_i \sigma_{j-1}, \ i < j, \\
\sigma_j \delta_j &= \sigma_j \delta_{j+1} = id, \\
\sigma_j \delta_i &= \delta_{i-1} \sigma_j, \ i > j+1.
\end{aligned} \tag{A.2}$$

Furthermore, all elements of $\mathrm{Hom}_\Delta(-,-)$ can be written as compositions of these face and degeneracy maps. It thus turns out that a simplicial object in a category $\mathcal{C}$ is simply a contravariant functor from the simplicial category Δ to $\mathcal{C}$, i.e.

$$S\mathcal{C} = \{\Delta^{op} \to \mathcal{C}\},$$

where Δ^{op} denotes the opposite category of the category Δ. By a simplicial group (resp. ring, abelian group, topological space, etc.) we shall mean a simplicial object in the category of groups (resp. the corresponding category).

Example A.3

For a given topological space X, the *total singular complex* $S(X)$ of X is the simplicial set defined as follows.
For $n \ge 0$, let

$$\Delta_n = \{(x_0,\dots,x_n) \in \mathbb{R}^{n+1} \mid 0 \le x_i \le 1, \ \sum_{i=0}^{n} x_i = 1\}.$$

Define

$$e_i : \Delta_{n-1} \to \Delta_n, \ 0 \le i \le n,$$
$$f_j : \Delta_{n+1} \to \Delta_n, \ 0 \le i \le n,$$

by

$$e_i : (x_0,\dots,x_n) \mapsto (x_0,\dots,x_{i-1},0,x_i,\dots,x_n),$$
$$f_j : (x_0,\dots,x_{n+1}) \mapsto (x_0,\dots,x_{j-1},x_j+x_{j+1},x_{j+2},\dots,x_{n+1}).$$

A *singular n-simplex* of X is a continuous map $\sigma : \Delta_n \to X$. The family $\{S(X)_n\}_{n \ge 0}$ of sets is a simplicial set with the face and degeneracy maps given by

$$d_i : S(X)_n \to S(X)_{n-1}, \ 0 \le i \le n,$$
$$s_j : S(X)_n \to S(X)_{n+1}, \ 0 \le j \le n,$$

defined by

$$d_i(\sigma) = \sigma \circ e_i, \ s_j(\sigma) = \sigma \circ f_j, \ \sigma \in S(X)_n.$$

Let R be a simplicial ring. Then a simplicial abelian group M is called a left *simplicial R-module* (or, simply an R-module), if there exists a simplicial map $f : R \times M \to M$ such that, for each i, f_i defines an R_i-module structure on M_i. Similarly, if G is a simplicial group, we can define *simplicial G-set* (resp. *simplicial G-space*) to be a simplicial set X (resp. topological space) with a simplicial map $G \times X \to X$.

A.3 Geometric Realization Functor

Let X be a simplicial set. The *geometric realization* $|X|$ of X is the topological space obtained from the disjoint union

$$\bigcup_n (X_n \times \Delta_n),$$

where the set X_n is viewed as a topological space with discrete topology, by making the following identifications:

$$(d_i x, p) \sim (x, e_i p), \ (x, p) \in X_n \times \Delta_{n-1},$$
$$(s_i x, p) \sim (x, f_i p), \ (x, p) \in X_{n-1} \times \Delta_n.$$

This construction defines the *geometric realization functor*

$$| \ | : \mathcal{S}\mathsf{Set} \to \mathsf{Top}$$

from the category $\mathcal{S}\mathsf{Set}$ of simplicial objects in the category Set of sets to the category Top of topological spaces.

The geometric realization can also be described as the coequalizer

$$\bigsqcup_{\phi:[n]\to[m]} (X_m \times \Delta_n) \rightrightarrows \bigsqcup_n (X_n \times \Delta_n) \longrightarrow |X|.$$

A.4 Skeleton and Coskeleton Functors

Let Δ_k be the full sub-category of the category Δ consisting of sets of cardinality at most $k + 1$, $k \ge 0$. Then any element from

$$\mathcal{S}_k \mathcal{C} := \{\Delta_k^{op} \to \mathcal{C}\}$$

is called a k-*truncated simplicial object in* $\mathcal{C}$. Clearly, for any $k \geq 0$, we have a functor

$$\mathrm{Tr}^k : \mathcal{SC} \to \mathcal{S}_k\mathcal{C},$$

which "truncates" the simplicial object at level k, i.e., forgets the part of simplicial object which appears in dimensions greater than k. It is known that, in case $\mathcal{C}$ has finite colimits, the functor Tr^k has a left adjoint functor sk^k, called the k-*skeleton* functor. Similarly, if $\mathcal{C}$ has finite projective limits, then Tr^k admits a right adjoint functor $cosk^k$, called the k-*coskeleton* functor. The k-skeleton functor can be constructed precisely by iterating the process of taking the so-called simplicial cokernels. For the detailed description of this construction, see [Dus75].

Example A.4

Let F be a free group with generators $\{x_i\}_{i \in I}$ and R its normal subgroup generated, as a normal subgroup, by the set $\{r_j\}_{j \in J}$. Consider the free product $F_1 = F * F_R$, where F_R is the free group with basis $\{y_j\}_{j \in J}$. Then we have the following three homomorphisms between free groups:

$$d_0 : F_1 \to F, \ x_i \mapsto x_i, \ i \in I, \ y_j \mapsto 1, \ j \in J,$$
$$d_1 : F_1 \to F, \ x_i \mapsto x_i, \ i \in I, \ y_j \mapsto r_j, \ j \in J,$$
$$s_0 : F \to F_1, \ x_i \mapsto x_i, \ i \in I.$$

It easy to see that the simplicial identities are satisfied for these maps and we have the 1-truncated simplicial group

$$S(X, \mathcal{R}) = F_1 \overset{\overset{d_0}{\underset{d_1}{\rightrightarrows}}}{\underset{s_0}{\leftarrow}} F.$$

We describe the 1-skeleton of this simplicial group. We have

$\mathrm{sk}^1 S(X, \mathcal{R})_0 = F = F(x_i, \ i \in I)$,
$\mathrm{sk}^1 S(X, \mathcal{R})_1 = F_1 = F(s_0(x_i), r_j, \ i \in I, \ j \in J)$,
$\mathrm{sk}^1 S(X, \mathcal{R})_2 = F(s_1 s_0(x_i), s_0(r_j), s_1(r_j), \ i \in I, \ j \in J)$,
$\mathrm{sk}^1 S(X, \mathcal{R})_3 = F(s_2 s_1 s_0(x_i), s_1 s_0(r_j), s_2 s_1(r_j), s_2 s_0(r_j), \ i \in I, \ j \in J)$,
$\ldots$

where, for a set X, by $F(X)$ we mean the free group generated by X. One can easily write the simplicial maps in $\mathrm{sk}^1(S(X, \mathcal{R}))$ in a natural way.

In a similar way, we can define the simplicial Lie algebra $\mathrm{sk}^1 S(X, \mathcal{R})$ for the case of a free Lie algebra F generated the the set $\{x_i\}_{i \in I}$ and its ideal R, which is a smallest ideal containing the set of elements $\{r_j \in F\}$, $j \in J$. Then we get the 1-truncated simplicial Lie algebra $S(X, \mathcal{R})$ and its 1-skeleton, viewed as a simplicial Lie algebra.

A.5 Moore Complex and Homotopy Groups

Let $\mathcal{C}$ be one of the following categories:

Gr:= the category of groups,
Lie:= the category of Lie algebras,
$_R$Mod:= the category of R-modules for some commutative ring R with identity.

For a given simplicial object $X \in S\mathcal{C}$, define a complex $(N_*(X), \bar{d}_*)$, called the *Moore complex* of X, by setting

$$N_n(X) = \bigcap_{0 \le i < n} \ker(d_i : X_n \to X_{n-1}), \qquad (A.3)$$

and the homomorphism $\bar{d}_n$ to be the restriction of $d_n : X_n \to X_{n-1}$ on $N_n(X)$.

The *homotopy groups* $\pi_i(X)$, $i \ge 0$, of a given simplicial object $X_* \in S\mathcal{C}$ are defined as the homologies of its Moore complex:

$$\pi_i(X) := H_i(N_*(X), \bar{d}_*), \quad i \ge 0. \qquad (A.4)$$

It is easy to show that, for any $X \in S\mathcal{C}$, $\pi_i(X)$ is an abelian group for $i \ge 1$.

For a given simplicial group G, denote by $Z_n(G)$ the nth chain subgroup of G_n, i.e.,

$$Z_n(G) = \ker(\bar{d}_n) = \bigcap_{0 \le i \le n} \ker(d_i),$$

and by $B_n(G)$ the nth boundary subgroup of G_n, i.e.,

$$B_n = \operatorname{im}(\bar{d}_{n+1}).$$

Thus we have, by definition,

$$\pi_n(G) = Z_n(G)/B_n(G), \quad n \ge 0.$$

In the case of an abelian simplicial group G, there is an equivalent way to compute the homotopy groups. Consider the chain complex $\{G_n, d_n\}$, where

$$d_n = \sum_{i=0}^{n} (-1)^i d_i : G_n \to G_{n-1}.$$

It can be checked directly that $d_n \circ d_{n+1} = 0$, and

$$\pi_i(G) = H_i(G_n, d_n).$$

The following Proposition follows directly from the definition of homotopy groups of simplicial groups (resp. R-modules).

Proposition A.5 *Let* $1 \to H \to G \to K \to 1$ *be a short exact sequence of simplicial groups (resp. simplicial R-modules). Then there exists an induced long exact sequence of homotopy groups:*

$$\ldots \to \pi_{i+1}(H) \to \pi_{i+1}(G) \to \pi_{i+1}(K) \to \pi_i(H) \to \ldots$$

It is easy to see that, for any simplicial group G, the π_0-functor coincides with the coequilizer functor:

$$\pi_0(G) = \text{coeq}(G_1 \overset{d_0, d_1}{\rightrightarrows} G_0).$$

A similar formula holds for simplicial R-modules.

For a given element $f \in G_0$, there is a simplicial automorphism $F_f : G \to G$ defined by

$$F_f : x \mapsto (s_0^n f)^{-1} x s_0^n f, \quad x \in G_n. \tag{A.5}$$

Let $f \in B_0(G)$, that is $f = d_1 f_1$, where $d_0 f_1 = 1$. Then

$$s_0^n d_1 f_1 = s_0^{n-1} d_2 s_0 f_1 = \ldots = d_{n+1} s_0^n f_1, \; n \geq 1$$

and

$$d_i s_0^n f_1 = s_0 d_{i-1} s_0^{n-1} f_1 = \ldots = s_0^n d_0 f_1 = 1, \; 0 \leq i \leq n;$$

hence

$$s_0^n f = s_0^n d_1 f_1 \in B_n(G).$$

Therefore, the map F_f defines an action of the group $\pi_0(G)$ on the abelian group $\pi_n(G)$, $n \geq 1$, i.e., $\pi_n(G)$ *can be viewed as a* $\mathbb{Z}[\pi_0(G)]$-*module*.

The computation of $\pi_1(G)$, even for the case of quite simple simplicial groups G, can turn out to be nontrivial. We present the computation for the case of simplicial groups which generalizes Example A.4.

Example A.6

(Brown-Loday [Bro87]) Let G be a simplicial group, such that G_2 is generated by degeneracy elements, i.e.,

$$G_2 = \langle s_0(G_1), \, s_1(G_1) \rangle. \tag{A.6}$$

Then

$$\text{im}(\bar{d}_2) = [\ker(d_1), \, \ker(d_2)]. \tag{A.7}$$

Hence, we have

$$\pi_1(G) = \frac{\ker(d_0) \cap \ker(d_1)}{[\ker(d_0), \, \ker(d_1)]}. \tag{A.8}$$

In Example A.4, the condition (A.6) clearly holds for the 1-skeleton $\mathrm{sk}^1 S(X, \mathcal{R})$. Therefore, the formula (A.8) holds. Clearly, we have

$$\ker(d_0) = \langle y_j, \ j \in J \rangle^{F_1},$$
$$\ker(d_1) = \langle y_j r_j^{-1}, \ j \in J \rangle^{F_1},$$

and

$$\pi_1(\mathrm{sk}^1 S(X, \mathcal{R})) = \frac{\langle y_j, \ j \in J \rangle^{F_1} \cap \langle y_j r_j^{-1}, \ j \in J \rangle^{F_1}}{[\langle y_j, \ j \in J \rangle^{F_1}, \langle y_j r_j^{-1}, \ j \in J \rangle^{F_1}]}. \qquad (A.9)$$

The action of $\pi_0(G)$ on $\pi_1(\mathrm{sk}^1 S(X, \mathcal{R}))$ is given by conjugation:

$$fR \circ x[\ker(d_0), \ker(d_0)] = f^{-1} x f[\ker(d_0), \ker(d_1)],$$
$$x \in \ker(d_0) \cap \ker(d_1), \ x \in F(X).$$

For the case of Lie algebras one can get an analogous result. First note, that for a simplicial Lie algebra G in which G_2 is generated by the degeneracy elements, the relation (A.7) again holds (where the bracket $[.,.]$ denotes the product in a Lie algebra) [Akc02]. Therefore, for a free Lie algebra F with generating set $\{x_i\}_{i \in I}$ and a subset $\{r_j \in F\}_{j \in J}$, one has

$$\pi_1(\mathrm{sk}^1 S(X, \mathcal{R})) = \frac{(y_j, \ j \in J)F \cap (y_j - r_j, \ j \in J)F}{[(y_j, \ j \in J)F, (y_j - r_j, \ j \in J)F]}. \qquad (A.10)$$

We say that the Moore complex $N_*(X)$ of a simplicial object X in a category $\mathcal{C}$ is of length $\leq k$ if $N_n(X) = 0$ for all $n \geq k + 1$. The simplicial objects with Moore complex of length $\leq n$ form a category; we denote this category by $\mathcal{SC}(n)$.

Proposition A.7 *If R is a principal ideal domain and X is a projective simplicial R-module, then $N(X)$ is a complex of projective R-modules.*

A.6 Dold-Kan Correspondence

The following result is the key to the construction of derived functors.

Theorem A.8 *The functor $N : \mathcal{S}_R\mathsf{Mod} \to \mathcal{C}h(R)$ is an equivalence of categories.*

To prove this result it is clearly enough to construct an inverse map

$$N^{-1} : \mathcal{C}h(R) \to \mathcal{S}_R\mathsf{Mod},$$

which is constructed by setting

$$(N^{-1}C)_n = \bigoplus_{f:[n] \twoheadrightarrow [m]} f^*(C_m), \ C \in Ch(R).$$

For example, for a given chain complex C of R-modules, the first few terms of $N^{-1}C$ can be written as

$$(N^{-1}C)_0 = C_0,$$
$$(N^{-1}C)_1 = C_1 \oplus s_0(C_0),$$
$$(N^{-1}C)_2 = C_2 \oplus s_0(C_1) \oplus s_1(C_1) \oplus s_0 s_0(C_0).$$

A.7 Eilenberg-Zilber Equivalence

For $(A, \partial_1), (B, \partial_2) \in Ch(R)$, the *tensor product* $(A \otimes_R B, \partial) \in Ch(R)$ is defined as follows:

$$(A \otimes_R B)_n = \oplus_{p+q=n} A_p \otimes_R B_q,$$
$$\partial(a \otimes b) = \partial_1 a \otimes b + (-1)^{dim(a)} a \otimes \partial_2 b.$$

For $X, Y \in S_R \mathsf{Mod}$, the *tensor product* $X \otimes_R Y \in S_R \mathsf{Mod}$ is defined as follows:

$$(X \otimes_R Y)_n = X_n \otimes_R Y_n,$$
$$\partial_i(x \otimes y) = \partial_i x \otimes \partial_i y, \ 0 \leq i \leq n,$$
$$s_i(x \otimes y) = s_i x \otimes s_i y, \ 0 \leq i \leq n.$$

For $x \in X_n, y \in Y_n$, the map

$$f : x \otimes y \mapsto \sum_{p+q=n} \partial_{n-p+1} \ldots \partial_{n-1} \partial_n x \otimes \partial_0^q y,$$

called the *Alexander-Whitney map*, induces the homomorphism of the normalized complexes

$$\bar{f} : N(X \otimes_R Y) \to N(X) \otimes_R N(Y). \tag{A.11}$$

The converse map

$$\bar{\nabla} : N(X) \otimes_R N(Y) \to N(X \otimes_R Y) \tag{A.12}$$

is induced by the 'shuffle-map', defined as follows. For $p, q \geq 1$, let

$$(a; b) = (a_1, \ldots, a_p; b_1, \ldots, b_q)$$

be a permutation of $(0, \ldots, p+q-1)$, such that $a_1 < \ldots < a_p$, $b_1 < \ldots < b_q$. We will refer to such $(a; b)$ as a $(p; q)$-shuffle. Denote $sign(a; b)$ to be the sign of the permutation $(a; b)$. For $x \in X_p$, $y \in Y_q$, define

$$\nabla : x \otimes y \mapsto \sum_{(p;q)-shuffles\ (a;b)} (-1)^{sign(a;b)} s_{a_p} \ldots s_{a_1} x \otimes s_{b_q} \ldots s_{b_1} y.$$

The maps (A.11) and (A.12) define the isomorphism of the chain complexes:

$$N(X \otimes_R Y) \simeq N(X) \otimes_R N(Y), \qquad (A.13)$$

called the *Eilenberg-Zilber equivalence*.

In the case when R is a principal ideal domain, and X is a free R-simplicial module, Künneth formula implies that there exists the following split exact sequence of R-modules

$$0 \to \oplus_i H_n(N(X)) \otimes_R H_{n-i}(N(Y)) \to H_n(N(X) \otimes_R N(Y)) \to$$
$$\oplus_i \operatorname{Tor}_1^R(H_i(N(X)), H_{n-i-1}(N(Y))) \to 0, \quad (A.14)$$

which can be written as

$$0 \to \oplus_i \pi_n(X) \otimes_R \pi_{n-i}(Y) \to \pi_n(X \otimes_R Y) \to$$
$$\oplus_i \operatorname{Tor}_1^R(\pi_i(X), \pi_{n-i-1}(Y)) \to 0, \quad (A.15)$$

due to the Eilenberg-Zilber equivalence (A.13).

A.8 Classifying Functor $\overline{W}$ and Homology

Let G be a simplicial group. Define the simplicial set WG by setting:

$$WG_n = G_n \times G_{n-1} \times \ldots \times G_0, \ n \geq 0 \qquad (A.16)$$

with face and degeneracy maps

$$d_i(g_n, \ldots, g_0) = (d_i g_n, d_{i-1} g_{n-1}, \ldots, (d_0 g_{n-i}) g_{n-i-1}, g_{n-i-2}, \ldots, g_0), \ i < n,$$
$$d_n(g_n, \ldots, g_0) = (d_n g_n, d_{n-1} g_{n-1}, \ldots, d_1 g_1),$$
$$s_i(g_n, \ldots, g_0) = (s_i g_n, s_{i-1} g_{n-1}, \ldots, s_0 g_{n-1}, 1, g_{n-i-1}, \ldots, g_0).$$

The simplicial set WG has a natural structure as a G-set, namely the one where the left G-action is given by

$$g \circ (g_n, \ldots, g_0) \mapsto (g g_n, g_{n-1}, \ldots, g_0), \ g \in G_n.$$

The space $\overline{W}G$ is the quotient of WG by the left G-action. Let $q : WG \to \overline{W}G$ be the quotient map.

A *reduced simplicial set* is a simplicial set having only one vertex i.e., a simplicial set X in which X_0 is a singleton. Denote the sub-category of the reduced simplicial sets by $\mathsf{r}\mathcal{S}\mathsf{Set}$. Then the construction $\overline{W}G$ defines a functor

$$\overline{W} : \mathsf{Gr} \to \mathsf{r}\mathcal{S}\mathsf{Set}, \tag{A.17}$$

called a *classifying space functor* on the category Gr of groups. Clearly, the components of $\overline{W}G$ can be written as

$$\overline{W}G_0 = 1, \ \overline{W}G_n = G_{n-1} \times G_{n-2} \times \cdots \times G_0, \ n > 0.$$

The face and degeneracy maps are defined as

$$d_0(g) = 1, \ d_1(g) = 1, \ g \in \overline{W}G_1, \ s_0(1) = 1,$$
$$d_0(g_n, \ldots, g_0) = (g_{n-1}, \ldots, g_0),$$
$$d_{i+1}(g_n, \ldots, g_0) = (d_i g_n, \ldots, d_1 g_{n-i+1}, \ g_{n-i-1} d_0 g_{n-i}, \ g_{n-i-2}, \ldots, g_0);$$
$$s_0(g_{n-1}, \ldots, g_0) = (1, g_{n-1}, \ldots, g_0);$$
$$s_{i+1}(g_{n-1}, \ldots, g_0) = (s_i g_n, \ldots, s_0 g_{n-i}, 1, g_{n-i-1}, \ldots, g_0).$$

Let M be a simplicial G-module. Following Quillen, define a graded abelian group, called *homology of G with coefficients in M* by setting

$$H_*(G, M) := \pi_*(\mathbb{Z}[WG] \otimes_{\mathbb{Z}[G]} M),$$

where $\mathbb{Z}[WG]$ and $\mathbb{Z}[G]$ are free abelian simplicial groups obtained by applying the group ring functor to WG and G respectively.

For a group G, consider the constant simplicial group with $G_i = G$ and all face and degeneracy maps equal to the identity map. Let M be a G-module, then, clearly, $H_*(G, M)$ is the same as ordinary group homology of G with coefficients in M. Clearly, $\mathbb{Z} \otimes_{\mathbb{Z}[G]} \mathbb{Z}[WG] = \mathbb{Z}[\overline{W}G]$, where $\mathbb{Z}$ is viewed as a constant simplicial G-module. Hence,

$$H_*(G, \mathbb{Z}) = \pi_*(\mathbb{Z}[\overline{W}G]).$$

A.9 Bisimplicial Groups

For a given category $\mathcal{C}$, a *bisimplicial object* in $\mathcal{C}$ is a functor

$$\Delta^{op} \times \Delta^{op} \to \mathcal{C}.$$

Clearly, in analogy with simplicial objects, any bisimplicial object can be viewed as a set of objects $X_{m,n} \in \mathcal{C}$, connected by certain maps. In this way, fixing the first index m, one gets the simplicial object $X_{m,*}$ and fixing the second index n, the simplicial object $X_{*,n}$.

Let G be a bisimplicial group, i.e., a bisimplicial object in the category Gr. It can be viewed as the data

$$G = \{G_{m,n},\ d_j^h,\ s_j^h,\ d_j^v,\ s_j^v\},$$

where each $G_{m,n}$ $(m,\ n \geq 0)$ is a group and

$$d_j^h : G_{m,n} \to G_{m-1,n},\ 0 \leq j \leq m,$$
$$s_j^h : G_{m,n} \to G_{m+1,n},\ 0 \leq j \leq m,$$
$$d_j^v : G_{m,n} \to G_{m,n-1},\ 0 \leq j \leq n,$$
$$s_j^v : G_{m,n} \to G_{m,n+1},\ 0 \leq j \leq n.$$

are homomorphisms satisfying appropriate relations. Define the *diagonal* $\mathcal{D}G$ of the bisimplicial group G to be the simplicial group given by setting

$$(\mathcal{D}G)_n = G_{n,n},\ d_j = d_j^h \circ d_j^v,\ s_j = s_j^h \circ s_j^v.$$

Theorem A.9 (Quillen [Qui66]). *For a bisimplicial group G there are two spectral sequences:*

$$E_{p,q}^2 = \pi_p^h \pi_q^v(G) \Longrightarrow \pi_{p+q}(\mathcal{D}G),$$
$$E_{p,q}^2 = \pi_p^v \pi_q^h(G) \Longrightarrow \pi_{p+q}(\mathcal{D}G),$$

where $\pi_p^h \pi_q^v(G)$ (resp. $\pi_p^v \pi_q^h(G)$) is the p-th homotopy group of the "horizontal" (resp. "vertical") simplicial group obtained by taking the q-th homotopy group of each of the "vertical" (resp. "horizontal") simplicial groups.

For a simplicial group $X_* \in S\mathsf{Gr}$, there is a natural first quadrant spectral sequence

$$E_{p,q}^1 = H_q(X_p, R) \Longrightarrow H_{p+q}(\overline{W}X_*, R), \tag{A.18}$$
$$d_r : E_{p,q}^r \to E_{p-r,q+r-1}^r. \tag{A.19}$$

A.10 Certain Simplicial Constructions

Kan's Construction

Let us recall the Kan's *loop group construction*. This is the functor

$$G : rS\mathsf{Set} \to S\mathsf{Gr}, \tag{A.20}$$

with the properties:

$$(i) \quad \pi_{n-1}(GX) \simeq \pi_n(X), \ X \in \mathcal{S}\mathsf{Gr},$$
$$(ii) \quad (GX)_n \text{ is a free group, } n \geq 0.$$

For a given $X \in \mathsf{r}\mathcal{S}\mathsf{Set}$, define the loop group $GX \in \mathcal{S}\mathsf{Gr}$ as follows:

Let $(GX)_n$ be the free group on the elements of X_{n+1} modulo the n-degenerate elements, i.e., elements of the form s_0x, $x \in X_n$; thus $(GX)_n$ can be presented as

$$(GX)_n = F(X_{n+1})/F(s_0(X_n)), \ n \geq 0.$$

The face and degeneracy maps are defined by

$$\tau(d_0x)d_0\tau(x) = \tau(d_1x),$$
$$d_i\tau(x) = \tau(d_{i+1}x), \ i > 0,$$
$$s_i\tau(x) = \tau(s_{i+1}x), \ i \geq 0,$$

where $\tau(x)$ denotes the class of the element $x \in X_{n+1}$ in $(GX)_n$. Obviously, for any $X \in \mathsf{r}\mathcal{S}\mathsf{Set}$, $(GX)_n$ is the free group on the set $X_{n+1}\backslash s_0X_n$.

The loop functor G is a left adjoint functor to the classifying space functor $\overline{W}$. These functors have the following main property:

For any reduced simplicial set X and a simplicial group Γ, the canonical maps

$$G(\overline{W}\Gamma) \mapsto \Gamma, \quad X \mapsto \overline{W}(GX)$$

are weak homotopy equivalences (see, for example, [Goe99, Section V]).

A simplicial group $X \in \mathcal{S}\mathsf{Gr}$ is called *free* if, for all $n \geq 0$, the components X_n are free groups, and there are sets of free generators $B_n \subset X_n$, $n \geq 0$, such that $s_i(B_n) \subseteq B_{n+1}$ for all $n \geq 0$, $0 \leq i \leq n$.

It is easy to see that free simplicial groups and CW-complexes have a lot of similar properties. To handle these similarities, Kan introduced the concept of *CW-basis* for a given free simplicial group [Kan59].

Let F be a free simplicial group. A family $\{\mathfrak{B}_n\}_{n\geq 0}$, $\mathfrak{B}_n \subset F_n$, is called a *CW-basis* of F if

(i) $\mathfrak{B}_n$ is a basis of F_n;

(ii) $s_i(x) \in \mathfrak{B}_{n+1}$ for all $x \in \mathfrak{B}_n$, $n \geq 0$, $0 \leq i \leq n$;

(iii) if $x \in \mathfrak{B}_n$ is non-degenerate, then $d_i(x) = 1$, $n \geq 1$, $0 \leq i \leq n-1$.

The non-degenerate elements of a *CW*-basis of F are called *generators* of F. For a given generator $x \in \mathfrak{B}_n$, the element $d_n(x) \in \mathfrak{B}_{n-1}$ is called the *attaching element* of x.

It is shown in [Kan58b] that *any free simplicial group has a CW-basis*. It follows directly from the definitions that a free simplicial group is determined by the generators of its CW-basis together with their attaching elements. Similar is the case for CW-complexes: every CW-complex is determined by its cells and attaching maps.

Let X be a reduced simplicial set, Γ a simplicial group. Then a *twisting function* $\tau : X \to \Gamma$ is a function which lowers dimension by one and is such that, for every $n > 0$ and $\sigma \in X_n$, the following conditions are satisfied:

$$d_i(\tau(\sigma)) = \tau(d_i(\sigma)), \ 0 \leq i < n - 1,$$

$$d_{n-1}(\tau(\sigma)) = \tau(d_{n-1}(\sigma))\tau(d_n(\sigma))^{-1},$$

$$s_i(\tau(\sigma)) = \tau(s_i(\sigma)), \ 0 \leq i < n,$$

$$\tau(s_n(\sigma)) = 1.$$

The twisting functions have a strong relation with the functor G (A.20) discussed above [Kan59]. Namely, for any reduced simplicial set X and a twisting function $\tau : X \to \Gamma$, there exists a simplicial homomorphism $g : GX \to \Gamma$, such that $g(\bar{\sigma}) = \tau(\sigma)$, $\sigma \in X$, where $\bar{\sigma}$ is the image of σ in GX. Therefore, there is a one to one correspondence between twisting functions $X \to \Gamma$ and simplicial homomorphisms $GX \to \Gamma$.

Let K be a CW-complex with a single 0-cell. The total singular complex $S(K)$ contains *the first Eilenberg sub-complex* $S^1(X)$, which consists of n-simplices $\sigma : \Delta_n \to K$ of $S(K)$, such that $\sigma(A_i)$ is the 0-cell in K for $0 \leq i \leq n$, where A_i, $0 \leq i \leq n$ are vertices of Δ_n. The Kan's construction is a simplicial map

$$\tau : S^1(K) \to B_K,$$

where B_K is a free simplicial group, τ a twisting function, which satisfies the following conditions:

(i) the elements $\tau(\sigma_c)$ are distinct and form the generators of a CW-basis of B_K, where c runs through the cells of K of dimension at least one;

(ii) for every sub-complex $L \subset K$, $\tau(S(L)) \subset B(L)$, where $B(L) \subset B_K$ denotes the simplicial subgroup of B_K, generated by elements $\tau(\sigma_c)$, $\sigma_c \in L$.

A twisting functor satisfying the above conditions defines a combinatorial homotopical model for based loops ΩK over the complex K. Namely, it defines a weak homotopical equivalence

$$\Omega K \simeq |B_K|.$$

In the reverse direction, for any free simplicial group B, there exists a reduced CW-complex K_B, together with a twisting function $\tau : S(K_B) \to B$, which satisfies the above conditions (i)-(ii) and is such that there are weak homotopy equivalences $K_{B_K} \simeq K$ and $B_{K_B} \simeq B$.

Example A.10

Let F be a free group with generators $\{x_i\}_{i \in I}$ and R a normal subgroup of F generated, as a normal subgroup, by the set $\{r_j\}_{j \in J}$. Form the standard two-dimensional complex K corresponding to the group presentation

$$\langle x_i, \ i \in I \mid r_j, \ j \in J \rangle.$$

The complex K is reduced, it has a single 0-cell, 1-cells are in one-to-one correspondence with the basis elements $\{x_i\}_{i \in I}$, 2-cells are in one-to-one correspondence with the elements $\{r_j\}_{j \in J}$. Then Kan's construction is the 1-skeleton of the truncation, described above:

$$B_K = \mathrm{sk}^1(S(X, \mathcal{R})).$$

Hence, the second homotopy module of K can be described as

$$\pi_2(K) \simeq \frac{\langle y_j, \ j \in J \rangle^{F_1} \cap \langle y_j r_j^{-1}, \ j \in J \rangle^{F_1}}{[\langle y_j, \ j \in J \rangle^{F_1}, \langle y_j r_j^{-1}, \ j \in J \rangle^{F_1}]},$$

with the action of $\pi_1(K) = \pi_0(\mathrm{sk}^1 S(X, \mathcal{R}))$, defined via the degeneracy map s_0.

Milnor's F[K]-construction

For a given pointed simplicial set K, the $F[K]$-construction [Mil56] is the simplicial group with $F[K]_n = F(K_n \setminus *)$, where $F(-)$ is the free group functor. There is the following weak homotopy equivalence:

$$|F[K]| \simeq \Omega \Sigma |K|.$$

Consider the simplicial circle $S^1 = \Delta[1]/\partial \Delta[1]$:

$$S_0^1 = \{*\}, \ S_1^1 = \{*, \sigma\}, \ S_2^1 = \{*, s_0\sigma, s_1\sigma\}, \ \dots, S_n^1 = \{*, x_0, \dots, x_n\},$$

where $x_i = s_n \dots \hat{s}_i \dots s_0 \sigma$. The $F[S^1]$-construction then clearly has the following terms:

$$F[S^1]_0 = 0,$$
$$F[S^1]_1 = F(\sigma), \text{ free abelian group generated by } \sigma,$$
$$F[S^1]_2 = F(s_0\sigma, s_1\sigma),$$
$$F[S^1]_3 = F(s_i s_j \sigma \mid 0 \le j \le i \le 2),$$
$$\dots$$

The face and degeneracy maps are determined naturally (with respect to the standard simplicial identities) for these simplicial groups. For example, the first nontrivial maps are defined as follows:

$$\partial_i : F[S^1]_2 \to F[S^1]_1, \ i = 0, 1, 2,$$
$$\partial_0 : s_0\sigma \mapsto \sigma, \ s_1\sigma \mapsto 1,$$
$$\partial_1 : s_0\sigma \mapsto \sigma, \ s_1\sigma \mapsto \sigma,$$
$$\partial_2 : s_0\sigma \mapsto 1, \ s_1\sigma \mapsto \sigma.$$

The above construction gives a possibility to define the homotopy groups $\pi_n(S^2)$ combinatorially, in terms of free groups. Since the geometrical realization of $F[S^1]$ is weakly homotopically equivalent to the loop space ΩS^2, the homotopy groups $\pi_n(S^2)$ are naturally isomorphic to the homotopy groups of the Moore complex of $F[S^1]$: $\pi_{n+1}(S^2) \simeq Z_n(F[S^1])/B_n(F[S^1])$. Here Z_n and B_n denote the cycles and the boundaries of the Moore complex of the corresponding simplicial group.

The explicit structure of the cycles and boundaries for $F[S^1]$ can be given in terms of certain normal subgroups in $F[S^1]$. This was realized by Jie Wu [Wu,01]. In fact, we have

$$\pi_{n+1}(S^2) \cong \frac{\langle y_{-1}\rangle^F \cap \langle y_0\rangle^F \cap \ldots \cap \langle y_{n-1}\rangle^F}{[[y_{-1}, y_0, \ldots, y_{n-1}]]}, \qquad (A.21)$$

where F is a free group with generators $y_0, \ldots, y_{n-1}$, $y_{-1} = (y_0 \cdots y_{n-1})^{-1}$, the group $[[y_{-1}, y_0, \ldots, y_{n-1}]]$ is the normal closure in F of the set of left-ordered commutators

$$[z_1^{\varepsilon_1}, \ldots, z_t^{\varepsilon_t}] \qquad (A.22)$$

with the properties that $\varepsilon_i = \pm 1$, $z_i \in \{y_{-1}, \ldots, y_{n-1}\}$ and all elements in $\{y_{-1}, \ldots, y_{n-1}\}$ appear at least once in the sequence of elements z_i in (A.22).

Hurewicz Theorem

The following simplicial analog of Hurewicz Theorem was obtained by Kan (see [Kan58a, Theorems 17.5, 17.6]):

Let F be a simplicial free group with $\pi_i(F) = 0$, $i \leq n$ ($n \geq 0$), then $\pi_i(F/[F, F]) = 0$, $i \leq n$, and the natural homomorphism $\pi_{n+1}(F) \to \pi_{n+1}(F/[F, F])$ is an isomorphism.

A.11 Free Simplicial Resolutions

For a given category $\mathcal{C}$, an *augmented simplicial object* in $\mathcal{C}$ is a pair

$$(X, X_{-1}), \ X \in \mathcal{SC}, \ X_{-1} \in \mathcal{C}$$

together with a morphism $d_0 \in \mathrm{Hom}_\mathcal{C}(X_0, X_{-1})$, such that

$$d_0 d_0 = d_0 d_1 : X_1 \to X_{-1}.$$

The augmented simplicial objects in $\mathcal{C}$ naturally form a category, which we denote by $\mathsf{a}\mathcal{SC}$; the m-truncated augmented simplicial category can also be defined in the obvious way; we denote it by $\mathsf{a}\mathcal{SC}_m$.

An augmented simplicial group (X, X_{-1}) is called a *resolution* of X_{-1} if $\pi_n(X) = 0$, $n > 0$ and $\pi_0(X) = X_{-1}$. The resolution (X, X_{-1}) of a group X_{-1} is called *free simplicial resolution* of X_{-1} if X is free.

A method of construction of free simplicial resolutions in the category of commutative algebras was described first by Andre [And70]. We start with the description of general construction of free simplicial resolutions from [Keu].

For a given $(X, X_{-1}) \in \mathsf{a}\mathcal{SC}$, define the nth *simplicial kernel* to be

$$Z_n(X, X_{-1}) = \{(x_0, \ldots, x_{n+1}) \in X_n^{n+2} \mid d_i x_j = d_{j-1} x_i, \ i < j\}.$$

Analogically the group $Z_n(X, X_{-1})$ can be defined for any $(X, X_{-1}) \in \mathsf{a}\mathcal{SC}_m$ for $m \geq n$. For such an object, there are $n + 2$ natural morphisms

$$p_i : Z_n(X, X_{-1}) \to X_n,$$
$$p_i : (x_1, \ldots, x_{n+1}) \mapsto x_i, \ i = 0, \ldots, n + 1$$

and $n + 1$ morphisms

$$q_i : X_n \to Z_n(X, X_{-1})$$
$$q_i : x \to (s_{i-1} d_0 x, \ldots, s_{i-1} d_{i-1} x, x, x, s_i d_{i+1} x, \ldots, s_i d_n x), \ x \in X_n.$$

The free simplicial resolution of a given object $X_{-1} \in \mathcal{C}$ can be constructed inductively. Firstly we choose a set E of generators of X_{-1}. Define X_0 to be a free object in $\mathcal{C}$ with basis E, which we denote by $F(E_0)$, and define $d_0 : F(E_0) \to X_{-1}$ to be the natural surjection. Now suppose we have defined an object $(X, X_{-1}) \in \mathsf{a}\mathcal{SC}_n$, where X_n is a free object in $\mathcal{C}$ over some set E_n. Complete the set $\bigcup_{i=0}^n q_i(E_n)$ to the set Y_n of generators of $Z_n(X, X_{-1})$. Then put $E_{n+1} = \{e_n \mid e_n \in Y_n\}$, and define X_{n+1} to be a free object $F(E_{n+1})$ in $\mathcal{C}$ on the generating set E_{n+1}. The maps d_i, s_i are defined by setting

$$d_i = p_i, \ s_i = q_i, \ i = 0, \ldots, n + 1,$$

where we use the identification of generators of $Z_n(X, X_{-1})$ with elements in X_{n+1}.

Similarly one can construct a free simplicial resolution starting with a free simplicial group X which is aspherical up to a fixeddimension, say n.

The above method gives an algorithm for converting X into a free simplicial resolution without changing X_i, $i \leq n$.

The free simplicial resolution of a given group can be constructed as an inductive limit of skeleton filtration, also constructed by step-by-step procedure [Mut99]. Let G be a group, given as a quotient $G = F/R$, where F is a free group with basis $\{x_i\}_{i \in I}$ and R its normal subgroup generated, as a normal subgroup, by the set $\{r_j\}_{j \in J}$. We use the notation from Example A.4. The first step in the construction of a free simplicial resolution of G is the 1-skeleton $F^1 = \mathrm{sk}^1 S(X, \mathcal{R})$. Clearly,

$$\pi_0(F^1) = G.$$

Now, in general, $\pi_1(F^1)$, which is generated by cosets

$$a_i[\ker(d_0), \ker(d_1)], \ i \in T,$$

is nontrivial. We can then define a new simplicial group F^2:

$$F_0 = F_0, \ F_1 = F_1, \ F_2 = \mathrm{sk}^1(F^1) * F(a_i \mid i \in T), \ \ldots$$

with obvious face and degeneracy maps, where all terms in dimensions ≥ 3 are free groups generated by degeneracy elements. Clearly,

$$\pi_0(F^2) = G, \ \pi_1(F^2) = 0.$$

Continuing this process by induction, "killing" homotopy groups at each dimension, we get the required free simplicial resolution of a given group.

A functorial construction for a free simplicial resolution of a given group can be given by the composition of the classifying space functor and Kan's loop group construction. Every group Γ can be viewed as a simplicial group X with $X_n = \Gamma$ and all face and degeneracy maps an identity. The composition of the functors G and $\overline{W}$ defines the free loop construction over a group Γ:

$$G\overline{W} : \mathsf{Gr} \to \mathcal{S}\mathsf{Gr},$$

which has the property:

$$\pi_0(G\overline{W}\Gamma) = \Gamma, \ \pi_i(G\overline{W}\Gamma) = 0, \ i > 0.$$

Clearly, $G\overline{W}\Gamma$ is a free simplicial group; thus it can be viewed as a free simplicial resolution of Γ.

Theorem A.11 (Comparison theorem [Keu]). *Let (X, X_{-1}) be a free simplicial resolution of X_{-1}, (Y, Y_{-1}) a resolution of Y_{-1} and $\alpha : X_{-1} \to Y_{-1}$ a group homomorphism. Then*

 (i) there exists a simplicial map $\gamma : X \to Y$, such that $\pi_0(\gamma) = \alpha$;

(*ii*) *for any simplicial map* $\gamma' : X \to Y$ *with* $\pi_0(\gamma') = \alpha$ *there exists a simplicial homotopy* $h : \gamma \to \gamma'$.

Proof. (i) Let $\{B_n \subset X_n\}_{n \geq 0}$ be the sets of free generators with $s_i(B_n) \subset B_{n+1}$, $n \geq 0$, $0 \leq i \leq n$. We construct γ by induction. Let

$$\gamma_{-1} : X_{-1} \to Y_{-1}$$

be the map α. Now, since $\pi_0(Y) = Y_{-1}$, the map $d_0 : Y_0 \to Y_1$ is an epimorphism. Hence, for any $x_0 \in B_0$, there exists an element $y \in Y_0$, such that

$$d_0 y = \gamma_{-1} d_0 x. \tag{A.23}$$

Define the map

$$\gamma_0 : X_0 \to Y_0$$

by setting $\gamma_0 : x \mapsto y$, $x \in B_0$ where y is defined by (A.23). Now suppose we have defined maps $\gamma_i : X_i \to Y_i$, $-1 \leq i \leq n-1$, $n \geq 1$, which satisfy the standard simplicial identities. We proceed to define $\gamma_n : X_n \to Y_n$. Clearly, for $x = s_i(x_0) \in B_n$, $x_0 \in B_{n-1}$, $0 \leq i \leq n-1$, we can define $\gamma_n(x) = s_i \gamma_{n-1}(x_0)$. Now let $x \in B_n \setminus \bigcup_{i=0}^{n-1} s_i(B_{n-1})$. Then

$$(\gamma_{n-1} d_0 x, \ldots, \gamma_{n-1} \gamma_n x) \in Z_{n-1} X.$$

The asphericity of X implies that there exists an element $y \in Y_n$, such that

$$d_i y = \gamma_{n-1} d_i y, \ 0 \leq i \leq n;$$

define $\gamma_n(x) = y$. This completes the construction of γ_n.

(ii) The inductive construction of the simplicial homotopy $h : \gamma \to \gamma'$ is standard; one can find all details in [Keu]. $\square$

A.12 Functorial Properties

Let R be a ring and n a non-negative integer. Define the *strong truncation functor*

$$\mathrm{str}_n : \mathcal{C}h(R) \to \mathcal{C}h(R)$$

in the category of chain complexes over the ring R by setting

$$\mathrm{str}_n(K)_i = K_i, \ i < n,$$
$$\mathrm{str}_n(K)_n = K_n / \ker(d_n),$$
$$\mathrm{str}_n(C)_i = 0, \ i \geq n+1,$$

for $K \in \mathcal{C}h(R)$ with obvious boundary maps.

The following results, which are due to Gruenenfelder, generalize the results from [Dol58a]. We follow the treatment in [Gru80].

Theorem A.12 *Let R be a principal ideal domain and X a projective simplicial R-module. Let $X' \in \mathcal{S}_R\mathsf{Mod}$ and $f_i : \pi_i(X) \to \pi_i(X')$, $i \geq 0$. Then there exists a map $f : X \to X'$, which induces the maps f_i.*

Proof. We construct the required map as $N^{-1}g : X \to X'$, where $g : N(X) \to N(X')$ is the map between Moore complexes which induces the maps $f_i : H_i(N(X)) \to H_i(N(X'))$.

Since X is a projective simplicial R-module, $N(X)_i$ are projective R-modules for $i \geq 0$ by Proposition A.7. The component $N(X)_i$ can be presented as $N(X)_i = \ker(\bar{d}_i) \oplus \operatorname{im}(\bar{d}_i)$. Therefore, the R-modules $\ker(\bar{d}_i)$ and $\operatorname{im}(\bar{d}_i)$, $i \geq 0$, are projective. Thus, there exist maps g_i', g_i'' such that the following diagram is commutative:

$$
\begin{array}{ccccccccc}
0 & \longrightarrow & \operatorname{im}(\bar{d}_{i+1}) & \longrightarrow & \ker(\bar{d}_i) & \longrightarrow & H_i(N(X)) & \longrightarrow & 0 \\
& & \downarrow{\scriptstyle g_i''} & & \downarrow{\scriptstyle g_i'} & & \downarrow{\scriptstyle f_i} & & \\
0 & \longrightarrow & \operatorname{im}(\bar{d}_{i+1}') & \longrightarrow & \ker(\bar{d}_i') & \longrightarrow & H_i(N(X')) & \longrightarrow & 0,
\end{array}
$$

and we have the required map

$$
g_i'' \oplus g_{i+1}' : \operatorname{im}(\bar{d}_{i+1}) \oplus \ker(\bar{d}_{i+1}) \to \operatorname{im}(\bar{d}_{i+1}') \oplus \ker(\bar{d}_{i+1}'). \quad \square
$$

Theorem A.13 *Let S be a ring, R a principal ideal domain and $X \in \mathcal{S}_R\mathsf{Mod}$, $X' \in \mathcal{S}_S\mathsf{Mod}$ projective simplicial modules. Let $F : {}_R\mathsf{Mod} \to {}_S\mathsf{Mod}$ be a covariant functor. Then*

(i) Any sequence of homomorphisms $\{f_i : \pi_i(X) \to \pi_i(X')\}$, $i \leq n$ induces homomorphisms $\{f_{F,i} : \pi_i(F(X)) \to \pi_i(F(X'))\}$, $i \leq n$.

(ii) If f_i is bijective for $i < n$ and surjective for $i = n$, then the same is true for $f_{F,i}$, $i \leq n$.

Proof. The homomorphisms f_i can be extended to homomorphisms between homotopy groups of strong truncations of the given simplicial modules:

$$
f_i : \pi_i(\operatorname{str}_{n+1}(X)) \to \pi_i(\operatorname{str}_{n+1}(X')), \quad i \geq 0.
$$

By Theorem A.12, we can assume that the sequence f_i, $i \geq 0$, is induced by a map

$$
f : \operatorname{str}_n(X) \to \operatorname{str}_n(X').
$$

Since $\operatorname{str}_{n+1}(X)$ is a direct summand of X, such that $\operatorname{str}_{n+1}(X)_i = X_i$, $i \leq n$, it follows that $F(\operatorname{sk}_{n+1}(X))$ is a direct summand of $F(X)$ with

$$F(\mathrm{str}_{n+1}(X))_i = F(X)_i, \ i \geq n.$$

Thus

$$\pi_i(F(X)) = \pi_i(F(\mathrm{str}_{n+1}(X))), \ i \leq n.$$

The sequence of maps $\{f_{F,i}\}$, $i \leq n$, can be defined as

$$f_{F,i} : \pi_i(F(X)) = \pi_i(F(\mathrm{str}_{n+1}(X))) \to \pi_i(F(\mathrm{str}_{n+1}(X'))) = \pi_i(F(X')).$$

(ii) Consider the following decomposition of the functor $N \circ f : N(\mathrm{str}_n(X)) \to N(\mathrm{str}_n(X'))$:

$$
\begin{array}{ccccccc}
N(\mathrm{str}_n(X)): & \mathrm{im}(d_{n+1}) & \longrightarrow & N(X)_n & \longrightarrow & N(X)_{n-1} & \longrightarrow & \cdots \\
& \downarrow & & \downarrow & \| & & \| & \\
C: & P & \longrightarrow & N(X)_n & \longrightarrow & N(X)_{n-1} & \longrightarrow & \cdots \\
\beta \downarrow & & \downarrow & & \downarrow & & \downarrow & \\
N(\mathrm{str}_n(X')): & \mathrm{im}(d'_{n+1}) & \longrightarrow & N(X')_n & \longrightarrow & N(X')_{n-1} & \longrightarrow & \cdots,
\end{array}
$$

where the left hand square is a pullback. First note, that the natural pullback properties imply that the map β induces isomorphisms

$$H_i(\beta) : H_i(C) \to H_i(N(\mathrm{str}_n(X')))$$

for all $i \geq 0$. Since P is projective R-module, the map

$$N^{-1} \circ \beta \circ N : N^{-1}(C) \to \mathrm{str}_n(X')$$

is a homotopy equivalence of *strongly* truncated simplicial groups. Obviously, the functor F preserves homotopy; hence

$$F(N^{-1} \circ \beta \circ N) : F(N^{-1}(C)) \to F(\mathrm{str}_n(X'))$$

is a homotopy equivalence. We have $C_i = N(\mathrm{str}_n(X))_i$, $i \leq n$, and a monomorphism $C_{n+1} \to N(\mathrm{str}_n(X))_{n+1}$. Hence $F(C)_i = F(N(\mathrm{str}_n(X'))_i$, $i \leq n$, therefore

$$\pi_i(F(N^{-1} \circ \beta \circ N)) : \pi_i(F(N^{-1}(C))) \to \pi_i(F(\mathrm{str}_n(X')))$$

is an isomorphism for $i < n$. The fact that the simplicial R-modules under consideration are projective implies that $\pi_n(F(N^{-1} \circ \beta \circ N))$ is an epimorphism. Therefore the map

$$\pi_i(F(X)) \simeq \pi_i(F(\mathrm{str}_n(X)) \to \pi_i(F(N^{-1}(C)) \to \pi_i(F(\mathrm{str}_n(X')) \simeq \pi_i(X')$$

is an isomorphism for $i < n$, and an epimorphism for $i = n$. $\square$

A.13 Derived Functors

Derived functors play a fundamental role in different areas of mathematics. The reader can find in [Ina98] an exposition of the general technique of derived functors. Here we mention only some examples of derived functors on the category of groups. However, it may be noted that the same construction is possible for other *algebraic* categories, like Lie algebras, crossed modules, cat^1-groups etc.

Let $T : \mathsf{Gr} \to \mathsf{Gr}$ be a functor on the category Gr of groups, $T(1) = 1$. Theorem A.11 clearly implies the following

Proposition A.14 *Let (X, X_{-1}) and (Y, Y_{-1}) be two free simplicial resolutions of $X_{-1} = Y_{-1}$. Then, for all $n \geq 0$,*

$$\pi_n(T(X)) = \pi_n(T(Y)).$$

From the above Proposition the definition of *left derived functors* of the functor T, denoted $\mathcal{L}_i T$, $i \geq 0$, follows naturally:

$$\mathcal{L}_0 T : \mathsf{Gr} \to \mathsf{Gr}, \ \mathcal{L}_i T : \mathsf{Gr} \to \mathsf{Ab}, \ i \geq 1,$$

are defined as

$$\mathcal{L}_i T : \Gamma \mapsto \pi_i(T(F_*)), \ i \geq 0,$$

where $F_* \in \mathcal{S}\mathsf{Gr}$ is a free simplicial resolution of $\Gamma \in \mathsf{Gr}$, and Ab is the category of abelian groups. Proposition A.14 implies that the resulting groups are independent of the choice of the free simplicial resolution F_*.

Example A.15

Let

$$Z_2 : \mathsf{Gr} \to \mathsf{Ab}, \quad \Gamma \mapsto \Gamma/\gamma_2(\Gamma), \ \Gamma \in \mathsf{Gr},$$

be the abelianization functor. Let X be a free simplicial resolution of Γ. By definition, X_i is free for all $i \geq 0$,

$$\pi_0(X) = \Gamma, \ \pi_i(X) = 0, \ i \geq 1.$$

Form the following bisimplicial group:

$$Y_{i,j} = \mathbb{Z}[(\overline{W}X_i)_j] = \mathbb{Z}[X_i^{\times(j-1)}].$$

Computing first homotopy groups of $Y_{i,j}$ for a fixed j, we get

$$\pi_0(\mathbb{Z}[(\overline{W}X_i)_j]) = \mathbb{Z}[\overline{W}_j(\Gamma)],$$
$$\pi_n(\mathbb{Z}[(\overline{W}X_i)_j]) = 0, \ n \geq 1.$$

On the other hand, computing first homotopy groups with fixed indices i, we get

$$\pi_n(\mathbb{Z}[(\overline{W}X_i)_j]) = H_n(X_i), \ n \geq 0.$$

Since X_i are free, we conclude that the last term is $\mathbb{Z}$ in dimension zero, abelianization $X_i/\gamma_2(X_i)$ in dimension one, and trivial in all other dimensions. Thus, by Theorem A.9,

$$\mathcal{L}_n Z_2(\Gamma) := \pi_n(X/\gamma_2(X)) = \pi_{n+1}(\mathbb{Z}[\overline{W}(\Gamma)]) = H_{n+1}(\Gamma), \ n \geq 0.$$

Example A.16

Let $Z_n : \mathsf{Gr} \to \mathsf{Gr}$, $n \geq 2$, be the functor which maps every group Γ to its quotient by the nth term of its lower central series:

$$Z_n : \Gamma \mapsto \Gamma/\gamma_n(\Gamma).$$

Let $\Gamma = F/R$ and X a free simplicial resolution of Γ with first two terms F_0, F_1 as in Example A.4, i.e.,

$$F_0 = F, \quad F_1 = F(y_j \mid j \in J) * F,$$

where y_j's are in one to one correspondence with a normal basis of R. Then, clearly,

$$\mathcal{L}_0 Z_n(\Gamma) = \Gamma/\gamma_n(\Gamma), \ n \geq 2.$$

The short exact sequence of simplicial groups

$$1 \to \gamma_n(X) \to X \to X/\gamma_n(X) \to 1,$$

gives the following long exact sequence:

$$\ldots \to \pi_1(X) \to \pi_1(X/\gamma_n(X)) \to \pi_0(\gamma_n(X)) \to \pi_0(X) \to \pi_0(X/\gamma_n(X)) \to 1.$$

Since X is aspherical and $\mathcal{L}_0 Z_n = Z_n$, we have

$$\mathcal{L}_1 Z_n(\Gamma) = \pi_1(X/\gamma_n(X)) = \ker\{\pi_0(\gamma_n(X)) \to \gamma_n(\Gamma)\}.$$

The group $\pi_0(\gamma_n(X))$ is a coequalizer of the diagram

$$\gamma_n(F_1) \ \underset{d_1}{\overset{d_0}{\rightrightarrows}} \ \gamma_n(F),$$

where the maps d_0, d_1 are restrictions of degeneracy maps in X. Clearly,

$$\ker(d_0) = [\langle F(y_j \mid j \in J)\rangle^{F_1}, \ _{n-1}F_1]$$

and

$$\mathrm{im}(d_1|_{\ker(d_0)}) = [R, \ _{n-1}F].$$

Hence,

$$\pi_0(\gamma_n(X)) = \frac{\gamma_n(F)}{[R, \ _{n-1}F]}$$

and therefore,

$$\mathcal{L}_1 Z_n(\Gamma) = \ker\{\frac{\gamma_n(F)}{[R, \ _{n-1}F]} \rightarrow \frac{\gamma_n(F)}{R \cap \gamma_n(F)}\} = \frac{R \cap \gamma_n(F)}{[R, \ _{n-1}F]}.$$

This is the same as the nth Baer invariant of Γ, which we denote by $M^{(n)}(\Gamma)$ (see Section 1.4); that is, $\mathcal{L}_1 Z_n(\Gamma) = M^{(n)}(\Gamma)$, $n \geq 2$.

Example A.17

For a given group F and its normal subgroup R, define the series $\{\delta_n(R, \ F)\}_{n \geq 1}$ by setting:

$$\delta_1(R, \ F) = [R, \ F], \ \delta_{n+1}(R, \ F) = [\delta_n(R, \ F), \ \delta_n(F)].$$

Let Γ be a group given by a free presentation $\Gamma = F/R$. Consider the functor

$$\mathfrak{D}_n : \mathsf{Gr} \rightarrow \mathsf{Gr}, \quad \Gamma \mapsto \Gamma/\delta_n(\Gamma), \ n \geq 1.$$

Then the same method as in the previous example gives the following:

$$\mathcal{L}_0 \mathfrak{D}_n = \mathfrak{D}_n, \ \mathcal{L}_1 \mathfrak{D}_n = \frac{R \cap \delta_n(F)}{\delta_n(R, \ F)}.$$

For example, the first derived functor of the metabelianization functor is

$$\mathcal{L}_1 \mathfrak{D}_2 = \frac{R \cap [[F, F], [F, F]]}{[[R, F], [F, F]]}.$$

Example A.18

Consider the commutator subgroup functor $T : G \mapsto [G, G]$. Then the foregoing arguments easily imply that for a given group $G = F/R$, one has the following:

$$\mathcal{L}_i T(G) = \begin{cases} \frac{[F, F]}{[R, F]}, \ i = 0, \\ H_{i+2}(G), \ i > 0. \end{cases} \tag{A.24}$$

A.14 Quadratic Functors

A functor $F : \mathsf{Ab} \rightarrow \mathsf{Ab}$ is quadratic, i.e., has degree ≤ 2, if $F(0) = 0$ and if the cross effect

$$F(A|B) = \ker(F(A \oplus B) \rightarrow F(A) \oplus F(B)), \quad A, \ B \in \mathsf{Ab},$$

is biadditive. We then have a binatural isomorphism

$$F(A \oplus B) = F(A) \oplus F(B) \oplus F(A|B)$$

given by $(F_{i_1}; F_{i_2}; i_{12})$, where $i_1 : A \hookrightarrow A \oplus B$; $i_2 : A \hookrightarrow A \oplus B$ and $i_{12} : F(A|B) \hookrightarrow F(A \oplus B)$ are the inclusions. Moreover, for any $A \in$ Ab one gets the diagram [Bau00]

$$F\{A\} := (F(A) \xrightarrow{H} F(A|A) \xrightarrow{P} F(A)). \tag{A.25}$$

Here $P = F(p_1 + p_2)i_{12} : F(A|A) \hookrightarrow F(A \oplus A) \to F(A)$ is given by the codiagonal $p_1 + p_2 : A \oplus A \to A$, where p_1 and p_2 are the projections. Moreover, H is determined by the equation $i_{12}H = F(i_1 + i_2) - F(i_1) - F(i_2)$ where $i_1 + i_2 : A \to A \oplus A$ is the diagonal map.

We recall from [Eil54] the definitions of certain quadratic functors.

Tor$(\mathbf{A}, \mathbf{C})$:

For abelian groups A and C, the abelian group $\mathrm{Tor}(A, C)$ ([Eil54], §11, p. 85) has generators

$$(a, m, c), \ a \in A, \ c \in C, \ 0 < m \in \mathbb{Z}, \ ma = mc = 0$$

and relations

$$(a_1 + a_2, m, c) = (a_1, m, c) + (a_2, m, c), \text{ if } ma_1 = ma_2 = mc = 0$$
$$(a, m, c_1 + c_2) = (a, m, c_1) + (a, m, c_2), \text{ if } ma = mc_1 = mc_2 = 0$$
$$(a, mn, c) = (na, m, c), \text{ if } mna = mc = 0$$
$$(a, mn, c) = (a, m, nc), \text{ if } ma = mnc = 0.$$

In particular, we have the functor $A \mapsto \mathrm{Tor}(A, A))$ on the category Ab. We denote the class of the triple (a, m, c) by $\tau_m(a, c)$.

$\Omega(\mathbf{A})$: Let A be an Abelian group. Then the group $\Omega(A)$, defined by Eilenberg-MacLane ([Eil54], p. 93), is the Abelian group generated by symbols $w_n(x)$, $0 < n \in \mathbb{Z}, x \in A$, $nx = 0$ with defining relations

$w_{nk}(x) = kw_n(x), \ nx = 0,$

$kw_{nk}(x) = w_n(kx), \ nkx = 0,$

$w_n(kx + y) - w_n(kx) - w_n(y) = w_{nk}(x + y) - w_{nk}(x) - w_{nk}(y), \ nkx = ny = 0,$

$w_n(x + y + z) - w_n(x + y) - w_n(x + z) - w_n(y + z) + w_n(x) + w_n(y) + w_n(z) = 0,$

$nx = ny = nz = 0.$

We continue to denote the class of the element $w_n(x)$ in $\Omega(A)$ by $w_n(x)$ itself.

Eilenberg-MacLane ([Eil54], p. 94) constructed a map

$$E : \mathrm{Tor}(A,\,A) \to \Omega(A)$$

by setting

$$\tau_n(a,\,c) \mapsto w_n(a+c) - w_n(a) - w_n(c).$$

A natural map

$$T : \Omega(A) \to \mathrm{Tor}(A,\,A) \tag{A.26}$$

can be defined by setting

$$w_n(x) \mapsto \tau_n(x,\,x), \quad x \in A, \quad nx = 0.$$

Clearly the composite map

$$E \circ T : \Omega(A) \to \Omega(A)$$

is multiplication by 2; for, as a consequence of the defining relations the elements $w_n(x)$ satisfy

$$w_n(mx) = m^2 w_n(x)$$

for all $m \in \mathbb{Z}$, $0 < n \in \mathbb{Z}$, $x \in A$.

Whitehead functor $\Gamma_2(A)$: The homogeneous component $\Gamma_2(A)$ of the graded functor $\Gamma(A)$ (see (5.36)-(5.39)) of degree 2 can be identified with the Whitehead functor ([Whi50], [Eil54], pp. 92 & 110) $\Gamma_2 : \mathsf{Ab} \to \mathsf{Ab}$, $A \mapsto \Gamma_2(A)$, where the group $\Gamma_2(A)$ is defined for $A \in \mathsf{Ab}$ to be the group given by generators $\gamma(a)$, one for each $x \in A$, subject to the defining relations

$$\gamma(-x) = \gamma(x), \tag{A.27}$$

$$\gamma(x+y+z) - \gamma(x+y) - \gamma(x+z) - \gamma(y+z) + \gamma(x) + \gamma(y) + \gamma(z) = 0 \tag{A.28}$$

for all x, y, $z \in A$.

$R(A)$: For $A \in \mathsf{Ab}$, let $_2A$ denote the subgroup consisting of elements x satisfying $2x = 0$. Define $R(A)$ ([Eil54]. p. 120) to be the quotient group of $\mathrm{Tor}(A,\,A) \oplus \Gamma_2(A)$ by the relations

$$\tau_m(x,\,x) = 0, \quad mx = 0, \tag{A.29}$$

$$\gamma_2(s+t) - \gamma_2(s) - \gamma_2(t) = \tau_2(s,\,t), \quad s,\,t \in_2 A. \tag{A.30}$$

The functors $A \mapsto \Gamma_2(A)$, $\mathrm{Tor}(A,\,A)$, $\Omega(A)$, $R(A)$ are all quadratic functors on the category of Abelian groups, i.e., all these functors have degree ≤ 2. Furthermore, $R(A) = H_5K(A;2)$; $\Omega(A) = H_7K(A;3)/(\mathbb{Z}/3\mathbb{Z} \otimes A)$, $R(A|B) = \Omega(A|B) = \mathrm{Tor}(A,\,B)$ and $R(\mathbb{Z}) = \Omega(\mathbb{Z}) = 0$ and $R(\mathbb{Z}/n) = \mathbb{Z}/(2,\,n)$; $\Omega(\mathbb{Z}/n) = \mathbb{Z}/n$.

The square functor. Let Ab_g denote the category of graded Abelian groups. For A, $B \in \mathsf{Ab}$, let $A \otimes B$ and $A * B := \mathrm{Tor}(A, B)$ be respectively the tensor product and the torsion product of A, B. The notion of tensor product of Abelian groups extends naturally to that of *tensor product $A \otimes B$ of graded Abelian groups A, B* by setting

$$(A \otimes B)_n = \bigoplus_{i+j=n} A_i \otimes B_j.$$

We also need the *ordered tensor product $A \overset{>}{\otimes} B$ of graded Abelian groups*, which is defined by setting

$$(A \overset{>}{\otimes} B)_n = \bigoplus_{i+j=n,\ i>j} A_i \otimes B_j$$

for A, $B \in \mathsf{Ab}_g$. In an analogous manner, we can define, for A, $B \in \mathsf{Ab}_g$, *torsion product $A * B$* and *ordered torsion product $A \overset{>}{*} B$* as

$$A * B = \bigoplus_{i+j=n} A_i * B_j, \quad (A \overset{>}{*} B)_n = \bigoplus_{i+j=n,\ i>j} A_i * B_j.$$

The tensor product, torsion product and the ordered tensor and torsion product are, in an obvious way, bifunctors on the category Ab_g.

Let $\wedge^2$ be the exterior square functor on the category Ab. The *weak square functor*

$$sq^\otimes : \mathsf{Ab}_g \to \mathsf{Ab}_g$$

is defined by

$$sq^\otimes(A)_n = \begin{cases} \Gamma_2(A_m), & \text{if } n = 2m,\ m \text{ odd}, \\ \wedge^2(A_m), & \text{if } n = 2m,\ m \text{ even}, \\ 0, & \text{otherwise.} \end{cases}$$

Let $(\mathbb{Z}_2)_{odd}$ be the graded Abelian group which is $\mathbb{Z}_2$ in odd degree ≥ 1 and trivial otherwise; thus $(\mathbb{Z}_2)_{odd}$ is the reduced homology of the classifying space $\mathbb{R}P_\infty = K(\mathbb{Z}_2, 1)$. The *square functor* $Sq^\otimes : \mathsf{Ab}_g \to \mathsf{Ab}_g$ is defined as follows:

$$Sq^\otimes(A) = A \overset{>}{\otimes} (A \oplus (\mathbb{Z}_2)_{odd}) \oplus sq^\otimes(A).$$

Define next the *torsion square functor*

$$Sq^\star(A) : \mathsf{Ab}_g \to \mathsf{Ab}_g$$

by setting

$$Sq^\star(A) = (A \overset{>}{*} (A \oplus (\mathbb{Z}_2)_{odd})) \oplus sq^\star(A),$$

where
$$sq^\star(A)_n = \begin{cases} \Omega(A_m), & n = 2m, \ m \text{ even} \\ R(A_m), & n = 2m, \ m \text{ odd} \\ 0, & \text{otherwise} \end{cases}$$

Now we are ready to formulate so-called universal coefficient theorem for the functor $\wedge^2$ due to Baues and Pirashvili [Bau00]. Let X be a simplicial group which is free Abelian in each degree. Then there exists [[Bau00], (4.1)] a natural short exact sequence of graded Abelian groups

$$0 \to Sq^\otimes(\pi_*(X)) \to \pi_*(\wedge^2 X) \to Sq^\star(\pi_*(X))[-1] \to 0 \qquad (A.31)$$

where $\pi_*(X)$ and $\pi_*(\wedge^2 X)$ are the graded homotopy groups of X and $\wedge^2 X$ respectively.

The exact sequence (A.31) leads to the description of the derived functors of the second lower central quotient functor. Since for a free group F, there is a natural isomorphism

$$\gamma_2(F)/\gamma_3(F) \simeq \wedge^2(F_{ab}),$$

for every free simplicial resolution $F_* \to G$, we obtain the following natural exact sequences:

$$0 \to H_2(G) \otimes H_1(G) \to \pi_1(\gamma_2(F_*)/\gamma_3(F_*)) \to \Omega(H_1(G)) \to 0$$
$$0 \to H_3(G) \otimes H_1(G) \oplus \Gamma_2(H_2(G)) \to \pi_2(\gamma_2(F_*)/\gamma_3(F_*))$$
$$\to \text{Tor}(H_2(G), H_1(G)) \to 0$$

Similar descriptions exist for other quadratic functors (see [Bau00]). Consider the functor
$$\bar{\Gamma} : \mathsf{Gr} \to \mathsf{Ab},$$

which is the composition of the abelianization and the functor Γ_2. For every group G there exists the following exact sequence of groups (see [Bau00]):

$$0 \to H_2(G) \otimes (H_1(G) \oplus \mathbb{Z}_2) \to \mathcal{L}_1\bar{\Gamma}(G) \to R(H_1(G)) \to 0.$$

A.15 Derived Functors in the Sense of Dold and Puppe

Let $T : \mathsf{Ab} \to \mathsf{Ab}$ be a functor with $T(0) = 0$ and A an abelian group. For $n \geq 0$, consider a free simplicial abelian group P_* with the following properties:

(i) $P_i = 0, \ i < n$,
(ii) $\pi_n(P_*) = A$,
(iii) $\pi_i(P_*) = 0, \ i \neq n$.

Define the ith derived functor of T (in the sense of Dold and Puppe [Dol61]) for the pair (A, n) as

$$\mathcal{L}_i T(A, n) := \pi_i(T(P_*)).$$

Standard arguments, similar to ones given in Proposition A.14, show that this definition is independent of a choice of P_*. As an example of P_*, we can choose the free abelian simplicial group

$$P_* = N^{-1}((A_1 \hookrightarrow A_0)[n]),$$

where N^{-1} is the inverse map to the Dold-Kan map (see A.6), A_1 and A_0 are free abelian and the sequence

$$0 \to A_1 \to A_0 \to A \to 0$$

is exact. For $n = 0$, we will use the notation $\mathcal{L}_i T(A) := \mathcal{L}_i T(A, 0)$.

The derived functors in the sense of Dold and Puppe play a fundamental role in topology in view of the following fact. For every abelian group A and $n \geq 1$, there exists a natural spectral sequence

$$E^2_{p,q} = \mathcal{L}_{p+q} \mathrm{SP}^q(A, n) \Rightarrow H_{p+q} K(A, n) \qquad (A.32)$$

which converges to the homology of the Eilenberg-MacLane space $K(A, n)$. This sequence degenerates [Bre99] and, therefore, the derived functors $\mathcal{L}_{p+q} \mathrm{SP}^q(A, n)$ define a canonical filtration of $H_{p+q} K(A, n)$.

Clearly, the sequence (A.31) provides a method of computing the derived functors of the exterior square. The universal coefficient theorem for quadratic functors SP^2 and Γ given in [Bau00] imply the following description of derived functors:

$$\mathcal{L}_i \mathrm{SP}^2(A, n) = \begin{cases} \mathrm{SP}^2(A), \ i = 0, \ n = 0, \\ A \overset{\wedge}{*} A, \ i = 1, \ n = 0 \\ \Gamma_2(A), \ i = 2n, \ n \neq 0 \ \text{even}, \\ \Lambda^2(A) \oplus \mathrm{Tor}(A, \mathbb{Z}_2), \ i = 2n, \ n \neq 1 \ \text{odd}, \\ R(A), \ i = 2n + 1, \ n \neq 0 \ \text{even}, \\ \Omega(A), \ i = 2n + 1, \ n \ \text{odd}, \\ A \otimes \mathbb{Z}_2, \ i = n + 2, \ n + 4, \ \ldots, \ n + 2[\frac{n-1}{2}], \\ \mathrm{Tor}(A, \mathbb{Z}_2), \ i = n + 3, \ n + 5, \ \ldots, \ n + 2[\frac{n-1}{2}] + 1, \ i \neq 2n, \\ 0, \ \text{otherwise}; \end{cases}$$

$$\mathcal{L}_i \Lambda^2(A, n) = \begin{cases} \Lambda^2(A), \ i = 0, \ n = 0, \\ \Gamma_2(A), \ i = 2n, \ n \ \text{odd}, \\ \Lambda^2(A) \oplus \mathrm{Tor}(A, \mathbb{Z}_2), \ i = 2n, \ n \neq 0 \ \text{even}, \\ R_2(A), \ i = 2n + 1, \ n \ \text{odd}, \\ \Omega_2(A), \ i = 2n + 1, \ n \ \text{even}, \\ A \otimes \mathbb{Z}_2, \ i = n+1, \dots, n + 2[\frac{n-1}{2}] + 1, \ i \neq 2n, \\ \mathrm{Tor}(A, \mathbb{Z}_2), \ i = n+2, \ n+4, \dots, n + 2[\frac{n-1}{2}], \\ 0, \ \text{otherwise}; \end{cases}$$

$$\mathcal{L}_i \Gamma_2(A, n) = \begin{cases} \Gamma_2(A), \ i = 2n, \ n \ \text{even}, \\ \Lambda^2(A) \oplus \mathrm{Tor}(A, \mathbb{Z}_2), \ i = 2n, \ n \ \text{odd}, \\ R_2(A), \ i = 2n + 1, \ n \ \text{even}, \\ \Omega_2(A), \ i = 2n + 1, \ n \ \text{odd}, \\ A \otimes \mathbb{Z}_2, \ i = n, n+2, \dots, n + 2[\frac{n-1}{2}], \ n > 0, \\ \mathrm{Tor}(A, \mathbb{Z}_2), \ i = n+1, n+3, \dots, n + 2[\frac{n-1}{2}] + 1, \ n > 0, \ i \neq 2n, \\ 0, \ \text{otherwise}. \end{cases}$$

For polynomial functors of higher degrees the functorial description of the derived functors is a deep problem. For example, consider the tensor cube. It follows from the work of MacLain [Mac60] that the derived functors can be described as follows.

$$\mathcal{L}_i \otimes^3 (A) = \begin{cases} \otimes^3(A), \ i = 0, \\ (\mathrm{Tor}(A, A) \otimes A)^{\oplus 3} / Jac_\otimes, \ i = 1, \\ \mathrm{Tor}(\mathrm{Tor}(A, A), A), \ i = 2 \end{cases}$$

where $Jac_\otimes$ is the subgroup in $(\mathrm{Tor}(A, A) \otimes A)^{\oplus 3}$, generated by elements

$$(a, n, b) \otimes c + (c, n, a) \otimes b + (b, n, c) \otimes a, \ a, b, c \in A, \ na = nb = nc = 0.$$

The derived functor $\mathcal{L}_1 \otimes^3 (A)$ is a part of the following short exact sequence [Mac60]:

$$0 \to \mathrm{Tor}(A, A) \otimes A \to \mathcal{L}_1 \otimes^3 (A) \to \mathrm{Tor}(A \otimes A, A) \to 0.$$

For an analogous description of the derived functors of the symmetric and the exterior powers see [Bre99] and [Jea02].

A.16 Derived Limits and Fibration Sequence

Let

$$\cdots \xrightarrow{f_n} G_n \xrightarrow{f_{n-1}} G_{n-1} \xrightarrow{f_{n-2}} \cdots \xrightarrow{f_1} G_1 \xrightarrow{f_0} G_0 \qquad (\text{A.33})$$

be an inverse system of groups in the category Gr. Then the group $\Pi_{\mathsf{Gr}} = \prod_i G_i$, defined as the unrestricted product, acts on the set $\Pi_{\mathsf{Set}} = \prod_i G_i$ by setting

$$(g_0, \ldots, g_n, \ldots) \circ (x_0, \ldots, x_n, \ldots) = (g_0 x_0 f_0(g_1^{-1}), \ldots, g_n x_n f_n(g_{n+1}^{-1}), \ldots),$$

x_i, $g_i \in G_i$. Let $\mathbf{1} = (1, 1, \ldots, 1, \ldots)$ be the identity element in $\Pi_i G_i$. Then, by definition,

$$\varprojlim_i G_i = \{g \in \Pi_{\mathsf{Gr}} G_i \mid g \circ \mathbf{1} = \mathbf{1}\}.$$

The *derived limit* $\varprojlim_i^1 G_i$ of the system (A.33) is defined to be the set of orbits of Π_{Set} under the above action of Π_{Gr}:

$$\varprojlim_i^1 G_i := \Pi_{Set} G_i / \{x \sim g \circ x : \ g \in \Pi_{\mathsf{Gr}}, \ x \in \Pi_{Set} G_i\}.$$

In general, $\varprojlim_i^1 G_i$ is a pointed set, but in case the group G_i are all abelian, it has a natural structure of an abelian group. Clearly, we can define the inverse and derived limit in the case of an inverse system of abelian groups by the exact sequence

$$1 \to \varprojlim_i G_i \to \prod_i G_i \xrightarrow{f} \prod_i G_i \to \varprojlim_i^1 G_i \to 1,$$

where the homomorphism $f : \prod_i G_i \to \prod_i G_i$ is defined by

$$(g_0, g_1, \ldots, g_n, \ldots) \mapsto (g_0 f_0(g_1^{-1}), g_1 d_1(g_2^{-1}), \ldots, g_n f_n(g_{n+1}^{-1}), \ldots).$$

The following result is well-known.

Proposition A.19 *Let*

$$1 \to \{G_n'\} \to \{G_n\} \to \{G_n''\} \to 1 \tag{A.34}$$

be a short exact sequence of inverse systems of groups. Then there is a sequence

$$1 \to \varprojlim_n G_n' \to \varprojlim_n G_n \to \varprojlim_n G_n'' \to$$

$$\varprojlim_i^1 G_n' \to \varprojlim_i^1 G_n \to \varprojlim_i^1 G_n'' \to 1. \tag{A.35}$$

of groups and pointed spaces which is exact as a sequence of groups at the first three terms, as a sequence pointed sets at the last three terms, and the set map

$$\varprojlim_n G_n'' \to \varprojlim_i^1 G_n'$$

extends to a natural action of $\varprojlim_i G_n''$ on $\varprojlim_i^1 G_n'$ such that elements of $\varprojlim_i^1 G_n'$ are in the same orbit if and only if they have the same image in $\varprojlim_i^1 G_n$.

In case (A.34) is a short exact sequence of abelian groups, then the sequence (A.35) is a long exact sequence of abelian groups.

It is of interest to note a characterization for the vanishing of the derived limit of a given inverse system of abelian groups.

An inverse system (A.34) of abelian groups is said to satisfy the *Mittag-Leffler condition* if, for every $m \geq 0$, the chain

$$\text{im}(f_m) \supseteq \text{im}(f_m \circ f_{m+1}) \supseteq \text{im}(f_m \circ f_{m+1} \circ f_{m+2}) \supseteq \cdots$$

is stationary.

Proposition A.20 (Gray [Gra66]). *Let (A.34) be an inverse system of countable abelian groups. Then $\varprojlim_i^1 G_i = 0$ if and only if this inverse system satisfies the Mittag-Leffler condition.*

Thus, in particular, for any inverse system of finite abelian groups, or for any inverse system of epimorphisms of abelian groups, its derived limit vanishes always.

Example A.21

Let p be a prime. Consider the inverse system

$$\cdots \to \mathbb{Z} \xrightarrow{p} \mathbb{Z} \xrightarrow{p} \mathbb{Z} \to \cdots \to \mathbb{Z}$$

of monomorphisms of the additive group of integers, defined by p-multiplication, i.e.,

$$z \mapsto pz, \ z \in \mathbb{Z}.$$

Then one has

$$\varprojlim_n^1 \mathbb{Z} = (\varprojlim_n \mathbb{Z}_{p^n})/\mathbb{Z},$$

i.e., p-adic integers modulo the rational integers.

Proposition A.22 (Harlap [Har75]). *Let (A.34) be an inverse system of finitely generated abelian groups. Then*
either *(A.34) satisfies the Mittag-Leffler condition and $\varprojlim_i^1 G_i = 0$,*
or $\varprojlim_i^1 G_i$ *is uncountable.*

Theorem A.23 [Bou72]. *Let $\mathcal{C}$ be a category with initial object $* \in Ob(\mathcal{C})$ and*

$$\cdots \to X_n \to X_{n-1} \to \cdots \to X_0 \to *$$

a tower of fibrations in $\mathcal{SC}$ with compatible base points. Then for any $i \geq 0$, there exists the following natural exact sequence:

$$* \to \varprojlim_n^1 \pi_{i+1}(X_n) \to \pi_i(\varprojlim_n X_n) \to \varprojlim_n \pi_i(X_n) \to *.$$

References

[Akc02] Akca, I. and Arvasi, Z. Simplicial and crossed Lie algebras. *Homology, Homotopy and Applications*, 4:43–57, 2002.

[And65] Andrews, J. J. and Curtis, M. L. Free groups and handlebodies. *Proc. Amer. Math. Soc.*, 16:192–195, 1965.

[And68] Andreev, K. K. Nilpotent groups and Lie algebras. *Algebra and Logic*, 7:206–211, 1968.

[And69] Andreev, K. K. Nilpotent groups and Lie algebras II. *Algebra and Logic*, 8:353–358, 1969.

[And70] André, M. *Homologie des Algébres Commutatives*, volume 206 of *Lecture Notes in Mathematics*. Springer-Verlag, 1970.

[Art25] Artin, E. Theorie der Zöpfe. *Abhandlungen Hamburg*, 4:47–72, 1925.

[Bae45] Baer, R. Representations of groups as quotient groups. I-III. *Trans. Am. Math. Soc.*, 58:295–419, 1945.

[Bak00] Bak, A. and Vavilov, N. Presenting powers of augmentation ideals and Pfister forms. *K-Theory*, 20:299–310, 2000.

[Bak04] Bak, A. and Tang, G. Solution to the presentation problem for powers of the augmentation ideal of torsion free and torsion abelian groups. *Advances in Math.*, 189:1–37, 2004.

[Bar69] Barr, M. and Beck, J. *Seminar on triples and categorical homology theory*, volume 80 of *Lecture Notes in Math.*, chapter Homology and standard constructions, pages 245–335. Springer, 1969.

[Bar07] Bardakov, V. and Mikhailov, R. On residual properties of link groups. *Sib. Mat. J.*, 48:485–495, 2007.

[Bau]   Baues, H.-J. and Mikhailov, R. Intersection of subgroups in free groups and homotopy groups. *Int. J. Algebra Comp.*, 18:803–823, 2008.

[Bau59] Baumslag, G. Wreath products and p-groups. *Proc. Cambridge Philos. Soc.*, 55:224–231, 1959.

[Bau67] Baumslag, G. Groups with the same lower central sequences as a relatively free group I. The groups. *Trans. Amer. Math. Soc.*, 129:308–321, 1967.

[Bau69] Baumslag, G. Groups with the same lower central sequences as a relatively free group II. Properties. *Trans. Amer. Math. Soc.*, 142:507–538, 1969.

[Bau77] Baumslag, G. and Stammbach, U. On the inverse limit of free nilpotent groups. *Comment. Math. Helv.*, 52:219–233, 1977.

[Bau90] Baues, H.-J. and Conduche, D. The central series for Peiffer commutators in groups with operators. *J. Algebra*, 133:1–34, 1990.

[Bau91] Baues, H. J. *Combinatorial homotopy and 4-dimensional complexes. With a preface by Ronald Brown.* Number 2 in Expositions in Mathematics. Walter de Gruyter & Co., Berlin, 1991.

[Bau97] Baues, H.-J. and Conduche, D. On the 2-type of an iterated. loop space. *Forum Math.*, 9:721–738, 1997.

[Bau00] Baues, H.-J. and Pirashvili, T. A universal coefficient theorem for quadratic functors. *J. Pure and Appl. Algebra*, 148:1–15, 2000.

[Ber79] Berridge, P. H. and Dunwoody, M. J. Non-free projective modules for torsion-free groups. *J. London Math. Soc.*, 19:433–436, 1979.

[Ber00] Berrick A. J. and Dwyer, W. The spaces that define algebraic K-theory. *Topology*, 39, 2000.

[Bha92a] Bhandari, A. K. and Passi, I. B. S. Lie nilpotency indices of group algebras. *Bull. London Math. Soc.*, 24:68–70, 1992.

[Bha92b] Bhandari, A. K. and Passi, I. B. S. Residually Lie nilpotent group rings. *Arch. Math.*, 58:1–6, 1992.

[Bog93] Bogley, W. J. H. C. Whitehead's asphericity question. In *Two-dimensional Homotopy and Combinatorial Group Theory*, volume 197 of *London Math. Soc. Lect. Note Ser.*, chapter X, pages 309–334. Cambridge University Press, Cambridge, 1993.

[Bou72] Bousfield, A. K. and Kan, D. M. *Homotopy limits, completions and localizations. Lecture Notes in Mathematics. 304.* Springer-Verlag, Berlin-Heidelberg-New York: Springer-Verlag., 1972.

[Bou77] Bousfield, A.K. Homological localization towers for groups and Π-modules. *Mem. Am. Math. Soc.*, 186:68 p., 1977.

[Bou92] Bousfield, A. K. On the *p*-adic completions of nonnilpotent spaces. *Trans. Amer. Math. Soc.*, 331:335–359, 1992.

[Bra81] Brandenburg J. and Dyer M. On J.H.C. Whitehead's aspherical question I. *Comm. Math. Helv.*, 56:431–446, 1981.

[Bre99] Breen, L. On the functorial homology of Abelian groups. *J. Pure Appl. Algebra*, 142(3):199–237, 1999.

[Bro75] Brown, K. S. and Dror, E. The Artin-Rees property and homology. *Israel J. Math.*, 22:93–109, 1975.

[Bro84] Brown, R. Coproducts of crossed *P*-modules: applications to second homotopy groups and to the homology of groups. *Topology*, 23:337–345, 1984.

[Bro87] Brown, R. and Loday, J.-L. Van Kampen theorems for diagrams of spaces. *Topology*, 26:311–335, 1987.

[Bru84] Brunner A. M., Frame M. L. and Lee Y. W., Wielenberg N. J. Classifying torsion-free subgroups of the Picard group. *Tranc. Amer. Math. Soc.*, 282:205–235, 1984.

[Bur97] Burns, J. and Ellis, G.. On the nilpotent multipliers of a group. *Math. Z.*, 226:405–428, 1997.

[Bur98] Burns, J. and Ellis, G. Inequalities for Baer invariants of finite groups. *Can. Math. Bull.*, 41:385–391, 1998.

[Car56] Cartan, H. and Eilenberg, S. *Homological Algebra.* Prineton University Press, 1956.

[Car02] Carrasco, P., Cegarra, A. M. and Grandjean, A. R. (Co)Homology of crossed modules. *J. Pure Appl. Algebra*, 168:147–176, 2002.

[Cli87] Cliff, G. and Hartley, B. Sjögren's theorem on dimension subgroups. *J. Pure Appl. Algebra*, 47:231–242, 1987.

[Coca] Cochran, T. and Harvey, S. Homological stability of series of groups. *Preprint arXiv:0802.2390.*

[Cocb] Cochran, T. and Harvey, S. Homology and derived p-series of groups. *Preprint arXiv:math/0702894.*

[Coc54] Cockcroft, W. On two-dimensional aspherical complexes. *Proc. London. Math. Soc.*, 4:375–384, 1954.

[Coc91] Cochran, T. *k*-cobordism for links in S^3. *Trans. Amer. Math. Soc.*, 327:641–654, 1991.

[Coc98] Cochran, T. and Orr, K. Stability of lower central series of compact 3-manifold groups. *Topology*, 37:497–526, 1998.

[Coc05] Cochran, T. and Harvey, S. Homology and derived series of groups. *Geom. Topol.*, 9:2159–2191, 2005.

[Coc08] Cochran, T. and Harvey, S. Homology and derived series of groups. *Geom. Topol.*, 12:199–232, 2008.

[Coh63] Cohn, P. M. A remark on the Birkhoff-Witt theorem. *J. London Math. Soc.*, 38:197–203, 1963.

[Con96] Conduché, D. Question de Whitehead et modules précroisés. *Bull. Soc. Math. France*, 124:401–423, 1996.

[Cur63] Curtis, E. Lower central series of semi-simplicial complexes. *Topology*, 2:159–171, 1963.

[Cur71] Curtis, E. B. Simplicial homotopy theory. *Adv. Math.*, 6:107–209, 1971.

[Dar92] Darmon, H. A refined conjecture of Mazur-Tate type for Heegner points. *Invent. Math.*, 110:123–146, 1992.

[Dol58a] Dold, Λ. Homology of symmetric products and other functors of complexes. *Ann. Math.*, 68:54–80, 1958.

[Dol58b] Dold, Albrecht and Thom, René. Quasifaserungen und unendliche symmetrische Produkte. *Ann. Math.*, 67:239–281, 1958.

[Dol61] Dold, A. and Puppe, D. Homologie nicht-additiver Funtoren; Anwendugen. *Ann. Inst. Fourier (Grenoble)*, 11:201–312, 1961.

[Du,92] Du, Xiankun. The centers of a radical ring. *Canad. Math. Bull.*, 35:174–179, 1992.

[Dun72] Dunwoody, M. J. Relation modules. *Bull. London Math. Soc.*, 4:151–155, 1972.

[Dus75] Duskin J. *Simplicial methods and the interpretation of "triple" cohomology*, volume 163 of *Mem. AMS.* Amer. Math. Soc., 1975.

[Dwy75] Dwyer, W. G. Homology, Massey products and maps between groups. *J. Pure Appl. Algebra*, 6:177–190, 1975.

[Dwy04] Dwyer, W. G. *Axiomatic, Enriched and Motivic Homotopy Theory (J. P. C. Greenlees, Ed.)*, chapter Localizations, pages 3–28. Proceedings of the NATO ASI. Kluwer, 2004.

[Dwy78] Dwyer, W.G. Homological localization of π-modules. 10:135–151, 1977/78.

[Dye75] Dyer, M. On the 2-realizability of 2-types. *Trans. Amer. Math. Soc.*, 204:229–243, 1975.

[Dye93] Dyer, M. N. Crossed modules and π_2 homotopy modules. In Wolfgang Metzler & Allan J. Sierdski Cynthia Hog-Angeloni, editor, *Two dimensional homotopy and combinatorial group theory*, volume 197 of *London Math. Soc. Lect. Note series*, pages 125–156. Cambridge University Press, Cambridge, 1993.

[Eil54] Eilenberg, S. and MacLane, S. On the groups $H(\Pi, n)$, II. *Ann. Math.*, 70:49–139, 1954.

[Ell89] Ellis G. An algebraic derivation of a certain exact sequence. *J. Algebra*, 127:178–181, 1989.

[Ell91] Ellis G. On the higher universal quadratic functors and related computations. *J. Algebra*, 140:392–398, 1991.

[Ell92] Ellis, G. Homology of 2-types. *J. London Math. Soc.*, 46:1–27, 1992.

[Ell02] Ellis, G. A Magnus-Witt type isomorphism for non-free groups. *Georgian Mathematical Journal*, 9:703–708, 2002.

[Ell08] Ellis, G. and Mikhailov, R. A colimit of classifying spaces. arXiv:0804.3581. 2008.

[Fal88] Falk, M. and Randell, R. Pure braid groups and products of free groups. *Contemp. Math.*, 78:217–228, 1988.

[Fox53] Fox, R. H. Free differential calculus. I: Derivations in the free group ring. *Ann. Math.*, 57:547–560, 1953.

[Fre95] Freedman, M. H. and Teichner, P. 4-manifold topology. II: Dwyer's filtration and surgery kernels. *Invent. Math.*, 122:531–557, 1995.

[Gas54]   Gaschütz, W. Über modulare Darstellungen endliche Gruppen, die von frei Gruppen induziert werden. *Math. Z*, 60:274–286, 1954.

[Gil]   Gilbert, N.D. Cockcroft complexes and the plus construction. Kim, A. C. (ed.) et al., Groups - Korea '94. Proceedings of the international conference, Pusan, Korea, August 18-25, 1994. Berlin: Walter de Gruyter. 119-125 (1995).

[Goe99]   Goerss, P. G. and Jardine, J. F. *Simplicial Homotopy Theory*, volume 174 of *Progress in Mathematics*. Birkhäuser, Basel-Boston-Berlin, 1999.

[Gor81]   Gordon, C. Ribbon concordance of knots in the 3-sphere. *Math. Ann.*, 257:157–170, 1981.

[Gou99]   Goussarov, M. Finite type invariants and n-equivalence for 3-manifolds. *C.R.A.S.P*, 329:517–522, 1999.

[Gra66]   Gray, B.I. Spaces of the same n-type, for all n. *Topology*, 5:241–243, 1966.

[Gra00]   Grandjean, A. R., Ladra, M. and Pirashvili, T. CCG-homology of crossed modules via classifying spaces. *J. Algebra*, 229:660–665, 2000.

[Gri04]   Grievel, P.-P. Une histoire de théorèm de Poincare-Birkhoff-Witt. *Expo. Math.*, 22:145–184, 2004.

[Gru]   Gruenberg, K.W. Free abelianised extensions of finite groups. Homological group theory, Proc. Symp., Durham 1977, Lond. Math. Soc. Lect. Note Ser. 36, 71-104 (1979).

[Gru36]   Gruen, O. Über die Faktogruppen freier Gruppen I. *Deutsche Math. (Jahrgang 1)*, 6:772–782, 1936.

[Gru57]   Gruenberg, K. W. Residual properties of infinite soluble groups. *Proc. Lond. Math. Soc., III. Ser.*, 7:29–62, 1957.

[Gru62]   Gruenberg, K. W. The residual nilpotence of certain presentations of finite groups. *Arch. Math.*, 13:408–417, 1962.

[Gru70]   Gruenberg, K. W. *Cohomological Topics in Group Theory*, volume 143 of *LNM*. Springer-Verlag, 1970.

[Gru72]   Gruenberg, K. W. and Roseblade, J. E. The augmentation terminals of certian locally finite groups. *Canad. J. Math.*, 24:221–238, 1972.

[Gru80]   Gruenenfelder, L. Lower central series, augmentation quotients and homology of groups. *Comment. Math. Helv.*, 55:159–177, 1980.

[Gup73]   Gupta, C. K. The free centre-by-metabelian groups. *Austral. math. Soc.*, 16:294–299, 1973.

[Gup78]   Gupta, C. K. and Gupta, N. D. Generalised Magnus embeddings and some applications. *Math. Z.*, 160:75–87, 1978.

[Gup82]   Gupta, N. On the dimension subgroups of metabelian groups. *J. Pure Appl. Algebra*, 24:1–6, 1982.

[Gup83]   Gupta, N. D. and Levin, F. On the lie ideals of a ring. *J. Algebra*, 81:225–231, 1983.

[Gup86]   Gupta, C. K. and Levin, F. Dimension subgroups of the free center-by-metabelian groups. *Illinois J. Math.*, 30:258–273, 1986.

[Gup87a] Gupta, C. K. and Passi, I. B. S. Magnus embeddings and residual nilpotence. *J. Algebra*, 106:105–113, 1987.

[Gup87b] Gupta, C. K., Gupta, N. D. and Levin, F. On dimension subgroups relative to certian product ideals. In *Group Theory (Bressanone, 1986)*, volume 1281 of *LNM*, pages 31–35, Springer, Berlin, 1987.

[Gup87c] Gupta, N. *Free Group Rings*, volume 66 of *Contemporary Math.* Amer. Math. Soc., 1987.

[Gup87d] Gupta, N. Sjögren's theorem for dimension subgroups-the metabelian case. In *Combinatorial group theory and topology (Alta, Utah), (1984)*, volume 111 of *Ann. Math. Studies*, pages 197–211, Princeton, NJ, 1987. Princeton University Press.

[Gup90]   Gupta, N. The dimension subgroup conjecture. *Bull. London Math. Soc.*, 22:453–456, 1990.

[Gup91a] Gupta, N. On groups without dimension property. *Int. J. Algebra Comput.*, 1(2):247–252, 1991.

[Gup91b] Gupta, N. A solution of the dimension subgroup problem. *J. Algebra*, 138:479–490, 1991.

[Gup91c] Gupta, N. and Srivastava, J. B. Some remarks on Lie dimesnion subgroups. *J. Algebra*, 143:57–62, 1991.

[Gup92] Gupta, N. and Kuz'min, Y. On varietal quotients defined by ideals generated by Fox derivatives. *J. Pure Appl. Algebra*, 78:165–172, 1992.

[Gup93] Gupta, N. D. and Tahara, K. The seventh and eighth Lie dimnsion subgroups. *J. Pure Appl. algebra*, 88:107–117, 1993.

[Gup94] Gupta, C. K., Gupta, N. D. and Passi, I. B. S. Dimension subgroups of cenre-by-metabelian groups. *Algebra Colloq.*, 1:317–322, 1994.

[Gup02] Gupta, N. The dimension subgroup conjecture holds for odd order groups. *J. Group Theory*, 5:481–491, 2002.

[Gup07] Gupta, N. and Passi, I. B. S. Commutator subgroups of free nilpotent groups. *Int. J. Algebra Comput.*, 17:1021–1031, 2007.

[Gut81] Gutierrez, M. A. and Ratcliffe, J. G. On the second homotopy group. *Quart. J. Math. (Oxford)*, 32:45–55, 1981.

[Hab00] Habiro, K. Claspers and finite type invariants of links. *Geom. Topol.*, 4:1–83, 2000.

[Hal85] Hales, A. W. Stable augmentation quotients of Abelian groups. *Pacific J. Math.*, 118:401–410, 1985.

[Har] Hartl, M. On the fourth integer dimension subgroup. *Preprint.*

[Har70] Hartley, B. The residual nilpotence of wreath products. *Proc. London Math. Soc.(3)*, 20:365–392, 1970.

[Har75] Harlap, A.E. Lokale Homologien und Kohomologien, homologische Dimension und verallgemeinerte Mannigfaltigkeiten. *Mat. Sb., N. Ser.*, 96(138):347–373, 1975.

[Har82a] Hartley, B. Dimension and lower central subgroups-Sjögren's theorem revisited. Lecture Notes 9, National University, Singapore, 1982.

[Har82b] Hartley, B. An intersection theorem for powers of the augmentation ideal in group rings of certain nilpotent p-groups. *J. London Math. Soc.*, 25:425–434, 1982.

[Har82c] Hartley, B. Powers of the augmentation ideal in group rings of infinite nilpotent groups. *J. London Math. Soc.*, 25:43–61, 1982.

[Har84] Hartley, B. *Group Theory: Essays for Philip Hall; ed. K. W. Gruenberg and J. E. Roseblade*, chapter Topics in the Theory of Nilpotent Groups, pages 61–120. Academic Press, 1984.

[Har91a] Hartley, B. and Kuzmin, Yu V. On the quotient of a free group by the commutator of two normal subgroups. *J. Pure Appl. Algebra*, 74:247–256, 1991.

[Har91b] Hartley, B. and Stöhr, R. Homology of higher relation modules and torsion in free central extensions of groups. *Proc. London Math. Soc.*, 62:325–352, 1991.

[Har95] Hartl, M. Some successive quotients of group ring filtrations induced by N-series. *Comm. Algebra*, 23:3831–3853, 1995.

[Har96a] Hartl, M. Polynomiality properties of group extensions with torsion-free abelian kernel. *J. Algebra*, 179:380–415, 1996.

[Har96b] Hartl, M. The nonabelian tensor square and Schur multiplicator of nilpotent groups of class 2. *J. Algebra*, 179:416–440, 1996.

[Har98] Hartl, M. Structures polynomiales en théorie des groupes nilpotents, Mémoire d'habilitation à diriger des recherches, Institut de Recherche Mathématique Avancée, Strasbourg, 94p. 1998.

[Har08] Hartl, M., Mikhalov, R. and Passi, I. B. S. Dimension quotients. Preprint. arXiv:0803.3290. 2008.

[Hau] Hausmann, J. C. Acyclic maps and the Whitehead aspherical problem *Preprint.*

[Hil71]   Hilton, P. J. and Stammbach, U. *A Course in Homological Algebra*, volume 4 of *GTM*. Springer-Verlag, 1971.

[How83]   Howie, J. Some remarks on a problem of J.H.C. Whitehead. *Topology*, 22:475–485, 1983.

[Ill71]   Illusie, Luc. *Complexe cotangent et déformations. I. (The cotangent complex and deformations. I.).* Lecture Notes in Mathematics. 239. Berlin-Heidelberg-New York: Springer-Verlag., 1971.

[Ina98]   Inassaridze H. *Non-abelian Homological Algebra and Its Applications*. Kluwer Acad. Publ., 1998.

[Jea02]   Jean, F. *Foncteurs dérivés de l'algèbres symétrique: Application au calcul de certains groupes d'homologie fonctorielle des espaces $K(B, n)$*. PhD thesis, University of Paris 13, 2002.

[Jen41]   Jennings, S.A. The structure of the group ring of a p-group over a modular field. *Trans. Amer. Math. Soc.*, 50:175–185, 1941.

[Jen55]   Jennings, S. A. The group ring of a class of infinite nilpotent groups. *Canad. J. Math.*, 7:169–187, 1955.

[Kan58a]   Kan, D. A combinatorial definition of homotopy groups. *Ann. Math.*, 67:288–312, 1958.

[Kan58b]   Kan, D. Minimal free c.s.s. groups. *Illinois J. Math.*, 2:537–547, 1958.

[Kan59]   Kan, D. A relation between CW-complexes and free c.s.s. groups. *Amer. J. Math.*, 81:512–528, 1959.

[Keu]   Keune, F. Derived functors and algebraic K-theory. Algebr. K-Theory I, Proc. Conf. Battelle Inst. 1972, Lect. Notes Math. 341, 166-176 (1973).

[Kru86]   Krushkal', S.L. and Apanasov, B.N. and Gusevskij, N.A. *Kleinian groups and uniformization in examples and problems. Transl. from the Russian by H. H. McFaden, ed. by Bernard Maskit*. Translations of Mathematical Monographs, 62. Providence, R.I.: American Mathematical Society (AMS), 1986.

[Kru03]   Krushkal, V. S. Dwyer's filtration and topology of 4-manifolds. *Math. Res. Lett.*, 10:247–251, 2003.

[Kuz87]   Kuzmin, Yu. V. Homology of groups of the form F/N'. *Dokl. Akad. Nauk SSSR*, 296:267–270, 1987.

[Kuz96]   Kuz'min, Yu. V. Dimension subgroups of extensions with an abelian kernel. *Mat. Sb.*, 187:65–70, 1996. translation in Sb. Math. 187 (1996), 685-691.

[Kuz06]   Kuz'min, Yu. V. *Homological Group Theory*. Factorial Press, Moscow, 2006.

[Lam00]   Lam, T. Y. and Leung, K. H. On vanishing sums of roots of unity. *J. Algebra*, 224:91–109, 2000.

[Laz54]   Lazard, M. Sur les algèbres enveloppantes universelles de certaines algèbres de Lie. *Publ. Sci. Univ. Alger. Sér A*, 1:281–294, 1954.

[Le 88]   Le Dimet, J.-Y. Cobordisme d'enlacements de disques. *Mémoires de la Société Mathématique de France Sér. 2*, 32:1–92, 1988.

[Lev40]   Levi, F. W. The commutator group of a free product. *J. Indian Math. Soc.*, 4:136–144, 1940.

[Lev91]   Levine, J. Finitely-presented groups with long lower central series. *Isr. J. Math.*, 73:57–64, 1991.

[Lic77]   Lichtman, A. I. The residual nilpotence of the augmentation ideal and the residual nilpotence of some classes of groups. *Israel J. Math.*, 26:276–293, 1977.

[Lic78]   Lichtman, A. I. Necessary and sufficient conditions for the residual nilpotence of free products of groups. *J. Pure Appl. Algebra*, 12:49–64, 1978.

[Lod82]   Loday, J.-L. Spaces with finitely many non-trivial homotopy groups. *J. Pure Appl. Algebra*, 24:179–202, 1982.

[Los74]   Losey, G. N-eries and filtrations of the augmentation ideal. *Canad. J. Math.*, 26:962–977, 1974.

[Luf96]   Luft, E. On 2-dimensional aspherical complexes and a problem of J.H.C. Whitehead. *Math. Proc. Cambridge Phil. Soc.*, 119:493–495, 1996.

[Lut02]   Luthar, I. S. and Passi, I. B. S. *Algebra Vol. 3 Modules*. Alpha Science International Ltd., Pangbourne UK, 2002.

[Lyn50]   Lyndon, R. C. Cohomology theory of groups with a single defining relation. *Ann. Math.*, 52:650–665, 1950.

[Mac50]   MacLane, S. and Whitehead, J. H. C. On the 3-type of a complex. *Proc. Nat. Acad. Sci. USA*, 36:41–48, 1950.

[Mac60]   Maclane, S. Triple torsion products and multiple Künneth formulas. *Math. Ann.*, 140:51–64, 1960.

[Mag35]   Magnus, W. Beziehungen zwischen Gruppen und Idealen in einem speziellen Ring. *Math. Ann.*, 111:259–280, 1935.

[Mag37]   Magnus, W. Über Beziehungen zwieschen höheren Kommutatoren. *J. reine angew. Math.*, 177:105–115, 1937.

[Mag66]   Magnus, W., Karrass, A. and Solitar, D. *Combinatorial Group Theory: Presentations of groups in terms of genrators and relations*. Interscience Publishers, New York-London-Sydney, 1966.

[Mal56]   Mal'cev, A. I. On certain classes of infinite soluble groups. *Mat. Sb.*, 28:567–588, 1951; Amer. Math. Soc. Transl. (2) 2, 1–21 (1956).

[Mal65]   Mal'cev, A. I. On faithful representations of infinite groups of matrices. *Mat. Sb.*, 8:405–422, 1940; Amer. Math. Soc. Transl. (2) 45, 1–18 (1965).

[Mal68]   Mal'cev, A. I. Generalized nilpotent algebras and their associated groups. *Mat. Sbornik N.S.*, 25:347–366, 1949; Amer. Math. Soc. Transl. (2) 69 (1968).

[Mal70]   Mal'cev, A. I. *Algebraicheskie sistemy (Russian)*. Nauka, Moscow, 1970. Posthumous edition, edited by D. Smirnov and M. Taĭclin. Translated from the Russian by B. D. Seckler and A. P. Doohovskoy. Die Grundlehren der mathematischen Wissenschaften, Band 192. Springer-Verlag, New York-Heidelberg, 1973.

[Mar]   Martins, Joao Faria. On 2-dimensional homotopy invariants of complements of knotted surfaces. arxiv:math/0507239.

[Mas67]   Massey, W. *Algebraic Topology: An introduction*. Harcourt, Brace and World, Inc., New York, 1967.

[Mas03]   Massuyeau, J.-B. Characterization of Y_2-equivalence for homology cylinders. *J. Knot Th. Ramifications*, 12:493–522, 2003.

[Mas07]   Massuyeau, G. Finite-type invariants of three-manifolds and the dimension subgroup problem. *J. London Math. Soc.*, 75:791–811, 2007.

[May67]   May, J. P. *Simplicial Objects in Algebraic Topology*. Van Nostrand, Princeton, 1967.

[Maz87]   Mazur, B. and Tate, J. Refined conjectures of the Birch and Swinnerton-Dyre type. *Duke Math. J.*, 54:711–750, 1987.

[McC96]   McCarron, James. Residually nilpotent one-relator groups with nontrivial centre. *Proc. Am. Math. Soc.*, 124(1):1–5, 1996.

[Mer87]   Merzlyakov, Yu. I. *Rational groups. (Ratsional'nye gruppy). 2n ed.* "Nauka". Glavnaya Redaktsiya Fiziko-Matematicheskoj Literatury, Moskva, 1987.

[Mik02]   Mikhailov, R. Transfinite lower central series in groups: parafree conditions and topological applications. *Proc. Steklov Math. Institute*, 239:236–252, 2002.

[Mik04]   Mikhailov, R. and Passi, I. B. S. Augmentation powers and group homology. *J. Pure Appl. Algebra*, 192:225–238, 2004.

[Mik05a] Mikhailov, R. On residual nilpotence and solvability of groups. *Sb. Math.*, 196:1659–1675, 2005.

[Mik05b] Mikhailov, R. and Passi, I. B. S. A transfinite filtration of schur multiplicator. *Int. J. Algebra Comput.*, 15:1061–1073, 2005.

[Mik06a] Mikhailov, R. On residual nilpotence of projective crossed modules. *Comm. Algebra*, 34:1451–1458, 2006.

[Mik06b] Mikhailov, R. and Passi, I. B. S. Faithfulness of certain modules and residual nilpotence of groups. *Int. J. Algebra Comput.*, 16:525–539, 2006.

[Mik06c] Mikhailov, R. and Passi, I. B. S. The quasi-variety of groups with trivial fourth dimension subgroup. *J. Group Theory*, 9:369–381, 2006.

[Mik07a] Mikhailov, R. V. Baer invariants and residual nilpotence of groups. *Izv. Math.*, 71(2):371–390, 2007.

[Mik07b] Mikhailov, R. Asphericity and residual properties of crossed modules. *Mat. Sb.*, 198:79–94, 2007.

[Mil56] Milnor, J. *On the construction F[K]. In Algebraic Topology. A Student's Guide.* London Mathematical Society Lecture Notes Series. 4. London: Cambridge University Press, 119–136 1972.

[Mil57] Milnor, J. *Algebraic Geometry and Topology: A Symposium in Honor of S. Lefschetz (R. H. Fox, D. Spencer, J. W. Tucker, eds.)*, chapter Isotopy of links, pages 208–306. Princeton Univ. Press, 1957.

[Mit71] Mital, J. *Residual Nilpotence.* PhD thesis, Kurkshetra University, Kurukshetra (India), 1971.

[Mit73] Mital, J. N. and Passi, I. B. S. Annihilators of relation modules. *J. Australian Math. Soc.*, 16:228–233, 1973.

[Mor70] Moran, S. Dimension subgroups modulo n. *Proc. Cambridge Philos. Soc.*, 68:579–582, 1970.

[Mus82] Musson, I. and Weiss, A. Integral group rings with residually nilpotent unit groups. *Arch. Math.*, 38:514–530, 1982.

[Mut99] Mutlu, A. and Porter, T. Free crossed resolutions from simplicial resolutions with given CW-basis. *Cah. Topologie Gom. Diffr. Catg.*, 40:'261–283, 1999.

[Nis73] Nishida, G. The nilpotency of elements of the stable homotopy groups of spheres. *J. Math. Soc. Japan*, 25:707–732, 1973.

[Nou67] Nouazé, Y. and Gabriel, P. Idéaux premiers de l'algèbre enveloppante d'une algèbre de Lie. *J. Algebra*, 6:77–99, 1967.

[Ols91] Olshanskii, A. Yu. *Geometry of defining relations in groups.* Kluwer Academic Publishers Group, Dodrecht, 1991.

[Par01] Parmenter. M. M. A basis for powers of the augmentation ideal. *Algebra Colloq.*, 8:121–128, 2001.

[Pas68a] Passi, I. B. S. Dimension subgroups. *J. Algebra*, 9:152–182, 1968.

[Pas68b] Passi, I. B. S. Polynomial maps on groups. *J. Algebra*, 9:121–151, 1968.

[Pas73] Passi, I. B. S., Sehgal, S. K. and Passman, D. S. Lie solvable group rings. *Can. J. Math.*, 15:748–757, 1973.

[Pas74] Passi, I. B. S. and Stammbach, U. A filtration of Schur multiplicator. *Math. Z.*, 135:143–148, 1974.

[Pas75a] Passi, I. B. S. Annihilators of relation modules-II. *J. Pure and Appl. Algebra*, 6:235–237, 1975.

[Pas75b] Passi, I. B. S. and Sehgal, S. K. Lie dimension subgroups. *Comm. Algebra*, 3:59–73, 1975.

[Pas77a] Passi, I. B. S. and Vermani, L. R. The associated graded ring of an integral group ring. *Math. Proc. Cambridge Philos. Soc.*, 82:25–33, 1977.

[Pas77b] Passman, D. S. *The Algebraic Structure of Group Rings.* Interscience, New York, 1977.

[Pas78] Passi, I. B. S. The associated graded ring of a group ring. *Bull. London Math. Soc.*, 10:241–255, 1978.

[Pas79] Passi, Inder Bir S. *Group rings and their augmentation ideals.* Lecture Notes in Mathematics. 715. Berlin-Heidelberg-New York: Springer-Verlag., 1979.

[Pas83] Passi, I. B. S. and Vermani, L. R. Dimension subgroups and Schur multiplicator. *J. Pure Appl. Algebra*, 30:61–67, 1983.

[Pas87a] Passi, I. B. S. and Sucheta. Dimension subgroups and schur multiplicator. *Topology and its Applications*, 25:121–124, 1987.

[Pas87b] Passi, I. B. S., Sucheta and Tahara, Ken-Ichi. Dimension subgroups and schur multiplicator. *Japan. J. Math. (N.S.)*, 13:371–379, 1987.

[Pas94] Passi, I. B. S. and Vermani, L. R. Schur multiplicator and Sjögren's theorem. *Res. Bull. Panjab Univ.*, 44:263–269, 1994.

[Pie74] Pietrowski, A. The isomorphism problem for one-relator groups with non-trivial centre. *Math. Z.*, 136, 1974.

[Plo71] Plotkin, B. I. The varieties and quasi-varieties that are connected with group representations (russian). *Dokl. Akad. Nauk SSSR*, 196:527–530, 1971. [Soviet Math. Dokl. 12 (1971), 192-196].

[Plo73] Plotkin, B. I. Remarks on stable representations of nilpotent groups. *Trans. Moscow Math. Soc.*, 29:185–200, 1973.

[Plo83] Plotkin, B. I. and Vovsi, S. M. *Varieties of Group Representations: General Theory, Connections and Applications.* Zinatne, Riga, 1983.

[Pri] Priddy, Stewart B. On $\Omega^\infty S^\infty$ and the infinite symmetric group. Algebraic Topology, Proc. Sympos. Pure Math. 22, 217-220 (1971).

[Pri91] Pride, S. J. *Group Theory fom a Geometric Point of View. Ed. E. Ghys, A. Haefliger and A. Verjovsky*, chapter Identitites among relations of group presentations, pages 687–717. World Scientific, Singapore, 1991.

[Qui] Quillen, D. Letter from Quillen to Milnor on Im $(\pi_i 0 \to \pi_i^s \to K_i \mathbb{Z})$. Algebr. K-Theory, Proc. Conf. Evanston 1976, Lect. Notes Math. 551, 182-188 (1976).

[Qui66] Quillen D. Spectral sequences of a double semi-simplicial group. *Topology*, 5:155–157, 1966.

[Qui68] Quillen, D. *Notes on the homology of commutative rings.* Mimiographed Notes, M. I. T., 1968.

[Qui69] Quillen, D. Rational homotopy theory. *Ann. Math.*, 90:205–295, 1969.

[Rap89] Raptis, E. and Varsos, D. Residual properties of HNN-extensions with base group an Abelian group. *J. Pure Appl. Algebra*, 59:285–290, 1989.

[Rap91] Raptis, E. and Varsos, D. The residual nilpotence of HNN-extensions with base group a finite or a f.g. abelian group. *J. Pure Appl. Algebra*, 76:167–178, 1991.

[Rat80] Ratcliffe, J. G. Free and projective crossed modules. *J. London Math. Soc.*, 22:66–74, 1980.

[Rec66] Rector, D. An unstable Adams spectral sequence. *Topology*, 5:343–346, 1966.

[Rei50] Reidemeister, K. Complexes and homotopy chains. *Bull. Amer. Math. Soc.*, 56:297–307, 1950.

[Ril91] Riley, D. M. Restricted Lie dimension subgroups. *Comm. Algebra*, 19:1493–1499, 1991.

[Rip72] Rips, E. On the fourth dimension subgroups. *Israel J. Math.*, 12:342–346, 1972.

[Rob95] Robinson, Derek J.S. *A course in the theory of groups. 2nd ed.* Graduate Texts in Mathematics. 80. New York, NY: Springer-Verlag., 1995.

[Rod04] Rodreguez, J. L. and Scevenels, D. Homology equivalences inducing an epimorphism on the fundamental group and Quillen's plus construction. *Proc. Amer. Math. Soc.*, 132:891–898, 2004.

[Röh85] Röhl, F. Review and some critical comments on a paper of Grün concerning the dimension subgroup conjecture. *Bol. Soc. Bras. Mat.*, 16(2):11–27, 1985.

[Ros79] Roseblade, J. E. and Smith, P. F. A note on the Artin-Rees property of certain polycyclic group algebras. *Bull. Lond. Math. Soc.*, 11:184–185, 1979.

[Rot88] Rotman, J. J. *An Introduction to Algebraic Topology.* Springer, 1988.

[San72a] Sandling, R. Dimension subgroups over arbitrary coefficient rings. *J. Algebra*, 21:250–265, 1972.

[San72b] Sandling, R. Subgroups dual to dimension subgroups. *Proc. Cambridge Philos. Soc.*, 71:33–38, 1972.

[San79] Sandling, Robert and Tahara, Ken-ichi. Augmentation quotients of group rings and symmetric powers. *Math. Proc. Camb. Philos. Soc.*, 85:247–252, 1979.

[Sch66] Schlesinger, J. The semi-simplicial free Lie ring. *Trans. Amer. Math. Soc.*, 122:436–442, 1966.

[Sco91] Scoppola, C. M. and Shalev, A. Applications of dimension subgroups to the power structure of *p*-groups. *Israel J. Math.*, 73:45–56, 1991.

[Ser51] Serre, J.-P. Homologie singulire des espaces fibrs. Applications. *Ann. Math.*, 54:425–505, 1951.

[Sha90a] Shalev, Λ. Dimension subgroups, nilpotency indices, and the number of genera-
          tors of ideals in p-group algebras. *J. Algebra*, 129:412–438, 1990.

[Sha90b] Sharma, R. K. and Srivastava, J. B. Lie ideals in group rings. *J. Pure Appl.
          Algebra*, 63:67–80, 1990.

[Sha91]   Shalev, A. Lie dimension subgroups, Lie nilpotency indices, and the exponent of
          the group of normalized units. *J. London Math. Soc.*, 1991.

[Shm65]   Shmel'kin, A. L. Wreath products and varieties of groups. *Izv. Akad. Nauk SSSR
          Ser. Mat.*, 29:149–170 (Russian), 1965.

[Shm67]   Shmel'kin, A. L. A remark on my paper "Wreath products and varieties of
          groups". *Izv. Akad. Nauk SSSR Ser. Mat.*, 31:443–444, 1967.

[Shm73]   Shmel'kin, A. L. Wreath products of Lie algberas and their applications to group
          theory. *Trudy Mosk. Mat. Obsc.*, 29:247–260, 1973; Trans. Moscow Math. Soc.
          29: 239-252 (1973).

[Sie93]   Sieradski, A. J. *Algebraic topology for two dimensional complexes*, volume 197
          of *London Mathematical Society Lecture Notes Series*, chapter II, pages 51–96.
          Cambridge University Press, 1993.

[Sjo79]   Sjogren, J. A. Dimension and lower central subgroups. *J. Pure Appl. Algebra*,
          14, 1979.

[Smi70]   Smith, M. On group algebras. *Bull. Amer. Math. Soc.*, 76:780–782, 1970.

[Smi82]   Smith, P. F. The Artin-Rees property. *Semin. d'algebre Paul Dubreil et Marie-
          Paule Malliavin, 34eme Annee, Proc., Paris 1981, Lect. Notes Math.*, 924:197–
          240, 1982.

[Sta65]   Stallings, J. Homology and central series of groups. *J. Algebra*, 2:170–181, 1965.

[Sta75]   Stallings, J. *Knots, Groups and 3-Manifolds*, volume 84 of *Annals of Math.
          Studies*, chapter Quotients of powers of the augmentation ideal in a group ring.
          Princeton University Press, Princeton, 1975.

[Sta98]   Stanford, T. B. Vassiliev invariants and knots modulo pure braidsubgroups.
          arXiv.org math/9805092, 1998.

[Ste98]   Stevenson, K. F. Conditions related to π_1 of projective curves. *J. Number Theory*,
          69:62–79, 1998.

[Stö87]   R. Stöhr. On torsion in free central extensions of some torsion-free groups. *J.
          Pure Appl. Algebra*, 46:249–289, 1987.

[Str74]   Strebel, R. Homological methods applied to the derived series of groups. *Com-
          ment. Math. Helv.*, 49:302–332, 1974.

[Sul71]   Sullivan, D. *Geometric topology. Part I. Localization, periodicity, and Galois
          symmetry. Revised version.* Massachusetts Institute of Technology, Cambridge,
          Mass., 1971.

[Tah77a] Tahara, Ken-Ichi. The fourth dimension subgroups and polynomial maps. *J.
          Algebra*, 45:102–131, 1977.

[Tah77b] Tahara, Ken-Ichi. On the structure of $Q_3(G)$ and the fourth dimension subgroup.
          Japan J. Math. (N.S.), 3:381–396, 1977.

[Tah78a] Tahara, Ken-Ichi. The fourth dimension subgroups and polynomial maps. II.
          Nagoya Math. J., 69:1–7, 1978.

[Tah78b] Tahara, Ken-Ichi. The fourth dimension subgroups and polynomial maps on
          groups, II. *Nagoya Math. J.*, 69:1–7, 1978.

[Tah81]   Tahara, Ken-Ichi. The augmentation quotients of group rings and the fifth di-
          mension subgroups. *J. Algebra*, 71:141–173, 1981.

[Tan95]   Tan, Ki-Seng. Refined theorems of the Birch and Swinnerton-Dyre type. *Annales
          de L'Institut Fourier*, 45:317–374, 1995.

[Tas93]   Tasić, V. A simple proof of Moran's theorem on dimension subgroups. *Comm.
          Algebra*, 21:355–358, 1993.

[Tod62]   Toda, H. Composition methods in homotopy groups of spheres. *Annals of Math-
          ematics Studies*, 49, 1962.

[Val80]   Valenza, Robert J. Dimension subgroups of semi-direct products. *J. Pure Appl.
          Algebra*, 18:225–229, 1980.

[Weh73] Wehrfritz, B.A.F. *Infinite linear groups. An account of the group-theoretic properties of infinite groups of matrices.* Ergebnisse der Mathematik und ihrer Grenzgebiete. Band 76. Berlin-Heidelberg-New York: Springer-Verlag, 1973.

[Whi41] Whitehead, J. H. C. On adding relations to homotopy groups. *Ann. Math.*, 42:409–428, 1941.

[Whi50] Whitehead, J. H. C. A certain exact sequence. *Ann. Math.*, 52:51–110, 1950.

[Wie78] Wielenberg, N. The structure of certain subgroups of the Picard group. *Math. Proc. Camb. Phil. Soc.*, 84:427–436, 1978.

[Wil03] Wilson, L. E. Dimension subgroups and p-th powers in p-groups. *Israel J. Math.*, 138:1–17, 2003.

[Wit37] Witt, E. Treue Darstellung Lieschen Ringe. *J. Reine angew. Math.*, 177:152–160, 1937.

[Wu,01] Wu, J. Combinatorial description of homotopy groups of certain spaces. *Math. Proc. Camb. Phil. Soc.*, 130:489–513, 2001.

Index

Lecture Notes in Mathematics

For information about earlier volumes
please contact your bookseller or Springer
LNM Online archive: springerlink.com

Vol. 1867: J. Sneyd (Ed.), Tutorials in Mathematical Biosciences II. Mathematical Modeling of Calcium Dynamics and Signal Transduction. (2005)
Vol. 1868: J. Jorgenson, S. Lang, $Pos_n(R)$ and Eisenstein Series. (2005)
Vol. 1869: A. Dembo, T. Funaki, Lectures on Probability Theory and Statistics. Ecole d'Eté de Probabilités de Saint-Flour XXXIII-2003. Editor: J. Picard (2005)
Vol. 1870: V.I. Gurariy, W. Lusky, Geometry of Müntz Spaces and Related Questions. (2005)
Vol. 1871: P. Constantin, G. Gallavotti, A.V. Kazhikhov, Y. Meyer, S. Ukai, Mathematical Foundation of Turbulent Viscous Flows, Martina Franca, Italy, 2003. Editors: M. Cannone, T. Miyakawa (2006)
Vol. 1872: A. Friedman (Ed.), Tutorials in Mathematical Biosciences III. Cell Cycle, Proliferation, and Cancer (2006)
Vol. 1873: R. Mansuy, M. Yor, Random Times and Enlargements of Filtrations in a Brownian Setting (2006)
Vol. 1874: M. Yor, M. Émery (Eds.), In Memoriam Paul-André Meyer - Séminaire de Probabilités XXXIX (2006)
Vol. 1875: J. Pitman, Combinatorial Stochastic Processes. Ecole d'Eté de Probabilités de Saint-Flour XXXII-2002. Editor: J. Picard (2006)
Vol. 1876: H. Herrlich, Axiom of Choice (2006)
Vol. 1877: J. Steuding, Value Distributions of L-Functions (2007)
Vol. 1878: R. Cerf, The Wulff Crystal in Ising and Percolation Models, Ecole d'Eté de Probabilités de Saint-Flour XXXIV-2004. Editor: Jean Picard (2006)
Vol. 1879: G. Slade, The Lace Expansion and its Applications, Ecole d'Eté de Probabilités de Saint-Flour XXXIV-2004. Editor: Jean Picard (2006)
Vol. 1880: S. Attal, A. Joye, C.-A. Pillet, Open Quantum Systems I, The Hamiltonian Approach (2006)
Vol. 1881: S. Attal, A. Joye, C.-A. Pillet, Open Quantum Systems II, The Markovian Approach (2006)
Vol. 1882: S. Attal, A. Joye, C.-A. Pillet, Open Quantum Systems III, Recent Developments (2006)
Vol. 1883: W. Van Assche, F. Marcellàn (Eds.), Orthogonal Polynomials and Special Functions, Computation and Application (2006)
Vol. 1884: N. Hayashi, E.I. Kaikina, P.I. Naumkin, I.A. Shishmarev, Asymptotics for Dissipative Nonlinear Equations (2006)
Vol. 1885: A. Telcs, The Art of Random Walks (2006)
Vol. 1886: S. Takamura, Splitting Deformations of Degenerations of Complex Curves (2006)
Vol. 1887: K. Habermann, L. Habermann, Introduction to Symplectic Dirac Operators (2006)
Vol. 1888: J. van der Hoeven, Transseries and Real Differential Algebra (2006)
Vol. 1889: G. Osipenko, Dynamical Systems, Graphs, and Algorithms (2006)
Vol. 1890: M. Bunge, J. Funk, Singular Coverings of Toposes (2006)
Vol. 1891: J.B. Friedlander, D.R. Heath-Brown, H. Iwaniec, J. Kaczorowski, Analytic Number Theory, Cetraro, Italy, 2002. Editors: A. Perelli, C. Viola (2006)
Vol. 1892: A. Baddeley, I. Bárány, R. Schneider, W. Weil, Stochastic Geometry, Martina Franca, Italy, 2004. Editor: W. Weil (2007)
Vol. 1893: H. Hanßmann, Local and Semi-Local Bifurcations in Hamiltonian Dynamical Systems, Results and Examples (2007)
Vol. 1894: C.W. Groetsch, Stable Approximate Evaluation of Unbounded Operators (2007)

Vol. 1895: L. Molnár, Selected Preserver Problems on Algebraic Structures of Linear Operators and on Function Spaces (2007)
Vol. 1896: P. Massart, Concentration Inequalities and Model Selection, Ecole d'Été de Probabilités de Saint-Flour XXXIII-2003. Editor: J. Picard (2007)
Vol. 1897: R. Doney, Fluctuation Theory for Lévy Processes, Ecole d'Été de Probabilités de Saint-Flour XXXV-2005. Editor: J. Picard (2007)
Vol. 1898: H.R. Beyer, Beyond Partial Differential Equations, On linear and Quasi-Linear Abstract Hyperbolic Evolution Equations (2007)
Vol. 1899: Séminaire de Probabilités XL. Editors: C. Donati-Martin, M. Émery, A. Rouault, C. Stricker (2007)
Vol. 1900: E. Bolthausen, A. Bovier (Eds.), Spin Glasses (2007)
Vol. 1901: O. Wittenberg, Intersections de deux quadriques et pinceaux de courbes de genre 1, Intersections of Two Quadrics and Pencils of Curves of Genus 1 (2007)
Vol. 1902: A. Isaev, Lectures on the Automorphism Groups of Kobayashi-Hyperbolic Manifolds (2007)
Vol. 1903: G. Kresin, V. Maz'ya, Sharp Real-Part Theorems (2007)
Vol. 1904: P. Giesl, Construction of Global Lyapunov Functions Using Radial Basis Functions (2007)
Vol. 1905: C. Prévôt, M. Röckner, A Concise Course on Stochastic Partial Differential Equations (2007)
Vol. 1906: T. Schuster, The Method of Approximate Inverse: Theory and Applications (2007)
Vol. 1907: M. Rasmussen, Attractivity and Bifurcation for Nonautonomous Dynamical Systems (2007)
Vol. 1908: T.J. Lyons, M. Caruana, T. Lévy, Differential Equations Driven by Rough Paths, Ecole d'Été de Probabilités de Saint-Flour XXXIV-2004 (2007)
Vol. 1909: H. Akiyoshi, M. Sakuma, M. Wada, Y. Yamashita, Punctured Torus Groups and 2-Bridge Knot Groups (I) (2007)
Vol. 1910: V.D. Milman, G. Schechtman (Eds.), Geometric Aspects of Functional Analysis. Israel Seminar 2004-2005 (2007)
Vol. 1911: A. Bressan, D. Serre, M. Williams, K. Zumbrun, Hyperbolic Systems of Balance Laws. Cetraro, Italy 2003. Editor: P. Marcati (2007)
Vol. 1912: V. Berinde, Iterative Approximation of Fixed Points (2007)
Vol. 1913: J.E. Marsden, G. Misiołek, J.-P. Ortega, M. Perlmutter, T.S. Ratiu, Hamiltonian Reduction by Stages (2007)
Vol. 1914: G. Kutyniok, Affine Density in Wavelet Analysis (2007)
Vol. 1915: T. Bıyıkoğlu, J. Leydold, P.F. Stadler, Laplacian Eigenvectors of Graphs. Perron-Frobenius and Faber-Krahn Type Theorems (2007)
Vol. 1916: C. Villani, F. Rezakhanlou, Entropy Methods for the Boltzmann Equation. Editors: F. Golse, S. Olla (2008)
Vol. 1917: I. Veselić, Existence and Regularity Properties of the Integrated Density of States of Random Schrödinger (2008)
Vol. 1918: B. Roberts, R. Schmidt, Local Newforms for GSp(4) (2007)
Vol. 1919: R.A. Carmona, I. Ekeland, A. Kohatsu-Higa, J.-M. Lasry, P.-L. Lions, H. Pham, E. Taflin, Paris-Princeton Lectures on Mathematical Finance 2004.

Editors: R.A. Carmona, E. Çinlar, I. Ekeland, E. Jouini, J.A. Scheinkman, N. Touzi (2007)

Vol. 1920: S.N. Evans, Probability and Real Trees. Ecole d'Été de Probabilités de Saint-Flour XXXV-2005 (2008)

Vol. 1921: J.P. Tian, Evolution Algebras and their Applications (2008)

Vol. 1922: A. Friedman (Ed.), Tutorials in Mathematical BioSciences IV. Evolution and Ecology (2008)

Vol. 1923: J.P.N. Bishwal, Parameter Estimation in Stochastic Differential Equations (2008)

Vol. 1924: M. Wilson, Littlewood-Paley Theory and Exponential-Square Integrability (2008)

Vol. 1925: M. du Sautoy, L. Woodward, Zeta Functions of Groups and Rings (2008)

Vol. 1926: L. Barreira, V. Claudia, Stability of Nonautonomous Differential Equations (2008)

Vol. 1927: L. Ambrosio, L. Caffarelli, M.G. Crandall, L.C. Evans, N. Fusco, Calculus of Variations and Non-Linear Partial Differential Equations. Cetraro, Italy 2005. Editors: B. Dacorogna, P. Marcellini (2008)

Vol. 1928: J. Jonsson, Simplicial Complexes of Graphs (2008)

Vol. 1929: Y. Mishura, Stochastic Calculus for Fractional Brownian Motion and Related Processes (2008)

Vol. 1930: J.M. Urbano, The Method of Intrinsic Scaling. A Systematic Approach to Regularity for Degenerate and Singular PDEs (2008)

Vol. 1931: M. Cowling, E. Frenkel, M. Kashiwara, A. Valette, D.A. Vogan, Jr., N.R. Wallach, Representation Theory and Complex Analysis. Venice, Italy 2004. Editors: E.C. Tarabusi, A. D'Agnolo, M. Picardello (2008)

Vol. 1932: A.A. Agrachev, A.S. Morse, E.D. Sontag, H.J. Sussmann, V.I. Utkin, Nonlinear and Optimal Control Theory. Cetraro, Italy 2004. Editors: P. Nistri, G. Stefani (2008)

Vol. 1933: M. Petkovic, Point Estimation of Root Finding Methods (2008)

Vol. 1934: C. Donati-Martin, M. Émery, A. Rouault, C. Stricker (Eds.), Séminaire de Probabilités XLI (2008)

Vol. 1935: A. Unterberger, Alternative Pseudodifferential Analysis (2008)

Vol. 1936: P. Magal, S. Ruan (Eds.), Structured Population Models in Biology and Epidemiology (2008)

Vol. 1937: G. Capriz, P. Giovine, P.M. Mariano (Eds.), Mathematical Models of Granular Matter (2008)

Vol. 1938: D. Auroux, F. Catanese, M. Manetti, P. Seidel, B. Siebert, I. Smith, G. Tian, Symplectic 4-Manifolds and Algebraic Surfaces. Cetraro, Italy 2003. Editors: F. Catanese, G. Tian (2008)

Vol. 1939: D. Boffi, F. Brezzi, L. Demkowicz, R.G. Durán, R.S. Falk, M. Fortin, Mixed Finite Elements, Compatibility Conditions, and Applications. Cetraro, Italy 2006. Editors: D. Boffi, L. Gastaldi (2008)

Vol. 1940: J. Banasiak, V. Capasso, M.A.J. Chaplain, M. Lachowicz, J. Miękisz, Multiscale Problems in the Life Sciences. From Microscopic to Macroscopic. Będlewo, Poland 2006. Editors: V. Capasso, M. Lachowicz (2008)

Vol. 1941: S.M.J. Haran, Arithmetical Investigations. Representation Theory, Orthogonal Polynomials, and Quantum Interpolations (2008)

Vol. 1942: S. Albeverio, F. Flandoli, Y.G. Sinai, SPDE in Hydrodynamic. Recent Progress and Prospects. Cetraro, Italy 2005. Editors: G. Da Prato, M. Röckner (2008)

Vol. 1943: L.L. Bonilla (Ed.), Inverse Problems and Imaging. Martina Franca, Italy 2002 (2008)

Vol. 1944: A. Di Bartolo, G. Falcone, P. Plaumann, K. Strambach, Algebraic Groups and Lie Groups with Few Factors (2008)

Vol. 1945: F. Brauer, P. van den Driessche, J. Wu (Eds.), Mathematical Epidemiology (2008)

Vol. 1946: G. Allaire, A. Arnold, P. Degond, T.Y. Hou, Quantum Transport. Modelling, Analysis and Asymptotics. Cetraro, Italy 2006. Editors: N.B. Abdallah, G. Frosali (2008)

Vol. 1947: D. Abramovich, M. Mariño, M. Thaddeus, R. Vakil, Enumerative Invariants in Algebraic Geometry and String Theory. Cetraro, Italy 2005. Editors: K. Behrend, M. Manetti (2008)

Vol. 1948: F. Cao, J-L. Lisani, J-M. Morel, P. Musé, F. Sur, A Theory of Shape Identification (2008)

Vol. 1949: H.G. Feichtinger, B. Helffer, M.P. Lamoureux, N. Lerner, J. Toft, Pseudo-Differential Operators. Quantization and Signals. Cetraro, Italy 2006. Editors: L. Rodino, M.W. Wong (2008)

Vol. 1950: M. Bramson, Stability of Queueing Networks, Ecole d'Eté de Probabilités de Saint-Flour XXXVI-2006 (2008)

Vol. 1951: A. Moltó, J. Orihuela, S. Troyanski, M. Valdivia, A Non Linear Transfer Technique for Renorming (2008)

Vol. 1952: R. Mikhailov, I.B.S. Passi, Lower Central and Dimension Series of Groups (2009)

Vol. 1953: K. Arwini, C.T.J. Dodson, Information Geometry (2008)

Vol. 1954: P. Biane, L. Bouten, F. Cipriani, N. Konno, N. Privault, Q. Xu, Quantum Potential Theory. Editors: U. Franz, M. Schürmann (2008)

Vol. 1955: M. Bernot, V. Caselles, J.-M. Morel, Optimal transportation networks (2008)

Vol. 1956: C.H. Chu, Matrix Convolution Operators on Groups (2008)

Vol. 1957: A. Guionnet, On Random Matrices: Macroscopic Asymptotics, Ecole d'Eté de Probabilités de Saint-Flour XXXVI-2006 (2008)

Vol. 1958: M.C. Olsson, Compactifying Moduli Spaces for Abelian Varieties (2008)

Recent Reprints and New Editions

Vol. 1702: J. Ma, J. Yong, Forward-Backward Stochastic Differential Equations and their Applications. 1999 – Corr. 3rd printing (2007)

Vol. 830: J.A. Green, Polynomial Representations of GL_n, with an Appendix on Schensted Correspondence and Littelmann Paths by K. Erdmann, J.A. Green and M. Schoker 1980 – 2nd corr. and augmented edition (2007)

Vol. 1693: S. Simons, From Hahn-Banach to Monotonicity (Minimax and Monotonicity 1998) – 2nd exp. edition (2008)

Vol. 470: R.E. Bowen, Equilibrium States and the Ergodic Theory of Anosov Diffeomorphisms. With a preface by D. Ruelle. Edited by J.-R. Chazottes. 1975 – 2nd rev. edition (2008)

Vol. 523: S.A. Albeverio, R.J. Høegh-Krohn, S. Mazzucchi, Mathematical Theory of Feynman Path Integral. 1976 – 2nd corr. and enlarged edition (2008)

Vol. 1764: A. Cannas da Silva, Lectures on Symplectic Geometry 2001 – Corr. 2nd printing (2008)

LECTURE NOTES IN MATHEMATICS  Springer

Edited by J.-M. Morel, F. Takens, B. Teissier, P.K. Maini

Editorial Policy (for the publication of monographs)

1. Lecture Notes aim to report new developments in all areas of mathematics and their applications - quickly, informally and at a high level. Mathematical texts analysing new developments in modelling and numerical simulation are welcome.

   Monograph manuscripts should be reasonably self-contained and rounded off. Thus they may, and often will, present not only results of the author but also related work by other people. They may be based on specialised lecture courses. Furthermore, the manuscripts should provide sufficient motivation, examples and applications. This clearly distinguishes Lecture Notes from journal articles or technical reports which normally are very concise. Articles intended for a journal but too long to be accepted by most journals, usually do not have this "lecture notes" character. For similar reasons it is unusual for doctoral theses to be accepted for the Lecture Notes series, though habilitation theses may be appropriate.

2. Manuscripts should be submitted either to Springer's mathematics editorial in Heidelberg, or to one of the series editors. In general, manuscripts will be sent out to 2 external referees for evaluation. If a decision cannot yet be reached on the basis of the first 2 reports, further referees may be contacted: The author will be informed of this. A final decision to publish can be made only on the basis of the complete manuscript, however a refereeing process leading to a preliminary decision can be based on a pre-final or incomplete manuscript. The strict minimum amount of material that will be considered should include a detailed outline describing the planned contents of each chapter, a bibliography and several sample chapters.

   Authors should be aware that incomplete or insufficiently close to final manuscripts almost always result in longer refereeing times and nevertheless unclear referees' recommendations, making further refereeing of a final draft necessary.

   Authors should also be aware that parallel submission of their manuscript to another publisher while under consideration for LNM will in general lead to immediate rejection.

3. Manuscripts should in general be submitted in English. Final manuscripts should contain at least 100 pages of mathematical text and should always include

   – a table of contents;
   – an informative introduction, with adequate motivation and perhaps some historical remarks: it should be accessible to a reader not intimately familiar with the topic treated;
   – a subject index: as a rule this is genuinely helpful for the reader.

   For evaluation purposes, manuscripts may be submitted in print or electronic form, in the latter case preferably as pdf- or zipped ps-files. Lecture Notes volumes are, as a rule, printed digitally from the authors' files. To ensure best results, authors are asked to use the LaTeX2e style files available from Springer's web-server at:

   ftp://ftp.springer.de/pub/tex/latex/svmonot1/ (for monographs).

Additional technical instructions, if necessary, are available on request from: lnm@springer.com.

4. Careful preparation of the manuscripts will help keep production time short besides ensuring satisfactory appearance of the finished book in print and online. After acceptance of the manuscript authors will be asked to prepare the final LaTeX source files (and also the corresponding dvi-, pdf- or zipped ps-file) together with the final printout made from these files. The LaTeX source files are essential for producing the full-text online version of the book (see www.springerlink.com/content/110312 for the existing online volumes of LNM).

The actual production of a Lecture Notes volume takes approximately 12 weeks.

5. Authors receive a total of 50 free copies of their volume, but no royalties. They are entitled to a discount of 33.3% on the price of Springer books purchased for their personal use, if ordering directly from Springer.

6. Commitment to publish is made by letter of intent rather than by signing a formal contract. Springer-Verlag secures the copyright for each volume. Authors are free to reuse material contained in their LNM volumes in later publications: a brief written (or e-mail) request for formal permission is sufficient.

Addresses:

Professor J.-M. Morel, CMLA,
École Normale Supérieure de Cachan,
61 Avenue du Président Wilson, 94235 Cachan Cedex, France
E-mail: Jean-Michel.Morel@cmla.ens-cachan.fr

Professor F. Takens, Mathematisch Instituut,
Rijksuniversiteit Groningen, Postbus 800,
9700 AV Groningen, The Netherlands
E-mail: F.Takens@math.rug.nl

Professor B. Teissier, Institut Mathématique de Jussieu,
UMR 7586 du CNRS, Équipe "Géométrie et Dynamique",
175 rue du Chevaleret
75013 Paris, France
E-mail: teissier@math.jussieu.fr

For the "Mathematical Biosciences Subseries" of LNM:

Professor P.K. Maini, Center for Mathematical Biology,
Mathematical Institute, 24-29 St Giles,
Oxford OX1 3LP, UK
E-mail: maini@maths.ox.ac.uk

Springer, Mathematics Editorial I, Tiergartenstr. 17
69121 Heidelberg, Germany,
Tel.: +49 (6221) 487-8259
Fax: +49 (6221) 4876-8259
E-mail: lnm@springer.com

Printing: Krips bv, Meppel, The Netherlands
Binding: Stürtz, Würzburg, Germany